Toward Functional Nanomaterials

Lecture Notes in Nanoscale Science and Technology

Volume 5

Volumes Published in This Series:

Volume 1: Self-Assembled Quantum Dots, Wang, Z.M. (Ed.), 2007
Volume 2: Nanoscale Phenomena: Basic Science to Device Applications, Tang, Z., and Sheng, P. (Eds.), 2007
Volume 3: One-Dimensional Nanostructures, Wang, Z.M. (Ed.), 2008
Volume 4: Nanoparticles and Nanodevices in Biological Applications, Bellucci, S. (Ed.), 2009
Volume 5: Toward Functional Nanomaterials, Wang, Z.M. (Ed.), 2009

Forthcoming Titles:

B-C-N Nanotubes and Related Nanostructures, Yap, Y.K. (Ed.), 2008
Epitaxial Semiconductor Nanostructures, Wang, Z.M., and Salamo, G., 2009

Zhiming M. Wang (Ed.)

Toward Functional Nanomaterials

Springer

Zhiming M. Wang
Institute of Nanoscale Science and Engineering
University of Arkansas
835 W. Dickson St.
Fayetteville AR 72701
zmwang@uark.edu

ISBN 978-0-387-77716-0 e-ISBN 978-0-387-77717-7
DOI 10.1007/978-0-387-77717-7
Springer Dordrecht Heidelberg London New York

Library of Congress Control Number: 2009921350

Printed on acid-free paper

Springer is part of Springer Science+Business Media (www.springer.com)

Preface

Advanced probes and new fabrication techniques enable nanomaterials to pervade multiple disciplines, including physics, chemistry, engineering and biology. Nanomaterials have been extensively investigated with various kinds of morphologies (nanoparticles, nanowhiskers, nanorods, nanowires, nanoclusters, quantum dots, etc.) and compositions (semiconductor, metal, polymer, etc.). Impressive progress has been made on directed assembly and synthesis, structure, and property characterization, as well as nanoscale device concepts and performance by a diverse group of experts. However, in spite of continued advancements in various aspects of functional nanomaterials, numerous challenges must still be overcome at different stages for practical applications to be realized.

It seems that there is a need for a book in which individual research groups comprehensively review their up-to-date efforts and simulate further developments in other laboratories. Therefore, I believe that this book, which consists of twelve chapters from nine countries, is a timely undertaking.

"Fabrication of Oxide Nanoparticles by Ion Implantation and Thermal Oxidation" is an in-depth review of the formation of oxide nanoparticles by metal ion implantation and subsequent thermal oxidation. Amekura and Kishimoto believe that there is a potential for an "embedded" breakthrough in the field of oxide nanoparticles similar to "Hache's finding of a breakthrough" in the field of metal nanoparticles. In "Design of Solution-Grown ZnO Nanostructures", Pauporté reviews significant progress toward the growth of well-controlled ZnO nanostructures in solution. He argues that solution-based methods are cost-effective and that the resulting nanostructures are easy to scale up for applications. In "Self-Assembled Metal Nanostructures in Semiconductor Structures", Ruffino et al. describe the self-assembly of metal nanostructures in semiconductor structures and demonstrate the consistency between structural and electrical measurements. In "Nanocrystal - Based Polymer Composites as Novel Functional Materials", Striccoli et al. focus on the incorporation of luminescent nanocrystals in plastic and structural polymers to obtain composite materials for integration in optoelectronic, photonic, and sensing devices.

While the first four chapters emphasize the fabrication of functional nanomaterials, chapters written by J. Li and S.-H. Wei, K. Miura et al., Sh. Michaelson et al., J.

Wei, and M. Gmitra and J. Barnaś 5–9 focus attention on their properties, behavior, and applications.

"Large-Scale *Ab Initio* Study of Size, Shape, and Doping Effects on Electronic Structure of Nanocrystals" covers the recent development of large-scale *ab initio* pseudopotential methods for calculating electronic structures of nanocrystals. Li and Wei demonstrate significant effects of size, shape, and doping on the properties of nanostructures. In "Chaotic Behavior Appearing in Dynamic Motions of Nanoscale Particles", Miura et al. reveal the dynamical motion of nanoscale particles with observations of chaotic behaviors. In "Hydrogen Concentration, Bonding Configuration and Electron Emission Properties of Polycrystalline Diamond Films: From Micro- to Nanometric Grain Size", Hoffman et al. summarize the studies of diamond films of varying grain size and thickness carried out in their laboratory. They reveal the impact of diamond grain size on hydrogen concentration, the shape of the Raman spectra, and the results of high resolution electron energy loss spectroscopy. In "Super-Resolution Optical Effects of Nanoscale Nonlinear Thin Film Structure and Ultrahigh -Density Information Storage", Wei introduces super-resolution optical effects of nanoscale nonlinear thin films and their application to ultrahigh density optical information storage. "Spin-Transfer and Current-Induced Spin Dynamics in Spin Valves: Diffusive Transport Regime" provides a demonstration by Gmitra and Barnaś of spin-transfer and current-induced spin dynamics in spin valves.

Gago et al. present a comprehensive overview of nanostructure evolution during nanopatterning by ion beam sputtering in "Self-Organized Surface Nanopatterning by Ion Beam Sputtering". While previous reviews are mainly devoted to ripple patterns, the focus of this chapter is on nanodot patterns due to their novelty and variety of applications. This is a must-have reference chapter for any researcher in the field of ion beam sputtering with an interest in nanotechnology. "Area-Selective Depositions of Self-assembled Monolayers on Patterned SiO_2/Si Surfaces" demonstrates area-selective depositions of self-assembled monolayers on patterned SiO_2/Si surfaces. Wang and Urisu believe that such a technique offers the potential for many practical applications, such as biosensor fabrication, cell studies, and tissue engineering.

While the above chapters focus on experimental manipulation of functional nanomaterials, virtual synthesis of electronic nanomaterials is the subject of the last chapter. Pozhar and Mitchel create atomic clusters and artificial molecules by computing the minimum total system energy. Perhaps these virtual nanomaterials exemplify possible outcomes of future experimental synthesis.

Last but not least, I am delighted to dedicate this book to my daughters, CC and KK. CC is ten years old and knows that nanomaterials are something small. The impact of nanotechnology to CC is nothing more than the iPod nano. KK is two years old and loves to read books loudly without knowing any words. I will record how she reads this book once it is published. I believe that nanotechnology will shape a better future for CC, KK, and all the children of the world!

Fayetteville, AR Zhiming M. Wang

Contents

Contributors

R. Akhvlediani Schulich Faculty of Chemistry, Technion – Israel Institute of Technology, Haifa 32000, Israel

Hiroshi Amekura Quantum Beam Center, National Institute for Materials Science (NIMS), 3-13 Sakura, Tsukuba, Ibaraki 305-0003, Japan

Józef Barnaś Department of Physics, Adam Mickiewicz University, Umultowska 85, 61-614 Poznan, Poland; Institute of Molecular Physics, Polish Academy of Sciences, M. Smoluchowskiego 17, 60-179 Pozna, Poland

Mario Castro Escuela Técnica Superior de Ingeniería and GISC, Universidad Pontificia Comillas de Madrid, E-28015 Madrid, Spain

R. Comparelli CNR-IPCF Institute for Physical Chemistry Processes c/o Dep. Chemistry, University of Bari, Via Orabona 4, 70126 Bari, Italy

Rodolfo Cuerno Departamento de Matem'aticas and Grupo Interdisciplinar de Sistemas, Complejos (GISC), Universidad Carlos III de Madrid, E-28911 Leganés, Spain

M.L. Curri CNR-IPCF Institute for Physical Chemistry Processes c/o Dep. Chemistry, University of Bari, Via Orabona 4, 70126 Bari, Italy

Raul Gago Instituto de Ciencia de Materiales de Madrid, Consejo Superior de Investigaciones Científicas, E-28049 Madrid, Spain

Javier Muñoz-García Departamento de Matem'aticas and Grupo Interdisciplinar de Sistemas, Complejos (GISC), Universidad Carlos III de Madrid, E-28911 Leganés, Spain

José A. Sánchez-García Instituto de Ciencia de Materiales de Madrid, Consejo Superior de Investigaciones Científicas, E-28049 Madrid, Spain

Filippo Giannazzo Consiglio Nazionale delle Ricerche-Istituto per la Microelettronica e Microsistemi (CNR-IMM), Stradale Primosole 50, 95121, Catania, Italy

Martin Gmitra Institute of Physics, P.J. Safarik University, Kosice, Park Angelinum 9, 040 01 Kosice, Slovak Republic

Maria Grazia Grimaldi Dipartimento di Fisica ed Astronomia and MATIS CNR-INFM, Università di Catania, via Santa Sofia 64, 95123 Catania, Italy

R. Harada Department of Physics, Aichi University of Education, Hirosawa 1, Igaya-cho, Kariya 448-8542, Japan

Alon Hoffman Schulich Faculty of Chemistry, Technion – Israel Institute of Technology, Haifa 32000, Israel, choffman@tx.technion.ac.il

M. Ishikawa Department of Physics, Aichi University of Education, Hirosawa 1, Igaya-cho, Kariya 448-8542, Japan

M. Kato Department of Physics, Aichi University of Education, Hirosawa 1, Igaya-cho, Kariya 448-8542, Japan

N. Kishimoto Quantum Beam Center, National Institute for Materials Science (NIMS), 3-13 Sakura, Tsukuba, Ibaraki 305-0003, Japan

Jingbo Li State Key Laboratory for Supterlattices and Microstructures, Instutite of Semiconductor, Chinese Academy of Science, P. O. Box 912, Beijing 100083, China

Sh. Michaelson Schulich Faculty of Chemistry, Technion – Israel Institute of Technology, Haifa 32000, Israel

William C. Mitchel Air Force Research Laboratory, Materials and Manufacturing Directorate, Wright-Patterson Air Force Base, OH 45433, USA

Kouji Miura Department of Physics, Aichi University of Education, Hirosawa 1, Igaya-cho, Kariya 448-8542, Japan

Thierry Pauporté CNRS, Laboratoire d'Electrochimie et Chimie Analytique, UMR7575, Ecole Nationale Supérieure de Chimie de Paris, 11 rue P. et M. Curie, 75231 Paris cedex 05, France

Liudmila A. Pozhar Center for Materials for Information Technologies, University of Alabama, Box 890209, Tuscaloosa, AL 35487-0209, USA

Vito Raineri Consiglio Nazionale delle Ricerche-Istituto per la Microelettronica e Microsistemi (CNR-IMM), Stradale Primosole 50, 95121, Catania, Italy

Fabrizio Roccaforte Consiglio Nazionale delle Ricerche-Istituto per la Microelettronica e Microsistemi (CNR-IMM), Stradale Primosole 50, 95121, Catania, Italy

Francesco Ruffino Dipartimento di Fisica ed Astronomia and MATIS CNR-INFM, Università di Catania, via Santa Sofia 64, 95123 Catania, Italy; Consiglio Nazionale delle Ricerche-Istituto per la Microelettronica e Microsistemi (CNR-IMM), Stradale Primosole 50, 95121, Catania, Italy

N. Sasaki Department of Applied Physics, Faculty of Science and Engineering, Seikei University, 3-3-1 Kichijoji Kitamachi, Musashino-shi, Tokyo 180-8633, Japan

Marinella Striccoli CNR- IPCF Institute for Physical Chemistry Processes c/o Dep. Chemistry, University of Bari, Via Orabona 4, 70126 Bari, Italy

O. Ternyak Schulich Faculty of Chemistry, Technion – Israel Institute of Technology, Haifa 32000, Israel

Tsuneo Urisu Department of Vacuum UV Photoscience, Institute for Molecular Science, Myodaiji, Okazaki, 444-8585, Japan

Luis Vázquez Instituto de Ciencia de Materiales de Madrid, Consejo Superior de Investigaciones Científicas, E-28049 Madrid, Spain

Changhsun Wang Department of Physics, Shanghai Jiao Tong University, 800 Dong Chuan Road, Shanghai 200240, China

Jingsong Wei Shanghai Institute of Optics and Fine Mechanics, Chinese Academy of Sciences, Shanghai 201800, China

Su-Huai Wei National Renewable Energy Laboratory, Golden, CO 80401, USA

Fabrication of Oxide Nanoparticles by Ion Implantation and Thermal Oxidation

H. Amekura and N. Kishimoto

Abstract Fabrication of oxide nanoparticles (NPs) in transparent insulators by metal ion implantation and subsequent thermal oxidation (II&TO) is reviewed. After a short historical review of the II&TO method, fundamental issues concerning two important processes in the II&TO method, i.e., (i) formation of metal NPs by ion implantation and (ii) thermal oxidation of the metal NPs, are described. Then the highlights of this chapter, i.e., the formation of oxide NPs by the II&TO method, are reviewed. Oxide NP systems of NiO, CuO, and ZnO have been formed by the conventional II&TO method, i.e., the II&TO method using atmospheric pressure of oxygen gas. Each of the NP system shows different characteristics. While NiO, CuO (and Cu_2O) NP systems show the oxide formation with little redistribution of the depth profiles, i.e., the oxide NPs are retained inside the SiO_2 substrate, ZnO NPs are formed *on the surface of* the SiO_2 substrate after prominent depth redistribution. Furthermore, recent developments in the II&TO method, i.e., the second generation of the II&TO method, are shown. ZnO NPs *embedded in* SiO_2 substrate are formed by low temperature and long-term oxidation. Cu_2O NPs, which are not most stable under atmospheric pressure of oxygen, are formed by two-step annealing. Consequently, selective formation of CuO and Cu_2O NPs is possible using the conventional II&TO and the two-step II&TO method, respectively. Finally, some remaining aspects of the oxide NP formation by the II&TO method are discussed.

Contents

H. Amekura (✉)
National Institute for Materials Science (NIMS), 3-13 Sakura, Tsukuba,
Ibaraki 305-0003, Japan
e-mail: amekura.hiroshi@nims.go.jp

Z.M. Wang (ed.), *Toward Functional Nanomaterials*, Lecture Notes in Nanoscale Science and Technology 5, DOI 10.1007/978-0-387-77717-7_1,

1 Introduction

While various oxides have been used as functional materials since ancient days [132], recently nano-sized oxides have been drawing much attention because of furthermore functionalities in peculiar optical [69, 30, 82], magnetic [139, 94], and catalytic [39] properties, etc. Ion implantation, which is now established as one of the most important processes in the semiconductor industry, is recently recognized as one of the promising methods in fabricating metal nanoparticles (NPs) in insulators with good controllability in fluence, spatial position, and depth [52, 31, 88, 47, 56, 86]. Furthermore, as proven in the semiconductor industry, ion implantation (and also thermal oxidation) is one of the high-purity processes (the purity of many nines, i.e., 99.99...%). Since ions are introduced in target materials in very high vacuum at ambient temperatures after (magnetic) mass separation, products are contamination-free. This is a big contrast to the wet chemical-based nanomaterial synthesis where the purity is not high (e.g., the purity of ~80%). The high purity of NPs fabricated by ion implantation is advantageous for the compatibility with the current silicon technology. Recently, extensive studies were conducted for development of nanomaterials for information technology. Someday, all the silicon technologies may be replaced by nanomaterials. However, what kinds of implementation plans are assumed? Since it is impossible to replace all the present silicon technologies with nanotechnology at once, some transition period should be assumed. In this period, new nanomaterials should co-exist and co-work with the silicon technologies. The compatibility of nanomaterials with the silicon technologies will be a very important issue. Ion implantation (and thermal oxidation) is one of the limited nanofabrication processes that meets this requirement.

While the first fabrication of metal NPs by ion implantation was reported in the mid-1970s [25, 26], extensive studies have started in the 1990s. Recently, more extensive studies are going on. However, not only *metal* NPs but also *oxide* NPs are attractive because of their peculiar functionalities.

Some research groups have partially succeeded in fabricating oxide NPs of VO_2 [82], ZnO [77], and a mixture of Cu_2O and CuO [61], etc., using sequential (positive) ion implantation. In some cases, however, secondary implantation of the sequence induces significant diffusion of the primarily implanted atoms, and hinders the compound formation. Furthermore, products of the sequential implantation strongly depend on the sequence of the implantations, i.e., the products of implantation sequence of element *A* and then *B* are different from the products of implantation sequence of *B* and then *A*. Ikeyama et al. [61] showed implantation sequence-dependent optical absorption of SiO_2 implanted with Cu ions then O ions and O ions then Cu ions. Hosono [60] compared products formed by sequential implantations of F ions and then Cu ions to SiO_2, and the reverse process. While only Cu NPs were formed after the former sequence, core-shell NPs of Cu and Cu_2O were formed in the latter sequence. Applying this phenomenon, Wuhan University group [115] formed Zn–ZnO core-shell NPs in SiO_2 by sequential implantation of Zn ions and F ions. Since various physical and chemical processes are expected under sequential implantation, it is generally difficult to predict the products without attempts.

In this chapter, an alternative method, which is more predictable than the sequential implantations and is applicable to oxides, i.e., metal ion implantation and subsequent thermal oxidation, is presented. As the name of this method has not been commonly recognized, we tentatively call this method as ion implantation and thermal oxidation (II&TO) method in this chapter. The principle of the II&TO method is summarized in Fig. 1. First, metal ions are introduced into a transparent insulator substrate, such as silica glass (SiO_2), sapphire (Al_2O_3), etc., by ion implantation of a few tens to a few hundreds kiloelectron volts up to much higher concentration than the solubility limit of the atoms in the substrate. Since the metal atoms cannot be dissolved in the substrate as isolated atoms in such a high concentration, precipitation is induced, i.e., a lot of the secondary phases of the metal aggregates in nanometer sizes are formed in the substrate. In other words, they are metal NPs.

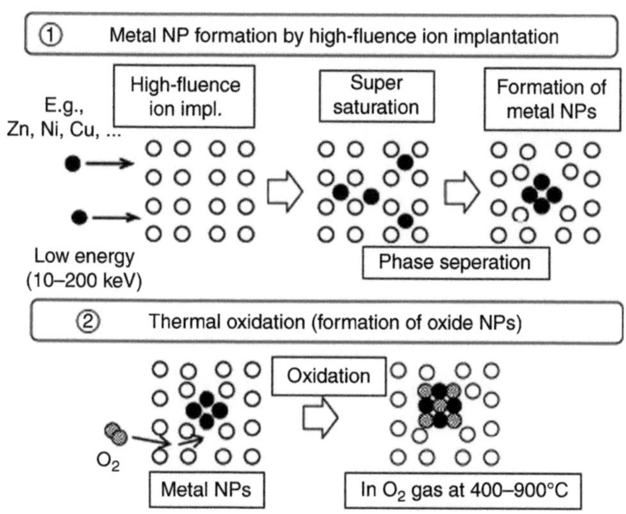

Fig. 1 Schematically depicted processes of oxide nanoparticle formation by the II&TO method

After this, the substrate containing metal NPs is annealed in oxygen gas typically at 700–800°C for 1 h. During the annealing, oxygen species migrate inside the substrate and oxidize the metal NPs. In the case of SiO_2 substrate, the oxygen atoms migrate as O_2 molecules [51]. For successful oxidation of the metal NPs, the migration depth of oxygen molecules should be comparable to the projected range of the ions. The energy of metal ions of 60 keV corresponds to the projected range of ~50 nm in SiO_2, which matches the migration depth of oxygen molecules of ~60 nm in SiO_2 at 800°C for 1 h. Consequently, megaelectron volt ion implantation is not practically adequate for this method. The migration depth of oxygen comparable to the projected range of the megaelectron volt ions, i.e., much larger than 1 μm, is unattainable at moderately high temperatures within several tens of hours. It may be attainable at very high temperatures, but some species of metal NPs become unstable at such high temperatures.

In this chapter, the formation of oxide NPs by the II&TO method are reviewed. After a short historical review of the II&TO method, the fundamental issues of two important processes in the II&TO method, i.e., (i) formation of metal NPs by ion implantation and (ii) thermal oxidation of the metal NPs, are described. Then the highlights of this chapter, i.e., the formation of oxide NPs by the II&TO method, are reviewed. Up to now, we have fabricated oxide NP systems of NiO, CuO, and ZnO by the II&TO method using atmospheric pressure of oxygen gas. Each of the NP system shows different characteristics: While NiO and CuO (and Cu_2O) NP systems show the oxide formation with little redistribution of the depth profiles, i.e., the oxide NPs are formed inside the SiO_2 substrate, ZnO system shows large depth redistribution, i.e., ZnO–NPs are formed on the surface of the SiO_2 substrate. The NP systems of NiO and ZnO show drastic transmittance changes in the visible region accompanied with the oxidation, while CuO (and Cu_2O) shows only a little change in the transparency in the visible region, indicating that the NiO and ZnO NP can be used as a building block for transparent devices.

Furthermore, recent developments in the II&TO method, i.e., the second generation of the II&TO method, are shown. ZnO NPs *embedded* in SiO_2 substrate are formed by low temperature and long-term oxidation. Cu_2O NPs, which are not most stable under atmospheric pressure of oxygen, are formed by the two-step annealing. Consequently, selective formation of CuO and Cu_2O NPs is possible using the conventional II&TO and the two-step II&TO method, respectively. Finally, some other aspects of the oxide NP formation by the II&TO method, including the drastic redistribution induced by the ZnO NP formation, are discussed.

2 A Short Historical Review of the II&TO Method

Table 1 summarizes a series of studies on the formation of oxide nanostructures in/on transparent insulators, e.g., SiO_2, Al_2O_3, etc, by the II&TO method. Although some researchers reported on the formation of oxide aggregates by ion implantation and annealing in the 1980s and the 1990s (e.g., [87, 45, 99, 111]), the oxide

Table 1 A list of representative references concerning the formation of oxide NPs by the ion implantation and thermal oxidation method up to 2007

First Author	R	Affiliation	E (keV)	Dose (cm^{-2})	Substrate	T_{Anneal} (°C)	Product(s)	TEM
In the 1980s–1990s, the formation of oxide NPs has been known, but was an unintended or unwanted process.								
Chen J	[33]	Fisk & ORNL	Zn 160	~1E17	SiO_2	~Ox700	mainly Zn	–
White CW	[138]	Fisk & ORNL	Zn 320	0.5-1E17	Al_2O_3	Ox, red 1000	$ZnAl_2O_4$	o
			Zn 320 + O 90	0.5-1E17	Al_2O_3	Ox, red 1000	$ZnAl_2O_4$	o
Liu XY	[79,80]	Changchun & Fisk	Zn 160	3E17	SiO_2	~Ox700	ZnO NP layer on surf.	–
	[81]							o
Liu YC	[78]	Changchun & Fisk	Zn 160	1E17	CaF_2	Ox300–700	ZnO (NPs?)	–
Amekura H	[2]	NIMS	Ni 60, Cu 60	~5E16	SiO_2	~Ox800	NiO?, CuO?	–
	[7]		Ni 60	~5E16		~Ox800	NiO NPs embedded	o
	[11]		Cu 60	~5E16		~Ox800	CuO NPs embedded	o
	[10]	NIMS & FZ-Juelich	Zn 60	1E17		~Ox700°C for 1 h	ZnO NPs on surf	o
	[14]	NIMS	Cu 60	~5E16		Ox800+Ar900°C	Cu_2O NPs embedded	o
	[19]	NIMS & FZ-Juelich	Zn 60	1E17		500°C for ~70 h	ZnO embedded	o
van Huis MA	[137]	Delft Univ. of Tech.	Zn 140	1E17	MgO	Vac650–1250°C (internal ox.)	Zn metal NPs embedded	o
Lee JK	[77]	Los Alamos NL	Zn100 + O 57	1E17	SiO_2	500–700	ZnO (NPs?)	–
Xiang X	[142]	Chengdu & Michigan	Ni 64	1E17	Al_2O_3	Ox600–900	NiO NPs	o
	[143]		Ni 64	1E17	$Y\text{-}ZrO_2$	Ox200–900	NiO NPs	o
	[144]		Zn 48	1E17	Al_2O_3	~Ox600	ZnO NPs	o
de Julian C	[35]	Univ. di Padova	Co 180	2E17	SiO_2	Ox800°C	$Co_3\,O_4$ on surf?	–
Muntele I	[95]	Alabama A&M	Zn 20	3E16	SiO_2	Ox, red 300–700°C	ZnO	–
Marques C	[83]	Inst. Tech. Nucl. PT	Zn 150	~1E17	$m\text{-}Al_2O_3$	Vac 1000°C	ZnO NPs?	–

aggregates were of secondary importance. They were by-products. A more intentional application of these processes was firstly proposed by a group from Fisk University and Oak Ridge National Laboratory (ORNL) [33]. The proposal was concerning the fabrication of ZnO NPs in SiO_2 by Zn ion implantation and subsequent thermal oxidation. They implanted Zn ions of 160 keV to SiO_2 up to a fluence of 1×10^{17} ions/cm^2, and carried out annealing both in reducing and oxidizing atmospheres up to 700°C. However, they did not obtain any clear evidence of ZnO formation. Though the annealing duration was not written in the paper, possibly the duration was too short to induce the migration of oxygen gas deeply enough inside the substrate.

In the next year, 1999, the ORNL and Fisk University group applied this method to Al_2O_3 substrates [138]. They implanted Zn ions of 320 keV to Al_2O_3 substrates up to fluences of 0.5–1×10^{17} ions/cm^2 at an implantation temperature of 550°C, and annealed at 1000°C for 1 h in oxidizing (O_2 gas) or reducing (4% H_2/Ar gas) atmosphere. They also carried out sequential implantations of Zn ions of 320 keV and O ions of 90 keV to fluences of 0.5–1×10^{17} ions/cm^2 for each ion at the implantation temperature of 550°C and annealing at 1000°C for 1 h. However, they observed precipitates of $ZnAl_2O_4$ and have never observed ZnO formation. The implantation at elevated temperature and annealing at high temperature around 1000°C were suitable for recovery of radiation damage in Al_2O_3 substrate but induced reactions between synthesized ZnO and the Al_2O_3 substrate to $ZnAl_2O_4$ phase.

The first success was reported from China in 2002. Changchun and Fisk University group [79, 80] implanted Zn ions to SiO_2 up to a very a high fluence of 3×10^{17} ions/cm^2 and carried out annealing in oxidizing atmosphere (air) at 700°C for 2 h. They observed x-ray diffraction (XRD) peaks from ZnO, which showed strong c-plane preferential orientation and ultraviolet (UV) photoluminescence around 377 nm. Furthermore, they carried out surface composition analysis using x-ray photoelectron spectroscopy (XPS). Although only very weak Zn signal and strong Si signal were observed in the as-implanted state, very strong ZnO signal and very weak Si signal were observed after annealing in air at 700°C. From the XPS results and the preferential XRD signal, they concluded that the SiO_2 surface was fully covered by ZnO nanocrystalline layer. The formation of the ZnO layer on SiO_2 surface was later confirmed by cross-sectional transmission electron microscopy (XTEM) [81].

The very high fluence of 3×10^{17} ions/cm^2 which they used probably made the Zn depth profile much shallower due to surface sputtering. The shallower depth profile of Zn NPs is favorable for oxidation. This may be one of the differences between the former challenges by Chen et al. [33].

On the other hand, starting from studies of annealing effects of Cu metal NPs [3] and Ni magnetic metal NPs [4, 5, 6], Amekura et al. have proposed and demonstrated the II&TO method by the formation of NiO NPs in SiO_2 [2, 7]. After this, they have applied this method to ZnO NP formation [10]. In their studies, SiO_2 was implanted with Zn ions of 60 keV up to a fluence of 1×10^{17} ions/cm^2, and was annealed in oxygen gas at 400–900°C for 1 h. Judging from changes in the absorption spectra and the XRD results, the Zn metal NPs transform to ZnO NPs around 700°C. Formation of droplet-like ZnO NPs on the SiO_2 surface and small NPs embedded in SiO_2 was confirmed from XTEM observation.

Encouraged by successes in SiO_2 substrates, some challenges have been extended toward other substrates. Liu et al. applied this method to ZnO NP formation in CaF_2 [78]. Xiang et al. applied it to NiO NP formation in Al_2O_3 [142] and Y-stabilized ZrO_2 [143], and ZnO NP formation in Al_2O_3 [144]. Comparing with the success of ZnO NP formation in Al_2O_3 by Xiang et al. and the past failure by White et al. [138], it seems that the key factor is the annealing temperature. Very high annealing temperatures lead to undesired reactions between ZnO and Al_2O_3 substrates. However, samples annealed at low temperatures are not free from radiation damage [83].

Alabama A&M University group formed ZnO by low-energy implantation of 20 keV Zn^+ to SiO_2 and annealing in oxygen gas [95]. Los Alamos group reported ZnO NP formation in SiO_2 by sequential implantation of Zn ions of 100 keV and O ions of 50 keV and post-implantation annealing [77].

Delft University of Technology group [137] implanted Zn ions of 140 keV to MgO and carried out high-temperature annealing in vacuum up to 1250°C, to intend internal oxidation of Zn NPs by O atoms in MgO lattice, but did not obtain ZnO phase. However, this does not mean that the internal oxidation is impossible. In fact, Marques et al. reported partial formation of ZnO NPs in m-cut Al_2O_3, but not in c-cut Al_2O_3 after vacuum annealing at 1000°C [83]. Amekura et al. observed partial formation of Zn_2SiO_4 phase in as-implanted state [22]. The Zn_2SiO_4 formation has been ascribed to sequential formation of ZnO by internal oxidation by oxygen librated from SiO_2 network by radiation damage and further reactions with SiO_2 substrate enhanced by radiation effects.

University of Padova group [35] reported the formation of Co_3O_4 nanophase by Co ion implantation to SiO_2 and oxygen annealing at 800°C for 4 h.

Recently, some modifications have been made in the II&TO method to form nanophases that are not formed by the conventional II&TO method. Although ZnO NPs are formed on the surface of SiO_2 substrate by the conventional II&TO method, ZnO NPs embedded in SiO_2 are formed by the II&TO method with low temperature and long-time annealing [19]. Cu_2O NPs that are not formed by the conventional II&TO method are formed by the II&TO method and two-step annealing [14]. These modifications can be called the second generation of the II&TO method.

3 Elemental Processes of the II&TO Method

3.1 Formation of Metal Nanoparticles by Ion Implantation

As for the formation of metal NPs by ion implantation, many excellent review papers have been already published (e.g., [52, 31, 88, 47, 56, 86]). Here, we describe only some important aspects of the metal NP formation by ion implantation briefly. Although the optical properties of metal NPs are of interest, we do not describe them here because of the limited pages of this chapter. As for metal NPs in general, reading of these references [110, 74] is suggested.

3.1.1 Nucleation and Growth of Metal Nanoparticles

As shown in Fig. 1, the formation of metal NPs in insulators by (high-fluence) ion implantation is regarded as secondary phase separation in a supersaturated solid solution. Since the depth distribution of ions implanted with a single energy approximately shows a Gaussian shape, i.e., not homogenous, the simple nucleation theory which assumes a homogenous concentration system cannot be exactly applicable. However, for simplicity, an explanation based on the simple nucleation theory is given here.

Since the driving force of doping by ion implantation is due to the acceleration voltage of implanter, i.e., a nonequilibrium physical process, the concentration of the implanted atoms, i.e., monomers, in an insulating substrate increases up to a very high level, higher than the solubility limit of the metal atoms in the substrate. Once the monomer concentration exceeds the solubility limit, the implants tend to aggregate if nucleation sites are available. However, this change is described by the first order phase transition, i.e., the solid solution is in a meta-stable state to some extent even exceeding the solubility limit. According to the elementary nucleation theory, the Gibbs energy change, ΔG, due to precipitation of spherical NP of radius R is given as

$$\Delta G = -\frac{4\pi}{3} R^3 \varepsilon + 4\pi R^2 \gamma \tag{1}$$

where ε and γ denote the energy gain per volume due to precipitation and the interface energy per surface area, respectively. Although ΔG becomes negative (i.e., the precipitation is thermodynamically favorable) when R is large enough, the ΔG has a positive maximum at $R^* = 2\gamma/\varepsilon$. In the case of very small clusters like monomers, dimers, trimers, etc., whose radius R is smaller than R^*, an increase of the size corresponds to an increase of ΔG, i.e., thermodynamically unfavorable processes. Microscopically, these small clusters continue attachment and detachment processes of monomers or possibly aggregation between the small clusters and the reverse processes. Once accidentally the size overcomes the critical value, R^*, with the help of thermal fluctuation or extrinsic nucleation centers, the size steadily increases because the size increase is thermodynamically favorable. Consequently, the nucleation, i.e., the formation of stable precipitates of the minimum size, is of low efficiency and time consuming in homogenous nucleation systems.

Although it has not been clarified whether the *homogenous* nucleation or the *heterogeneous* nucleation (i.e., with extrinsic nucleation centers) is dominant under ion implantation, introduction of defect sites in the substrate by ion implantation is favorable for efficient nucleation. In fact, Kaiser et al. [68] have succeeded in in situ observation of the aggregation processes of implanted ions at a defect site using high-angle annular dark-field (HAADF) imaging in a scanning transmission electron microscopy (STEM).

Once the nucleation sites form, the monomer concentration steeply decreases due to steady growth of the nuclei. From Gibbs–Thomson relation, the monomer concentration in NP of radius R is given as

$$c^{\mathrm{GT}}(R) = c_{\infty} \exp\{R_C/R\} \tag{2}$$

where c_{∞} denotes the monomer concentration at a flat surface. The capillary length R_{C} is given as

$$R_{\mathrm{C}} = 2\sigma V_{\mathrm{a}}/k_{\mathrm{B}}\, T \tag{3}$$

where σ, V_{a}, k_{B}, and T denote the surface tension, the atomic volume, the Boltzmann constant, and temperature, respectively. As long as the monomer concentration in the matrix is higher than $c^{\mathrm{GT}}(R)$ of any NPs, all the NPs increase the sizes at the expense of monomers. (For simplicity the diffusion limited aggregation is assumed.) Once the monomer concentration in the matrix is lower than the $c^{\mathrm{GT}}(R)$ of one of the NPs, the NP shrinks. Since smaller NP has higher $c^{\mathrm{GT}}(R)$ as shown in Equation (2), smaller NPs shrink and larger ones grow. The growth mode changes to the Ostwald ripening.

NIMS (National Institute for Materials Science) group [113] carried out in situ measurements of implantation-induced light emission from silica glass implanted with 60 keV Cu^- ions up to a few of 10^{17} ions/cm^2, with attaching spectrometers and CCD arrays to an implantation chamber. They also carried out in situ measurements of optical absorption under the implantation with changing the optical paths. The results are shown in Fig. 2(a) and (b). Figure 2(a) shows evolutions of the implantation-induced luminescence intensity from isolated Cu^+ ions (monomers) in silica glass with the fluence, which is proportional to the concentration of the Cu monomers in the substrate. Below the fluence of 0.5×10^{16} ions/cm^2, the monomer concentration steeply increases, indicating that the nucleation has not started. Around a fluence of 0.5×10^{16} ions/cm^2, the monomer concentration turns to decrease, which means the beginning of the nucleation. At fluence $2-4\times10^{16}$ ions/cm^2, the concentration approaches constant values which depend on the ion flux, indicating steady-state growth of NPs. The steady-state concentration is higher under higher flux implantation.

Figure 2(b) shows the fluence dependence of the optical absorption at the surface plasmon resonance (SPR) peak of Cu NPs in SiO_2. The absorption intensity is approximately proportional to the total number of Cu atoms forming Cu metal NPs. It should be noted again that Fig. 2(a) shows the fluence dependence of the Cu *monomer* concentration, and that Fig. 2(b) shows the dependence of Cu *NPs*, i.e., *aggregates of monomers*. An interesting feature is the incubation period below the fluence of 3×10^{16} ions/cm^2, where the absorption is kept approximately zero.

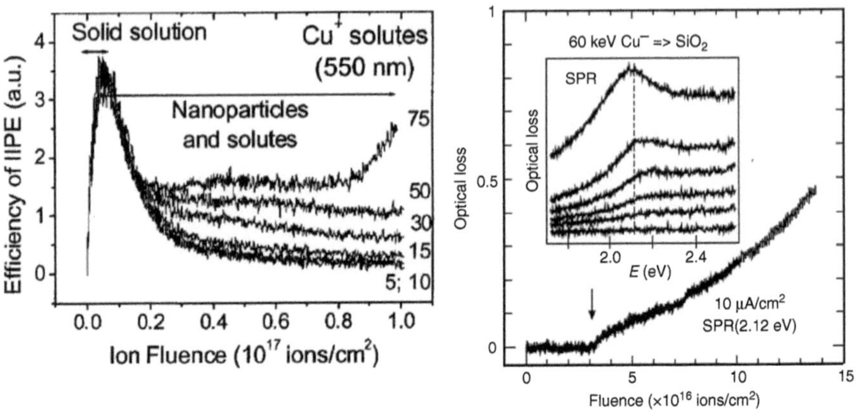

Fig. 2 Fluence dependences of (**a**) the ion-induced photo-emission (IIPE) intensity normalized by the ion flux, of the Cu^+ solute band at various ion fluxes and (**b**) the optical absorption intensity at the surface plasmon resonance (SPR) at an ion flux of 10 μA/cm^2, under the implantation of 60 keV Cu^- ions. The numbers in the figure (**a**) indicate the ion flux in the units of μA/cm^2. The inset in figure (**b**) indicates optical absorption spectra at various fluences (Fig. 2(a) is reprinted with permission from Plaksin et al. [113]. © (2006) American Institute of Physics)

Corresponding calculations were given by Strobel [124] using Kinetic 3-Dimension Lattice Monte Carlo (K3DLMC) method and shown in Fig. 3. The fluence dependence of the monomer concentration is similar with the experimental results shown in Fig. 2(a). It should be noted that Fig. 3 also shows the fluence dependence of the number of NPs (n), while Fig. 2(b) shows the SPR absorption, i.e., approximately the total number of Cu atoms forming NPs. While the number of NPs (n) becomes almost constant after finishing the nucleation stage, the total number of Cu atoms forming NPs increases almost linearly. However, these dependences are not inconsistent with each other. Also in the experiments, the number of NPs may be constant in the growth stage, e.g., $4 - 8 \times 10^{16}$ ions/cm^2. Since the monomer concentration is almost constant in this fluence region as shown in Fig. 2(a), all the implanted ions should be absorbed by NPs that have already formed. Consequently, the SPR absorption increases almost linearly.

The incubation period for the NP formation is clearly shown in the fluence dependence of the NP number in Fig. 3. The NPs start forming when the nucleation has started. It is contrary with the experimental results shown in Figs. 2(a) and (b). The nucleation starts around 0.5×10^{16} ions/cm^2 as shown in Fig. 2(a), while the SPR appears around 3×10^{16} ions/cm^2 as shown in Fig. 2(b). The difference will be discussed in Section 3.1.2.

As for the formation, growth, and Ostwald ripening processes of NPs, extensive studies have been carried out from numerical simulations (e.g., [125, 126, 127, 56]) and from experiments (e.g., [116, 136, 36, 42, 117]). To form a narrow size

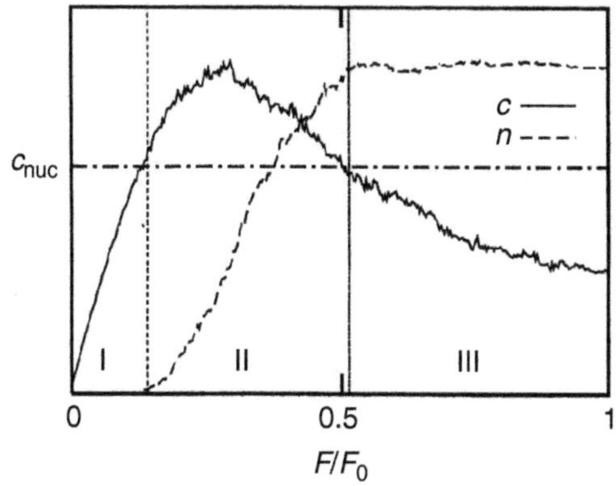

Fig. 3 Fluence dependence of the monomer concentration c and the NP density n of a finite homogenous system under a constant flux, simulated by a kinetic 3D lattice Monte Carlo (K3DLMC) method. Regions I, II, and III denote accumulation/saturation, nucleation, and growth regions, respectively. The effective concentration threshold for nucleation is indicated as c_{nuc} (Strobel [124])

distribution of NPs by ion implantation, a method to separate the nucleation and the growth stages was proposed, which consists of short-time implantation and spike annealing for the nucleation stage followed by long-time implantation for the growth stage [114].

3.1.2 Metal–Nonmetal (M–NM) Transitions of Nanoparticles

Metal NPs are formed by aggregation processes starting from isolated atoms (monomers). The monomer is an atom having discrete energy levels, which is different from metals which have continuous energy levels around the Fermi energy. By increasing the size, i.e., the dimer, the trimer, etc., the density of the discrete energy levels increases. Above a certain size of cluster, the energy levels around the Fermi level are practically regarded as continuous. This is called "metal–nonmetal (M–NM) transition" of NPs. According to a simplified criterion [148, 47], the M–NM transition occurs when the cluster includes 50, 39, and 34 atoms of Fe, Co, and Ni, respectively. These sizes approximately correspond to NPs with a diameter of $\sim$1 nm.

Too long incubation period for the SPR absorption comparing with the monomer concentration shown in Figs. 2(a) and (b) can be explained by the context of the M–NM transition of Cu NPs. Since the SPR is due to the metallic nature of NPs, nonmetallic NPs do not show the SPR absorption. Although the nucleation starts and the NPs are formed around 0.5×10^{16} ions/cm^2, the NPs are too small to show metallic characters, i.e., the SPR absorption. After increasing the sizes of NPs and going through the M–NM transition, the NPs show the SPR absorption. The incubation fluence of 3×10^{16} ions/cm^2 may correspond not to the nucleation of NPs but to the M–NM transition.

As well known, the ferromagnetism of bulk transition metals, Fe, Co, and Ni, is due to itinerant *d*-electrons, i.e., a metallic property. Another interesting issue concerning the M–NM transition is whether the magnetic–nonmagnetic transition occurs simultaneously with the M–NM transition of magnetic metal NPs. Nickel magnetic NPs were formed in SiO_2 by implantation of negative Ni ions of 60 keV to $\sim5\times10^{16}$ ions/cm^2 [4]. The number of Ni atoms included in the sample was evaluated by three different methods: Rutherford backscattering spectrometry (RBS) detects the total number of Ni atoms in the sample, irrespective of the existing forms whether isolated or aggregated. The optical absorption at the SPR peak detects the number of Ni atoms forming metal NPs that are larger than the critical size of the M–NM transition. The saturation magnetization detects the number of Ni NPs having the magnetization. Most (>85%) of the Ni atoms form *metallic* NPs even in as-implanted state. It is predicted from the calculation that the M–NM transition of Ni NP occurs around 1 nm of diameter [148]. This is consistent with TEM observation of the as-implanted sample, which shows the mean diameter and the standard deviation of Ni NPs are 2.9 nm and 1.0 nm, respectively, indicating that most of NPs are larger than 1 nm [6]. However, the saturation magnetization in the as-implanted state is ~50% of the value expected from the bulk state, and increases to ~70% after annealing at 800°C. The magnetization shows different size dependences from the metallic nature of the NPs. Although these results are neither direct nor clear evidence, the simultaneity between the M–NM and the magnetic–nonmagnetic transitions is not trivial.

3.1.3 High-Fluence Effects for Nanoparticle Formation

The first success of ion implantation in industrial applications was in the 1970s in impurity doping to semiconductor devices. In this application, ion implantation has shown excellent controllability of dopant concentration due to ion-current measurement, reliable precision of dopant depth profiles due to the acceleration voltage, spatial controllability with mask techniques, and the robustness against the surface states and the interactions with other impurities of the substrate. In this application, a fluence less than 10^{14} ions/cm^2 was enough, since the dopants were introduced as isolated atoms. It should be noted that later the applications of ion implantation in the semiconductor industry have been exploited up to high fluence, such as SIMOX, smart-cut, etc.

SRIM code, originally called TRIM [149], has been developed with assuming low implantation fluences, although nowadays SRIM code is one of the most common codes in the ion implantation community and is regarded as a basic tool to estimate the ion range, and so on. The SRIM code assumes that implanted ions are independent of each other: Every time when introducing a new ion for simulation, the implanted substrate is assumed to be virgin, i.e., including no previously implanted ions, while this procedure is repeated up to 100,000 times for statistical averaging. All the effects induced by previously implanted ions, such as sputtering removal of the surface layer, composition changes of the substrate due to previously implanted element, etc., are not included.

However, NP formation by ion implantation requires very high concentration of implants, which sometimes exceeds over 10 at.% (atomic%). The effects of the previously implanted ions cannot be neglected, and are clearly shown in Fig. 4(a) and (b). Figure 4(a) shows comparisons between calculated depth profiles by SRIM (dotted lines) and TRIDYN codes (solid lines) for implantation of Zn ions of 60 keV to SiO_2. TRIDYN code is less common but includes the sputtering loss of the surface layer and the composition changes induced by high-fluence implantation [91, 92]. While both the profiles are identical at 1×10^{16} ions/cm^2, shallowing of the TRIDYN profile starts at 3×10^{16} ions/cm^2, where the peak concentration comes to ~10 at.%. At 1×10^{17} ions/cm^2, the difference between both the profiles are clear. While SRIM code gives the mean projected range of 46 nm, TRIDYN gives 30 nm. By exceeding the fluence of 1×10^{17} ions/cm^2, the total content of implanted ions shows a saturation. Although the profile at 2×10^{17} ions/cm^2 slightly shifts to the surface side compared with that at 1×10^{17} ions/cm^2, both the contents are almost the same. An interesting feature is the profile at 3×10^{17} ions/cm^2, which is slightly smaller than that at 2×10^{17} ions/cm^2. The peak concentration has never exceeded 35 at.%. At high fluence, a steady state is attained under balance between outgoing Zn ions by sputtering and incoming Zn ions by implantation. Of course, the fluence for the saturation depends on the ion energy, the combination with the ion species and the substrate. Calculated fluence dependence ofthe areal density is shown in

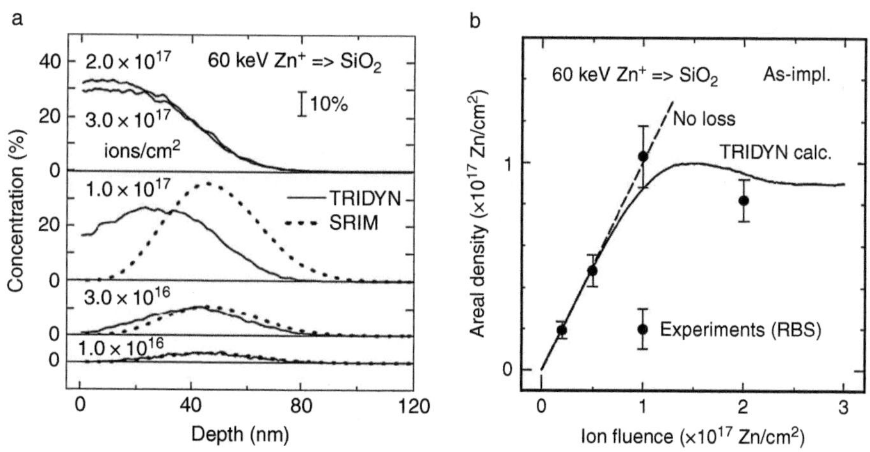

Fig. 4 (**a**) Depth profiles of Zn atoms in SiO_2 implanted with Zn^+ ions of 60 keV to various fluences and (**b**) the fluence dependence of Zn areal density (ions/cm^2), calculated by TRIDYN code are shown by *solid lines* in each figure. The profiles calculated by SRIM2003 code at the fluences of 0.10, 0.30, and 1.0×10^{17} ions/cm^2 are shown by *dotted lines* in figure (**a**). Experimental results of Zn areal density are plotted in figure (**b**) as *closed circles* (Amekura et al. [18].)

Fig. 4(b) with experimental data determined by RBS [18]. Whatever high fluence is used, it is shown that the areal density of Zn ions never exceeds $\sim 1\times10^{17}$ ions/cm^2.

When the II&TO method with high-fluence implantation is applied, these facts should be reminded: (1) the depth profile of implanted ions can be shallower than the value estimated by SRIM code. (The shallowing of the depth profile can be advantageous for oxidation of NPs.) (2) The atomic concentration saturates, and the depth profile approaches to the steady state one. Ion implantation up to very high fluences might be nothing but waste of time. However, it should be noted that high fluence of 1×10^{17} ions/cm^2 or higher is required for efficient ZnO NP formation because of other reasons (see Section 4.2.5).

It should also be noted that Maxwell–Garnett (MG) theory, which is often used for calculation of optical absorption spectra of insulators including metal NPs [84], is, *in principle*, applicable only for dilute limit of the NP concentration. Although the reason is not clear, the MG theory can be *practically* applicable up to several atomic percent (at.%) of the concentration or higher. It should be noted that the implantation of Zn ions of 60 keV up to a fluence of 1×10^{17} ions/cm^2 gives a peak concentration of $\sim$25 at.% as shown in Fig. 4(a).

3.1.4 Substrates

Ion implantation introduces a lot of defects in the substrates, and the NP formation requires high fluence. Selection of the substrate materials is important. Most of the crystalline materials are subjected to amorphization or at least severe damage by high-fluence implantation. Annealing is important for recovery of good crystallinity. Since the II&TO method includes the annealing step in oxidizing atmosphere, the annealing step may have two effects, i.e., the oxidation of metal NPs and the damage recovery of the substrate. As pointed out by Marques et al. [83], crystalline direction in anisotropic crystals may have a special importance.

Since we have mainly used silica glasses, i.e., amorphous SiO_2, we are free from problems of amorphization. However, for example, the bonding angle distribution of amorphous SiO_2 network changes with the ion fluence [37]. While the SiO_2 samples before and after the implantation are both amorphous, the structures may not be similar. Furthermore, SiO_2 has some special properties induced by ion implantation, i.e., radiation-induced compaction (densification), radiation-induced viscous flow, and anisotropic deformation [37, 105, 123, 29]. Fig. 5 shows surface morphology of an SiO_2 sample detected by AFM (atomic force microscopy) (a) before and (b) after Zn^+ ion implantation of 60 keV to a fluence of 1×10^{17} ions/cm^2 [13]. The Zn ion irradiation induces the surface smoothing, probably due to the radiation-induced viscous flow: The surface roughness R_q, i.e., the root mean square of height deviation, decreases from 1.51 nm to 0.39 nm by the irradiation. However, after increasing the fluence up to 2×10^{17} ions/cm^2, some textures, probably due to Zn NPs, appear on the surface. Long-time sputtering due to the high fluence removes the surface layer, and the embedded Zn NPs appear on the surface (Amekura et al. unpublished).

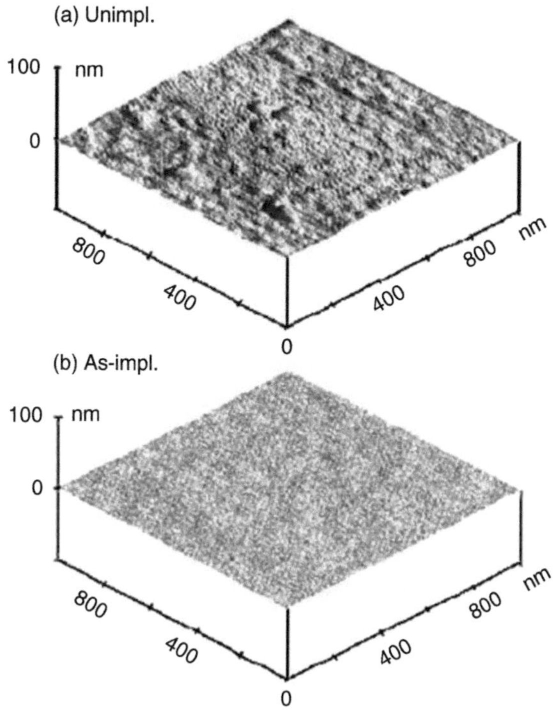

Fig. 5 Surface morphology change of a SiO_2 sample induced by Zn ion implantation of 60 keV to a fluence of 1.0×10^{17} ions/cm^2 detected by AFM, (**a**) before and (**b**) after the implantation

Also SiO_2 is very sensitive to electronic excitation like alkali halides. Defect formation via electronic excitation is often observed in this material (e.g., [48]). In fact, voids observed in Zn-implanted SiO_2 are very sensitive even to electron exposure for TEM observation, and disappear after long-time exposure [135]. Instabilities of substrate materials should be considered when selecting the substrates.

Responses of various insulators to ion irradiation have been studied and summarized by Townsend et al. [131] in order to fabricate the ion-implanted surface optical waveguides.

3.1.5 Other Issues

While implants are accelerated in the form of ions, i.e., charged particles, the charge state of the implants consisting metal NPs is neutral. It is worthwhile to discuss the charge states of ions (atoms) implanted in the insulating substrate. The question is whether the charge states of accelerated ions are preserved in the substrate to some extent, or not. In most cases, implanted metal ions show positively charged states, such as M^+, M^{2+}, or M^{3+}, etc., in insulators, when the fluence is low enough. However, the ions should be neutral to form NPs. Otherwise, the repulsive coulomb force hinders the NPs formation. McHargue et al. [87] studied the fluence dependence of the charge states of Fe implants in Al_2O_3, which were implanted as Fe^+ ions of

160 keV, using the conversion electron Moessbauer spectroscopy. It is interesting to note that Fe^{2+}, Fe^{3+}, Fe^{0} states are observed, while the Fe ions are implanted as the Fe^{+} state. By increasing the fluence, the Fe^{2+} state decreases and the Fe^{3+} state and Fe^{0} state (metal aggregates) increase. The results indicate that the charge states of implants change in accordance with the fluence, even when implanting with the same charge state of ions. The charge compensation between different charge states is readily acceptable for semiconducting or metal substrates, but not for the insulating substrates. Electronic excitation due to ion implantation may play an important role for the charge compensation.

Another example of the nonconservation of the charge state of the implanted ions is Er-implanted insulators. While luminescent Er in 1.5 μm region is in Er^{3+} state, in most of cases, Er ions are implanted at Er^{+} or Er^{2+} states.

As for implantation of negative ions, each of the negative ion which consists of a neutral atom and a loosely bounded electron, the ions are implanted into solids in a neutral state with losing the loosely bounded electrons at the surface of the substrate (Toyota, 1996, PhD Thesis, Kyoto University, unpublished).

3.2 Thermal Oxidation of Metal Nanoparticles in Matrix

3.2.1 Migration of Oxygen

Since we have mainly used SiO_2 as substrates for the II&TO method, migration processes of oxygen species in SiO_2 are very important. These processes have been well studied because of the industrial importance for thermal oxidation of Si substrate. According to thermal oxidation experiments of Si substrate covered by SiO_2 layer, using oxygen gas including O isotopes [51], the isotopes were only found close to the boundary between the SiO_2 layer and the Si substrate, but not in the SiO_2 layer, indicating that oxygen migrates in the SiO_2 layer in the form of O_2 molecules and is not dissociated inside the SiO_2 layer but at the boundary. From the first principle calculations [50, 66, 102], it was confirmed that O_2 molecules are stable in amorphous SiO_2 network, while O atoms are strongly bounded in SiO_2 network with forming Si–O–O–Si bonds. The diffusion constant of O_2 molecules in SiO_2 is given as

$$D(T) = D_{\mathrm{o}} \exp(-E/k_{\mathrm{B}}T) \tag{4}$$

where $D_{\mathrm{o}} = 2\times10^{-9}$ cm^2/s and $E = 1.3$ eV, respectively [140]. The diffusion length L of O_2 molecules is given as

$$L = \sqrt{4D(T)t} \tag{5}$$

where t denotes the annealing duration.

Temperature dependence of the diffusion length L is plotted in Fig. 6 for annealing duration $t = 1$ h. Even at 1000°C, the diffusion length L is shorter than 200 nm. For oxidation of implanted metal NPs, theprojected range of ions should be of the

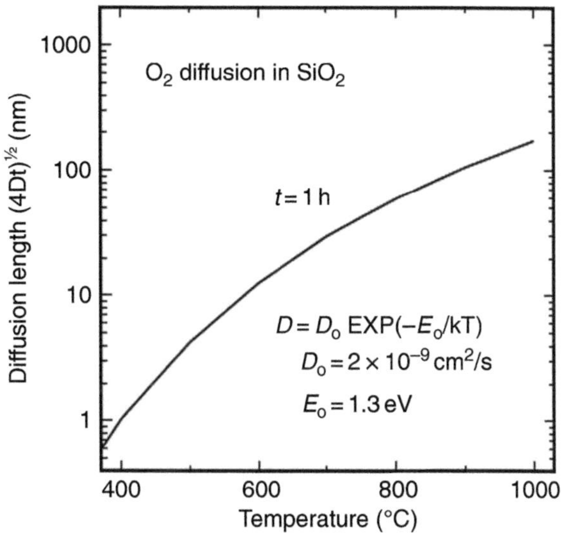

Fig. 6 Temperature dependences of diffusion length $(4Dt)^{1/2}$ of O_2 molecules in SiO_2 for annealing time $t = 1$ h. The diffusion constant $D(T) = D_o\exp(-E/kT)$ was used with values of $D_o = 2\times10^{-9}$ cm^2/s and $E = 1.3$ eV

same order of the oxygen diffusion length L. It is known that very high annealing temperatures make metal NPs in SiO_2 unstable [21, 3]. Implantation of a few tens to a few hundreds kiloelectron volts is appropriate for the II&TO method.

On the other hand, the projected range of megaelectron volts ions in SiO_2 is larger than 1 μm (e.g., 1.6 μm for 2 MeV Cu in SiO_2). Oxidation of metal NPs formed by megaelectron volts implantation requires very long annealing duration or very high temperature. It is difficult to attain such long diffusion lengths by oxidation around 1000°C for 1 h. However, atmosphere-dependent effects for 1 h annealing of SiO_2 implanted with Cu, Ag, or Au ions of 2 MeV between reducing (H_2/N_2) and oxidizing (air) atmospheres are reported by Universidad of Nacional Autonoma de Mexico [101, 118]. Roiz et al. ascribed the atmosphere effects to diffusion of H_2 but not O_2. According to them, the diffusion constant D of H_2 at 230°C is 10^{10} times larger than that of O_2 at the same temperature. In order to simplify the discussion, we did not use hydrogen annealing. In this chapter, we have compared annealing atmosphere effects between pure oxygen gas and vacuum, not in hydrogen.

Johannessen et al. observed partial oxidation of Cu NPs in SiO_2 implanted MeV Cu ions after annealing in H_2/N_2 gas. They ascribed it to oxidation by oxygen atoms or molecules liberated from SiO_2 substrate by radiation damage [67]. This process cannot be neglected in kiloelectron volt implantation, and examples are shown in Section 4.2.3.

3.2.2 Criteria for Reactions Between Implants and Substrate

When metal ions (M) are implanted to a substrate material, e.g., SiO_2, metal NPs are not the only products. In some cases, oxides, silicides, and silicates can be formed. Because of the nonequilibrium nature of ion implantation, the products could not be completely predicted by thermodynamics. This topic was already reviewed by Cattaruzza [31]. However, since it is important to understand the mechanisms of the II&TO method, a short summary is given here.

Hosono [58] found that the angle distribution of Si–O–Si bonds of implanted SiO_2 is similar with that of unimplanted SiO_2 at high fictive temperature T_f ~3000 K. He assumed that the implantation-induced reactions were similar to those in thermal equilibrium at the high fictive temperature. He proposed that the implantation-induced reactions in SiO_2 were explained by the Gibbs energy changes $\Delta G^{\circ}(T)$ at T ~3000 K. Later, Hosono et al. [59] approached this problem from another point of view based on the electronegativity (EN). Implants (M) with EN $<$ 2.5, e.g., Li, B, etc., pull oxygen atoms out from SiO_2 network forming M–O bonds and leaving Si–Si bonds. On the other hand, implants with EN $>$ 3.5, e.g., F, substitute the oxygen sites of SiO_2 network, leaving O_2 molecules or peroxy radicals. Only implants with $2.5 < EN < 3.5$ are chemically inactive with the SiO_2 network, forming aggregates, i.e., NPs.

On the other hand, Padova group (e.g., [31]) proposed the two-step model, which distinguishes two different energy steps in ion implantation process, i.e., high-energy ballistic processes and low-energy chemically guided processes. According to Cattaruzza, the two-step model gives the best prediction of the products in the three models. This means that thermodynamics is applicable to some extent by including the ballistic processes such as radiation damage.

3.2.3 "Thermodynamics" of the II&TO Method

As shown in Section 3.2.2, thermodynamics is still applicable to ion beam–based material syntheses such as metal NP formation, if the effects of the ballistic processes are included. Moreover, thermal oxidation, which is one of the most important processes of the II&TO method, is also governed by thermodynamics. Reactions and the Gibbs energy changes concerning three metal species, i.e., Ni, Zn, and Cu, which will demonstrate the oxide NP formation in this chapter, are summarized in Table 2. The reactions (i–iii), (iv–vii), (viii–ix), and (x–xii), respectively, correspond to the aggregation of metal atoms, and the formation of oxide phases, silicate phases, and silicide phases.

Next, we discuss about internal oxidation of metal NPs, i.e., the formation of oxide NPs by possible reaction with metal NPs fabricated by ion implantation and oxygen atoms from the substrate lattice. For example, metal species M are implanted in SiO_2, and then vacuum annealing is carried out to induce the oxidation of M by oxygen from SiO_2 network at the expense of the reduction of Si. (It should be noted that this process is different from the reaction with O species librated from SiO_2 network by radiation damage. This process assumes the reaction with oxygen atoms forming SiO_2 network, i.e., which includes breaking of SiO_2 network.) Figure 7 shows the Ellingham diagram (e.g., [128]). The ordinate means the Gibbs energy changes for the oxide formation. Most of the substrate materials for ion-implanted NPs, e.g., SiO_2, Al_2O_3, MgO, locate at the bottom part of Ellingham diagram, indicating that metal atoms (i.e., Si, Al, and Mg) and oxygen atoms are strongly coupled.

Table 2 Gibbs energy changes of related reactions for oxide nanoparticle formation. Gray columns indicate positive or nearly positive values of the Gibbs energy changes, i.e., unpreferable reactions

Reaction		ΔG° (kJ/mol)		
		$T = 298$ K	1000 K	1200 K
(i)	Ni (atom) → Ni (metal)	–386	–281	–252
(ii)	Zn (atom) → Zn (metal)	–95	–15	+7
(iii)	Cu (atom) → Cu (metal)	–298	–207	–182
(iv)	$Ni + 1/2O_2 \rightarrow NiO$	–212	–149	–132
(v)	$Zn + 1/2O_2 \rightarrow ZnO$	–320	–248	–225
(vi)	$Cu + 1/2O_2 \rightarrow CuO$	–128	–66	–50
(vii)	$Cu + 1/4O_2 \rightarrow Cu_2O$	–74	–48	–41
(viii)	$2ZnO + SiO_2 \rightarrow Zn_2SiO_4$	–31	–29	–28
(ix)	$2NiO + SiO_2 \rightarrow Ni_2SiO_4$	–11	–5	–4
(x)	$Ni + SiO_2 \rightarrow NiSi + O_2$	+768	+647	+614
(xi)	$2Ni + SiO_2 \rightarrow Ni_2Si + O_2$	+709	+595	+563
(xii)	$3Ni + SiO_2 \rightarrow NiSi + 2NiO$	+344	+349	+350

On the contrary, oxide phase of implants, e.g., Cu, Zn, Ni, locate at the upper part of Ellingham diagram, indicating that the oxide phases of Cu, Zn, and Ni are less stable than the substrate oxide materials. If the internal oxidation, i.e., vacuum annealing, is applied to, e.g., Zn-implanted SiO_2, the formation of ZnO at the expense of the reduction of Si cannot be expected because SiO_2 is more stable than ZnO as shown

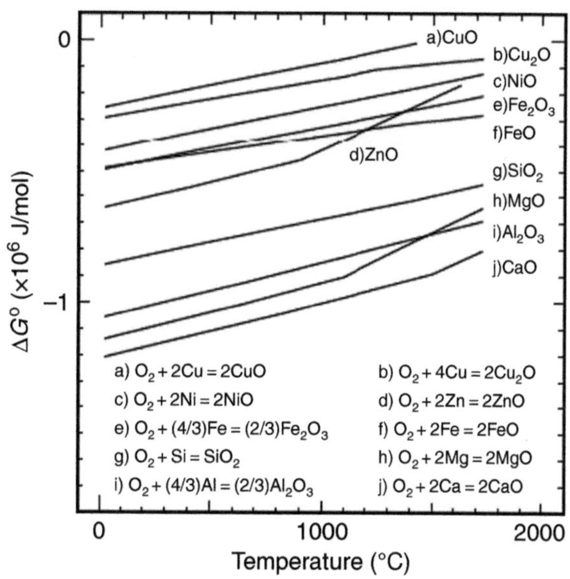

Fig. 7 Ellingham diagram of oxides, i.e., standard free energy changes of formation of oxides as a function of temperature, of some oxides related to this study

in the Ellingham diagram. On the other hand, internal oxidation may be possible in Si-implanted ZnO matrix where SiO_x NPs can be formed in ZnO at the expense of Zn reduction. However, the difference of the Gibbs energy change between ZnO and SiO_2 is smaller when compared with the pairs of CuO–SiO_2, Cu_2O–SiO_2, and NiO–SiO_2. This can be a reason why various reactions are observed in Zn-implanted SiO_2 system, which will be shown in Section 4.2.

As for thermodynamics data, the Kubaschewski's book [76] and the FACT web [44] were used.

4 Oxide Nanoparticle Formation

Oxidation behaviors and properties of four different oxide NP systems, i.e., NiO [7, 9], ZnO [10, 12, 13, 15, 16, 19, 20], CuO [11], and Cu_2O [14, 16], in SiO_2 are discussed in this section. Each NP system shows strong contrast with each other. While NiO, CuO, and Cu_2O NPs are formed inside the SiO_2 substrates after oxidation, ZnO NPs mainly form on the surface of SiO_2 substrate. Similar shallowing of the depth profiles are reported in Fe oxide NPs in Al_2O_3 [87], and Co_3O_4 NPs in SiO_2 [35]. While NiO and ZnO NPs show transparency in the visible region due to the bandgaps larger than ~3 eV, CuO and Cu_2O show color due to the relatively small bandgaps of ~1.5 and ~2.1 eV, respectively. Under atmospheric pressure of oxygen gas, ZnO, NiO, and CuO NPs are the most stable phase but not Cu_2O NPs. A special method is necessary for the formation of Cu_2O NPs.

4.1 NiO Nanoparticles in SiO_2: A Simple Embedded System

4.1.1 Fundamental Properties of NiO

Nickel oxide (NiO) is an antiferromagnetic insulator (Neel temperature $T_N = 525$ K) with a crystalline structure of NaCl type, although the band theory predicted the metallic natures of this material. This material has a historical importance, because the concept of Mott–Hubbard insulator, i.e., insulators due to strong electron correlation, was firstly introduced. The sample is transparent, and the absorption edge locates at ~4 eV. (In some cases, the crystals show weak green color due to forbidden *d–d* transitions.) Later photoelectron spectroscopy has shown the absorption edge is due to transitions from the oxygen *p*-band to the upper Hubbard *d*-band, i.e., a charge transfer gap, instead of the pure Mott–Hubbard gap, i.e., transitions between the lower and the upper Hubbard *d*-bands [132]. Recently, ultrafast optical nonlinearity has been observed in some quasi-one-dimensional Mott insulators, e.g., Sr_2CuO_3 [98]. Some interesting properties may be found in nano-sized zero-dimensional NiO.

4.1.2 Formation of NiO NPs and Discussion

Optical-grade silica glasses (SiO_2) of KU-1 type (OH 820 ppm) of 15 mm diameter and 0.5 mm thickness were implanted with negative ions of Ni (Ni^-) of 60 keV to a fluence of 5.5×10^{16} ions/cm^2. Figure 8 shows the optical absorption spectra of an implanted sample (a) in as-implanted state, after annealing at 800°C for 1 h in (b) vacuum or (c) oxygen gas [7]. Since the implanted sample is brownish but semi-transparent, multiple reflections in the sample cannot be excluded. Each absorption spectrum was determined from four different measurements of transmittance (T) incident from the implanted side T_f, incident from the rear side T_r, reflectance (R) incident from the implanted side R_f, and incident from the rear side R_r [5].

In the as-implanted state, the products are metal Ni NPs, and the spectrum agrees well with the one calculated from Mie theory [89]. The broad peaks at 3.3 and 5.8 eV are characteristic in the formation of Ni NPs in SiO_2 [5,6]. The sample shows brownish color. After annealing in vacuum at 800°C for 1 h, the spectrum is almost the same as the as-implanted state, indicating that the products are still metal Ni NPs. The sample still shows brownish color. However, after annealing in oxygen gas at the same temperature, 800°C, for the same duration, 1 h, the spectrum drastically changes. Low energy absorption ($E < 3.5$ eV) completely disappears, and a steep absorption edge appears at around ~4 eV. The color of the sample becomes transparent. This change is consistent with the formation of NiO NPs, because NiO is a transparent insulator with a band gap of ~4 eV [132].

Results of isochronal annealing in oxygen gas are shown in Fig. 9. Up to 400°C annealing for 1 h, the spectrum is similar with that of the as-implanted state. After

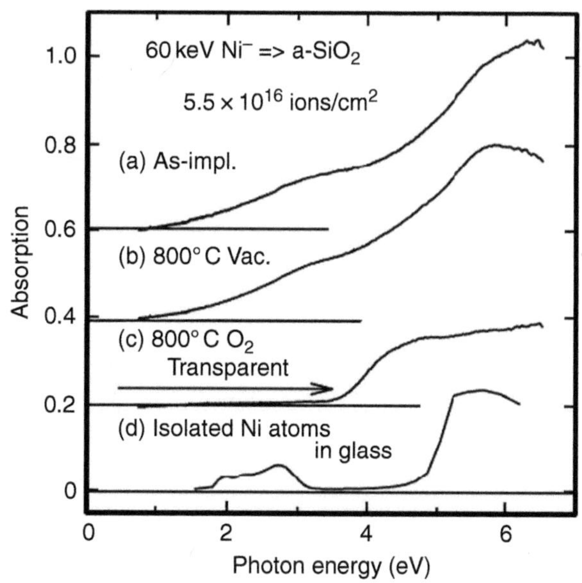

Fig. 8 Optical absorption spectra of SiO_2 implanted with Ni^- ions of 60 keV in (**a**) the as-implanted state and after annealing at 800°C for 1 h in (**b**) vacuum and in (**c**) O_2 gas flow. A spectrum of isolated Ni atoms in silicate glass from literature [43] is shown in (**d**). The spectra are vertically shifted from each other for clarity, with *horizontal lines* showing the zero level of each spectrum (Reprinted with permission from Amekura et al. [7]. © (2004) American Institute of Physics)

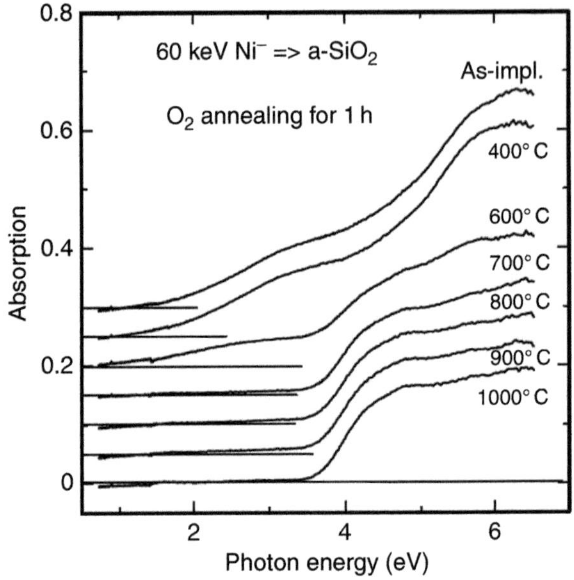

Fig. 9 Optical absorption spectra of Ni^--implanted SiO_2 during isochronal annealing in oxygen gas at 400–1000°C for 1 h each. The measurements were taken at room temperature (Reprinted with permission from Amekura et al. [7]. © (2004) American Institute of Physics). The white bar in (a) indicates the projectile Range Rp

annealing at 600°C, the intensity of the peak around 6 eV decreases. After annealing at 700°C, the absorption below ~3.5 eV disappears and a steep absorption edge appears at around 4 eV, indicating that the oxidation of Ni NPs has completed.

The formation of NiO NPs is also confirmed by grazing incident X-ray diffraction (GIXRD) as shown in Fig. 10. While the sample in the as-implanted state shows diffraction peaks due to an fcc lattice of Ni metal, the diffraction peaks change to an NaCl-type lattice of NiO after thermal oxidation.

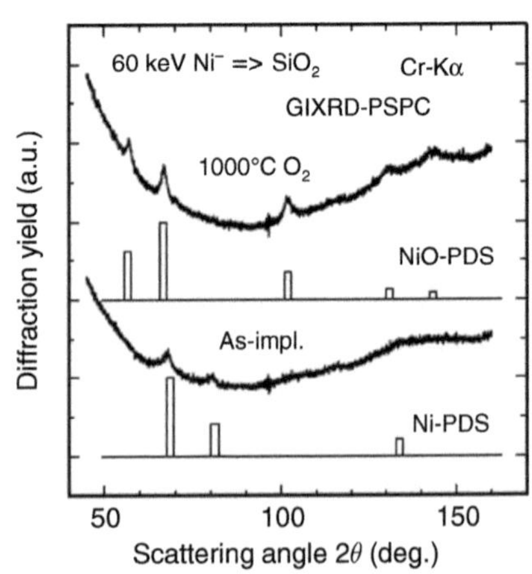

Fig. 10 Grazing incident x-ray diffraction (GIXRD) patterns of SiO_2 implanted with Ni^- ions of 60 keV in the as-implanted state and after annealing at 1000°C for 1 h in oxygen gas

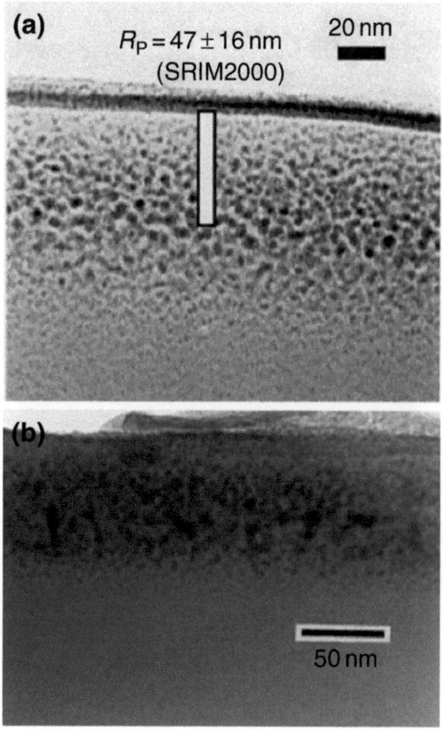

Fig. 11 Cross-sectional transmission electron microscopy (XTEM) images of SiO_2 implanted with 60 keV Ni^- to 5.5×10^{16} ions/cm^2 (**a**) in the as-implanted state (Fig. 11(a) is reprinted with permission from Amekura et al. [6]. © (2004) Elsevier) and (**b**) after annealing at 800°C for 1 h in oxygen gas flow (Fig. 11(b) is reprinted with permission from Amekura et al. [7]. © (2004) American Institute of Physics). Ni and NiO nanoparticles are observed as *black dots* in both (**a**) and (**b**). In (**a**), the surface is coated with a Cr-marker layer. Large structures on the sample surface shown in (**b**) are due to glue

Figure 11 shows cross-sectional transmission electron microscopy (XTEM) images of (a) as-implanted state (Ni metallic NPs) and (b) after annealing in oxygen gas at 800°C for 1 h (NiO NPs). The formation of NiO NPs has been directly confirmed. The mean diameters (d) of Ni NPs in the as-implanted state and of NiO NPs after the oxidation were 2.9 nm and 3.6 nm, respectively. The average number of Ni atoms included in one NP is estimated from the mean volumes $\langle V \rangle$ of NPs and the Ni atom density n in Ni and NiO. The mean Ni numbers included are 1.2×10^3 and 1.3×10^3 atoms for Ni and NiO NPs, respectively. Although the mean diameter of NiO NPs is larger than that of Ni NPs, the numbers of Ni atoms included are almost the same. This is because NiO NPs consist of Ni and O atoms but Ni NPs consist of Ni atoms only.

Comparing the results obtained before and after oxidation, shallowing of depth profiles is less observed. The depth profiles are more clearly detected by RBS [9]. Figure 12 shows the RBS spectra around Ni edge of SiO_2 implanted with 60 keV Ni^- ions in as-implanted state and after annealing at 800°C for 1 h in (b) oxygen gas and in (c) vacuum. After annealing in either of the atmospheres, the peak intensity slightly decreases. However, any distinct shallowing of the depth profile was not observed. (Compare this figure with Fig. 17 (ZnO).) The decrease in the intensity after annealing is a common phenomenon in Ni- or Cu-implanted SiO_2 but not in

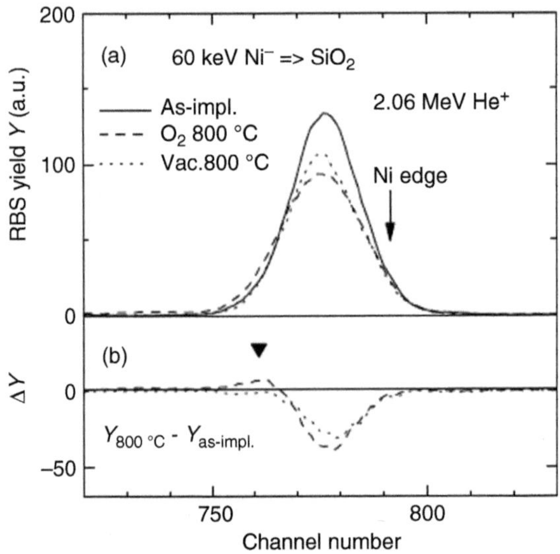

Fig. 12 RBS spectra of SiO_2 implanted by Ni^- of 60 keV around the Ni edge. The *solid*, *broken*, and *dotted lines* in figure (**a**) indicate the depth distributions of Ni atoms in SiO_2 in the as-implanted state, after oxygen annealing at 800°C and after vacuum annealing at 800°C, respectively. The broken (*dotted*) line in figure (**b**) shows the difference between the sample annealed at 800°C in oxygen gas (vacuum) and the as-implanted one. The *downward triangle* indicates the region where $\Delta Y > 0$ (Reprinted with permission from Amekura et al. [9]. © (2005) Elsevier)

Zn-implanted SiO_2. However, the mechanism has not been clarified yet. We speculate that the phenomenon is due to fast migration of Ni or Cu atoms in SiO_2, which have not aggregated into metal NPs but exist as isolated atoms. These isolated atoms in SiO_2 can have higher diffusivity than metal atoms included in NPs, because the atoms included in NPs are bounded to each other. In the case of Zn-implanted SiO_2, most of the Zn atoms are included in Zn NPs at room temperature (RT), because of higher diffusivity of Zn atoms in SiO_2.

In the as-implanted state and after annealing in vacuum, Ni NPs show large magnetization due to ferromagnetic, or more precisely superparamagnetic nature of Ni NPs [6,8]. After oxidation, the large magnetization was lost. Only very small magnetization, probably due to antiferromagnetic NiO NPs, remains.

4.2 ZnO Nanoparticles: NP Formation on the Substrate Surface

Zinc oxide (ZnO) has been known as a compound semiconductor of the II–VI group and has been used as transparent conductive films, varistors, acoustic wave devices, etc., for decades. Recently, ZnO is again receiving considerable attention because this material is a wide-gap semiconductor with a large exciton binding energy of 60 meV, which stabilizes the excitons even at RT and higher. Because of the concentrated oscillator strength with respect to exciton transitions, laser action with low threshold is attainable. One of the big difficulties in this material for electroluminescence (EL) application was p-type doping. However, a reproducible success of p-type doping and EL from p-n junctions was recently reported [133], although many irreproducible successes of the p-type doping were reported before. Recent progresses of studies on bulk ZnO are reviewed by [106].

Nanostructures of ZnO show more exciting optical properties. Because of the concentrated oscillator strength to exciton transitions, laser action is easily attainable even in nanostructures. Moreover, new lasing phenomena characteristic to nanostructures have been observed including random laser [30] self-formed cavity laser [69]. Nanostructures of ZnO are highly attractive and various attempts have been made to fabricate ZnO nanostructures of high quality. One of the successful processes is the II&TO method, which will be described here.

4.2.1 ZnO NP Formation by the Conventional II&TO Method

Formation of Metallic Zn NPs by Ion Implantation

As with Ni^- implantation described in Section 4.1.2, optical-grade silica glasses of KU-1 type (OH 820 ppm) were used for substrates. Zn^+ ions with a mass number of 64 were selected and were implanted with an energy of 60 keV up to a fluence of 1.0×10^{17} ions/cm^2. While negative ions were used for Ni and Cu implantation, positive ions were used for Zn because of the low electron affinity of Zn atoms, i.e., very low efficiency of Zn negative ions. The positive ion flux was limited to less than 2 μA/cm^2 in order to maintain the sample temperature below 100°C during the implantation. The pressure inside the implantation chamber was always less than 1×10^{-4} Pa.

According to SRIM2003 code [149], the projected range in SiO_2 (assuming the density of 2.2 g/cm^3) was 49 nm. However, because of the high fluence of 1.0×10^{17} ions/cm^2, the projected range can be shallower due to the sputtering loss of the surface layer of the substrate. TRIDYN code [91, 92], which includes the sputtering loss and the compositional changes due to high-fluence implantation, predicts a range of 30 nm.

Figure 13 shows the optical absorption spectra of a sample in the as-implanted state and after annealing in oxygen gas at various temperatures for 1 h each [16]. To exclude multiple reflections in the sample, each absorption spectrum was determined from measurements of four different quantities, T_f, T_r, R_f, and R_r as described in Section 4.1.2 [5].

The sample in the as-implanted state shows a very broad band peaked at ~4.8 eV, whose low-energy tail covers all over the visible region and extends to the near-infrared (NIR) region. The as-implanted sample shows brownish color due to the absorption tail. As for the 4.8 eV peak, similar peaks were observed at ~5 eV in Zn-implanted SiO_2 [33] and at ~4.2 eV in Zn-implanted MgO [78]. Both the peaks were ascribed to the surface plasmon resonance (SPR) of Zn NPs. However, both the assignments to the SPR were due to simple speculations. Since there were no other

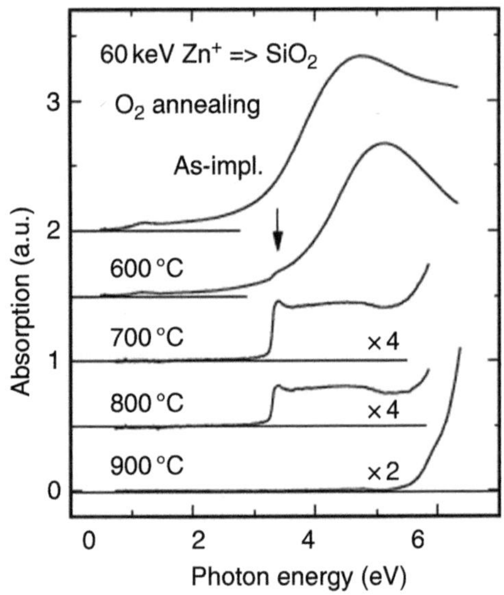

Fig. 13 Optical absorption spectra of SiO_2 implanted with Zn^+ ions of 60 keV up to 1.0×10^{17} ions/cm^2 in the as-implanted state and annealed in oxygen gas flow at 600, 700, 800, and 900°C for 1 h each. The spectra at 700, 800, and 900°C are magnified by a factor of 4, 4, and 2, respectively. The spectra are vertically shifted from each other for clarity, with *horizontal lines* showing the zero level of each spectrum. (Reprinted with permission from Amekura et al. [16]. © (2006) Elsevier)

peaks in the visible and the ultraviolet (UV) region, they ascribed the broad peak to the SPR. However, we have observed a new peak at ~1.2 eV as shown in Fig. 13, which has never been reported before [16]. Since the 1.2 eV peak is reproduced from Mie theory using the refractive index data of bulk Zn [147], the peak is not due to impurities or defects, but of intrinsic nature of Zn NPs. Tentatively, we ascribe the 1.2 eV peak to one of the SPRs of Zn NPs [23]. The relationship with the 4.8 eV band, which was believed as the SPR of Zn NPs, will be discussed elsewhere.

Figure 14 shows grazing incident x-ray diffraction (GIXRD) patterns of the sample in the as-implanted state and after annealing in oxygen gas at various temperatures for 1 h each. Using the Cr Kα line as an x-ray source, the x-ray was incident on the sample at a fixed angle of 3° from the sample surface, and diffracted x-ray peaks were detected by a multichannel position-sensitive proportional counter. The as-implanted sample shows a big broad band around 32° and five small peaks. The broad band is ascribed to the amorphous SiO_2 substrate. All the small five peaks agree well with the powder diffraction pattern of Zn metal, indicating that at least some portions of implanted Zn atoms form Zn NPs with the same crystalline structure as the bulk. Furthermore, the observed peaks show almost the same intensity ratios with the powder patterns, indicating no correlated alignment between Zn NPs.

It should be noted that the diffraction peak intensity of NPs shows a superlinear dependence on the concentration of the implants. By decreasing the concentration, the sizes of NPs decrease, i.e., the width of the peak steeply increases. Although the integrated peak intensity linearly decreases with the fluence, the peak intensity superlinearly decreases. The results shown in Fig. 14 determine the major phase(s),

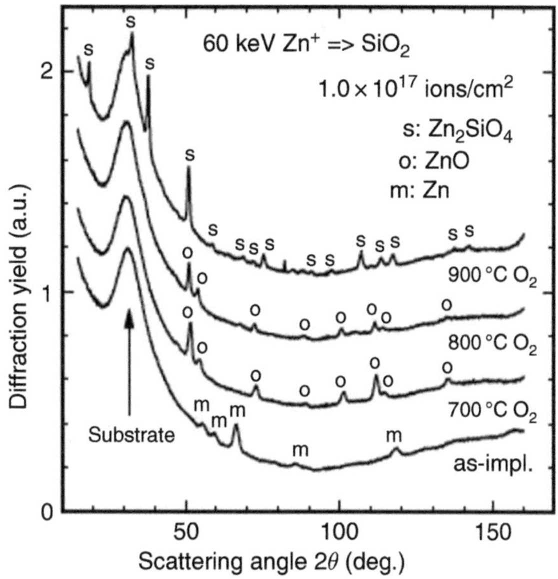

Fig. 14 Grazing incidence x-ray diffraction (GIXRD) spectra of silica glass implanted with Zn^+ ions of 60 keV up to 1.0×10^{17} ions/cm^2 in the as-implanted state and annealed in oxygen gas flow at 700, 800, and 900°C for 1 h each. Diffraction peaks labeled by "m," "o," and "s" are ascribed to Zn NPs, ZnO NPs, and Zn_2SiO_4, respectively (Reprinted with permission from Amekura et al. [16]. © (2006) Elsevier)

but do not exclude the existence of minor phases. As for the minor phases, XPS results will be discussed in Section 4.2.3.

A cross-sectional TEM (XTEM) image of the as-implanted sample is shown in Fig. 15(a) [12]. Spherical-shaped Zn metal NPs of 5–15 nm in diameter were observed in the depth region between 10 and 70 nm beneath the surface. NPs were hardly observed in the region shallower than 10 nm in depth. This is because the implanted ions are too energetic to stop on the surface or region shallower than 10 nm.

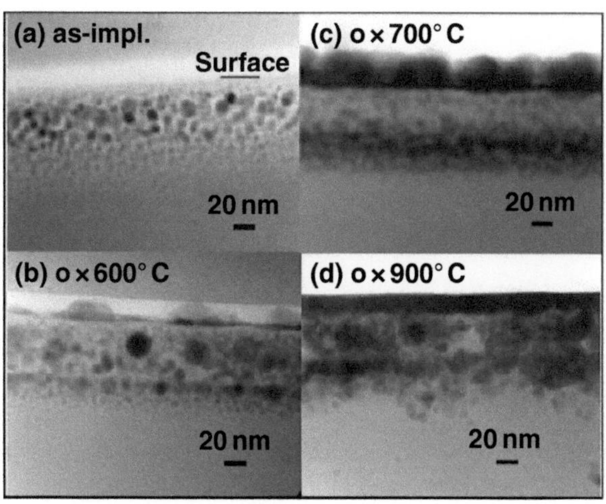

Fig. 15 Cross-sectional TEM (XTEM) images of SiO_2 samples implanted with Zn^+ ions of 60 keV to a fluence of 1.0×10^{17} ions/cm^2 (**a**) in the as-implanted state and after annealing in oxygen gas for 1 h at (**b**) 600°C, (**c**) 700°C, and (**d**) 900°C. In figure (**a**), the substrate surface is indicated by a *line* (Reprinted with permission from Amekura et al. [12]. © (2006) American Institute of Physics)

X-ray photoelectron spectroscopy (XPS) of the implanted samples was carried out using monochromatic Al Kα line (1486.7 eV) as an x-ray source, with an electron neutralizer [13]. For depth profiling, an Ar^+ ion beam of 1 kV was used for sputtering. To evaluate the chemical states of Zn species, XPS spectra at the Zn $L_3M_{45}M_{45}$ Auger transition were detected. This is because the chemical shift of the Auger transition between Zn metal and ZnO is much larger than that of Zn $2p_{3/2}$ transition [96]. Sometimes, XPS analysis using the Auger transition is called x-ray-excited Auger electron spectroscopy (XAES), as we called in our past literature. However, for simplicity, we call this method also XPS, while the spectra are plotted with the kinetic energy of photoelectrons, not the binding energy.

Figure 16(a) shows an XPS spectrum from the sample surface of the as-implanted sample. Only very weak signals are observed, which is consistent with the result of XTEM shown in Fig. 15(a), where no NPs are observed in the layer shallower than 10 nm. It was confirmed from the XPS spectra at other transitions (Si 2*p*, O 1*s*, C 1*s*, Zn 2*p*) that the surface mostly consisted of Si and O atoms, apart from a low concentration of Zn atoms and residual carbon contamination. After removing the surface layer of ~20 nm, a twin-peak structure due to the Zn metallic state appears at 922.2 eV and 995.6 eV [96] in the kinetic energy, which is consistent with the

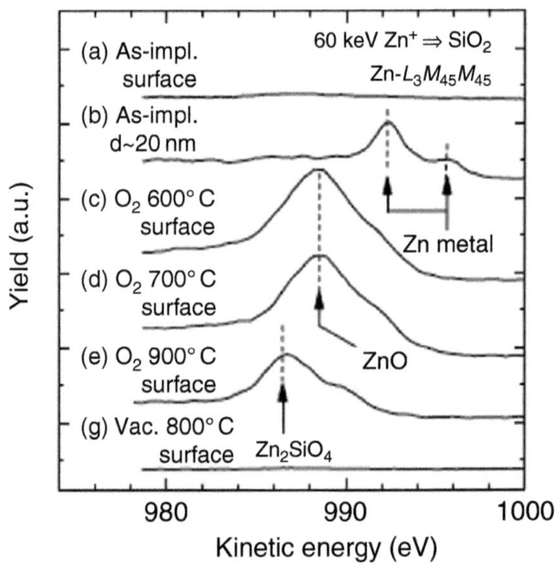

Fig. 16 XPS spectra around Zn-$L_3M_{45}M_{45}$ transitions from SiO_2 samples implanted with Zn ions of 60 keV to a fluence of 1.0×10^{17} ions/cm^2 in the as-implanted state (**a**) and (**b**) and after annealing for 1 h at 600°C in O_2 gas (**c**), 700°C in O_2 (**d**), 900°C in O_2 (**e**), and 800°C in vacuum (**g**). All of the spectra were detected at the surface without sputtering except (**b**), which was detected after sputtering out of the surface layer of ~20 nm thickness (Reprinted with permission from Amekura et al. [13]. © (2006) American Institute of Physics)

XTEM image shown in Fig. 15(a), confirming the formation of Zn metal NPs at the depth of ~20 nm beneath the surface.

ZnO Formation by Annealing in Oxygen Gas

Next, effects of the annealing in oxygen gas are described. It should be firstly noted that all the annealing effects are *evaluated by measurements at room temperature* (RT) after annealing at elevated temperatures. The implanted samples, which consequently include Zn NPs as shown in the Section "Formation of Metallic Zn NPs by Ion Implantation", were annealed in a tube furnace with oxygen gas flow (gas purity of higher than 99.9%, the pressure of $\sim 1.0 \times 10^5$ Pa, the flow of ~100 sccm) at elevated temperatures for 1 h each, where sccm denotes centimeter cube per minute at the standard condition. Up to 600°C annealing, the 4.8 eV absorption band due to Zn NPs is dominant, as shown in Fig. 13. However, after annealing at 700°C for 1 h, the 4.8 eV band disappears and a steep absorption edge appears at ~3.3 eV. The energy of the absorption edge agrees with that of bulk ZnO. The color of the sample changes transparent.

The GIXRD patterns also change drastically as shown in Fig. 14. The five peaks observed in the as-implanted state, all of which are ascribed to Zn metal, completely disappear, and new eight peaks, which are ascribed to ZnO, appear after the annealing at 700°C for 1 h. These facts indicate the transformation of Zn NPs to ZnO due to annealing in oxygen gas at 700°C for 1 h.

In Fig. 13, the spectrum after 700°C annealing is shown with four times magnification, indicating that the integrated absorption intensity drastically decreases after the transformation to ZnO. However, it does not mean the drastic loss of the total Zn content inside the sample, rather it only means much smaller oscillator strength of ZnO in this energy region. The total Zn content was monitored by RBS and the areal densities after annealing at different temperatures are shown in Fig. 17 [15]. Even after annealing at 900°C for 1 h, the Zn content is 1.0×10^{17} ions/cm^2, i.e., the same as the as-implanted state, although a shallowing of the depth profile is visible after annealing in oxygen gas at 700°C or higher for 1 h.

The shallowing is more clearly shown in the XTEM images of Figs. 15(b) and (c). After annealing at 600°C for 1 h, a lot of Zn NPs are still observed inside the SiO_2 substrate with slight coarsening of the sizes. A drastic change from the as-implanted state is the formation of droplet-like ZnO NPs on the sample surface. After annealing at 700°C, the number and the size of the surface ZnO NPs increases, and all the NPs, not only on the surface but also embedded Zn NP in the substrate, change to ZnO or other phases. This is because the sample becomes transparent in the visible region, indicating the disappearance of Zn metallic NPs which absorb photons in the visible region.

It should be noted again that the surface was free from Zn species in the as-implanted state as confirmed by XPS shown in Fig. 16. The formation of the surface NPs indicates the transportation of Zn species from inside of the substrate to the surface and the oxidation close to the surface. This transportation corresponds to the shallowing observed by RBS (Fig. 17).

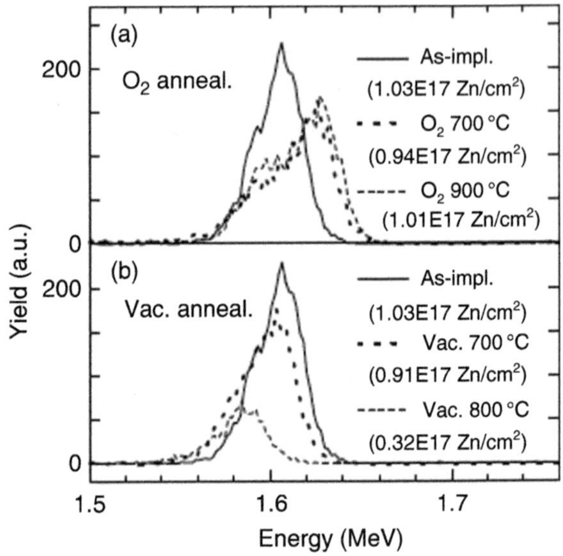

Fig. 17 RBS spectra at the Zn edge of SiO_2 implanted with Zn^+ ions of 60 keV up to 1.0×10^{17} ions/cm^2 after annealing in (**a**) oxygen gas of $\sim1\times10^5$ Pa pressure at 700 and 900°C for 1 h each and (**b**) vacuum of $\sim1\times10^{-3}$ Pa pressure at 700 and 800°C for 1 h each (Reprinted with permission from Amekura et al. [15]. © (2006) Elsevier)

The formation of the surface ZnO NPs is also confirmed by XPS from the sample surfaces. As shown in Fig. 16, a broad peak at ~988.1 eV is observed after annealing at 600°C and 700°C, which confirms that the surface NPs consist of ZnO phase [96].

The partial formation of ZnO NPs after annealing at 600°C for 1 h is also confirmed by the absorption spectra of Fig. 13. Although the spectrum is dominated by the broad absorption due to Zn metal NPs, a weak but distinct kink is visible at ~3.3 eV, indicating the partial formation of ZnO at 600°C.

The surface morphology changes are also detected by AFM (atomic force microscopy) and shown in Figs. 18, 19, and 20 [13]. Figure 18(a) shows the surface morphology of a sample before the implantation. The surface roughness R_q, i.e., the root mean square of height deviation, was 1.51 nm. After the implantation, R_q decreases to 0.39 nm, as shown in Fig. 18(b), indicating a consequence of radiation-induced surface smoothing (see Section 3.1.4). Until annealing at 500°C for 1 h, the surface was as smooth as the as-implanted state. After 600°C annealing for 1 h in oxygen gas, numerous domelike structures of ZnO NPs appear. By increasing the annealing temperature to 700 and 800°C, the size of the domelike structures increases. The corresponding values of R_q increase to 6.92, 13.0, and 16.5 nm at 600, 700, and 800°C, respectively. The annealing temperature dependence of the roughness R_q is plotted in Fig. 19.

On the other hand, the surface was much flatter for samples annealed in vacuum even at 800°C than those annealed in oxygen gas, as shown in Fig. 20, although the R_q slightly increases from the as-implanted value of 0.39 nm to 0.62 and 0.88 nm after annealing at 700 and 800°C, respectively. The small increase

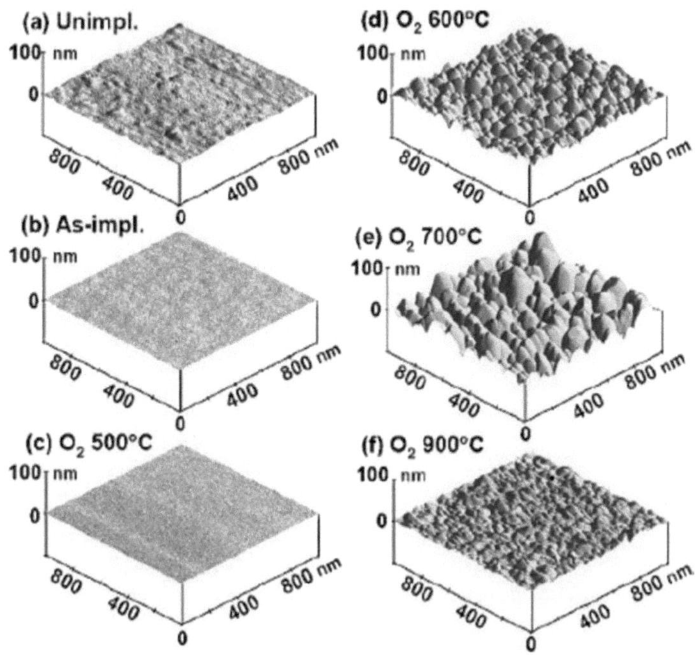

Fig. 18 Evolution of the surface morphology of SiO_2 samples induced by Zn ion implantation of 60 keV to a fluence of 1.0×10^{17} ions/cm^2 and subsequent annealing in *oxygen gas* of $\sim1\times10^5$ Pa pressure detected by AFM, before implantation (**a**), in the as-implanted state (**b**), and after annealing for 1 h at 500°C (**c**), 600°C (**d**), 700°C (**e**), and 900°C (**f**) (Reprinted with permission from Amekura et al. [13]. © (2006) American Institute of Physics)

of the R_q value after vacuum annealing is probably due to thermal roughening. Figure 16(g) shows an XPS spectrum from the surface of the sample annealed in vacuum at 800°C for 1 h. Of course no Zn-related signal is observed on the surface.

As shown in the GIXRD results (Fig. 14), Zn NPs show diffraction peaks whose intensity ratios are almost the same as the powder diffraction pattern. This fact indicates no correlated alignment between the Zn NPs. Contrarily, ZnO NPs show preferential intensity ratio. Intensities of ZnO (xx2) peaks, i.e., Zn (002), (102), (112), and (202), are more pronounced than other peaks. These facts indicate that ZnO NPs show preferential growth along the c-plan, and that each NP has correlated alignment. The non-correlated alignment of Zn NPs and the partial-correlated alignment of ZnO NPs can be explained from the XTEM images (Fig. 15). Since both Zn and ZnO are hexagonal crystals, the driving force comes not from the crystalline structures but rather from the matrix. Because Zn NPs are embedded in amorphous SiO_2, i.e., isotropic media, there are no special directions for growth. On the other hand, a major part of ZnO NPs form on the surface of the substrate, i.e., the growth directions are limited.

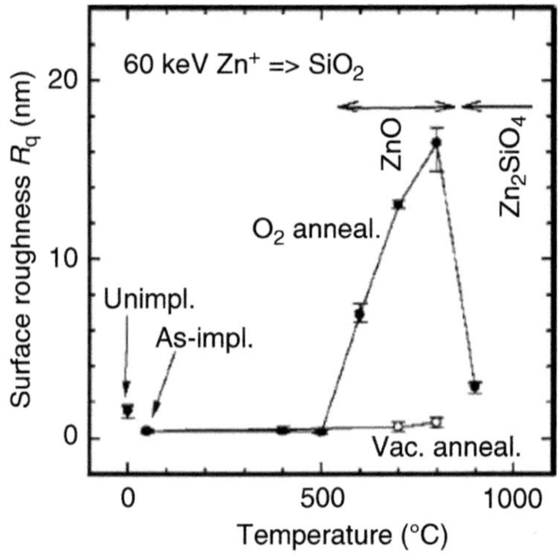

Fig. 19 Annealing temperature dependence of the surface roughness R_q of SiO_2 samples implanted with Zn ions of 60 keV to a fluence of 1.0×10^{17} ions/cm^2 and subsequent annealing for 1 h, in oxygen gas of $\sim 1 \times 10^5$ Pa (*closed circles*) and in vacuum of less than 1×10^{-3} Pa pressure (*open circles*). The values of R_q before implantation and in the as-implanted state are also shown (Reprinted with permission from Amekura et al. [13]. © (2006) American Institute of Physics)

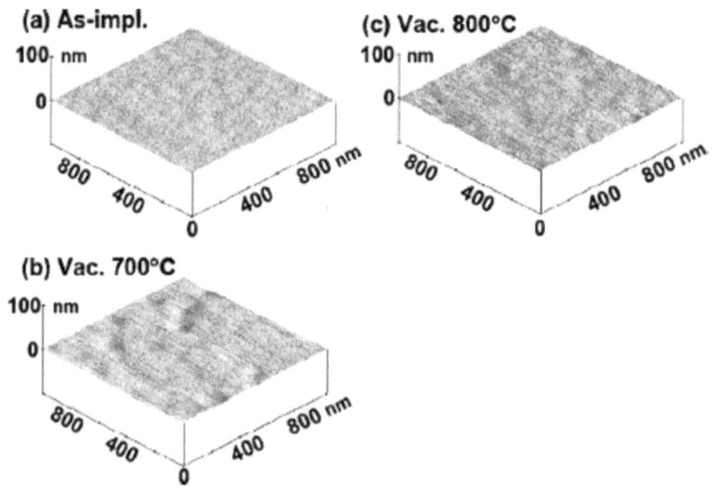

Fig. 20 Evolution of the surface morphology of SiO_2 samples induced by Zn ion implantation of 60 keV to a fluence of 1.0×10^{17} ions/cm^2 and subsequent annealing in *vacuum* of less than 1×10^{-3} Pa pressure detected by AFM, in the as-implanted state (**a**) and after annealing for 1 h at 700°C (**b**) and 800 °C (**c**) (Reprinted with permission from Amekura et al. [13]. © (2006) American Institute of Physics)

Further Reactions at Higher Oxidation Temperatures

As shown in Fig. 13, annealing at 800°C does not induce any drastic changes on the absorption spectrum, but the absorption intensity slightly decreases. A similar result is obtained in the GIXRD patterns shown in Fig. 14. The annealing at 800°C does not change the diffraction pattern, but slightly decreases the intensity of each peak.

Drastic changes are observed after annealing at 900°C for 1 h. In the absorption spectrum (Fig. 13), the absorption edge of ZnO NPs located at ~3.3 eV completely disappears, and a new absorption edge appears at ~5.3 eV. Drastic changes are also induced in the GIXRD pattern as shown in Fig. 14. All the eight diffraction peaks due to ZnO NPs completely disappear, and a lot of new peaks clearly appear. All the new peaks are ascribed to the Zn_2SiO_4 phase. The absorption edge energy of ~5.3 eV corresponds to the band-gap energy of Zn_2SiO_4 [32].

The Zn_2SiO_4 phase is formed by the following reactions. First, Zn atoms implanted into SiO_2 form Zn metal NPs. According to the Padova group, the formation process of Zn metal NPs

$$\text{Zn (isolated atoms) in SiO}_2 \rightarrow \text{Zn (metal NPs) in SiO}_2 \tag{6}$$

is approximated by

$$\text{Zn (gas, i.e, atom)} \rightarrow \text{Zn (bulk metal)} \tag{7}$$

$$\Delta G^\circ_{298\,\mathrm{K}} = -95\,\mathrm{kJ/mol} \tag{8}$$

$$\Delta G^\circ_{1200\,\mathrm{K}} = +7\,\mathrm{kJ/mol} \tag{9}$$

The Gibbs energy change of this reaction is negative at RT, but positive at high temperature of 1200 K. Following the two-step model, the Gibbs energy change around RT is applicable for ion implantation at RT. In fact, Zn NPs are formed even in the as-implanted state. The positive value of ΔG at 1200 K may indicate that Zn NPs are no longer stable at 1200 K and that Zn NPs prefer to dissolute as Zn atoms in SiO_2.

Annealing (oxidation) process in oxygen gas is approximated by

$$\mathrm{Zn} + 1/2\,\mathrm{O}_2 \rightarrow \mathrm{ZnO} \tag{10}$$

$$\Delta G^\circ_{298\,\mathrm{K}} = -320\,\mathrm{kJ/mol} \tag{11}$$

This process gives the Gibbs energy change with negative sign and large absolute value, contributing to the stabilization of ZnO NPs. The Gibbs energy changes at high temperatures of 1000 and 1200 K of the reaction (11) are given in Table 2, showing large negative values. Annealing in oxygen gas at higher temperatures induces further reaction between ZnO NPs and SiO_2 substrate:

$$2\,ZnO + SiO_2 \longrightarrow /Zn_2SiO_4 \tag{12}$$

$$\Delta G^{\circ}_{298\,K} = -31\,kJ/mol \tag{13}$$

Since the Gibbs energy change is negative, this reaction spontaneously proceeds. A coordinate diagram of these reactions is schematically shown in Fig. 21. Since all the Gibbs energy changes of the reactions are negative, once O_2 is introduced into the SiO_2 matrix containing Zn NPs, the system finally moves to the most stable phase, i.e., Zn_2SiO_4. However, the experimental observation of the relatively stable ZnO phase indicates the existence of a potential barrier between the ZnO and the Zn_2SiO_4 phase, which stabilize the ZnO phase in a meta-stable state. It should be noted that ZnO NPs are practically stable enough at RT while they are in a meta-stable state. ZnO NPs formed by the II&TO method survive for many years at RT.

As shown in the XTEM image, Fig. 15(d), the domelike structures of ZnO NPs change to layer-like structures which cover most of the surface of the substrate. This is also consistent with the AFM image of Fig. 18(f), where the surface structures become smaller and the roughness R_q decreases to 2.89 nm. The new surface layer is ascribed to the Zn_2SiO_4 phase from the XPS result shown in Fig. 16.

Figure 22 shows the UV-absorption spectra of the Zn-implanted samples after annealing in oxygen gas at 700, 800, and 900°C for 1 h each. Due to the bulk ZnO, the absorption starts at around 3.3 eV and stays almost a constant value up to 6.5 eV. Consequently, absorption increase observed above ~5.3 eV is mostly due to the Zn_2SiO_4 phase. Then the ZnO and the Zn_2SiO_4 contributions to the absorption spectra are separated and plotted as shown in the inset of Fig. 22. By increasing the annealing temperature, the Zn_2SiO_4 phase increases at the expense of the ZnO

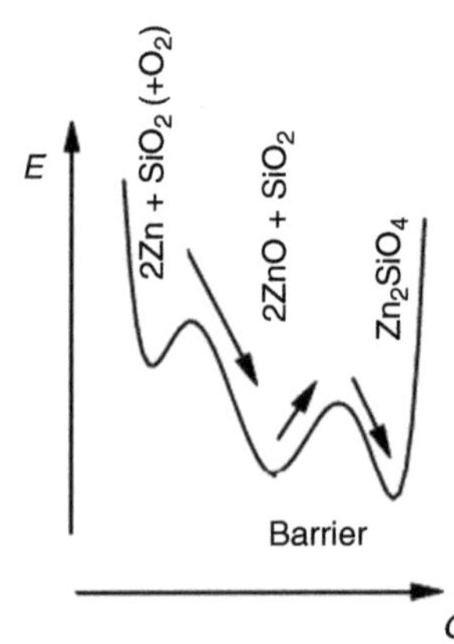

Fig. 21 Schematically depicted configuration curve of the reactions induced in Zn-implanted SiO_2 by additional O_2 molecule and heat treatment. While Zn_2SiO_4 is the most stable phase, ZnO (NPs) exist as a meta-stable phase separated from other phases by relatively high potential barriers

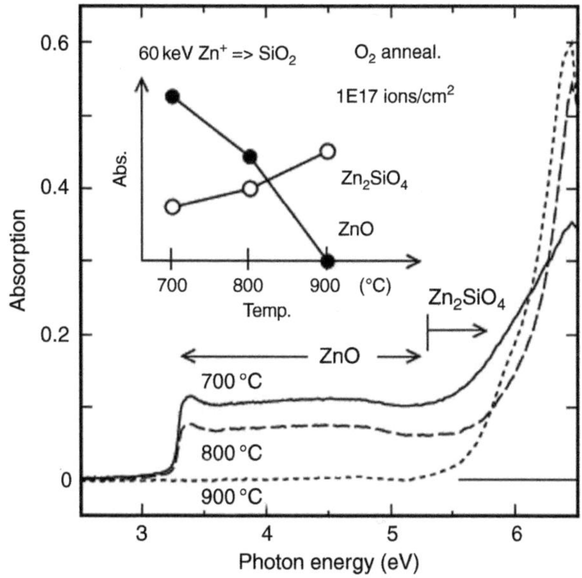

Fig. 22 Optical absorption spectra in the UV region of SiO_2 implanted with Zn^+ ions of 60 keV up to 1.0×10^{17} ions/cm^2 and annealed in oxygen gas at 700, 800, and 900°C for 1 h each. The inset shows the annealing temperature dependence of the integrated absorption intensities of ZnO and Zn_2SiO_4 components

phase. The Zn_2SiO_4 phase exists even after the oxidation at 700°C, and increases with the annealing temperature.

4.2.2 Comparison with Annealing in Vacuum

To distinguish the oxidation effect and pure thermal effects of annealing in oxygen gas, annealing effects in vacuum were observed and compared with those in oxygen gas. The experimental setup is the same as the annealing in oxygen gas except the annealing atmosphere. Instead of oxygen gas of $\sim1\times10^5$ Pa, vacuum of less than 1×10^{-3} Pa was applied.

Figure 23(a) shows the optical absorption spectra in the as-implanted state and after annealing for 1 h at elevated temperatures in vacuum [21]. As mentioned in Section "Formation of Metallic Zn NPs by Ion Implantation", the broad absorption peak around 4.8 eV is due to Zn metal NPs in SiO_2 (hereafter call the Zn-NP band). After vacuum annealing at 400°C for 1 h, the intensity of the Zn-NP band slightly increases with a peak shift of 0.1 eV to the higher energy side. These changes are probably due to growth of Zn NPs. After annealing at 600°C, the absorption spectra hardly change. After annealing at 700°C, the intensity of the Zn-NP band decreases. After annealing at 800°C, the optical absorption disappears completely, except a very weak tail above ~5.5 eV.

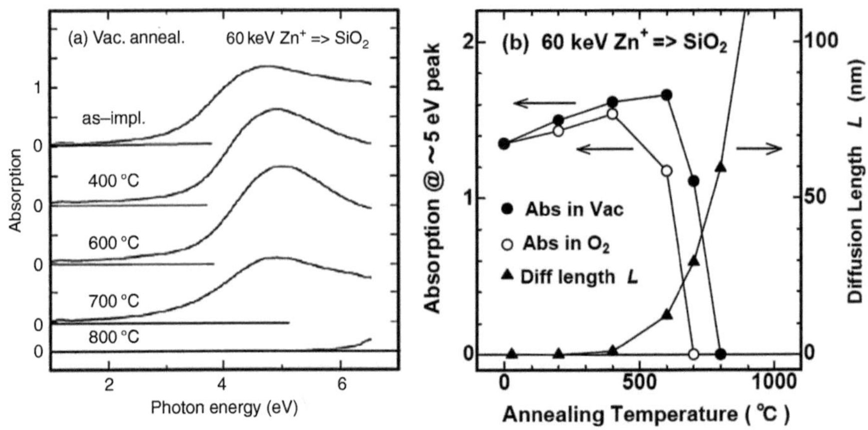

Fig. 23 (**a**) Optical absorption spectra of SiO_2 implanted with Zn ions of 60 keV to 1.0×10^{17} ions/cm^2 in the as-implanted state and after annealing in vacuum of 1×10^{-3} Pa pressure, for 1 h at 400, 600, 700, and 800°C. The spectra are shifted vertically for clarity. (**b**) Annealing temperature dependence in vacuum (*closed circles*) and in oxygen gas (*open circles*) of the intensity of the absorption band at ~5 eV, which is ascribed to Zn NPs in SiO_2. The annealing duration was 1 h each. Diffusion length L of oxygen which was defined in Equation (5) is also shown (Reprinted with permission from Amekura et al. [21]. © (2007) Elsevier)

Annealing temperature dependences of the peak intensity of the Zn-NP band both in vacuum and in oxygen gas are plotted in Fig. 23(b). Up to 400°C annealing, both the annealing atmospheres give almost the same intensity changes, indicating pure thermal effects. After 600°C annealing, the vacuum atmosphere increases the intensity, while the oxygen atmosphere decreases it. These behaviors are explained by the diffusion length of O_2 molecules in SiO_2, as introduced in Section 3.2.1. It should be noted again that the projected ranges calculated by SRIM2003 code [149] and by TRIDYN [91] are 49 and 30 nm, respectively. At 400°C, the diffusion length L is ~1 nm, much shorter than the projected range. Only pure thermal effects are observed in both the atmospheres. At 600°C, the diffusion length increases to ~10 nm. A part of Zn NPs are oxidized during oxygen annealing. Then the intensity of the Zn-NP band decreases during oxygen annealing, but not during vacuum annealing. At 700°C, the diffusion length increases to ~30 nm, inducing oxidation of almost all the Zn metal NPs. The intensity of the Zn NP band steeply decreases to zero during oxygen annealing at 700°C.

Also the intensity of the Zn NP band decreases during vacuum annealing at 700°C, and the peak completely disappears after 800°C annealing in vacuum. Depth profiles of Zn atoms after 700°C and 800°C annealing in vacuum are detected by RBS and shown in Fig. 17 [15]. In contrast with oxygen annealing, the depth profiles of Zn atoms do not show any shift toward the surface, but rather a weak shift to deeper region. The surface morphology observed by AFM is shown in Fig. 20. The surface was mostly flat, and no surface NPs were observed after vacuum annealing at 700°C and 800°C. The evolution of the roughness values are plotted in Fig. 19.

4.2.3 Detailed Formation Processes

Radiation-Induced Formation of Zn_2SiO_4 in As-Implanted State

Figure 24 shows XPS spectra at Zn-$L_3M_{45}M_{45}$ Auger transitions, i.e., XAES, of Zn-implanted sample in as-implanted state at various sputtering depths. There is little evidence of Zn signal on the surface (0 nm), which has already been shown in Fig. 16. By increasing the sputtering depth, two peaks appear at kinetic energies of 922.2 and 995.6 eV, both of which are ascribed to metallic Zn [96]. The metallic peaks grow larger with depth, reaching a maximum at a depth of ~15 nm and decreasing in deeper regions. A weak peak is visible at 986.1 eV at a depth of 6 nm, which is ascribed to the Zn_2SiO_4 phase. The weak peak coexists with the strong metallic signals. While the metallic signals disappear in the regions deeper than 40 nm, the Zn_2SiO_4 signal shows a maximum at ~42 nm. It should be noted that ZnO signal, which should appear at 988.1 eV [96], is not observed at any depth. The formation of the Zn_2SiO_4 phase is also confirmed in the Zn $2p_{3/2}$ transition of XPS.

The observation of Zn_2SiO_4 phase as a minor component in the as-implanted state is not inconsistent with the absorption spectrum and GIXRD as shown in Figs. 13 and 14. The Zn_2SiO_4 phase has absorption only in regions with energy higher than ~5.3 eV, and the oscillator strength of the Zn_2SiO_4 phase is smaller than the major component of metallic Zn. As mentioned in Section 4.2.1, the sensitivity of GIXRD shows a superlinear dependence with the concentration of the component, i.e., less sensitive to minor components.

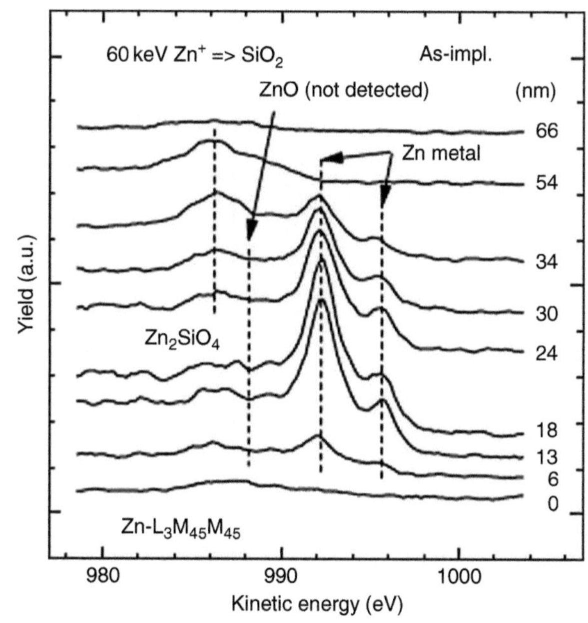

Fig. 24 XPS spectra around Zn-$L_3M_{45}M_{45}$ Auger transitions from as-implanted SiO_2 sample which was implanted with 60 keV Zn^+ ions to a fluence of 1.0×10^{17} ions/cm^2. The sputtering depth of each spectrum is indicated on the right-hand side of the figure. Each spectrum is shifted vertically for clarity. The energy positions of Zn, ZnO, and Zn_2SiO_4 species are shown by dashed lines (Reprinted with permission from Amekura et al. [22]. © (2007) American Institute of Physics)

The depth profiles of Zn atoms, which form the metallic Zn, ZnO, and Zn_2SiO_4 phases, were determined from the XPS spectra and plotted in Fig. 25 [22]. As already mentioned, the ZnO component was below the detection limit. The Zn and Zn_2SiO_4 phases show concentration maxima at ~15 nm and ~42 nm, respectively. A spatial separation is observed along the depth between the metallic Zn and Zn_2SiO_4 phases.

In the same figure, depth profiles of Si and O atoms are shown. Characteristic differences are observed between the Si and the O profiles. While the Si profile has two minima, which correspond to the metallic Zn maximum and the Zn_2SiO_4 maximum, the O profile has only one minimum corresponding to the metallic Zn maximum. The numerical ratio of O atoms to Si atoms is plotted in Fig 25. The region with an O/Si ratio larger than 2 (smaller than 2) corresponds to the O-rich (Si-rich) region in SiO_2. While the depth around the metallic Zn peak is Si-rich, the depth around the Zn_2SiO_4 peak is O-rich. The inhomogeneous O/Si ratio along the depth is probably a consequence of atomic collisions during the ion implantation in SiO_2. The formation of the Si-rich and O-rich regions is reproduced by both SRIM2003 [149] and TRIDYN [91,92] codes. The result of TRIDYN calculation is shown in Fig. 26. Si-rich and O-rich regions are formed in the shallow and deep regions, respectively, probably due to the difference in the atomic mass of Si and O atoms. Since an O atom is lighter than a Si atom, the O atoms are displaced to deeper regions than the Si atoms on average.

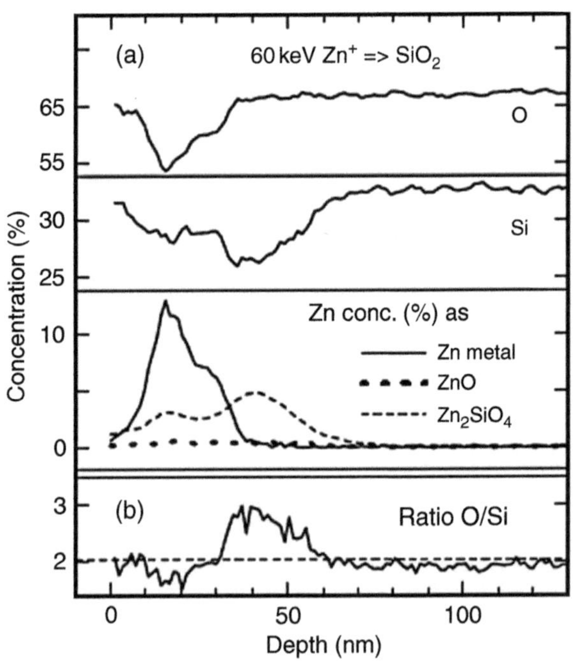

Fig. 25 (**a**) Concentration profiles of O, Si, and Zn atoms along the depth, determined by XPS. (**b**) Numerical ratio of O atoms to Si atoms (N_O/N_{Si}) in SiO_2 implanted with 60 keV Zn^+ ions to a fluence of 1.0×10^{17} ions/cm^2, plotted along the depth (Reprinted with permission from Amekura et al. [22]. © (2007) American Institute of Physics)

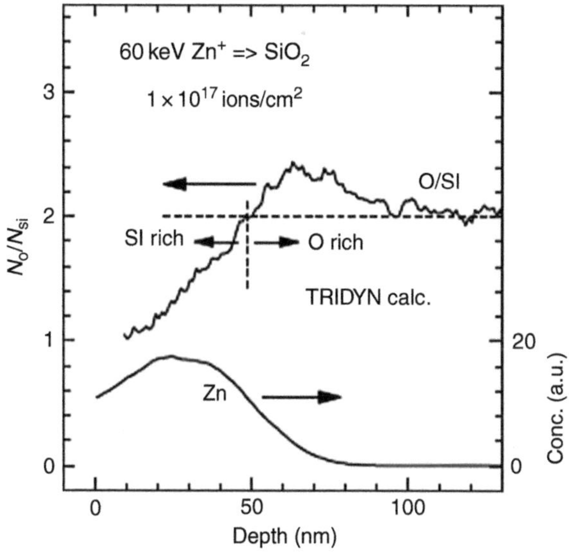

Fig. 26 Numerical ratio of O atoms to Si atoms (N_O/N_{Si}) in SiO_2 implanted with 60 keV Zn^+ ions to a fluence of 1.0×10^{17} ions/cm^2, calculated by TRIDYN code, plotted along the depth. The depth profile of the implanted Zn atoms is also shown. (Reprinted with permission from Amekura et al. [22]. © (2007) American Institute of Physics)

In the O-rich region, excess O atoms may exist as atomic oxygen, oxygen molecules [122], or oxygen-excessive defects in the SiO_2 network such as peroxy radicals [48]. These excess O species react with the implanted Zn atoms and SiO_2 matrix to form Zn_2SiO_4 via reactions as

$$2Zn + SiO_2 + O_2 \rightarrow Zn_2SiO_4 \quad (14)$$

$$2Zn + SiO_2 + 2O \rightarrow Zn_2SiO_4 \quad (15)$$

However, as already described in the Section "Further Reactions at Higher Oxidation Temperatures", the formation of Zn_2SiO_4 phase requires not only an excess of oxygen but also a temperature of ~900°C. The reaction is probably enhanced by the nonthermal energy conveyed by implanted ions into nanoscopic regions of the implanted layer via high-density electron excitation and/or the high-density of local atomic vibrations. This would be consistent with the lack of the ZnO phase. Because of the nonthermal energy deposition by the ions, once the ZnO phase is formed, most of it is rapidly transformed to the Zn_2SiO_4 phase via enhanced reaction with the SiO_2 matrix, even when the mean temperature of the implanted region is as low as RT.

Detailed Processes in Annealing in Oxygen Gas

Sputtering depth profiles by XPS of the samples annealed in oxygen gas for 1 h each at (a) 600°C, (b) 700°C, and (c) 900°C are shown in Figs. 27(a), (b), and (c). The sputtering depths, which are determined from the sputtering time and the

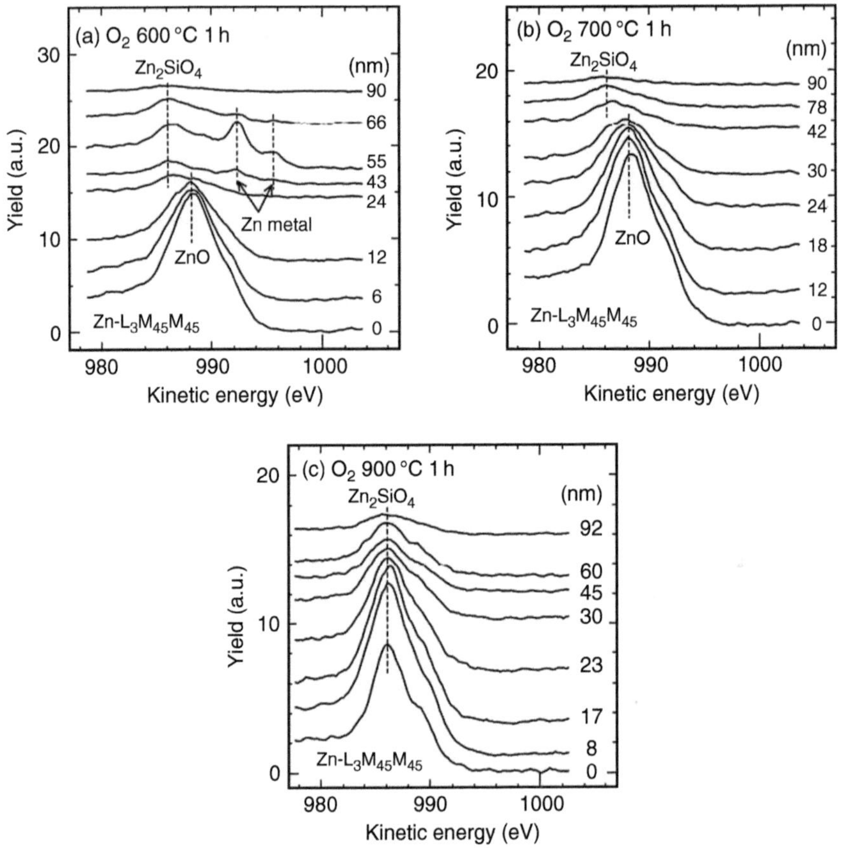

Fig. 27 XPS spectra around Zn-$L_3M_{45}M_{45}$ Auger transitions from Zn-implanted SiO_2 samples after annealing in oxygen gas for 1 h each at (**a**) 600°C (Fig. 27(a) is reprinted with permission from Amekura et al. [20]. © (2007) Elsevier), (**b**) 700°C, and (**c**) 900°C. The sputtering depth of each spectrum is indicated on the right-hand side of the figure. Each spectrum is shifted vertically for clarity. The energy positions of Zn, ZnO, and Zn_2SiO_4 species are shown by dashed lines (Fig. 27(b) in reprinted with permission from Amekura et al. [15]. © (2006) Elsevier)

efficiency, are shown in the right side of the figures. However, it should be noted that the surfaces of the SiO_2 substrates are partly covered by ZnO and Zn_2SiO_4 phases after annealing as shown in Fig. 15, i.e., the surface is neither flat nor homogeneous. The sputtering depth includes relatively large ambiguity.

After annealing at 600°C for 1 h, the surface is mainly covered by ZnO phase. However, after removing a layer of ~30 nm thickness, the metallic Zn phase appears. These results are consistent with the XTEM image shown in Fig. 15(b). Also, weak signals due to Zn_2SiO_4 phase are observed in the deeper region. After annealing at 700°C for 1 h, the ZnO phase is observed not only on the surface but also inside the SiO_2 matrix. No signal due to metallic Zn is observed in the whole of the depth region. While only signals from the Zn_2SiO_4 phase are observed in the region deeper than ∼40 nm, weak but certain signal due to Zn_2SiO_4 phase is observed in shallower region. After annealing at 900°C for 1 h, all the ZnO phase changes to Zn_2SiO_4 phase. At any depth, only the signal from the Zn_2SiO_4 phase is observed.

From these results, it is found that the small but certain amount of Zn_2SiO_4 phase always stays in deep region. Probably the Zn_2SiO_4 phase is formed in as-implanted state, and survives during the annealing because it is the most stable phase in the Zn–SiO_2 system as shown in Fig. 21.

Detailed Processes in Annealing in Vacuum at 800°C

As shown in Fig. 17, the total content of Zn atoms decreases to ∼90% and ∼30% of the as-implanted value after annealing in vacuum at 700°C and 800°C, respectively. However, if all the remaining Zn atoms after 800°C annealing (3.2×10^{16} ions/cm^2) formed metal NPs in SiO_2, the Zn-NP band should be as intense as ∼30% of the as-implanted value. However, the Zn-NP band completely disappeared, indicating the Zn atoms exist in a different form other than Zn metal NPs. The XPS spectra of the sample annealed in vacuum at 800°C for 1 h at various depths are shown in Fig. 28. Neither Zn metal nor ZnO phases are observed but only Zn_2SiO_4 phase is observed. From the comparison with the as-implanted state (Fig. 24), the migration of the Zn_2SiO_4 phase toward deeper region is clearly shown, which is also consistent with the RBS results shown in Fig. 12. XTEM images of samples (a) in as-implanted state and after annealing in vacuum for 1 h at (b) 700°C and (c) 800°C are shown in Fig. 29. After the annealing at 700°C, the sizes of the metallic Zn NPs

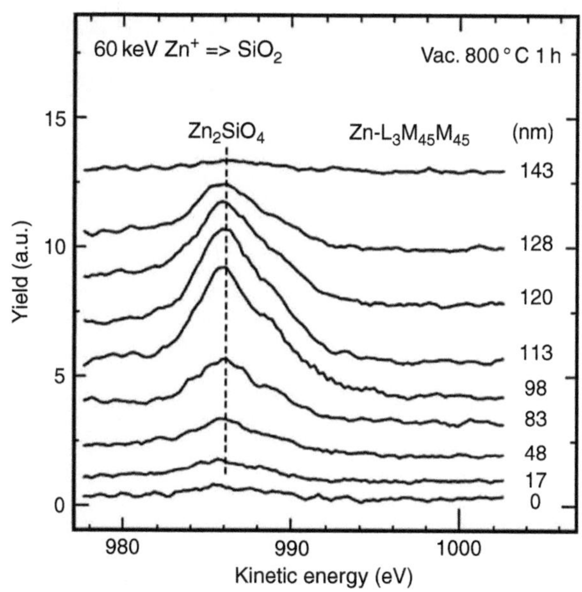

Fig. 28 XPS spectra around Zn-$L_3M_{45}M_{45}$ Auger transitions from Zn-implanted SiO_2 samples after annealing in *vacuum* at 800°C for 1 h at various depths. The sputtering depth of each spectrum is indicated on the right-hand side of the figure. Each spectrum is shifted vertically for clarity

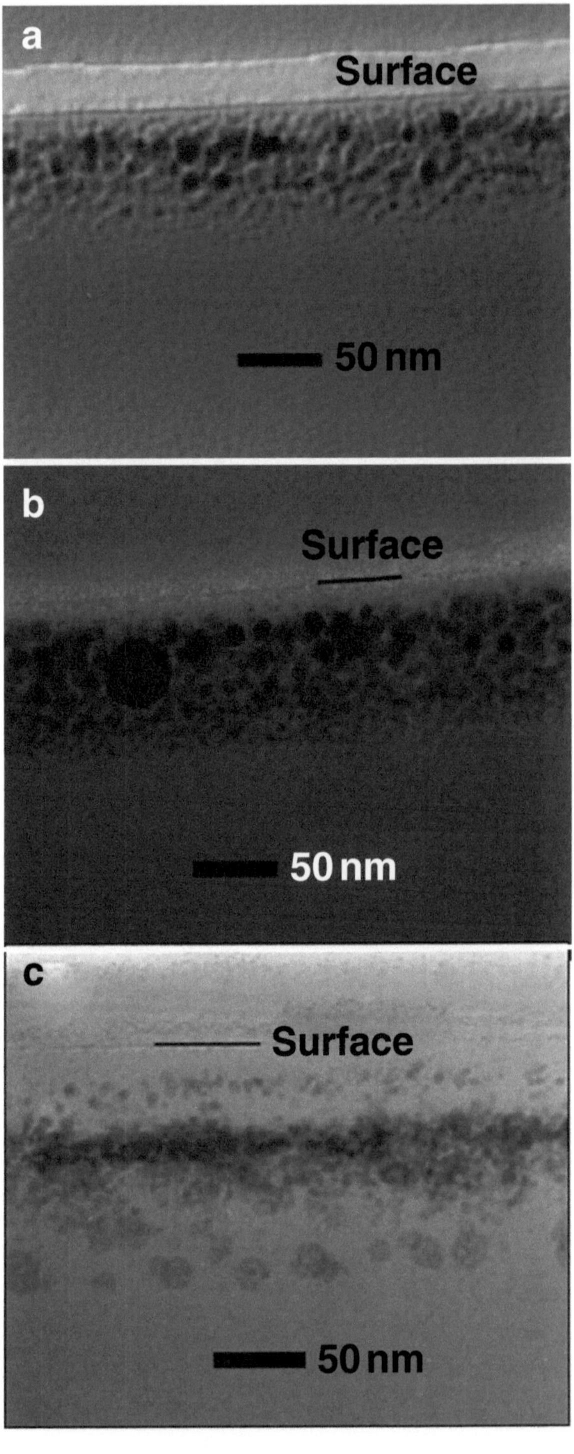

Fig. 29 Cross-sectional TEM (XTEM) images of SiO_2 samples implanted with Zn^+ ions of 60 keV to a fluence of 1.0×10^{17} ions/cm^2 (**a**) in the as-implanted state and after annealing in vacuum for 1 h at (**b**) 700°C and (**c**) 800°C. The substrate surface is indicated by a line in each figure

increase. After annealing at 800°C in vacuum, drastic changes are observed. Most of NPs shallower than ~40 nm disappear, and a lot of very small NPs are observed between ~40 nm and ~80 nm in depth, which are again consistent with RBS and XPS results.

The appearance of Zn_2SiO_4 phases after vacuum annealing at 800°C can be described, e.g.,

$$2Zn + 2SiO_2 \rightarrow 2ZnO + Si + SiO_2 \qquad (16)$$

$$\rightarrow Zn_2SiO_4 + Si$$

$$\Delta G^{\circ}_{298\,K} = +185\,kJ/mol \quad \Delta G^{\circ}_{1100\,K} = +209\,kJ/mol$$

Since the Gibbs energy change is positive, the reaction does not spontaneously proceed. One possible explanation is that the Zn_2SiO_4 phases exist from the beginning of the annealing, i.e, the phases were formed in as-implanted state, as discussed in Section 4.2.3. The Zn_2SiO_4 phases slightly move to deeper region by vacuum annealing at 800°C.

It should be noted again that RBS and XPS results show that ~70% of Zn atoms are missing from the SiO_2 matrix and ~30% of the Zn atoms stay in the SiO_2 as Zn_2SiO_4 phase in deeper region. As shown in Equation (7), Gibbs energy change of Zn NP formation is negative (i.e., preferable) at RT, but turns to positive (i.e., unpreferable) at 1200 K, i.e., ~900°C. When the temperature increases to ~800°C in vacuum, Zn NPs may shrink and disappear with emitting Zn monomers that show drastic long migration. Only the Zn_2SiO_4 NPs, i.e., slow diffusers, remain.

4.2.4 "Defect-Band-Free" Luminescence

Photoluminescence (PL) spectra of Zn-implanted SiO_2 samples after annealing are shown in Fig. 30 by solid lines, which were detected under excitation of the 325 nm (3.81 eV) line from a HeCd laser [12]. The excitation power density on the sample was ~2 W/cm^2. Corresponding absorption spectra are shown by dotted lines. After annealing at 700°C for 1 h where most of Zn NPs transform to ZnO NPs, a sharp peak and a broad band (the so-called green band) were observed in the PL spectrum at 375 and ~500 nm, respectively. The sharp peak and the green band are, respectively, due to free excitons and defects such as Zn interstitials or O vacancies [106]. The green band is a relatively common defect band and in many cases coexists with the exciton peak. Although the absorption spectrum does not change after annealing at 800°C for 1 h, large changes are observed in PL spectrum: the green band and the exciton peak become three times stronger and ~10 times weaker, respectively. The increase in the green band intensity corresponds to the degradation of the ZnO NPs.

One of the interesting observations is that the green (defect) band is drastically reduced after annealing at 600°C for 1 h. In Fig. 30, the spectrum around the defect

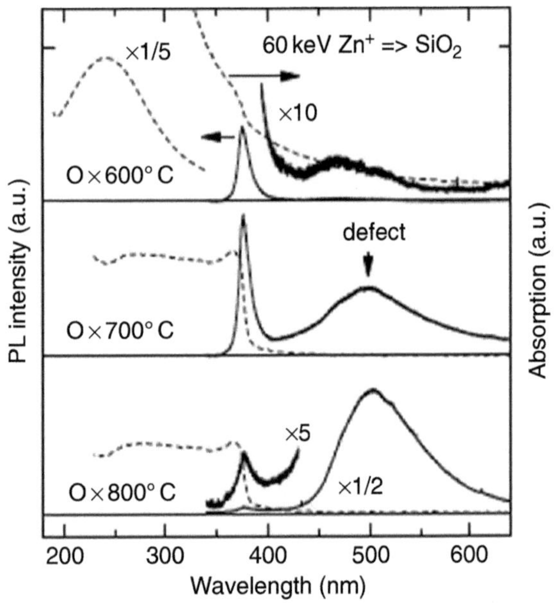

Fig. 30 Photoluminescence (*solid lines*) and optical absorption (*dotted lines*) spectra of SiO_2 samples implanted with Zn^+ ions of 60 keV to a fluence of 1.0×10^{17} ions/cm^2, and annealed in oxygen gas for 1 h at 600, 700, and 800°C, respectively. The spectra are shifted vertically for clarity and the horizontal lines indicate the base lines. A part of the spectrum at 600°C is also shown with magnification of ten times. (Reprinted with permission from Amekura et al. [12]. © (2006) American Institute of Physics)

band is shown with magnification of ten times. Comparing with the sample annealed at 700°C, the intensity of the defect band decreases to ∼1/20 and the peak shifts to ∼470 nm, which can be ascribed to other defect species. As for the exciton peak, the intensity after 600°C annealing is almost comparable to that after 700°C annealing. It should be noted again that the absorption spectrum of the sample annealed at 600°C is dominated by strong absorption due to Zn metallic NPs covering the whole of the visible region (see dotted lines in Fig. 30). In fact, the sample shows brownish color. It could be contrary to common sense if a homogenous semiconductor has absorption in the visible region but emits PL in the UV region only. In fact, the sample after annealing at 600°C is no longer homogenous. While the surface is covered by the droplet-like ZnO NPs, the Zn metallic NPs survive in the SiO_2 substrate as shown in Fig. 15(b). The absorption in the visible region and the PL in the UV region are due to Zn NPs in the deep region and ZnO NPs on the surface, respectively. One might consider that the disappearance of the green band was due to the absorption by the remaining Zn metallic NPs. However, it cannot happen because the absorption by Zn metallic NPs is weaker in the visible region than in the UV region.

PL spectra excited from the implanted surface side and those excited from the rear surface side were compared in the samples oxidized at 600 and 700°C. The results are shown in Fig. 31 [20]. It should be noted that the sample was excited from one side, and the PL was detected from the same side. In the sample oxidized at 700°C, the PL spectrum excited from the implanted side almost coincides with the spectrum excited from the rear side. However, in the sample oxidized at 600°C,

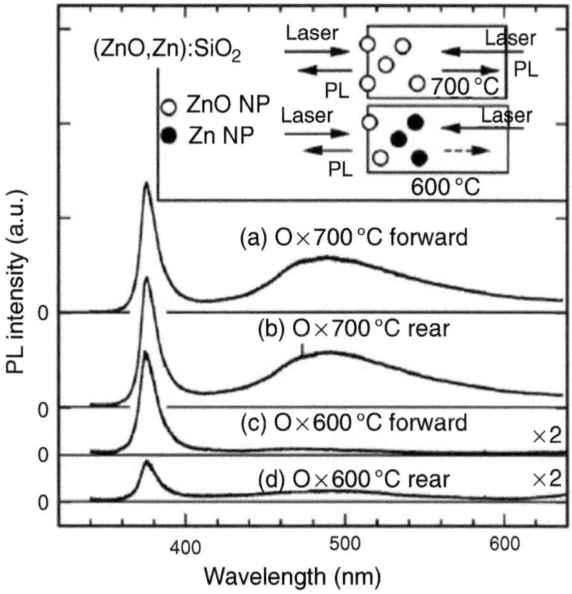

Fig. 31 Photoluminescence (PL) spectra of SiO_2 samples implanted with Zn^+ ions of 60 keV and annealed for 1 h at 700°C (**a**) and (**b**) and 600°C (**c**) and (**d**). The implanted surface (**a**) and (**c**) and the rear surface (**b**) and (**d**) were illuminated with the 325 nm line from a HeCd laser, and the PL was detected from the same surface side. The spectra are shifted vertically for clarity and the horizontal lines indicate the base lines. The inset explains why the free exciton line was weaker under condition (**d**) (Amekura et al. [20])

the PL spectrum excited from the implanted side shows a stronger exciton peak than the spectrum excited from the rear side.

While all the Zn NPs have transformed to ZnO in the sample annealed at 700°C for 1 h, both Zn and ZnO NPs coexist in the sample annealed at 600°C for 1 h. Since the oxidation of Zn NPs started from the (implanted) surface side, more ZnO NPs are present in the shallower region, and more Zn NPs survive in the deeper regions. When the sample is excited from the rear side, the excitation laser is partially absorbed by Zn NPs in the deeper region at first, then reaches ZnO on the implanted surface (see the inset of Fig. 31). The excitation intensity experienced by ZnO NPs is weaker than the direct excitation. The emitted luminescence from the ZnO NPs again goes through the Zn NP layer, and is partially absorbed again. When the sample is excited from the implanted side, the laser directly excites ZnO NPs on the surface, and the emitted luminescence is not absorbed by Zn NPs in the deep region. The luminescence from the rear excitation then becomes much weaker than the implanted surface excitation. If Zn and ZnO NPs had the same depth distributions, the excitation surface dependence would not be noticeable. This idea has been confirmed by sputter depth profiling by XPS shown in Fig. 27(a), which shows spatial separation between ZnO phase (close to the surface) and Zn phase (located ~35 nm deep).

As shown in Fig. 21, Zn_2SiO_4 phase is more stable than the mixture phase of ZnO and SiO_2. ZnO NPs in SiO_2 exist in a meta-stable state. While all the ZnO NPs in SiO_2 transform to the Zn_2SiO_4 phase after annealing in oxygen gas at 900°C for 1 h as shown in Fig. 13, even lower temperature annealing at 800°C for 1 h converts some portions of ZnO NPs to the Zn_2SiO_4 phase. This is supported by the UV absorption spectra shown in Fig. 22, where annealing temperature dependence of ZnO absorption and Zn_2SiO_4 absorption is shown. It is pointed out that an increase in the green band occurs during the transformation from ZnO to Zn_2SiO_4 because a lot of Zn interstitials and O vacancies are produced during the bond rearrangements between ZnO and Zn_2SiO_4 phases [53]. Consequently, larger intensity ratio of the green band to the exciton peak is observed after annealing at 800°C. By decreasing the annealing temperature, the transformation to Zn_2SiO_4 phase is reduced. The sample annealed at 700°C shows weaker green band than the sample annealed at 800°C. The "defect-band-free" PL at 600°C may be due to the reduction in the transformation to Zn_2SiO_4 because of low temperature.

Another possible origin of the defect-band-free PL is the remaining metallic Zn NPs in the deeper region at 600°C. When ZnO NPs form at the surface of SiO_2 substrate, the metallic Zn NPs may act as reservoirs to supply Zn atoms continuously to the surface. Then the formation of stoichiometric ZnO NPs is attained. However, more studies are necessary for identification of the mechanism of the defect-band-free PL at 600°C.

Detailed discussion about the mechanism of the "Defect-band-free PL" is given in reference [20].

4.2.5 Fluence Dependence: The Best Fluence for ZnO NP Formation

Also as described in Section 3.1.3, very high fluence might be meaningless, because the fluence dependence of the implant concentration in sample shows a saturation exceeding 1×10^{17} ions/cm^2 in the case of Zn implantation of 60 keV to SiO_2.

Optical absorption spectra of three different fluences of 2.0×10^{16}, 5.0×10^{16}, and 1.0×10^{17} ions/cm^2 are shown both in as-implanted state and after annealing at 700°C for 1 h in oxygen gas. Since the absorption intensity principally increases with the fluence at least in the as-implanted state, the absorption spectra are normalized by the fluence as shown in Fig. 32[18]. In the as-implanted state, the absorption intensity is approximately similar because the ordinate is normalized by the fluence, while the absorption peak shifts to the high energy side with decreasing fluence and becomes higher than 6.5 eV at 2×10^{16} ions/cm^2. However, after oxidation annealing at 700°C, the ZnO absorption, which starts from ~3.3 eV and extends toward the high-energy side, shows a strong fluence dependence even after the normalization by the fluence. The normalized (absolute) ZnO absorption of the 5×10^{16} ions/cm^2 sample decreases to ~40% (~20%) of the normalized (absolute) absorption of the 1×10^{17} ions/cm^2 sample. In the 2×10^{16} ions/cm^2 sample, the ZnO absorption is not observed but the Zn_2SiO_4 absorption starts at ~5.3 eV. From RBS, the Zn content in all the samples was confirmed to be almost constant up to 900°C annealing in oxygen gas. These results indicate that the branching ratio to ZnO (Zn_2SiO_4) phase at 700°C annealing decreases (increases) with decreasing the fluence. This is because the Zn_2SiO_4 phase is formed via a reaction between SiO_2 matrix and a part of tentatively formed ZnO as

$$2ZnO + SiO_2 \rightarrow Zn_2SiO_4 \quad (17)$$

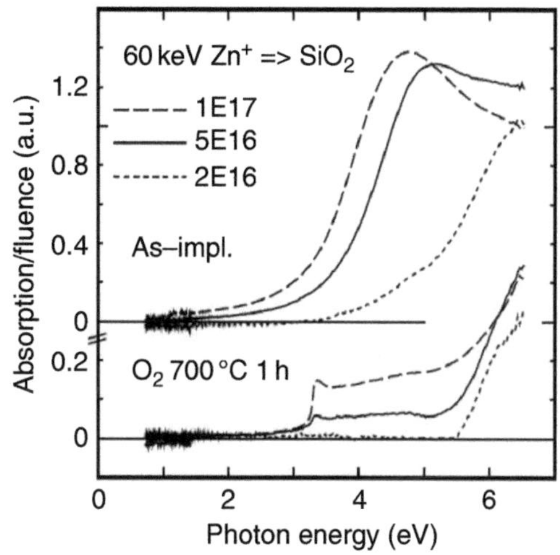

Fig. 32 Optical absorption spectra of SiO_2 implanted with Zn^+ ions of 60 keV up to three different fluences of 0.20×10^{17} (*dotted lines*), 0.50×10^{17} (*solid lines*), and 1.0×10^{17} ions/cm^2 (*broken lines*) in the as-implanted state and annealed in oxygen gas at 700°C for 1 h

By decreasing the fluence, the size of ZnO NPs formed at 700°C annealing becomes smaller in average. Consequently, the surface-to-volume ratio becomes larger, leading to more efficient reaction between ZnO NPs and SiO_2 substrate.

4.2.6 Embedment of ZnO NPs into SiO_2

As shown in the Section "ZnO Formation by Annealing in Oxygen Gas", the ZnO NPs are formed on the surface of the SiO_2 substrate. However, embedded NPs have some advantages against NPs formed on the surface. The embedded NPs are protected from attacks from outside, e.g., chemicals, force, etc., by the matrix. The matrix also inhibits, to some extent, the aggregations between NPs. It is a challenging task to form embedded ZnO NPs in a SiO_2 matrix using ion implantation and thermal annealing/oxidation [19].

Scenario for Embedment of ZnO NPs

Our strategy is the following: The formation of ZnO NPs on the surface is probably due to a much larger diffusion constant of Zn atoms in SiO_2 than that of oxygen. Our recent interpretation on the formation of ZnO NPs on the surface is slightly different. As shown in Section 3.2.1, the diffusion constant of oxygen in SiO_2 is written as

$$D^{\mathrm{ox}}(T) = D_{\mathrm{o}}^{\mathrm{ox}} \exp(-E_{\mathrm{ox}}/kT), \tag{18}$$

where $D_{\mathrm{o}}{}^{\mathrm{ox}} = 2 \times 10^{-9}$ cm^2/s and $E_{\mathrm{ox}} = 1.3$ eV [140]. The diffusion constant of Zn in SiO_2 has not been elucidated but is probably given by the Arrhenius form

$$D^{\mathrm{Zn}}(T) = D_{\mathrm{o}}^{\mathrm{Zn}} \exp(-E_{\mathrm{Zn}}/kT). \tag{19}$$

Even when $D^{Zn}(700°C) >> D^{ox}(700°C)$, annealing at a temperature much lower (or higher) than 700°C comparatively reduces the Zn diffusion and enhances the oxygen diffusion, if the activation energy E_{Zn} is larger (or smaller) than E_{ox}. In fact, our present study has shown that the reduced formation of surface NPs and enrichment of embedded NPs are attainable by low-temperature oxidation at 500°C for ~70 h, indicating $D^{Zn}(500°C) \sim D^{ox}(500°C)$.

Formation of the Embedded ZnO NPs

Figure 33 shows the absorption spectra of a Zn-implanted SiO_2 sample, in the as-implanted state and after annealing at 500°C in oxygen gas for various annealing time. In the as-implanted state, a broad band due to metallic Zn NPs is observed around 4.8 eV. After 1 h annealing at 500°C, the peak shifts to 4.9 eV, probably due to growth of Zn NPs and a partial recovery of damaged SiO_2 substrate. At 8 h and 14 h, the spectra show almost no changes except that the peak shifts to 4.95 eV. After 44 h annealing, a sharp peak appears at 3.3 eV, indicating the partial formation of ZnO NPs. However, the broad band due to Zn metal NPs still coexists with the ZnO peak. An absorption spectrum of a sample oxidized at 700°C for 1 h is plotted in Fig. 33 by a dotted line. After 70 h annealing at 500°C, almost the same spectrum as that observed at 700°C annealing for 1 h was obtained. A period of ~70 h was required to oxidize most of the Zn NPs at 500°C, while less than 1 h was required at 700°C. From Equation (18), the diffusion constants of oxygen in SiO_2 are given as

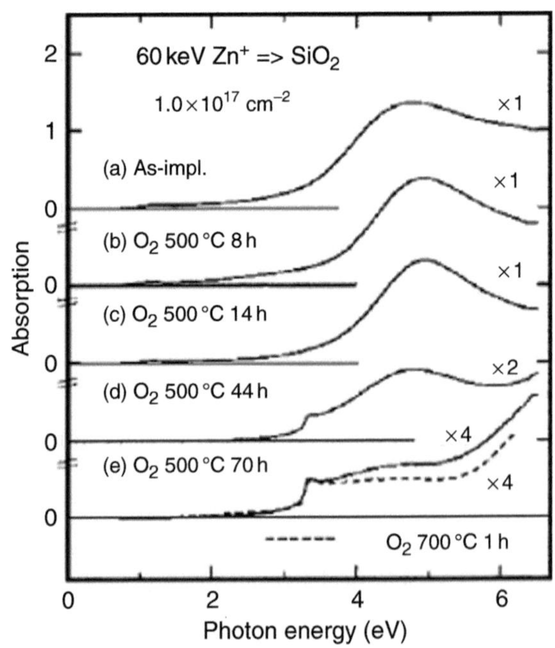

Fig. 33 Optical absorption spectra of SiO_2 samples implanted with 60 keV Zn^+ ions to a fluence of 1.0×10^{17} ions/cm^2 measured at room temperature in the as-implanted state (**a**) and after annealing in oxygen gas at 500°C for various durations (**b**–**e**). The dotted line shows the absorption spectrum after annealing at 700°C for 1 h. The spectra are shifted vertically for clarity and the horizontal lines indicate the base lines. (Reprinted with permission from Amekura et al. [19]. © (2007) American Institute of Physics)

1×10^{-17} and 6×10^{-16} cm^2/s at 500 and 700°C, respectively. To obtain the same diffusion length $L = (4Dt)^{1/2}$ at 700°C for 1 h, 6×10 h is required at 500°C. This shows good agreement with the experimental value of ~70 h, suggesting an oxygen diffusion-limited reaction.

The formation of ZnO NPs on the surface of the SiO_2 substrate and the morphology changes by oxygen annealing were evaluated by AFM and shown in Fig. 34. In the as-implanted state (Fig. 34(a)), the surface was very flat as indicated by the R_q value of 0.39 nm. After annealing at 700°C for 1 h, numerous ZnO NPs with domelike shapes overlapped each other and covered the whole surface, as shown in Fig. 34(b). The R_q value increased to 13.0 nm. The corresponding cross-sectional view was observed by XTEM, and is shown in Fig. 35(a). The XTEM image also confirms that the ZnO NPs were mainly formed on the surface of the SiO_2, while some NPs remained inside the SiO_2.

In the case of annealing at 500°C, the surface was as flat as in the as-implanted state up to ~5 h duration. Small hemispheres appeared on the surface at 8 h duration. After 14 h duration, many ZnO NPs whose shapes were not exactly domelike but rather flat island shapes appeared on the surface. After 70 h duration, when most of the Zn NPs had been transformed to ZnO as shown in Fig. 33, a large number of ZnO NPs having flat island shapes were formed on the surface. Some of the NPs

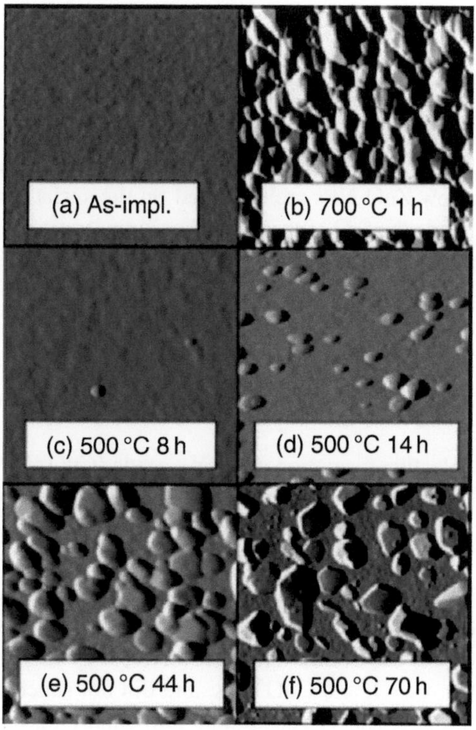

Fig. 34 AFM images of SiO_2 sample surfaces implanted with 60 keV Zn^+ ions to a fluence of 1.0×10^{17} ions/cm^2 in the as-implanted state (**a**) and after annealing in oxygen gas at 700°C for 1 h (**b**) and at 500°C for various durations ranging from 8 to 70 h (**c–f**). The size of each image is 1×1 μm^2. (Reprinted with permission from Amekura et al. [19]. © (2007) American Institute of Physics)

Fig. 35 Cross-sectional TEM (XTEM) images of SiO_2 samples implanted with 60 keV Zn^+ ions to a fluence of 1.0×10^{17} ions/cm^2 and annealed in oxygen gas at 700°C for 1 h (**a**) and at 500°C for 7×10 h (**b**). (Reprinted with permission from Amekura et al. [19]. © (2007) American Institute of Physics)

showed facets. However, comparing with Fig. 34(b) and (f), it is clear that ZnO NP formation on the surface was drastically reduced by low-temperature annealing at 500°C. An XTEM image corresponding to Fig. 34(f) is shown in Fig. 35(b). Only a few ZnO NPs are observed on the surface, and most of the ZnO NPs are formed inside the SiO_2 substrate. These results are confirmed by RBS measurements, where the migration of Zn species toward the surface direction was observed after annealing at 500°C for 70 h but much less than after annealing at 700°C for 1 h.

4.3 Selective Formation of CuO and Cu_2O Nanoparticles

Selective formation of cupric oxide (CuO) NPs and of cuprous oxide (Cu_2O) NPs is discussed in this section using the II&TO method and the extension.

Ikeyama et al. [61] carried out sequential implantation of 1.8 MeV Cu^+ and 1.0 MeV O^+ ions to SiO_2, and formed Cu, CuO, and Cu_2O NPs in SiO_2. By changing the dose ratio of Cu^+ to O^+ and the implantation sequence, i.e., Cu implantation followed by O implantation or vice versa, they changed the production ratio of Cu/Cu_2O/CuO NPs. However, they always obtained samples with mixtures of Cu, Cu_2O, and CuO NPs, and did not succeed in purification of any NP species in the samples. The sequential implantation is a nonequilibrium synthesis method with plenty of freedom. However, proper control of the products is sometimes difficult because of too much freedom. Contrarily, thermal oxidation/reduction is governed by thermodynamics. The controllability is limited to some extent, but the results are easily predictable using the accumulated knowledge of thermodynamics. For the selective formation of CuO and Cu_2O NPs, strategies based on thermodynamics are proposed in this section.

Early study on the oxidation of Cu NPs was motivated by the formation of core(Cu)-shell(Cu oxide, either CuO or Cu_2O) structure by thermal oxidation. The surface plasmon resonance (SPR) peak of metal NPs is enhanced when the dielectric constant of the matrix increases. Since the dielectric constants of both CuO and Cu_2O are higher than SiO_2, it could be expected that the SPR of the Cu NPs was enhanced when the outer layer of the Cu NPs was oxidized. This conclusion was supported by roughly calculated absorption spectra shown in Fig. 36(a), which does not include any size effects. In the calculation, the radius of the Cu core R_{core} decreases with increasing thickness of the oxide shell L_{shell} under keeping the sum $R_{core} + L_{shell}$ = constant. The SPR is enhanced with red shift when the Cu NPs are coated by the oxide layer of 1 nm thickness (L_{shell} = 1 nm) compared with the NPs without the oxide layer (L_{shell} = 0 nm). By including the mean-free-path confinement (MFPC) effect, the enhancement becomes less prominent as shown in Fig. 36(b). However, clear evidence of the formation of Cu/Cu oxide core/shell NPs by the II&TO method has not yet been reported.

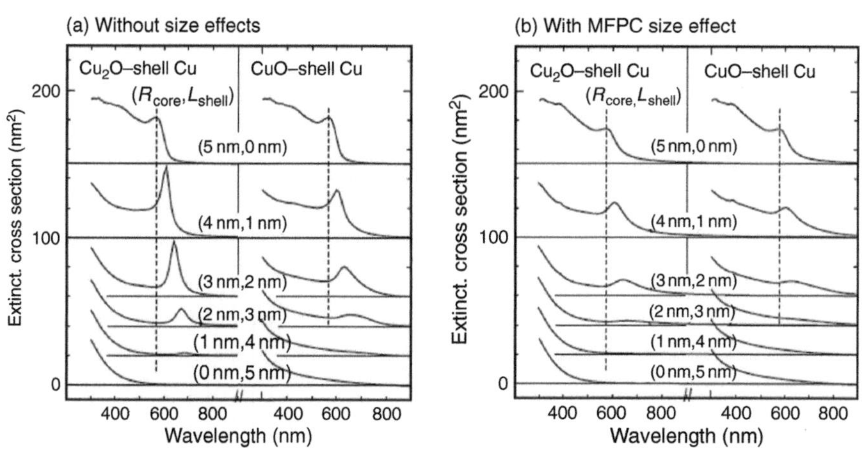

Fig. 36 Extinction cross sections (~ absorption) of Cu NPs coated with Cu_2O or CuO shell of various thickness L_{shell} are calculated using Mie theory without any size effects (**a**) and with the mean-free-path confinement effect (**b**). To simulate the core-shell structures formed by thermal oxidation, the sum of the radius of the core R_{core} and the thickness of the shell L_{shell} is kept constant (5 nm) in each calculation

4.3.1 Basic Properties of CuO and Cu_2O

Cuprous oxide (Cu_2O) is a semiconductor with a direct band gap of ~2.169 eV at 4.2 K [112], and shows an uncommon crystal structure called the cuprite structure [34], which is a type of three-dimensional cubic structure with an inversion symmetry (the space group $O_h{}^4$). In this structure, O atoms occupy the sites of a body-centered cubic lattice and each O atom has four covalent bonds extending in the tetrahedral directions. The adjacent O atoms are bridged with each other via a linearly coordinated Cu atom; i.e., each Cu atom is coordinated by only two adjacent O atoms. The four tetrahedral bonding of O atoms are also unusual.

Because of its uncommon crystalline structure, Cu_2O shows an unusual electronic structure [112], which may promise applications in solar cells, etc. A schematic band structure is shown in Fig. 37. In this material, both the extreme of the lowest conduction band and that of the highest valence band are located at the Γ point in the Brillouin zone, forming a direct energy gap. The highest valence band splits into an upper $\Gamma_7{}^+$ band and a lower $\Gamma_8{}^+$ band due to the spin-orbit interaction ($\Delta_{SO} = 0.124$ eV). Since the lowest conduction band ($\Gamma_6{}^+$) and the split valence bands ($\Gamma_7{}^+$ and $\Gamma_8{}^+$) mainly have a Cu 4*s* and Cu 3*d* character, respectively, the transitions between them ($\Gamma_7{}^+$-$\Gamma_6{}^+$ and $\Gamma_8{}^+$-$\Gamma_6{}^+$; i.e., the so-called yellow and green transitions) are all parity-forbidden. While the direct-forbidden band gap of this material has been known since the 1960s [41], an additional peculiarity of Cu_2O is that the second lowest conduction band having the allowed parity ($\Gamma_8{}^-$) is located very close to the lowest conduction band; i.e., only 0.558 eV higher. Since the transitions between the second lowest conduction band and the split valence bands ($\Gamma_7{}^+$-$\Gamma_8{}^-$ and $\Gamma_8{}^+$-$\Gamma_8{}^-$; i.e., the so-called blue and indigo transitions) are direct and allowed, Cu_2O shows strong exciton absorption peaks around ~2.6 eV with a weak absorption continuum starting from ~2.0 eV due to the parity-forbidden transitions.

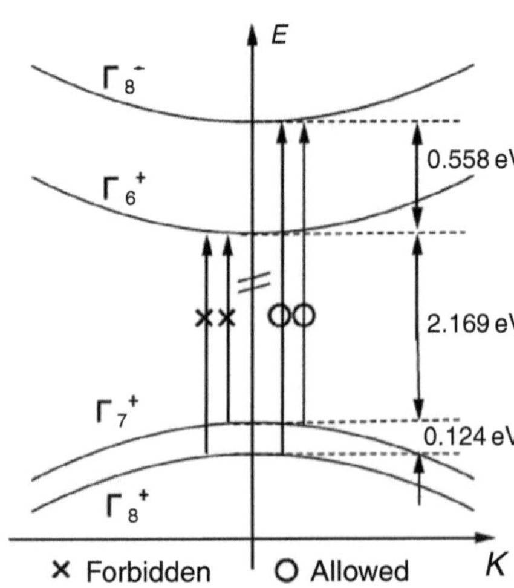

Fig. 37 A schematic band structure of Cu_2O. Arrows with circles and with crosses indicate parity-allowed and parity-forbidden transitions, respectively

The blue and indigo exciton states are regarded as autoionization states; i.e., resonant states in the weak continuous states due to the yellow and green transitions.

Although the bulk form of Cu_2O has been known for several decades, counterparts in nanophases have been recently under investigation. In fact, Cu_2O NPs have been fabricated using mainly chemical routes [145, 141], discharge [146], evaporation [27], and annealing [70], with considering applications mainly to catalysis [145, 141], and solar cells [145, 108]. Recently, Cu_2O has been pointed out as a basic material of high Tc oxide superconductor [150].

On the other hand, cupric oxide (CuO) is an anti-ferromagnetic insulator (semiconductor) with the Neel temperature of ~213 K. The crystal structure is a monoclinic tenorite structure. However, the electronic structures of CuO are still under confusion: While one source has ascribed the absorption edge at ~1.4 eV to an indirect gap of the semiconductor type [109], others have ascribed it to a charge-transfer gap of the Mott-insulator type [64]. In fact, band calculation [34] has predicted an open Fermi surface, i.e., a metallic nature, although CuO shows insulating behaviors. Recently the quasi-one-dimensional nature of the antiferromagnetism is pointed out in CuO, inspite of the three dimensional crystal structure [121].

4.3.2 Formation of CuO NPs by the Conventional II&TO Method

Figure 38 shows the phase diagram of copper–oxygen system plotted for oxygen partial pressure versus temperature [120]. Under oxygen partial pressure of ~760 Torr, where the conventional II&TO method has been carried out, the CuO phase is stable up to ~1100°C. The Cu_2O phase is only stable under much lower oxygen partial pressure than the atmospheric pressure, i.e., ~760 Torr. Consequently, the product of the II&TO method under the atmospheric oxygen partial pressure is the CuO phase, not the Cu_2O phase. In this section, the formation of CuO NPs by annealing in oxygen gas of the atmospheric pressure is described by comparison with annealing in vacuum.

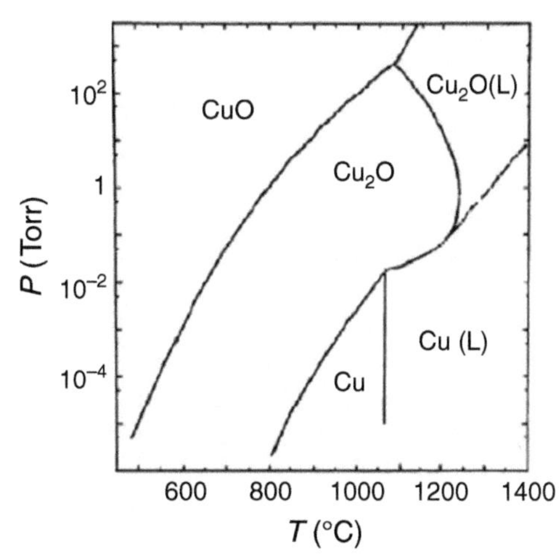

Fig. 38 Phase diagram of copper–oxygen system plotted for oxygen partial pressure versus temperature

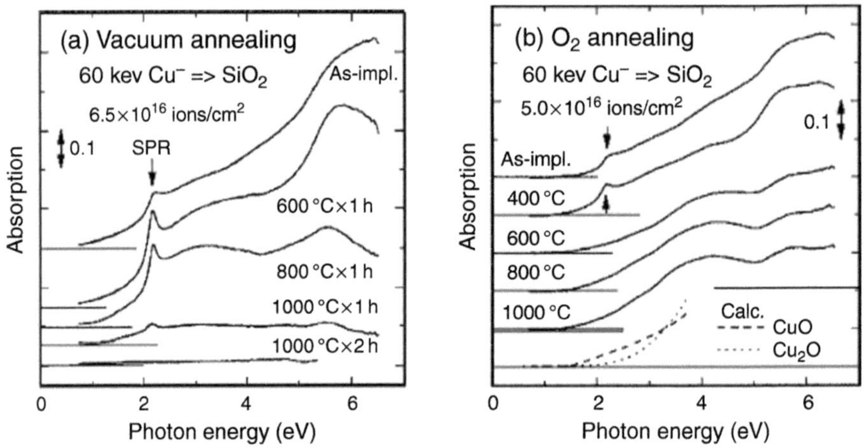

Fig. 39 Optical absorption spectra of SiO_2 implanted with Cu^- ions of 60 keV in the as-implanted state and after annealing at elevated temperatures for 1 h each in (**a**) vacuum and in (**b**) oxygen gas. The spectra are shifted vertically for clarity, with horizontal lines showing the zero level of each spectrum. Arrows indicate the surface plasmon resonance (SPR) of Cu NPs. Absorption spectra of CuO and Cu_2O NPs in SiO_2 are calculated by Mie theory using bulk dielectric functions of CuO and Cu_2O, and are shown in figure (**b**) by broken and dotted lines, respectively. (Reprinted with permission from Amekura et al. [11]. © (2005) American Institute of Physics)

Figure 39(a) shows vacuum annealing effects on the absorption spectra of SiO_2 implanted with Cu^- ions of 60 keV [11]. Even in the as-implanted state, the SPR peak at 2.15 eV and a broad absorption at the higher energy side are observed, which are evidences of Cu NP formation [119]. The Cu NP formation was also confirmed by XTEM [72]. After vacuum annealing at 600°C for 1 h, the SPR peak becomes sharper probably due to growth of Cu NPs. A decrease in the absorption around 6 eV is observed. A similar decrease is observed in SiO_2 implanted with Ni^- ions of 60 keV [6]. A similar observation in Cu-implanted SiO_2 confirms that the absorption is not due to implanted metals but due to radiation-induced defects, e.g., E' centers. After vacuum annealing at 800°C, absorption around 5.6 eV also decreases. This is due to the growth of Cu NPs, because a similar behavior is reproduced by Mie theory including higher order (Amekura 2007, unpublished). After vacuum annealing at 1000°C for 1 h, the whole absorption decreases drastically. This is due to the disappearance of Cu NPs from the sample. Although the detailed mechanism has not been clarified yet, this phenomenon has already been reported [134, 2] and recognized as a limit of thermal processing of Cu NPs in SiO_2. After 1000°C annealing for 2 h, no absorption is observed.

Figure 39(b) shows oxygen annealing (thermal oxidation) effects on the absorption spectra of Cu-implanted SiO_2. The spectrum after oxygen annealing at 400°C is almost the same as the as-implanted state, but the SPR peak slightly becomes clearer and moves to the lower energy side of about 0.02 eV. The low-energy shift is consistent with the core/shell formation. However, clear evidence was not obtained from TEM observation. The SPR peak disappears after annealing at 600°C in oxygen gas. After oxygen annealing at 800°C and even at 1000°C, the spectral shape fairly changes. These observations are in contrast with that of vacuum annealing, where the disappearance of Cu NPs is observed at 1000°C. The difference is due to the fact that oxides generally have higher thermal stability than metals. It should be noted that the band-gap energies are ~1.4 eV for bulk CuO [85] and ~2.0 eV for bulk Cu_2O [62] at RT. These band-gap energies are close to the absorption onset of the SPR of Cu NPs. (The SPR peak locates at ~2.15 eV.) The absorption spectra of CuO and of Cu_2O NPs in SiO_2 are calculated from complex refractive indices of the bulk materials [107] using the Mie theory [89] with neglecting size effects. These assumptions are not so bad for CuO and Cu_2O NPs larger than ~10 nm in diameter. The observed spectra after annealing at 600°C or higher agree better with the calculated spectrum of CuO NPs in SiO_2 than that of Cu_2O NPs.

Figure 40 shows the GIXRD spectra of as-implanted state and after oxygen annealing at 800°C and 1000°C for 1 h each. Powder diffraction patterns of Cu, CuO, and Cu_2O from JCPDS library (JCPDS) are also shown in Fig. 40 as rectangles. In the as-implanted state, a strong peak (~67 deg) and a weak peak (~78 deg) are observed, which are ascribed to Cu (111) and (200) diffraction, respectively. After oxidation annealing at 800°C for 1 h, the diffraction pattern is completely changed: Two strong peaks are observed at ~53° and ~58°, which are ascribed to CuO (11-1) and (111) diffraction, respectively. After oxygen annealing at 1000°C for 1 h, the peaks become stronger and sharper, as shown in Fig. 40. The GIXRD results indicate that the major component after the oxidation is CuO NPs. No evidence of the Cu_2O NP formation was obtained from the GIXRD measurements.

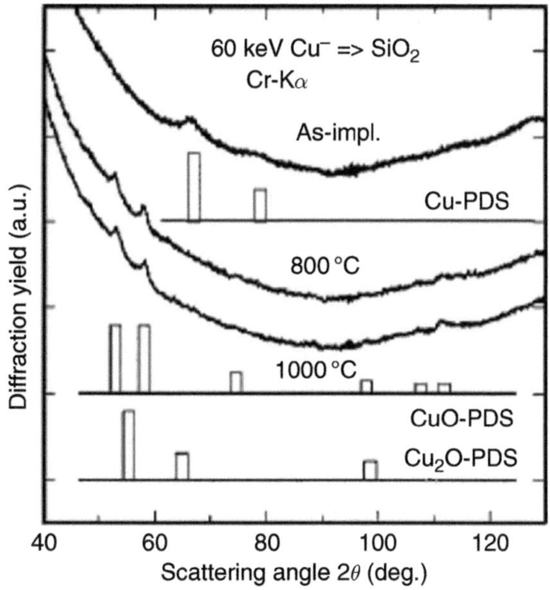

Fig. 40 Grazing incident XRD patterns of SiO_2 implanted with Cu^- ions of 60 keV to 6.5×10^{16} ions/cm^2 in the as-implanted state and after annealing at 800°C and 1000°C for 1 h each in O_2 gas flow. The incident angle of x-ray was 2°. Powder diffraction patterns of Cu, CuO, and Cu_2O from the JCPDS library are shown as rectangles. (Reprinted with permission from Amekura et al. [11]. © (2005) American Institute of Physics)

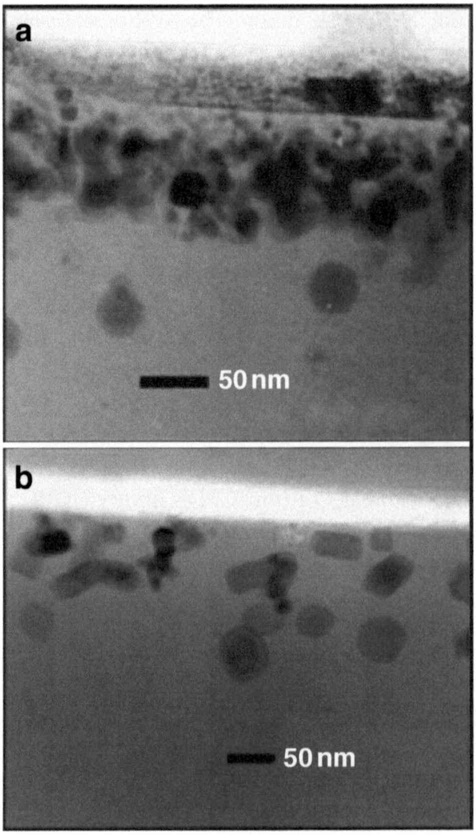

Fig. 41 Cross-sectional TEM images of (**a**) CuO NPs in SiO_2 fabricated by implantation of 60 keV Cu^- ions and annealing in oxygen gas at 800°C for 1 h, and (**b**) Cu_2O NPs in SiO_2 fabricated by soft reduction of CuO NPs in LOP Ar gas at 900°C for 5 h. The surface of sample (**a**) is covered with glue (Reprinted with permission from Amekura et al. [14]. © (2006) American Institute of Physics)

Figure 41(a) shows an XTEM image of the sample after annealing at 800°C for 1 h, which shows a direct evidence of the oxide NP formation. Various sizes of NPs of 5–30 nm in diameter are observed.

Figure 42 shows RBS spectra of Cu-implanted SiO_2 sample, which correspond to the depth profiles of Cu atoms in SiO_2. In the as-implanted state, the implanted Cu atoms locate within a depth layer of ~100 nm thickness. The 1000°C annealing in vacuum strongly enhances the diffusion of Cu atoms in SiO_2 and the Cu depth profile becomes very broader. The center of the depth profile moves to much deeper than the projectile range, and the tail extends until 600 nm deep from the surface after the annealing at 1000°C for 1 h. After the annealing at 1000°C for 2 h, almost no Cu signal was detected. On the other hand, after the oxygen annealing, the depth profile of Cu atoms, which exist as CuO NPs, does not show any significant changes up to 1000°C, except a slight decrease in the Cu peak. It is speculated that the detachments of Cu atoms from CuO NPs are much lower than the detachments from Cu NPs at the same temperatures. Once Cu NPs are converted to CuO NPs, the NPs are more stable than Cu NPs at elevated temperatures. The oxygen annealing

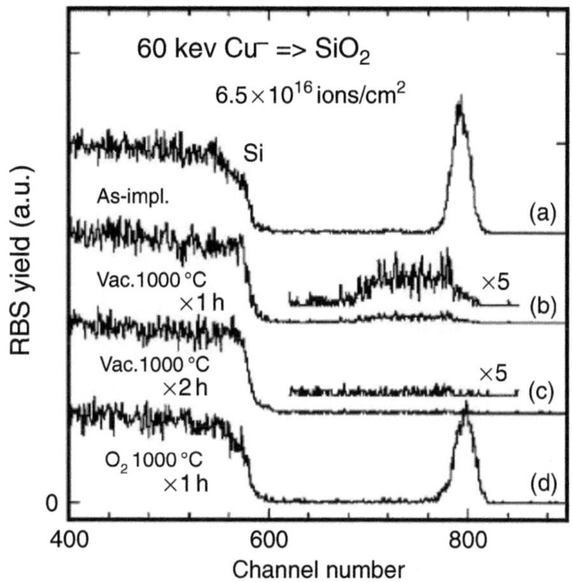

Fig. 42 RBS spectra of SiO_2 implanted with Cu^- ions of 60 keV to 6.5×10^{16} ions/cm^2 in (**a**) as-implanted state and after annealing at 1000°C in (**b**) vacuum for 1 h, (**c**) vacuum for 2 h, and (**d**) oxygen gas for 1 h. (Reprinted with permission from Amekura et al. [11]. © (2005) American Institute of Physics)

improves the thermal instability of Cu NPs in SiO_2 observed around 1000°C. We have succeeded in the selective formation of CuO NPs in SiO_2.

4.3.3 Formation of Cu_2O NPs by the Two-Step II&TO Method

Next, the selective formation of Cu_2O NPs is discussed. As shown in Fig. 38, the Cu_2O phase is only stable at less than a few hundreds torr and in a limited temperature region. With decreasing oxygen pressure, the stable region of Cu_2O moves to lower temperature. However, oxygen migration in the implanted layer requires another limitation for the adequate temperature range for the formation of Cu_2O NPs. Since the diffusion constant of oxygen in SiO_2 is given by Equation (4), an annealing at 700°C for 1 h is required for the diffusion length $L \sim 30$ nm, which is comparable to the TRIDYN ion range of 30 nm. On the other hand, a temperature in excess of 1000°C is not appropriate because dissolution of Cu NPs is induced as shown in Fig. 42. Consequently, a reasonable temperature window is between 700 and 900°C, which corresponds to oxygen pressure window of 10^{-4} and 10^{-1} Torr. Using high-purity Ar gas containing O_2 impurities of 10^{-3} Torr in partial pressure for annealing atmosphere, we have carried out two different methods of heat treatments to synthesize Cu_2O NPs: (1) low-oxygen-pressure (LOP) oxidation of Cu NPs which had previously been formed by ion implantation, and (2) LOP reduction of CuO NPs which had previously been formed by the conventional II&TO method, namely two-step annealing.

Low-Oxygen-Pressure Oxidation of Cu NPs to Cu_2O NPs

First, we intended to convert Cu NPs in SiO_2 to Cu_2O NPs by LOP oxidation in Ar gas. As-implanted samples were directly annealed in LOP Ar gas without any previous heat treatments. Figure 43 shows GIXRD spectra of Cu-implanted SiO_2 after LOP Ar gas annealing [17]. After annealing at 800°C for 1 h in LOP Ar gas, a weak peak of Cu_2O (111) diffraction appears at $2\theta \sim 55°$, in addition to more intense peaks of Cu (111) and Cu (200) diffraction at 66.5 and 78.6°, respectively. With increasing the annealing duration to 10 h while keeping the temperature at 800°C in LOP Ar gas, the Cu_2O (111) peak increases to $\sim 120\%$ of the value after 1 h annealing and the Cu (111) peak decreases to $\sim 70\%$. It should be noted that no peaks ascribed to the CuO phase were observed. This is in contrast to annealing in oxygen gas of ~ 760 Torr, where only CuO NPs form.

Although the Cu_2O phase increases with increasing annealing duration, the efficiency is poor. When increasing the annealing temperature to 900°C, the efficiency decreases as shown in curve (c) in Fig. 43. No Cu_2O peaks are observed. This behavior can be explained from thermodynamics. The formation of Cu_2O phase via oxidation of Cu NPs is described as

$$2\mathrm{Cu} + 1/2\,\mathrm{O}_2 \leftrightarrow \mathrm{Cu}_2\mathrm{O}. \tag{20}$$

The equilibrium constant K_1 of the reaction (20) is given by

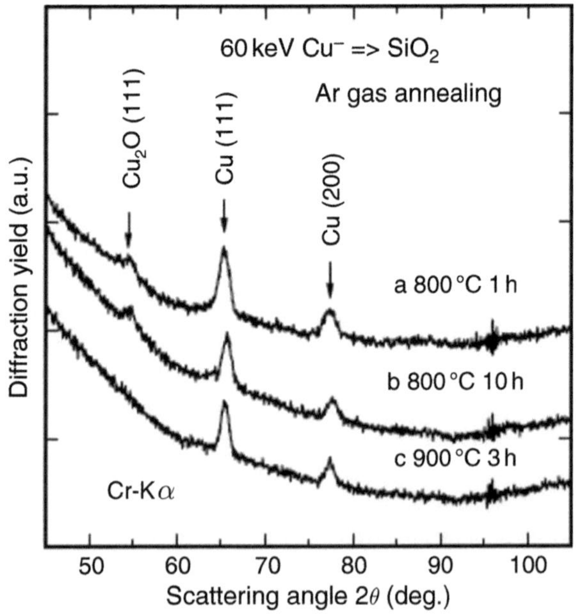

Fig. 43 GIXRD patterns of SiO_2 samples implanted with Cu^- ions of 60 keV, and annealed in Ar gas including oxygen impurities of 10^{-3} Torr in partial pressure. The diffraction yield is plotted on a linear scale

$$K_1 = \exp\left(-\frac{\Delta G_1^{\circ}}{RT}\right) \tag{21}$$

where $\Delta G_1{}^{\circ}$, R, and T denote the Gibbs energy change of the reaction (20), the gas constant, and temperature, respectively. The Gibbs energy change is given by

$$\Delta G_1^{\circ} = A_1 + B_1 T \log T + C_1 T \quad (\mathrm{J/mol}) \tag{22}$$

where $A_1 = -1.690\times10^5$ J/mol, $B_1 = -16.4$ J/Kmol and $C_1 = +123$ J/Kmol at 298 K $< T <$ 1356 K [76]. Temperature dependence of the equilibrium constant K_1 is shown in Fig. 44 as curve (a). With increasing temperature from 800°C to 900°C, the constant K_1, i.e., the Cu_2O phase, decreases. This result qualitatively explains the reason why the Cu_2O phase decreases with increasing temperature from 800°C to 900°C.

To understand the inapplicability of the LOP oxidation of Cu NPs in SiO_2 matrix, let's consider the conservation of amount of O_2 molecules migrated inside the SiO_2 matrix, i.e., the product of the oxygen concentration (~partial pressure) and the annealing. Since it takes ~1 h at 800°C for oxidation of most Cu NPs ($\sim5\times10^{16}$ ions/cm^2) to CuO in O_2 gas of ~760 Torr, the product becomes 760×1 Torr·h. To oxidize the same amount of Cu NPs to Cu_2O under low oxygen pressure of $\sim10^{-3}$ Torr, it could take 10^6 h (~760 Torr·h/10^{-3} Torr). The partial formation of Cu_2O NPs after LOP oxidation at 800°C for 1 h is rather surprising. Probably, the partial formation is due to oxygen atoms or molecules included in the SiO_2 substrate as impurities or those librated from the SiO_2 network by radiation damage [122]. This mechanism has been assumed by some authors: Johannessen et al. [67] implanted Cu ions of 0.8–5 MeV to SiO_2 and carried out annealing of the samples in forming gas of 5% H_2 + 95% N_2 at 500–1100°C for 1 h. They observed partial oxidation of

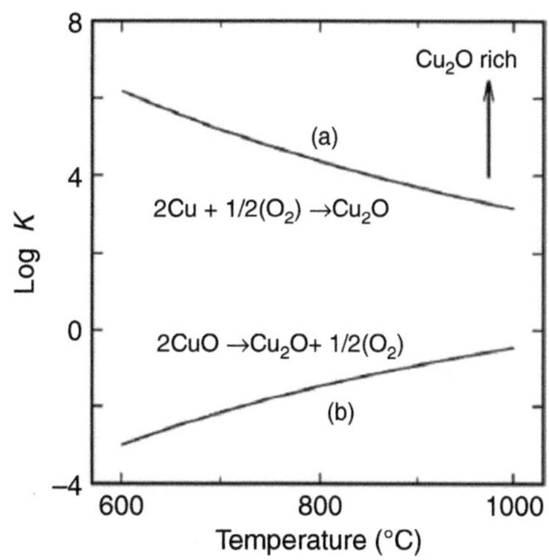

Fig. 44 Temperature dependences of the equilibrium constants of the reaction $2Cu + 1/2O_2 \rightarrow Cu_2O$ and the reaction $2CuO \rightarrow Cu_2O + 1/2O_2$ are shown as curves (**a**) and (**b**), respectively (Amekura et al. [17])

Cu NPs by EXAFS, which is due to the oxygen atoms or molecules included in the SiO_2 substrate as impurities or those librated from the SiO_2 network by radiation damage. Also, we have already assumed this mechanism to explain the formation of Zn_2SiO_4 phase in Zn-implanted SiO_2 in the as-implanted state.

This assumption explains the reason why the 10 h annealing at 800°C marginally increases the Cu_2O peak. Most of the oxygen in the depth region where most Cu NPs exist has been exhausted for the initial annealing of 1 h.

Low-Oxygen-Pressure Reduction of CuO NPs to Cu_2O NPs

Next, LOP reduction of CuO NPs to Cu_2O NPs, which is expressed by Equation (23),

$$2CuO \leftrightarrow Cu_2O + 1/2\ O_2 \tag{23}$$

was examined. The equilibrium constant K_2 of the reaction (23) is calculated using equations similar to Equations (21) and (22) with values of $A_2 = 1.462\times10^5$ J/mol, B_2 =25.5 J/Kmol and $C_2 = -185$ J/Kmol at 298 K $< T <$ 1300 K [76]. The calculated temperature dependence is shown as curve (b) in Fig. 44. The Cu_2O phase is found to increase with increasing temperature. Furthermore, the reaction (23) corresponds to the pulling out of oxygen atoms from CuO lattice. Very low oxygen density of the annealing atmosphere does not limit the reaction rate.

As previously mentioned, CuO NPs are readily formed in SiO_2 by the II&TO method. Using the CuO NPs as starting materials, the reduction of CuO NPs to Cu_2O was carried out in LOP Ar gas [17]. Figure 45 shows the GIXRD spectra of CuO NPs in the as-formed state and after reduction in LOP Ar gas at 800°C and 900°C. After reduction at 800°C for 1 h, the Cu_2O (111) diffraction appears at 55.3°, in addition to strong peaks of CuO (11-1) and CuO (111) at 53.9° and 59.0°, respectively. Even increasing the annealing duration up to 20 h at 800°C, the peak ratio Cu_2O (111)/CuO (11-1) did not significantly change. Furthermore the Cu (111) peak appeared after 1 h annealing at 800°C, and the peak marginally increased after the 20 h annealing at 800°C. It seems that not only the reduction of CuO to Cu_2O but also additional reduction of the Cu_2O phase to metallic Cu occurs.

When the annealing temperature was increased to 900°C, a drastic improvement was observed. The Cu_2O (111)/CuO (11-1) peak ratio increases to 1.34 after 1 h annealing at 900°C, and all the CuO peaks and Cu peaks disappear after 5 h annealing, as shown in Fig. 45, indicating that a pure phase of Cu_2O NPs is attained. The optical absorption spectroscopy, which has much higher sensitivity to Cu NPs through the SPR absorption than the GIXRD, supports the nonexistence of Cu NPs in the sample annealed at 900°C for 5 h as shown in Section 4.3.4. The conversion efficiency from CuO to Cu_2O phase at 900°C is higher than that at 800°C. This behavior is qualitatively consistent with the dependence shown as the curve (b) in Fig. 44. Moreover, the over-reduction of Cu_2O to Cu is suppressed. Consequently the Cu_2O phase is much more stable at 900°C than 800°C.

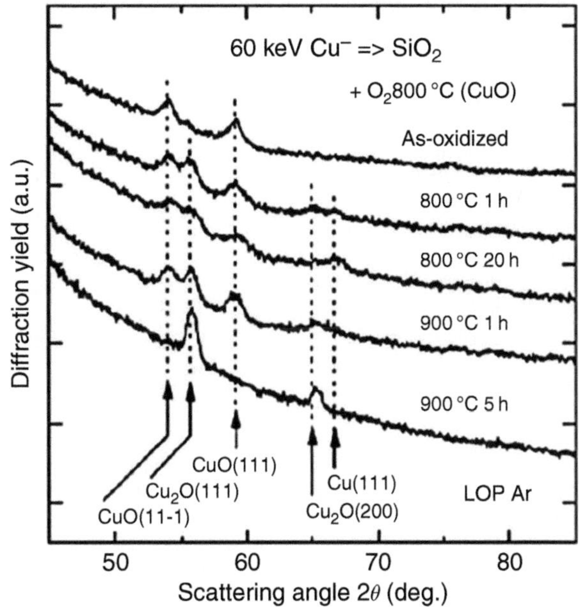

Fig. 45 GIXRD patterns of SiO_2 samples including CuO NPs, which were formed by ion implantation and oxidation, in the as-oxidized state and after annealing at 800°C and 900°C in Ar gas with oxygen impurities of 10^{-3} Torr partial pressure. The diffraction yield is plotted on a linear scale (Amekura et al. [17])

4.3.4 Selective Formation of Cu, CuO, and Cu_2O NPs and the Optical Absorption

Figure 46 shows the GIXRD results of samples #1, #2, and #3. An as-implanted sample (#1) shows broad peaks ascribed to fcc Cu metal (111) and (200) diffractions (JCPDS), indicating the formation of crystalline Cu NPs in the sample. After oxidation annealing at 800°C for 1 h (#2), sharp peaks due to CuO (11-1), (111), and (20-2) diffractions are observed, indicating transformation from Cu NPs to CuO NPs. After the soft reduction in LOP Ar gas at 900°C for 5 h (#3), two sharp peaks due to Cu_2O (111) and (200) diffractions are observed, indicating transformation from CuO NPs to Cu_2O NPs.

XTEM images of #2 and #3 are shown in Fig. 41(a) and (b), respectively. A corresponding XTEM image of Cu NPs in the as-implanted state can be found in a previous report from our group [134]. Although the CuO NPs have a roughly spherical shape, the Cu_2O NPs are rather rectangular in shape. The different shapes of the NPs are probably due to the different crystalline structures of CuO (tenorite structure, monoclinic) and Cu_2O (cuprite structure, cubic).

The XTEM image as shown in Fig. 41(b) indicates that Cu_2O NPs form inside the SiO_2 substrate, similar to NiO and CuO, but not on the surface of the substrate as ZnO. This fact is supported by the RBS results shown in Fig. 47, which shows that the depth profile of Cu_2O NPs is similar with the as-implanted Cu NPs, i.e., embedded in SiO_2 substrate.

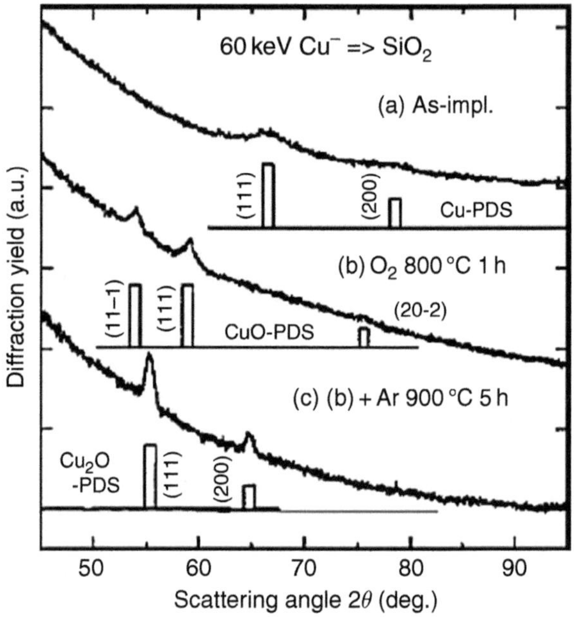

Fig. 46 GIXRD patterns of SiO_2 implanted with Cu^- ions of 60 keV. (**a**) In the as-implanted state, (**b**) after annealing in oxygen gas at 800°C for 1 h, and (**c**) after two-step annealing in oxygen gas at 800°C for 1 h and in LOP Ar gas at 900°C for 5 h. Powder diffraction patterns of Cu, CuO, and Cu_2O from the JCPDS Library are shown as *rectangles*. The spectra are plotted on a linear scale, and vertically shifted from each other for clarity. (Reprinted with permission from Amekura et al. [14]. © (2006) American Institute of Physics)

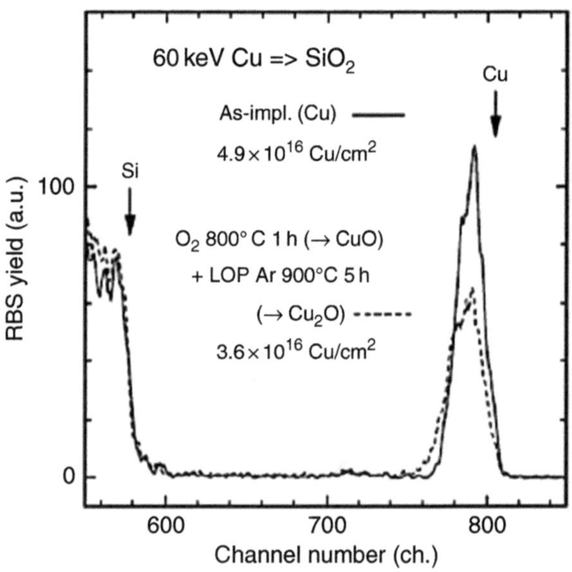

Fig. 47 RBS spectra of SiO_2 samples implanted with Cu ions of 60 keV. *Solid* and *broken lines* indicate the spectra in the as-implanted state and after two-step annealing, respectively. Arrows indicate Cu and Si edges. While Cu metallic NPs are formed in the as-implanted state, Cu_2O NPs are formed after the two-step annealing (Amekura et al. [17])

Figure 48 shows the optical absorption spectra of samples #1, #2, and #3 measured at RT [14]. The Cu NPs show an SPR peak at ~2.2 eV. The absorption increases with the photon energy at higher than ~2 eV, which is ascribed mainly to the *d*-band transitions of Cu NPs and partly to radiation damage [5]. The CuO NPs show an absorption edge at ~1.4 eV and a broad peak around ~4.2 eV. The

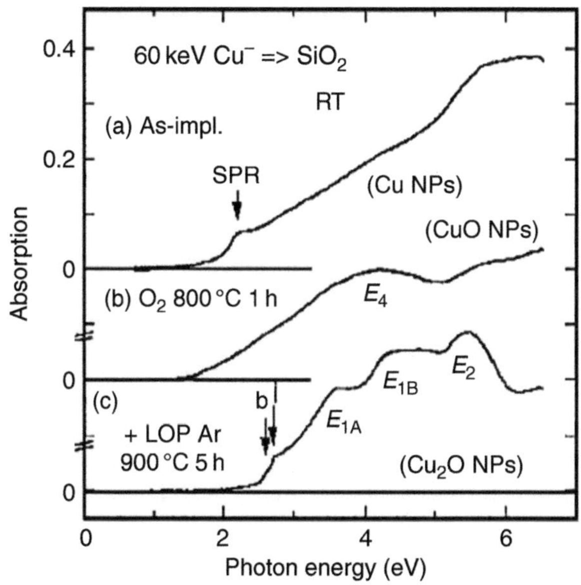

Fig. 48 Optical absorption spectra of (**a**) Cu NPs in SiO_2 fabricated by implantation of 60 keV Cu^- ions, (**b**) CuO NPs in SiO_2 fabricated by oxidation of Cu NPs in oxygen gas at 800°C for 1 h, and (**c**) Cu_2O NPs in SiO_2 fabricated by soft reduction of CuO NPs in LOP Ar gas at 900°C for 5 h. The spectra are vertically shifted from each other for clarity, with horizontal lines showing the zero level of each spectrum. Arrows indicate the surface plasmon resonance (SPR) of Cu NPs, the blue (**b**) and the indigo (i) transitions of Cu_2O NPs. (Reprinted with permission from Amekura et al. [14]. © (2006) American Institute of Physics)

Cu_2O NPs show an absorption edge around ~2.1 eV, two sharp peaks at 2.58 and 2.71 eV, and three broad peaks at ~3.65, ~4.5, and ~5.46 eV. No peaks were observed around ~2.15 eV or ~2.25 eV, where well-known exciton series of the yellow and green transitions were observed in the bulk crystals. This is because of the small oscillator strengths due to the forbidden transitions of the yellow and green transitions.

Ito et al. estimated the optical absorption spectra of bulk Cu_2O [63] and CuO [64] from their spectroscopic ellipsometry data. The absorption spectra shown in Fig. 48(b) and (c) agree well with their estimations except for small deviations of the peak energies. From a comparison with past literature [112], the sharp lines at 2.58 and 2.71 eV, which are indicated by "b" and "i" in Fig. 48(c), are ascribed to the parity-allowed exciton peaks of the blue (Γ_7^+-Γ_8^-) and indigo (Γ_8^+-Γ_8^-) transitions. The absorption edge at ~2.1 eV is due to the parity-forbidden Γ_7^+-Γ_6^+ transition. The broad bands at ~3.65 (3.45) and ~4.5 eV (4.25) are ascribed to $X_3 \rightarrow X_1$ and $M_1 \rightarrow M_1$ transitions in the Brillouin zone, respectively [73]. The values in parentheses are those estimated by Ito et al. [63]. Although the broad band at ~5.46 (5.7) eV was not assigned by Ito et al. to any transition in the Brillouin zone, it can be assigned to the $R'_{25} \rightarrow R_{15}$ transition.

As for CuO, there is still confusion over the basis of interpretation of the optical transitions as shown in Section 4.3.1. Probably, the Mott insulator concept is the majority opinion for CuO at present. Although Ito et al. [64] predicted a broad absorption peak at 3.35 eV from the ellipsometry data, we observed a broad peak at ~4.2 eV from CuO NPs in SiO_2. The deviation was as large as ~0.9 eV. Since the ellipsometry study determined the transition energies under many assumptions, the absorption spectroscopy, which we used, is a direct and more reliable method.

Figure 49(b) shows the temperature dependence of the optical loss (–log T) of Cu_2O NPs around the blue, indigo, and E_{1A} transitions. In the temperature-dependent optical measurements, we measured transmittance but not the reflectance. With decreasing temperature, all of these peaks become narrower, which are probably due to a decrease in phonon scattering. The widths of the blue and indigo peaks decrease down to 30 K, and show saturation in the line width below 30 K. Since the full width at half maximum (FWHM) of the blue and indigo peaks was ~0.03 eV at 2.8 K, much larger than the measurement resolution of 0.001 eV, the widths are not due to limitation of the measurements; i.e., the saturation of the width is not due to an experimental artifact. The blue and indigo lines show very narrow widths at low temperatures, indicating the semiconductor nature of the direct-allowed exciton transitions.

For comparison, the optical loss spectra of the SPR peak of Cu NPs at 2.9 and 300 K are shown in Fig. 49(a). The width of the SPR peak is slightly sharper at 2.9 K but the change is very small, indicating the nature of metallic NPs [54]. These results indicate that sharp exciton lines of the direct-allowed transitions (the blue and indigo lines), which exist as autoionization states over the direct-forbidden yellow and green transitions, are available in Cu_2O NPs, even at RT.

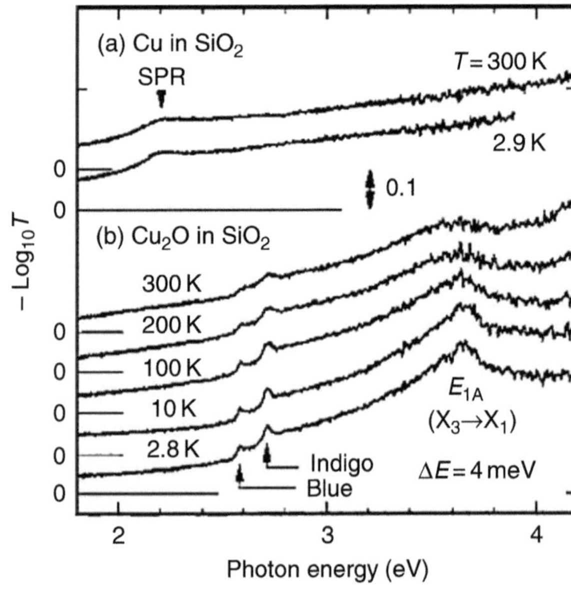

Fig. 49 Temperature dependence of optical loss spectra of (**a**) Cu NPs in SiO_2 fabricated by implantation of 60 keV Cu^- ions and (**b**) Cu_2O NPs in SiO_2 fabricated by two-step annealing of Cu NPs. The spectra are vertically shifted from each other for clarity, with horizontal lines showing the zero level of each spectrum (Reprinted with permission from Amekura et al. [14]. © (2006) American Institute of Physics)

5 Discussion

As shown by XTEM (Fig. 15), RBS (Fig. 17), AFM (Fig. 18), and XPS (Fig. 16), Zn implants show drastic migration toward the SiO_2 surface (i.e., shallowing of the depth profile) and form ZnO NPs on the surface of the SiO_2 substrate during annealing in oxygen gas at 700°C for 1 h. On the other hand, after annealing in *vacuum* at the same temperature and duration, Zn implants do not show any significant migration toward the surface and the peak of the depth profile maintains almost the same depth or moves slightly deeper as clearly shown by RBS (Fig. 17), XTEM (Fig. 29), and AFM (Fig. 20). The shallowing of the depth profile accompanied with the formation of oxide NPs has not been observed for NiO, CuO, and Cu_2O as shown in XTEM and RBS (Figs. 11, 12, 41, 42, and 47).

A similar annealing atmosphere dependence of shallowing of the implant depth profile was observed in Fe-implanted Al_2O_3 [87]. They implanted Fe^+ ions of 160 keV to Al_2O_3 up to 4×10^{16} ions/cm^2 at RT, and compared annealing atmosphere effects in oxygen gas with that in diluted hydrogen gas using RBS. The results are shown in Fig. 50. In the case of oxygen annealing, Fe signal close to the surface increased with increasing the annealing temperature, showing a bimodal distribution of Fe concentration along the depth. The formation of the oxide phase was confirmed by TEM close to the surface. On the other hand, diluted hydrogen

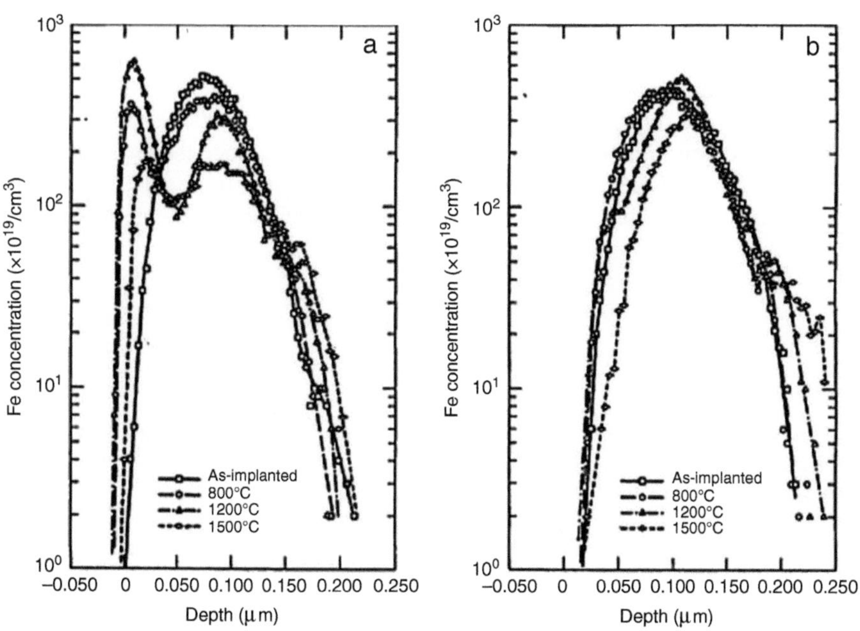

Fig. 50 Iron depth profiles determined by RBS in the α-Al_2O_3 sample implanted at room temperature with $^{56}Fe^+$ ions of 160 keV to 4×10^{16} ions/cm^2 and subsequently annealed in oxygen (**a**) and hydrogen (**b**) (Reprinted with permission from McHargue et al. [87]. © (1987) Elsevier)

annealing induced almost no change in the depth profile, or a little shift toward deeper direction. These behaviors are similar with Zn-implanted SiO_2.

The shallowing of the depth profile accompanied with the oxide phase formation was recently observed by Padova group [35]. They implanted SiO_2 glasses by 180 keV Co^+ ions to 2.0×10^{17} ions/cm^2 and annealed the samples in pure oxygen gas at 800°C for 4 h. The shallowing of the depth profile was confirmed by RBS as shown in Fig. 51. The oxide phase was ascribed to Co_3O_4 by GIXRD.

It seems that there are two kinds of oxide NP formation by the II&TO method, i.e., the oxide NP formation accompanied with and without the shallowing of the depth profile, which depends on the metal species and probably on the matrix. Implants of Zn (and probably Co) in SiO_2 belongs to the former and those of Ni and Cu to the latter.

In the late 1990s, Rossendorf group extensively studied strange annealing effects on Ge-implanted SiO_2 layers [55, 103]. Annealing in the *so-called inert gas* such as N_2 or Ar induces drastic changes on the Ge depth profile, from the Gaussian-like one in the as-implanted state to the bimodal one after annealing. Later, the origin of the strange annealing effect was ascribed to the in-diffusion of oxidants (such as a few parts per million of humidity or oxygen) from the annealing atmosphere.

In the course of the study, Rossendorf group also observed similar Ge redistribution and formation of amorphous GeO_x NPs under annealing in pure oxygen gas, and reproduced the experimental behaviors by numerical calculations using a kinetic 3D lattice Monte Carlo method [55]. The results are very suggestive when considering the mechanism of the oxide NP formation under annealing in oxygen gas.

The results are shown in Fig. 52. At elevated temperatures, Ge atoms are detached from Ge NPs and migrate in SiO_2 matrix performing nondirectional random walk. However, oxygen is supplied from the surface side only. Once the Ge atoms convert to the oxide phase by a reaction with oxygen supplied from atmosphere, they become almost immobile. Recent study shows that GeO and GeO_2 are mobile and

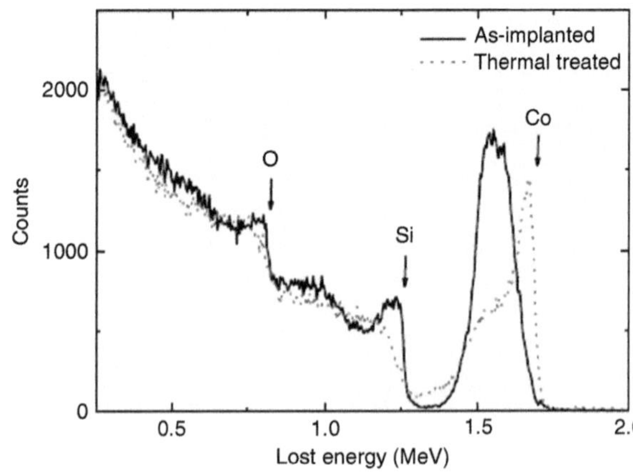

Fig. 51 RBS spectra of the silica glass sample implanted with Co ions of 180 keV to 2.0×10^{17} ions/cm^2 in the as-implanted state and after the oxygen thermal treatment for 4 h. Arrows correspond to the Co, Si, and O edges (Reprinted with permission from de Julian Fernandez et al. [35]. © (2006) Elsevier)

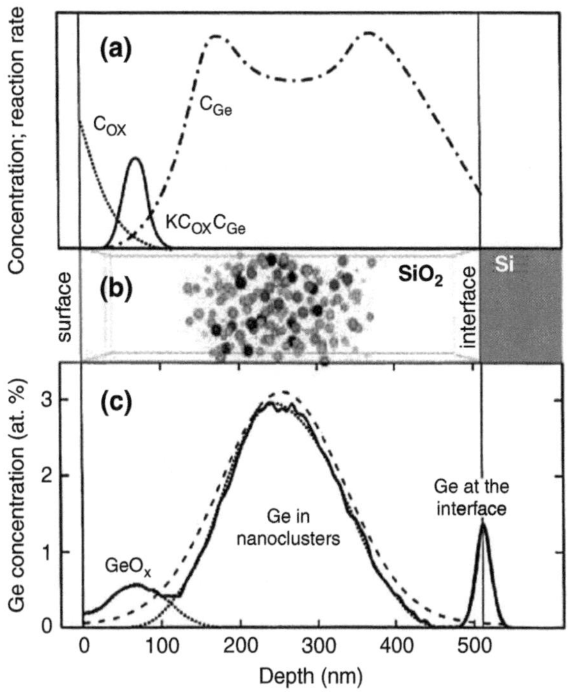

Fig. 52 A kinetic lattice Monte Carlo (KLMC) simulation of the Ge redistribution and cluster formation in Ge^+-implanted SiO_2 layer on a Si wafer during annealing. (**a**) Scheme of the oxidant concentration c_{ox}, the Ge monomer concentration c_{Ge} and the Ge oxidation rate $kc_{ox}c_{Ge}$ in SiO_2. (**b**) The spatial and size distribution of Ge nanocrystals from a KLMC simulation. (**c**) The calculated total Ge depth profile (*full curve*), which evolved from the as-implanted profile (*dashed curve*) during annealing, consists of the three Ge components (i) GeO_2 (subsurface peak), (ii) Ge in nanocrystals (central peak), and (iii) Ge accumulated at the interface (Reprinted with permission from Heinig et al. [55]. © (1999) Elsevier)

immobile, repectively [28]. The immobile GeO_x phase in shallow region gradually increases and the depth distribution of Ge gradually moves to shallower region, although the migration of Ge atoms itself is nondirectional one. This mechanism can be applicable to ZnO NP formation.

In the cases of Ni and Cu which have higher melting temperatures than Zn, the detachment of metal atoms from metal NPs is much less at the same annealing temperature. Consequently, oxygen molecules migrate sufficiently inside the SiO_2 matrix, and oxidize directly the metal NPs. In this case, oxide NPs form one by one at the same depth where metal NPs once located.

However, the difference between Ge case and Zn case should be pointed out. While the Ge oxide NPs form in a shallow region inside the SiO_2 matrix, ZnO NPs form on the surface of the substrate. Additional mechanisms such as stress effects might be included. Another problem was observed on Fe NPs in SiO_2. Although the atmosphere-dependent annealing behaviors were clearly observed in Fe NPs in Al_2O_3 as shown in Fig. 50, the behaviors in SiO_2 are rather complex. In annealing

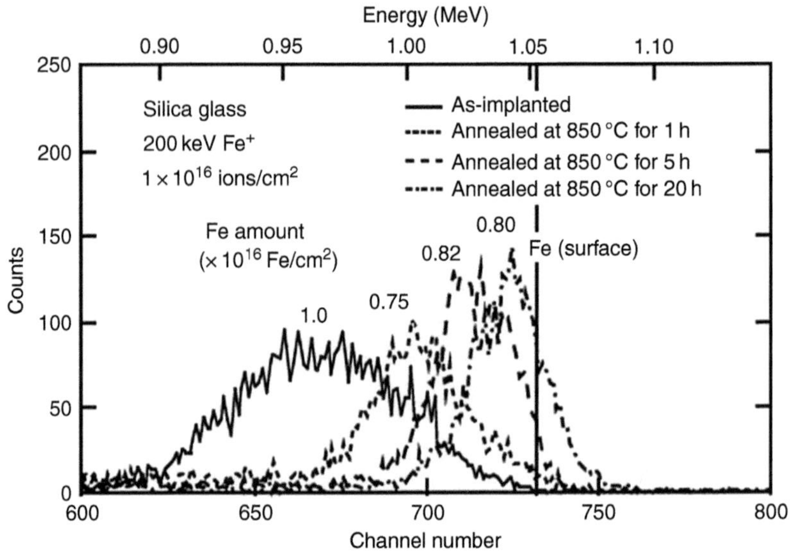

Fig. 53 RBS spectra of silica glasses implanted with 200 keV Fe^+ ions to a fluence of 1×10^{16} ions/cm^2 and subsequently annealed at 850°C for 1, 5, and 20 h in vacuum. The amount of Fe was estimated by area integration of the Fe signal (Reprinted with permission from Oyoshi [104]. © (2002) Institute of Pure and Applied Physics)

at 850°C in vacuum less than 1×10^{-6} Torr, Fe species show a shallowing of the depth profile as shown in Fig. 53 [104]. Using AFM, another group reported the NP formation on the surface of the SiO_2 implanted with Fe ions and annealed in N_2 atmosphere at 800°C [38] as shown in Fig. 54. From XPS, it was found that the surface of the NPs mainly consisted of γ-Fe_2O_3 but metallic Fe was included inside.

The formation processes of oxide NPs have not been clarified yet. Collaboration between numerical calculations and experiments might be very important.

6 Summary

The formation of oxide NPs by the II&TO method has been reviewed. Starting from the history of this method, physics of two important steps of this method, i.e., (1) metal NP formation by ion implantation and (2) thermal oxidation of metal NPs in SiO_2 matrix, have been described. After this, four different systems of oxide NPs in SiO_2 matrix, i.e., NiO, ZnO, CuO, and Cu_2O NPs, are reviewed. Although it is expected that the II&TO method is applicable to a lot of metal oxide systems, the number of oxide NP systems fabricated is still limited because of the limited number of the researchers working in this field.

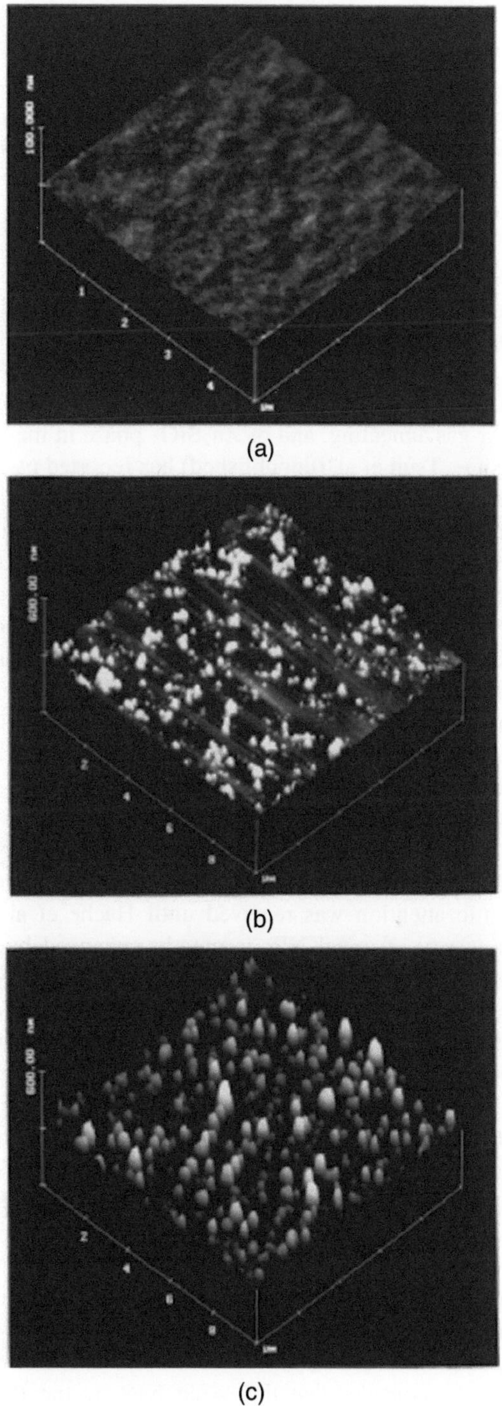

Fig. 54 AFM images of the surface morphology of the iron-implanted silica film in (**a**) the as-implanted state and after annealing in nitrogen gas with a duration of: (**b**) 800°C for 1 h and (**c**) 800°C 4 h (Reprinted with permission from Ding et al. [38]. © (1999) American Institute of Physics)

However, the II&TO method is coming to the second generation. While ZnO NPs are formed on the surface of the SiO_2 substrate when using the conventional II&TO method, the embedment of ZnO NPs in the SiO_2 substrate is possible using low temperature and long-duration annealing. The second example is Cu_2O NPs. While CuO NPs are formed by the conventional II&TO method, Cu_2O NPs are formed by the II&TO method with two-step annealing, where CuO NPs are firstly formed in atmospheric oxygen gas and then reduced to Cu_2O NPs under low oxygen pressure.

To extend the applicability of this method to oxide phases that are not the most stable under atmosphere pressure of oxygen, some modifications of the II&TO method can be important. However, the method applied to Cu_2O NP formation is not almighty. New ideas are important.

The role of oxygen species librated from SiO_2 network by radiation damage should be again stressed. They induce the partial formations of Cu_2O NPs under Ar gas annealing, and of Zn_2SiO_4 phase in the as-implanted state of Zn-implanted SiO_2. Tsuji et al. (unpublished) has reported partial formation of Ge oxide phase in the as-implanted state of Ge-implanted SiO_2 thin layer. They explained the experimental results by assuming the same mechanism that we assume.

As for the formation of ZnO NPs, the clarification of the formation mechanism of the surface NPs is important. Another issue for clarification is whether Zn NPs are liquid phase or solid phase when Zn NPs are oxidized, because the melting point of bulk Zn is 419.6°C, which is much lower than the oxidation temperature. Recently, this question has been almost clarified by expriments. However, NPs embedded in matrix are suffered by high-pressure due to embedment. In fact, Xe NPs in Al exist as solid phase at RT, while the melting point of bulk Xe is –111.9°C [90]. Another important issue is spatial controlled fabrication of oxide NPs in micro- and nanometric regions. The first step has been already given [24].

Looking back to the history of *metal* NPs in glass fabricated by ion implantation, probably the first door was opened by Arnold [25, 26]. However, only very little attention was received until Hache et al. [49] found that the optical nonlinearity of metal NPs is greatly enhanced by SPR. Since then, extensive studies have started and continued in the 1990s. Now topics of the metal NP studies are very widespread. Very fundamental issues such as the nucleation of metal NPs are extensively being studied by numerical calculations [127, 125], by in situ optical measurements [113], by TEM observations [117, 136], etc. The concept of SPR has been reconfirmed [5, 23]. An approach based on ion implantation techniques toward Plasmonics is proposed [93]. The single electron transport was attained using NPs by ion implantation [97]. Catalytic property of metal NPs fabricated by ion implantation has been applied for the fabrication of carbon nanotubes [1]. Morphology control of metal NPs, i.e., the formation of spheroidal NPs, is extensively studied using swift heavy ion irradiation (e.g., [40]). Spatial control of the NP formation is one of the most important issues, which is challenged by laser co-irradiation [100], by masked implantation [46], etc. Numerous studies are going on.

We consider that the oxide NPs by the II&TO method is in the era before "Hache's finding of a breakthrough" in metal NPs. Once a big breakthrough is made, avalanches of researchers will come into this field. We believe the breakthrough is "embedded" in this field. Since the first manuscript was submitted in August 2007, recent advances after then were not included. Briefly we would like to point out some new results: Tin oxide (SnO_2) NPs were formed in silica by the II&TO method by Kuiri et al [75]. Vanadium oxide (probably V_2O_5) NPs were formed in crystalline SiO_2 by the II&TO method [57]. Tagliente et al. reported important stress effects on the formation of ZnO NPs by the II&TO metod [129,130].

Acknowledgments The authors thank Drs. N. Umeda, Y. Takeda, K. Kono, M. Yoshitake, Y. Sakuma, M. Ohnuma, S. Hishita, M. Tanaka, H.-S. H. Boldyryeva (NIMS), Dr. Y. Katsya (Spring-8 service Co. Ltd.), Wang, O.A. Plaksin (SSC RF, A.I. Leypunsky Institute of Physics. & Power Engineering, Russia), Profs. Ch. Buchal and S. Mantl (Forschungszenturum Juelich, Germany) for collaborations. They appreciate the staffs of BL15XU, NIMS and of Spring-8 for their help at the beam line. The high temperature XRD measurements were performed under the approval of NIMS Beamline station (Proposal No. 2007A4501 and 2007B4502). Also the authors thank Profs. X.T. Zu (University of Electronic Science & Technology, China), Y.C. Liu (Northeast Normal University China), D. Ila (Alabama A&M University, USA), Dr. C. Marques (Instituto Tecnolo'gico e Nuclear, Portugal), Dr. P.K. Kuiri (Institute of Physics, India), Prof. G. Mattei (University of Padova, Italy), Dr. M.A. Tagliente (Centro Ricerche Brindisi, Italy), and Prof. Y. Saito (University of Yamanashi, Japan) for exchange of information.

Some parts of this study were financially supported by JSPS-Kakenhi (No. 18510102), the Budget for Nuclear Research of the MEXT based on the screening and counseling by the Atomic Energy Commission, Futaba Electronics Memorials Foundation, and Nippon Sheet Glass Foundation for Materials Science and Engineering.

References

1. Adhikari AR, Huang MB, Wu D, Dovidenko K, Wei BQ, Vajtai R, Ajayan PM (2005) Appl. Phys. Lett. 86, 053104.
2. Amekura H, Takeda Y, Kono K, Kitazawa H, Kishimoto N (2003) Rev. Adv. Mater. Sci. 5, 178.
3. Amekura H, Kishimoto N (2003) J. Appl. Phys. 94, 2585.
4. Amekura H, Kitazawa H, Kishimoto N (2004) Nucl. Instrum. Methods B219–220, 825.
5. Amekura H, Takeda Y, Kishimoto N (2004) Nucl. Instrum Methods B222, 96.
6. Amekura H, Kitazawa H, Umeda N, Takeda Y, Kishimoto N (2004) Nucl. Instrum. Methods B 222, 114.
7. Amekura H, Umeda N, Takeda Y, Lu J, Kishimoto N (2004) Appl. Phys. Lett. 85, 1015.
8. Amekura H, Fudamoto Y, Takeda Y, Kishimoto N (2005) Phys. Rev. B 71, 172404.
9. Amekura H, Umeda N, Takeda Y, Lu J, Kono K, Kishimoto N (2005) Nucl. Instrum. Methods B230, 193.
10. Amekura H, Umeda N, Sakuma Y, Kishimoto N, Buchal Ch (2005) Appl. Phys. Lett. 87, 013109.
11. Amekura H, Kono K, Takeda Y, Kishimoto N (2005) Appl. Phys. Lett. 87, 153105.
12. Amekura H, Umeda N, Sakuma Y, Plaksin OA, Takeda Y, Kishimoto, Buchal Ch (2006) Appl. Phys. Lett. 88, 153119.
13. Amekura H, Plaksin OA, Yoshitake M, Takeda Y, Kishimoto N, Buchal Ch (2006) Appl. Phys. Lett. 89, 023115.
14. Amekura H, Umeda N, Takeda Y, Kishimoto N (2006) Appl. Phys. Lett. 89, 223120.
15. Amekura H, Umeda N, Yoshitake M, Kono K, Kishimoto N, Buchal Ch (2006) J. Cryst. Growth 287, 2.
16. Amekura H, Kono K, Kishimoto N, Buchal Ch (2006) Nucl. Instrum. Methods B 242, 96.

17. Amekura H, Plaksin OA, Kono K, Takeda Y, Kishimoto N (2006) J. Phys. D: Appl. Phys. 39, 3659.
18. Amekura H, Ohnuma M, Kishimoto N, Buchal Ch, Mantl S (2008) J. Appl. Phys. 104, 114309.
19. Amekura H, Umeda N, Boldyryeva H, Kishimoto N, Buchal Ch, Mantl S (2007) Appl. Phys. Lett. 90, 083102.
20. Amekura H, Sakuma Y, Yoshitake M, Takeda Y, Kishimoto N, Buchal Ch (2007) Nucl. Instrum Methods B 257, 64.
21. Amekura H, Plaksin OA, Kishimoto N, Buchal Ch (2007) Surf. Coating Tech. 210, 8214.
22. Amekura H, Yoshitake M, Plaksin OA, Kishimoto N, Buchal Ch, Mantl S (2007) Appl. Phys. Lett. 91, 063113.
23. Amekura H, Umeda N, Kono K, Takeda Y, Kishimoto N, Buchal Ch, Mantl S (2007) Nanotechnology 18, 395707.
24. Amekura H, Wang HS, Hishita S, Pan J, Kishimoto N, Buchal Ch, Mantl S (2009) Nanotechnology 20, 065303.
25. Arnold GW (1975) J. Appl. Phys. 46, 4466.
26. Arnold GW and Borders JA (1977) J. Appl. Phys. 48, 1488.
27. Balamurugan B, Aruan I, Mehta BR (2004) Phys. Rev. B 69, 165419.
28. Beyer V, von Borany J (2008) Phys. Rev. B 77, 014107.
29. Brongersma ML, Snoeks E, van Dillen T, Polman A (2002) J. Appl. Phys. 88, 59.
30. Cao H, Zhao YG, Ho ST, Seelig EW, Wang QH, Chang RPH (1999) Phys. Rev. Lett. 82, 2278.
31. Cattaruzza E (2000) Nucl. Instrum Methods B 169, 141.
32. Chang H, Park HD, Sohn KS (1999) J. Korean Phys. Soc. 34, 545.
33. Chen J, Mu R, Ueda A, Wu MH, Tung YS, Gu Z, Henderson DO, White CW, Budai JD, Zuhr RA (1998) J. Vac. Sci. Technol. **A 16**, 1409.
34. Ching WY, Xu Y-N, Wong KW (1989) Phys. Rev. B 40, 7684.
35. de Julian Fernandez C, Mattei G, Sada C, Battaglin C, Mazzoldi P (2006) Mater. Sci. Eng. C 26, 987.
36. De Marchi G, Mattei G, Mazzoldi P, Sada C, Miotello A (2002) J. Appl. Phys. 92, 4249.
37. Devine RAB (1994) Nucl. Instrum Methods B 91, 378.
38. Ding X-Z, Chiah MF, Cheung WY, Wong SP, Xu JB, Wilson IH, Wang H-M, Chen L-Z, Liu X-H (1999) J. Appl. Phys. 86, 2550.
39. Diwald O, Thompson TL, Goralski EG, Walck SD, Yates JT (2004) J. Phys. Chem. B 180, 52.
40. D'Orleans C, Stoquert, JP, Estournes C, Cerruti C, Grob JJ, Guille JL, Haas F, Muller D, Richard-Plouet M (2003) Phys. Rev. B 67, 220101.
41. Elliott RJ (1961) Phys. Rev. 124, 340.
42. Espiau de Lamaestre R, Bernas H (2006) Phys. Rev. B 73, 125317.
43. Estourenes C, Lutz T, Guille JL (1996) J. Non-cryst. Solids 197, 192.
44. FACT (Facility for the Analysis of Chemical Thermodynamics) http://www.crct.polymtl.ca/fact/ http://www.crct.polymtl.ca/reacweb.htm
45. Farlow GC, Sklad PS, White CW, McHargue CJ (1990) J. Mater Res. 5, 1502.
46. Fromknecht R, Linker G, Sun K, Zhu S, Wang LM, van Veen A, van Huis MA, Weimann T, Wng J, Niemeyer J, Eichhorn F, Wang T (2004) Mat. Res. Soc. Symp. Proc. 792, 151.
47. Gonella F, Mazzoldi P (2002) In: Nalwa HS (de) Handbook of nanostructured materials and nanotechnology. Academic Press, San Diego, Vol. 4, Chap. 2.
48. Griscom DL (1991) J. Ceramic. Soc. Jpn. 99, 923.
49. Hache F, Ricard D, Flytzanis C (1986) J. Opt. Soc. Am. B 3, 1647.
50. Hamann DR (1996) Phys. Rev. Lett. 76, 660.
51. Han C-J, Hems CR (1988) J. Electrochem. Soc. 135, 1824.
52. Haglund, RF, Yang L, Magruder RH, White CW, Zuhr RA, Yang L, Dorsinville, Alfano RR (1994) Nucl. Instrum Methods B 91, 493.

53. He H, Wang Y, Zou Y (2003) J. Phys. D: Appl. Phys. 36, 2972.
54. Heilmann A, Kreibig U (2000) Eur. Phys. J: Appl. Phys. 10, 193.
55. Heinig K-H, Schmidt B, Markwitz A, Groetzschel R, Strobel M, Ostwald S (1999) Nucl. Instrum. Methods B 148, 969.
56. Heinig K-H, Mueller T, Schmidt B, Strobel M, Moeller W (2003) Appl. Phys. A 77, 17.
57. Ho AY, Mitsui N, Mochiduki K, Fujita T, Ijima K, Saito Y (2008) Jpn. J. Appl. Phys. 47, 4612.
58. Hosono H (1993) Jpn. J. Appl. Phys. 32, 3892.
59. Hosono H, Imagawa H (1994) Nucl. Instrum Methods B 91, 510.
60. Hosono H (1995) Phys. Rev. Lett. **74,** 110.
61. Ikeyama M, Nakao S, Tazawa S, Kadono K, Kamada K (2001) Nucl. Instrum. Methods B **175-177**, 652.
62. Ito T, Masumi T (1997) J. Phys. Soc. Jpn. 66, 2185.
63. Ito T, Kawashima T, Yamaguchi H, Masumi T (1998) J. Phys. Soc. Jpn. 67, 2125.
64. Ito T, Yamaguchi H, Masumi T, Adachi S (1998) J. Phys. Soc. Jpn. 67, 3304.
65. JCPDS library, No. 40836 for Cu, No. 481548 for CuO, and No. 50667 for Cu_2O, International Centre for Diffraction Data.
66. Jin Y-G, Chang KJ (2001) Phys. Rev. Lett. 86, 1793.
67. Johannessen B, Kluth P, Glover CJ, Azevedo GM, Llewllyn DL, Foran GJ, Ridgeway MC (2005) J. Appl. Phys. 98, 024307.
68. Kaiser U, Muller DA, Grazul JL, Chuvilin A, Kawasaki M (2002) Nature Materials 1, 102.
69. Kawasaki M, Ohtomo A, Ohkubo I, Koinuma H, Tang ZK, Yu P, Wong GKL (1998) Mater. Sci. Eng. B56, 239.
70. Kim JY, Rodriguez JA, Hanson JC, Frenkel AI, Lee PL (2003) J. Am. Chem. Soc.125, 10684.
71. Kishimoto N, Gritsyna VT, Kono K, Amekura H, Saito T (1997) Nucl. Instrum. Methods B **127/128**, 579.
72. Kishimoto N, Gritsyna VT, Kono K, Amekura H, Saito T (1997) Mater. Res. Soc. Symp. Proc. 438, 435.
73. Kleinman L, Mednick K (1980) Phys. Rev. 21, 1549.
74. Kreibig U, Vollmer M (1995) Optical properties of metal clusters, Spring, Berlin.
75. Kuiri PK, Lenka HP, Ghatak J, Sahu G, Joseph B, Mahapatra DP (2007) 102, 024315.
76. Kubaschewski O, Evans ELl, Alcock CB (1967) Metallurgical Thermochemistry, 4th ed. Pergamon, London.
77. Lee JK, Tewell CR, Schulze RK, Nastasi M, Hamby DW, Lucca DA, Jung HS, Hong KS (2005) Appl. Phys. Lett. **86**, 183111.
78. Liu YC, Xu HY, Mu R, Henderson DO, Lu YM, Zhang JY, Shen DZ, Fan XW, White CW (2003) Appl. Phys. Lett. 83, 1210.
79. Liu YX, Liu YC, Shen DZ, Zhong GZ, Fan XW, Kong XG, Mu R, Henderson DO (2002) J. Cryst. Growth **240**, 152.
80. Liu YX, Liu YC, Shen DZ, Zhong GZ, Fan XW, Kong XG, Mu R, Henderson DO (2002) Solid State Commun. 121, 531.
81. Liu YX, Liu YC, Shao CL, Mu R (2004) J. Phys. D: Appl. Phys. **37**, 30215.
82. Lopez R, Haynes TE, Boatner LA, Feldman, Haglund RF (2002) Opt. Lett. 27, 1327.
83. Marques C, France N, Alves LC, da Silva RC, Alves E, Safran G, McHargue CJ (2007) Nucl. Insturm. Methods 257, 515.
84. Maxwell-Garnett JC (1906) Philos. R. Soc. London 205, 237.
85. Masumi T, Yamaguchi H, Ito T, Shimoyama H (1998) J. Phys. Soc. Jpn. 67, 67.
86. Mazzoldi P, Mttei G (2005) Rivista del Nuovo Cimento 28, 1.
87. McHargue CJ, Farlow GC, Sklad PS, White CW, Perez A, Kornilios N, Marest G (1987), Nucl. Insturm. Methods B 19/20, 813.
88. Meldrum A, Haglund RF, Boatner LA, White CW (2001) Adv. Mater 13, 1431.
89. Mie G (1908) Ann. Phys. 25, 377 [in German].

90. Mitsuishi K, Song M, Furuya K, Allen CW, Birtcher RC, Dahmen U (2003) Nucl. Instrum. Methods B 206, 109.
91. Moeller W, Eckstein W (1984) Nucl. Instrum. Methods B 2, 814.
92. Moeller W, Eckstein W, Biersack JP (1988) Computer Phys. Commun. 51, 355.
93. Mueller T, Heinig K-H, Schmidt B (2002) Mater. Sci. Eng. C19, 209.
94. Muroi M, Street R, McCormick PG (2000) J. Appl. Phys. 87, 3242.
95. Muntele I, Muntel C, Thevenard P, Ila D (2007) Surf. Coating Tech., 201, 8557.
96. Nagoshi M (1998) -ray Photoelectron Spectroscopy, in The Surface Science Society of Japan (ed.), Maruzen, Tokyo p. 137 [in Japanese].
97. Nakajima A, Futatsugi T, Nakao H, Usuki T, Horiguchi N, Yokoyama N (1998) J. Appl. Phys. 84, 1316.
98. Ogasawara T, Ashida M, Motoyama N, Eisaki H, Uchida S, Tokura Y, Ghosh H, Shukla A, Mazumdar S, Kuwata-Gonokami M (2000) 85, 2204.
99. Ohkubo M, Hioki T, Kawamoto J (1986) J. Appl. Phys. 60, 1325.
100. Okubo N, Umeda N, Takeda Y, Kishimoto N (2003) Nucl. Instrum. Methods B 206, 610.
101. Oliver A, Cheang-Wong JC, Roiz J, Rodriguez-Fernandez L, Hernandez JM, Crespo-Sosa A, Munoz E (2002) Nucl. Instrum Methods B 191, 333.
102. Orellana W, da Slva AJR, Fazzio A (2001), Phys. Rev. Lett. 87, 155901.
103. Ostwald S, Schmidt B, Heinig K-H (2000) Surf. Interface Anal. 29, 249.
104. Oyoshi K (2002) Jpn. J. Appl. Phys. 41, 6145.
105. Oyoshi K, Tagami T, Tanaka S (1991) Jpn. J. Appl. Phys. 30, 1854.
106. Ozgur U, Alivov Y, Liu C, Teke A, Reshchikov MA, Dogan S, Avrutin V, Cho S-J, Morkoc H (2005) J. Appl. Phys. 98, 0431301.
107. Palik ED (1998) Handbook of Optical Constants of Solids, Academic Press, San Diego.
108. Paul GK, Nawa Y, Sato H, Sakurai T, Akimoto K (2006) Appl. Phys. Lett. 88, 141901.
109. Pena O, Rodriguez-Fernandez L, Cheang-Wong JC, Santiago P, Crespo-Sosa A, Munoz E, Oliver A (2006) J. Non-Cryst. Solids 352, 349.
110. Perenboom JAAJ, Wyder P, Meier F (1981) Phys. Reports 78, 173.
111. Perez A, Treilleux M, Capra T, Griscom DL (1987) J. Mater Res. 2, 910.
112. Peyghambarian N, Koch SW, Mysyrowicz A (1993) Introduction to Semiconductor Optics, Prentice-Hall, Englewood Cliffs, NJ, Chap. 7, p. 174.
113. Plaksin OA, Takeda Y, Amekura H, Kishimoto N (2006) J. Appl. Phys. 99, 044307.
114. Ramaswamy V, Haynes TE, White CW, Moberly-Chan WJ, Roorda S, Aziz MJ (2005) Nano Lett. 5, 373.
115. Ren F, Guo LP, Shi Y, Chen DL, Wu ZY, Jiang CZ (2006) J. Phys. D: Appl. Phys. 39, 488.
116. Rizza GC, Stobel M, Heinig K-H, Bernas H (2001) Nucl. Instrum Methods B 178, 78.
117. Rizza GC, Cheverry H, Gacoin T, Lamasson A, Henrry S (2007) J. Appl. Phys. 101, 014321.
118. Roiz J, Oliver A, Munoz E, Rodriguez-Fernandez L, Hernandez JM, Cheang-Wong JC (2004) J. Appl. Phys. 95, 1783.
119. Ruppin R (1986) J. Appl. Phys. 59, 1355.
120. Schmid-Whitley RD, Martinez-Clemente M, Revocolecschi A (1974) J. Cryst. Growth 23, 113.
121. Shimizu T, Matsumoto T, Goto A, Chandrasekar Rao TV, Yoshimura K, Kosuge K (2003) Phys. Rev. B 68, 224433.
122. Skuja L, Guttler B (1996) Phys. Rev. Lett. 77, 2093.
123. Snoeks E, Weber T, Cacciato A, Plolman A (1995) J. Appl. Phys. 78, 4723.
124. Strobel M (1999) Modeling and Computer Simulation of Ion Beam Synthesis of Nanostructures. Forschungszentrum Rossendorf, Wissenschaftlich-Technische Berichte FZR-277.
125. Strobel M, Heinig K-H, Moeller W, Meldrum A, Zhou DS, White CW, Zuhur RA (1999) Nucl. Instrum. Methods B 147, 343.
126. Strobel M, Heinig K-H, Moeller W (1999) Nucl. Instrum. Methods B 148, 104.
127. Strobel M, Heinig K-H, Moeller W (2001) Phys. Rev. B 64, 245422.

128. Swalin RA (1962) Thermodynamics of Solids, John Wiley and Sons, New York.
129. Tagliente MA, Massaro M (2008) Nucl. Instrum. Methods B 266, 1055.
130. Tagliente MA, Massaro M, Mattei G, Mazzoldi P, Bello V, Pellegrini G (2008) J. Appl. Phys. 104, 093505.
131. Townsend PD, Chandler PJ, Zhang L (1994) Optical Effects of Ion Implantation. Cambridge University Press, Cambridge.
132. Tsuda N, Nasu K, Fujimori A, Shiratori K (1993) Electron conduction in oxides. Springer, Berlin.
133. Tsukazaki A, et al. (2005) Nature-Mater. 4, 42.
134. Umeda N, Kishimoto N, Takeda Y, Lee CG, Gritsyna VT (2000) Nucl. Instrum. Methods B 166–167, 864.
135. Umeda N, Amekura H, Kishimoto N (2008) Vacuum 83, 645.
136. Valentin E, Bernas H, Ricolleau C, Creuzet F (2001) Phys. Rev. Lett. 86, 99.
137. van Huis MA, van Veen A, Schut H, Kooi BJ, De Hosson JThM, Du XS, Hibma T, Fromknecht R (2004) Nucl. Instrum. Methods B 216, 390.
138. White CW, Meldrum A, Sonder E, Budai JD, Zuhr RA, Withrow SP, Henderson DO (1999) Mat. Res. Soc. Symp. Proc. **540**, 219.
139. White CW, Withrow SP, Budai JD, Boatner LA, Sorge KD, Thompson JR, Beaty KS, Meldrum A (2002) Nucl. Instrum. Methods B 191, 437.
140. Williams EL (1965) J. Am Cer. Soc. 48, 190.
141. Wu H, Wang W, Zhu W (2006) J. Phys. Chem. B 110, 13829.
142. Xiang X, Zu XT, Bao JW, Zhu S, Wang LM (2005) J. Appl. Phys. 98, 073524.
143. Xiang X, Zu XT, Zhu S, Wang LM (2005) Physica B369, 88.
144. Xiang X, Zu XT, Zhu S, Wei QM, Zhang CF, Sun K, Wang LM (2006) Nanotechnology 17, 2636.
145. Yang H, Ouyang J, Tang A, Xiao Y, Li X, Dong X, Yu Y (2006) Mater. Res. Bull. 41, 1310.
146. Yao W-T, Yu S-H, Zhou Y, Jiang J, Wu Q-S, Zhang L, Jiang J (2005) J. Phys. Chem. 109, 14011.
147. Yarovaya RG, Shklyarevskii IN, El-Shazly AFA (1974) Sov. Phys. - JETP 38, 331.
148. Zhao J, Chen X, Wang G (1994) Phys. Rev. B 50, 15424.
149. Ziegler JF, Biersack JP, Littmark U (1985) The Stopping and Range of Ions in Solids, Pergamon, New York, Chap. 8. http://www.srim.org/
150. Zou JM, Kim M, O'Keeffe M, Spence JCH (1999) Nature 401, 49.

Design of Solution-Grown ZnO Nanostructures

Thierry Pauporté

Abstract The renewed interest of the scientific community in zinc oxide (ZnO) during the last decade has been mostly powered by the development of new low-temperature methods for the synthesis of ZnO nanostructures with a controlled morphology. The wide variety of morphology includes nanoparticles, nanowires, nanorods, nanotubes, nanosheets, as well as nanoporous films. The present chapter is a review of the most recent progresses made in the design of these structures by the use of different solution-based low-temperature preparation methods. The methods include chemical, sol-gel and hydrothermal synthesis, electrospinning, electroless deposition and electrodeposition. Special attention is paid to the preparation of organic/inorganic hybrid films, to patterning and to the doping of nanostructured ZnO layers. The interest in these nanostructures is illustrated by a large variety of applications, such as in solar cells, light emitting diodes (LED), photocatalysis and surfaces with controllable wettability.

Contents

T. Pauporté (✉)
Laboratoire d'Électrochimie et Chimie Analytique, UMR7575, ENSCP-CNRS-Univ.Paris6, École Nationale Supérieure de Chimie de Paris, 11 rue P. et M. Curie, 75231 Paris cedex 05 France
e-mail: thierry-pauporte@enscp.fr

Z.M. Wang (ed.), *Toward Functional Nanomaterials*, Lecture Notes in Nanoscale Science and Technology 5, DOI 10.1007/978-0-387-77717-7_2,

1 Introduction

Zinc oxide (ZnO) is not a newly discovered compound and research on its preparation and characterization has continued for several decades. However, a renewed interest of the scientific community in ZnO can be measured by the very large number of publications which have appeared on this compound in the past few years. ZnO is a large bandgap n-type semiconductor which combines several advantages over its competitors such as GaN used today for production of short-wavelength light-emitting devices. Its exciton energy (60 meV) is larger than the room-temperature thermal energy, and high-excitonic near-UV luminescence and laser emissions can be expected. The recent synthesis of p-type ZnO reported by several groups (e.g. [1–5]) is promising for the preparation of diodes with efficient charge injection and also for the preparation of efficient light-emitting and laser diodes in the near future. However, due to the present lack of reproducible and high-quality p-type ZnO, some alternative methods such as n–p heterojunction or metal-insulator-semiconductor diodes can be considered for exploiting the advantages of ZnO. ZnO is stable under high-energy radiation and can be easily etched in acidic or alkaline solutions. It is also a promising candidate for spintronic applications with a Curie temperature higher than 300°C for Mn-doped ZnO and also high Curie temperatures for Fe-, Co- or Ni-alloyed ZnO.

The preparation of nanostructured ZnO in a large variety of shapes is an emerging field. One-dimensional (1D) ZnO nanostructures such as nanorods, nanowires, nanobelts, nanotubes and nanowhiskers are important for their physical properties arising from quantum confinement (such as electronic quantum transport and enhanced radiative recombination of carriers). Nanowires have promising potentials in extensive applications and are the fundamental building blocks for fabricating short-wavelength nanolaser, field effect transistors (FET), ultrasensitive nanosized gas sensors, nanoresonators, transducers, actuators, nanocantilevers and field emitters [6–12]. These properties could have numerous applications in various areas such as nanoscale electronics, optoelectronic devices and high-density magnetic memories [13]. Two-dimensional (2D) nanostructures such as nanosheets and nanoplatelets have also been widely studied. Due to their large surface-to-volume ratio, they are interesting catalysts and photocatalysts. They may be useful in energy conversion application and data storage and memory devices [14, 15]. The formation of three-dimensional (3D) nanostructures by assembling 1D and 2D nanostructures would be desirable for advanced electronics and optoelectronics applications, as demonstrated with well-organized epitaxial nanorods grown at the junction of nanowalls [15]. Finally, mesoporous films are also widely studied nanostructures since they are of great interest for sensor, catalysis, photocatalysis, solar cells (such as dye-sensitized solar cells) and luminescence applications. In the case of ZnO, the preparation of mesoporous films with well-controlled morphologies has been generally achieved by wet synthetic methods [7].

Up to now, high-quality 1D and 2D ZnO nanostructures have been fabricated mainly by physical techniques such as molecular beam epitaxy, metal-organic chemical vapor deposition and vapor-phase transport and condensation processes [6, 11]. These techniques are expensive and energy consuming since they are operated under extreme conditions. In contrast, solution-based methods are emerging as soft, easy-to-operate, cost-effective processes which are suitable for the preparation of high-quality 1D or 2D ZnO nanostructures at low temperature. They are easy to scale up for the preparation of films on large surfaces at low cost.

In this chapter, we review some significant progresses made recently in the field of solution-grown well-controlled ZnO nanostructures. We also review the main synthetic methods for the preparation of ZnO nanoparticles and nanostructured thin films, with special attention paid to the growth mechanism. Their optical properties and the fabrication of functional optoelectronic devices such as solar cells or light-emitting diodes (LED) in which they act as the active layer are also discussed.

2 Structural and Optical Properties of ZnO

The thermodynamically stable crystal phase of ZnO is hexagonal wurtzite with two lattice parameters, a and c of 3.250 Å and 5.207 Å respectively [16]. It belongs to the point group 3 m and space group $P6_3mc$ or C^4_{6v}. A schematic representation of the hexagonal structure of ZnO is shown in Fig. 1. Its ionic and polar structure can be described as two interpenetrating hexagonal close-packed (hcp) sublattices, each of which consists of one type of atom (O or Zn) displaced with respect to each other by a factor $3/8 = 0.375$ in the [0001] direction. In the ZnO structure, each Zn^{2+} ion is surrounded by four O^{2-} ions and vice-versa. The hexagonal unit cell exhibits a polar, negatively charged, oxygen-terminated $(000\bar{1})$ low-index face and a polar, positively charged, Zn-terminated (0001) face [17]. The [0001] axis (noted c-axis) points from the face of the O plane to the Zn plane. $\{10\bar{1}0\}$ faces are nonpolar. The low-symmetry nonpolar faces with 3-fold coordinated atoms are the most stable ones. Due to the presence of a center of inversion in the wurtzite structure, there is an inherent asymmetry along the c-axis of ZnO which allows the growth of anisotropic crystallites.

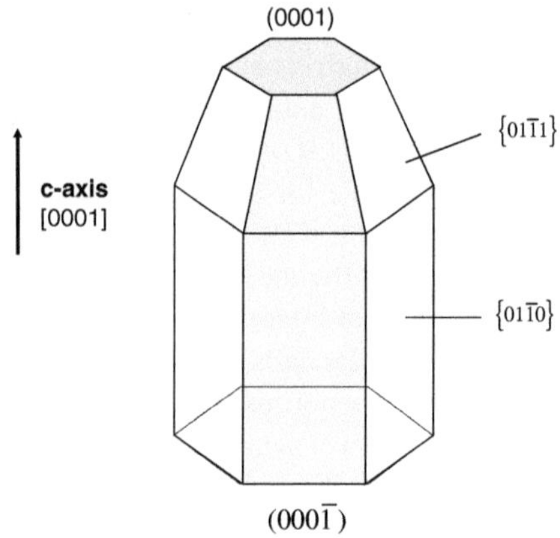

Fig. 1 Hexagonal wurtzite crystal structure of ZnO

Because of its noncentral symmetry, ZnO is piezoelectric, a key property for building electromechanical-coupled sensors and transducers [6]. It is also a II–VI semiconductor and is intrinsically n-type. P-type doping has proven extremely difficult and has been achieved recently by physical methods [1–5]. However, the reproducibility of the published data and the quality of these compounds remain controversial [6]. To the best of our knowledge, there is presently no report on the synthesis of p-type ZnO by a wet method. ZnO direct bandgap is ~3.3 eV at room temperature [11]. It presents a high exciton binding energy at 60 meV. To achieve efficient UV emission at room temperature, exciton energy must be much greater than the thermal energy at room temperature ($k_BT \sim 25$ meV, where k_B is the Boltzmann's constant). In this regard, ZnO is a very promising candidate compared to GaN (25 meV) or ZnSe (22 meV), two other well-studied wide-bandgap semiconductors.

The luminescence spectra of high-quality ZnO samples present at low temperature a succession of fine structures as shown in Fig. 2a [16]. At high energy (~3.38 eV), emission peaks are due to free excitons. Below, the luminescence spectra are dominated by several bound exciton lines with prominent lines from exciton bound to neutral donors (3.35–3.37 eV). In the spectral region of 3.32–3.34 eV, emission peaks are due to two electron transition satellites. These transitions involve radiative recombination of an exciton bound to a neutral donor, leaving the donor in an excited state, thereby leading to a transition energy which is less than the donor-bound exciton energy. At lower energies, the spectral region contains LO-phonon replicas of the main transitions. Excitonic emissions are observed in ZnO single-crystal nanorods at different temperatures [18]. At 10 K fine structures are observed (Fig. 2b). The high-energy emission (3.376 eV) is attributed to a free exciton peak (X_A), whereas at lower energies, three UV emission peaks (noted I_2-1, I_2-2, I_2-3) are ascribed to neutral donor-bound excitons. The presence of free exciton emission indicates that the ZnO compound is of high optical quality. With increasing temperature, the peak intensities decrease dramatically and are shifted toward

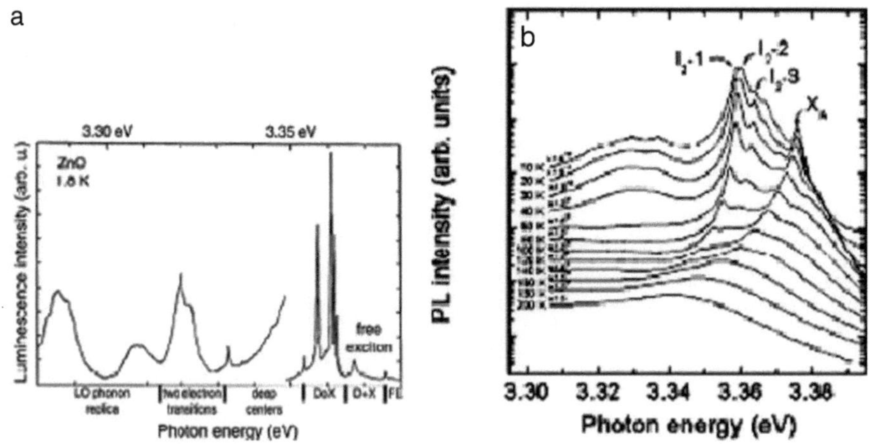

Fig. 2 (**a**) Typical UV luminescence spectrum of high-quality ZnO at 1.8 K (after [12]). (**b**) Effect of temperature on the PL spectrum of high-optical-quality ZnO nanorods (after [18])

lower energies. The I_2 peaks progressively disappear, and at 200 K the spectrum is dominated by the free exciton emission. This behavior is probably the result of the decomposition of bound excitons to free excitons due to the increased thermal energy.

Most photoluminescence (PL) spectra of undoped ZnO also show deep-level defect emissions in the visible wavelength region. The energy levels of different defects in ZnO extracted from different studies are reported in Fig. 3 [19–23]. In the

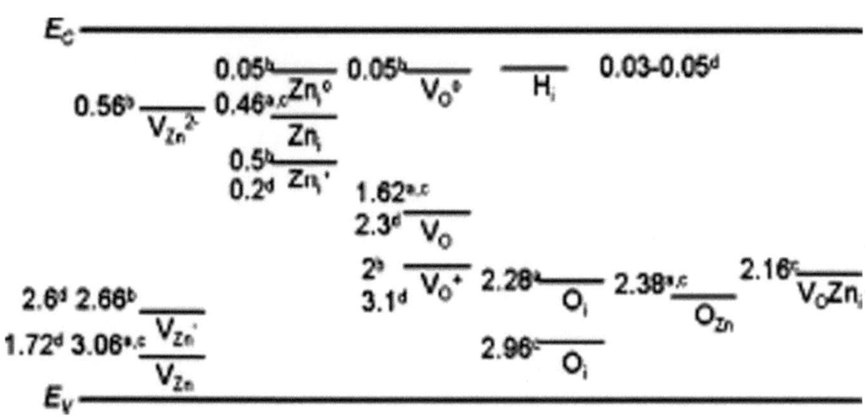

Fig. 3 Energy levels of different defects in ZnO from different studies in the literature: (**a**) [20], (**b**) [21], (**c**) [22] and (**d**) [23]. E_v and E_c are the energies of valence and conduction bands respectively. V_{Zn}, V_{Zn}^- and V_{Zn}^{2-} denote neutral, singly charged and doubly charged zinc vacancy respectively. Zn_i° and Zn_i indicate neutral zinc interstitial while Zn_i^+ denotes singly charged zinc interstitial. V_O° and V_O denote neutral oxygen vacancy while V_O^+ denotes singly charged oxygen vacancy. O_i and H_i represent oxygen and hydrogen interstitials, respectively. V_OZn_i denotes the complex of oxygen vacancy and zinc interstitial (after [19])

literature, green luminescence (GL), yellow luminescence (YL), orange-red luminescence (OL) and red luminescence (RL) have been reported, depending on the energy at which the broad visible emission is centered. The origin of the GL emission appearing at about 2.5 eV remains controversial. A number of hypotheses have been proposed but it is generally associated with oxygen vacancies [24–27] and in most cases with paramagnetic singly ionized oxygen vacancies, V_O^+. The presence of Cu in ZnO can also cause an emission band in the green spectral range [6]. In undoped ZnO, the yellow emission (~2.2 eV) has been related to the interstitial oxygen defect, O_i. In the case of ZnO nanostructures with high surface area, correlation between YL and surface OH group has been suggested [19].

3 Preparation of ZnO Nanosized Powders with Controlled Shape

The classical route for ZnO formation in solution is based on the oxide precipitation by reacting zinc salts with hydroxide ions. In the Zn system, the most stable solid phase in contact with the media will in general precipitate first. At room temperature the pK_s of ZnO and $Zn(OH)_2$ are 16.8 and 16.5 respectively [28]. Therefore, ZnO is more stable thermodynamically than $Zn(OH)_2$ and it should be theoretically possible to prepare ZnO at room temperature in aqueous solvent. However, the two pK_a are close and a competition occurs for the formation of the two compounds due to kinetic aspects and experimentally $Zn(OH)_2$ is clearly the dominant species formed at room temperature. A first approach to overcome this difficulty consists in substituting the aqueous solvent by an organic one which favors the formation of the dehydrated product at low temperature. The second solution consists in raising the preparation temperature above 50°C in order to accelerate the kinetic of the oxide formation reaction. In both cases, ZnO powders are obtained by homogeneous nucleation in the liquid phase.

3.1 Synthesis in Organic Solvents

The general approach for the preparation of ZnO quantum dots (QDs) was first described by Spanhel et Anderson in 1991 [29] and simplified by several others (e.g. [30–33]). Zinc precursor is prepared by dissolving zinc acetate dihydrate in boiling ethanol at atmospheric pressure and cooled at 0°C. A white powder of anhydrous acetate is precipitated close to the room temperature. Then a mixture of LiOH (or NaOH, $(CH_3)_4NOH$) in ethanol cooled at 0°C is added dropwise to the previous suspension which then turns transparent. In order to avoid particle aggregation, it is important to precisely control the amount of hydroxide ion added and the pH of the solution. The final pH must be close to the point of zero change (pzc ~ 9) of the colloidal particles. The particles synthesized are very small in size with diameter varying between 2.5 and 7 nm with small size dispersion. The size control is achieved by monitoring the storage time since ZnO nanoparticles continue to grow after the synthesis even when stored at 0°C [30]. Alternatively, octanethiol has been reported to inhibit the growth of ZnO nanoparticles in solution [32]. Due to a quantum confinement effect, absorption curves show a blue-shift with particle size decrease (Fig. 4a).

The bandgaps, E_g, of the spherical nanosized particles are approximately determined from the absorbance curves at $\lambda_{\frac{1}{2}}$ (Fig. 4a). The shift of the absorption onset reveals a decrease in the optical bandgap with particle size. The tendency is fitted based on an effective mass approximation (Fig. 4b) [33]:

$$E_g = E_g^b + \frac{\hbar^2\pi^2}{2er^2}\left(\frac{1}{m_e m_0} + \frac{1}{m_h m_0}\right) - \frac{1.8e^2}{4\pi\varepsilon\varepsilon_0 r} \tag{1}$$

where E_g^b is the bulk bandgap energy (~3.3 eV), r is the particle radius, m_e and m_h are the effective masses of the electrons and the holes respectively, m_0 is the mass of electron (9.1×10^{-31} kg), ϵ is the relative permittivity, ϵ_0 is the permittivity of free space (8.85×10^{-12} F m^{-1}), $\hbar$ is Planck's constant divided by 2π (1.05×10^{-34} J s) and e is the charge on the electron (1.6×10^{-19} C). For ZnO, it is accepted that $m_e = 0.26$, $m_h = 0.59$ and $\epsilon = 8.5$ [33]. Bandgap enlargement can be seen for particles of diameter less than 8 nm.

The effect of particle size, r, on photoluminescence (PL) was investigated in [31]. The green PL intensity was higher than the UV excitonic emission. Increasing r clearly decreased the excitonic emission energy but had no effect on the green emission energy. The near-band-edge-(NBE) to-deep-defect-intensity ratio was promoted by aging time, and then with r. Therefore, the optical quality of the particles increased with r.

By a similar method, multipod nanorods were synthesized in 2-propanol [34]. The key parameter for the particle structure change was the use of poly(N-vinyl-2-pyrrolidone) (PVP) as an additive. The particles were rather large in size (more than

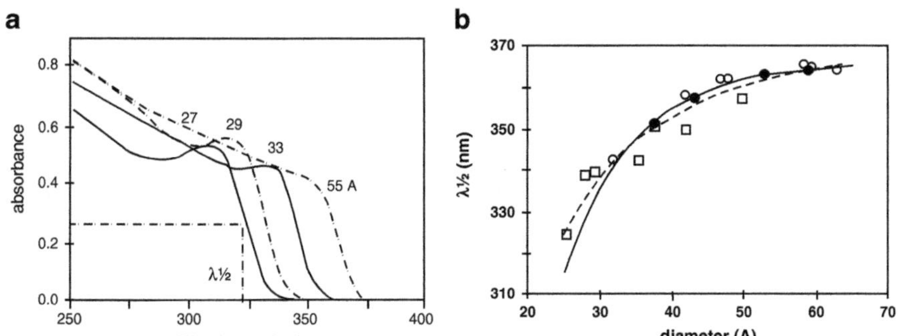

Fig. 4 (**a**) UV/vis absorbance spectra of ZnO sols of particles with various diameters. (**b**) Size dependence of ZnO optical bandgap. The dots represent experimental values and the lines are fits with Eq. (1). (after [30])

100 nm). Single-crystalline ZnO nanorods with an average diameter of 16 nm were prepared from zinc acetate dihydrate and hydrazine monohydrate in the presence of dodecylbenzene sulfonic acid sodium salt (DBS)[35]. Rod-shaped micelles were formed in a mixed organic solvent of chloroform and ethylene glycol and ZnO was grown in a controlled manner inside the micelles. Strong UV excitonic emission and quite weak deep-level emission were observed.

Another route for the formation of elongated ZnO nanoparticles with variable aspect ratio (length-to-diameter ratio, L/D) consists in adding water to the precipitation of ethanolic solutions [36]. The presence of water favors the growth of polar planes along the *c*-axis. These ZnO nanoparticles were synthesized in sealed hydrothermal bath at higher temperature, 75–190°C, to avoid the formation of zinc hydroxide. Generally, it was observed that the aspect ratio was promoted by increasing water content and at 50% (v/v) water content, L/D was higher than 50. The individual ZnO nanorods and nanowires were determined to be single crystals, elongated along their *c*-axes. Increasing L/D clearly enhanced deep-level PL emission of ZnO.

Pan et al. [14] studied the effect of different parameters on the growth of ZnO particles from $Zn(OH)_2$ seeds in closed bath at 190°C. The results are summarized in Fig. 5. When ZnO was grown in pure anhydrous ethanol, 2D nanosheets with 6 nm

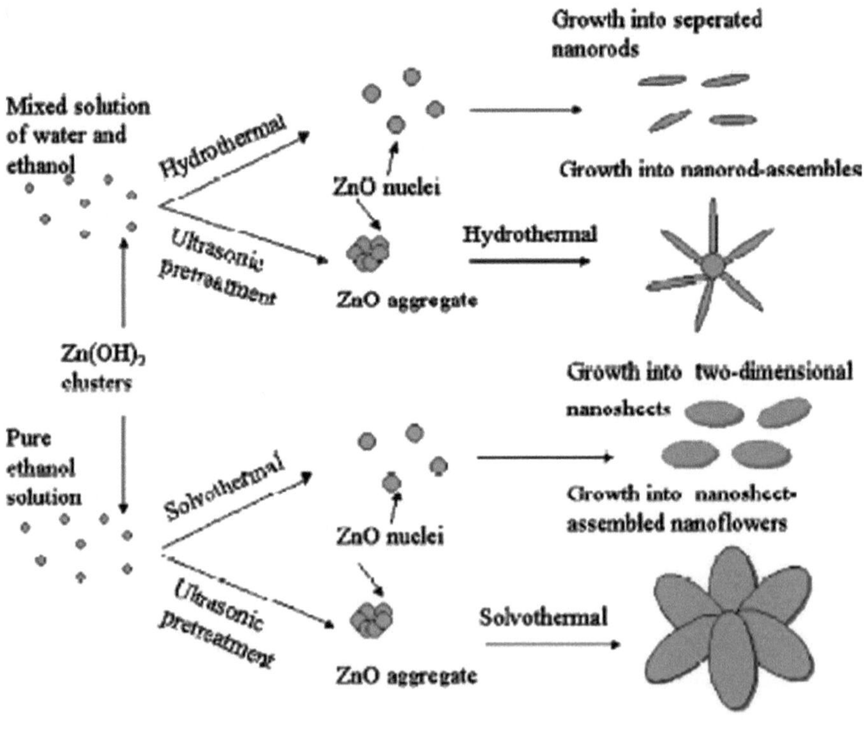

Fig. 5 Control of the shape of ZnO nanocrystals grown from $Zn(OH)_2$ seeds by solvothermal and hydrothermal methods (after [14])

thickness were formed from ZnO nucleus. After an ultrasonic treatment, aggregates of ZnO nuclei were formed and nanoflowers made of assembled nanosheets were obtained. When the particles were grown in a 50/50 (v/v) mixture of ethanol and water, nanorods were synthesized from well-separated ZnO nuclei, whereas nanorod assemblies were grown from ZnO nuclei aggregates formed by an ultrasonic treatment.

3.2 Nanoparticle Preparation in Aqueous Solvent

Large ZnO single crystals are classically grown from a ZnO seed by hydrothermal methods under extreme conditions [37–39]. The aqueous alkaline growth solutions are 2–10 molal, the temperature ranges between 200 and 500°C and the pressure between 500 and 2000 bars. However, aqueous media are also of interest in the growth of small particles of well-controlled shape. ZnO nanoparticles can be prepared by homogeneous nucleation and growth in aqueous solvent at temperatures higher than 50°C.

ZnO nanowhiskers were grown from $Zn(OH)_2$ seeds at 180–200°C under hydrothermal conditions [40]. The seeds were obtained by precipitation in ammonia medium and the whiskers were grown by the "Ostwald ripening" mechanism, in which the smaller whiskers dissolve and feed the growth of larger crystallites. The growth habit of ZnO single crystals under hydrothermal conditions was investigated by Li et al. [41]. Zinc acetate or zinc hydroxide colloids were mixed to ammonia hydroxide and microcrystals were grown at high temperature. A new rule was concluded concerning the growth habit of ZnO considering the relationship between the growth rate and the orientation of the coordination polyhedron at the interface: the direction of the crystal face with the corner of the coordination polyhedron occurring at the interface has the fastest growth rate; the direction of the crystal face with the edge of the coordination polyhedron occurring at the interface has the second fastest growth rate; the direction of the crystal face with the face of the coordination polyhedron occurring at the interface has the slowest growth rate. ZnO nanocrystals could be grown by simply mixing zinc acetate and ammonia at 50°C [42]. Various crystal shapes were obtained by varying $[NH_3]/[Zn] = R$ concentration ratio. At $R = 2$, hexagonal nanorods with sharp edges were formed. At $R = 4$, ellipsoidal particles were obtained, whereas at $R = 6$, multineedle shapes were formed (Fig. 6). The PL spectra of the particles depended on their shape: the spectrum of the hexagonal particles was dominated by the NBE UV emission, the ellipses and multineedles presented emissions from both excitons at 390 nm and deep levels centered at 530 nm.

Increase in pH of the solution and hydrolysis of Zn(II) can be monitored by the thermal decomposition of an ammonia precursor such as urea [43–45], hexamethylene tetramine (HMT) [46] or triethanolamine (TEA) [47]. By varying the growth conditions, different particle shapes have been reported such as nanorods [45], nanowires [43, 44], nanoellipses with controllable aspect ratio [47] and hexagonal microrods with sheetlike and platelike nanostructures [46]. Various ZnO

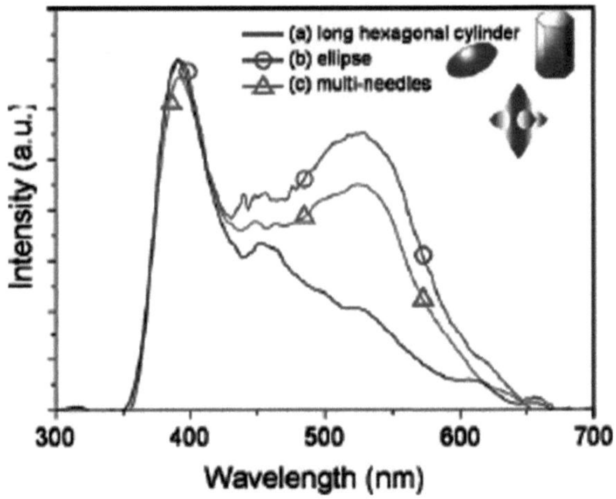

Fig. 6 Photoluminescence properties of ZnO crystals of various shapes prepared by precipitation in solution (after [42])

nanostructures of controlled morphology can be prepared in alkaline aqueous solution by controlling the concentration of a complexing agent, ethylenediamine, and the final pH for precipitation [48].

4 Chemical Deposition of Nanostructured ZnO Films in Aqueous Solutions

The growth of wurtzite ZnO in solution onto substrate requires the first step of heterogeneous nucleation, which is followed by the oxide growth. Indeed, the interfacial energy between crystal and substrates is usually smaller than the interfacial energy between crystal and solutions. Consequently heteronucleation takes place at a lower supersaturation onto a substrate than in homogeneous solution. Different combinations of zinc precursors and water soluble additives have been proposed to grow ZnO films. The synthesis temperature is higher than 50°C. Usually, the substrate is placed in a glass bottle filled with the deposition aqueous mixture and is sealed with an autoclavable screw cap. However, synthesis in open bath is also reported. The deposition rate is rather slow. Scheme 1 encompasses the main chemical routes for the preparation of ZnO films and summarizes the effect of substrate (F-doped SnO_2 coated glass, nanocrystalline ZnO template, Au-coated conducting glass, single-crystal sapphire) on the film morphology.

4.1 Film Preparation in Alkaline Solutions

There are several reports in the literature on the direct heterogeneous nucleation and growth of ZnO layer in ammonia solutions containing zinc salts [50–52]. Yamabi and Imai [50] studied the preparation of ZnO in various ammonium salts

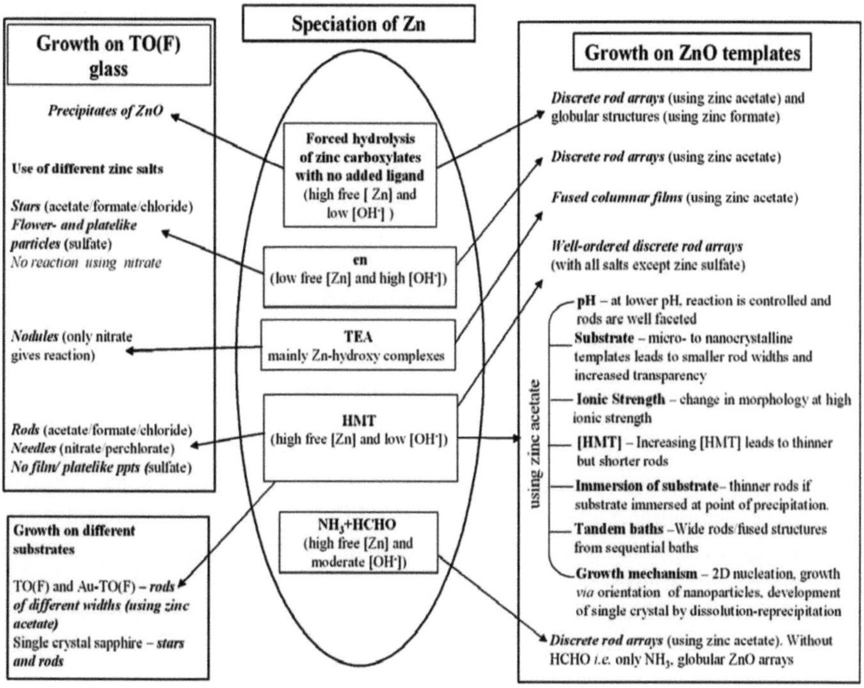

Scheme 1 Film morphologies obtained on different substrates by the main chemical growth methods of ZnO, except electroless deposition (after [49])

as complexing agent (NH_4F, NH_4Cl, NH_4NO_3, $(NH_4)_2SO_4$). The molar ratio $R = NH_4^+/Zn(II)$ ranged between 20 and 50. The pH of the solution was adjusted by adding NaOH. ZnO was produced in the pH range of 9–13 at 60°C. The pH of ZnO formation increased with the complexing agent concentration and then with the R value. The stability range of ZnO formation did not depend on the ammonium salt used. The authors also carried out an interesting investigation of the effect of the substrate on the film morphology. It was shown that the density of crystallite varies with the substrate, being low on glass slide and silicon wafer and being high on polyethylene terephthalate (PET) and F-doped tin oxide. High density vertically aligned hexagonal nanorods with diameters of 20–100 nm and lengths of 2 μm were formed on a buffer layer of nanocrystallized ZnO. Duan et al. [51] prepared arrays of ZnO nanotubes by precipitation of ZnO by ammonia in the presence of polyethylene glycol (PEG) templating agent (see Section 7).

ZnO films can be grown in the presence of Lewis bases such as ethane-1,2 diamine (ethyelenediamine) (*en*) [49, 53] or triethanolamine [49] which form a complex with zinc. The complexes act as a reservoir for the metal and release Zn(II) in alkaline medium upon ZnO precipitation. The deposition rate can be controlled by altering the bath temperature, the pH or the relative concentration of the reactants in solution (chelating agent or metal ion). Deposition *en* method consists in preparing a mixture of zinc salt and ethylenediamine. The chelating agent is in excess with a concentration ratio of about 2–2.5. The deposition temperature ranges between 50 and 80°C. The pH of the solution is increased and fixed to a final value with an aqueous solution of sodium hydroxide. The best quality films were formed when the adjusted final pH was above but close to the supersaturation line which describes the thermodynamical stability limit of ZnO (see Section 6.1) [53]. The substrate had a great influence on film morphology. ZnO nanorod arrays could be obtained using a ZnO nanocrystalline seed layer (SL) (Scheme 1). However, the rods were densely arrayed and not well hexagonal-faceted [49]. The preparation of well-designed nanostructured films seems rather difficult by the method.

4.2 Electroless Deposition of ZnO Films

Izaki and Omi have developed an electroless deposition method for the preparation of polycrystalline particulate ZnO films [54, 55]. In the first step, Pd particles are deposited on the surface by a simple chemical treatment. The conventional two-step activation process is composed of, first, a sensitizer treatment followed by an activator treatment. The activated substrate is subsequently immersed in an aqueous solution maintained at 50–60°C, which contains a zinc salt, nitrate ions and dimethylamine-borane (DMAB). The suggested deposition mechanism was described as follows:

$$(CH_3)_2NHBH_3 + H_2O \rightarrow BO_2{}^- + (CH_3)_2NH + 7H^+ + 6e^- \quad (2)$$

$$NO_3^- + H_2O + 2e^- \rightarrow NO_2{}^- + 2OH^- \quad (3)$$

$$Zn^{2+} + 2\,OH^- \rightarrow ZnO + H_2O \quad (4)$$

The reduction reaction of nitrate ion plays an important role in ZnO formation from solution and is driven by the oxidation of DMAB precursor. However, we can observe that the OH^- balance between reactions (2) and (3) should yield a pH decrease of the solution and not an increase as expected for ZnO precipitation.

The films prepared presented a good transparency in the visible and a bandgap measured at 3.3 eV [53, 54]. Atomic force microscopy (AFM)images showed that the film surface was composed of grains in the 100 nm range. Their size decreased with increasing DMAB concentration. Boron was incorporated in the film. It acted as a dopant and increased the ZnO film conductivity. The electroless deposition method was successfully employed for the low-temperature preparation of ZnO micropatterns using self-assembled monolayers (SAMs) of phenylsilane [56, 57] (see Section 8). The patterned deposit yielded an orange cathodoluminescence centered at 600 nm.

4.3 Thermal Decomposition of Hydroxide Precursors

4.3.1 Preparation of Nanostructured Films

The method consists in adding a precursor to the deposition aqueous solution which slowly decomposes with the elevation of temperature and generates ammonia. Ammonia is hydrolyzed and produces hydroxide ions:

$$NH_3 + H_2O \rightarrow NH_4{}^{+} + OH^{-} \tag{5}$$

The decomposition rate depends on the operating conditions (pH, temperature, concentration of the precursor, etc.). The method is simple and has been used by many researchers for the preparation of high-quality nanostructured ZnO films [17, 49, 58–66]. Urea is a first example of hydroxide precursor [66]. By thermal decomposition, aqueous solutions of urea produce ammonia. In the presence of Zn(II), ZnO is precipitated as a thin film by the following global reaction:

$$Zn(II) + (NH_2)_2CO + 2H_2O \rightarrow ZnO + 2NH_4{}^{+} + CO_2 \tag{6}$$

In recent works, hexamethylene tetramine (HMT), also called methenamine or hexamine, with formula $C_6H_{12}N_4$, has been a more employed precursor. HMT is a highly water soluble, nonionic tetradentate cyclic ternary amine. Classically, the synthesis is performed from a mixture of HMT and of Zn(II) salt heated at 50–95°C. The substrate is placed in a sealed bottle filled with the solution. HMT undergoes a slow thermal decomposition which produces H_2O, NH_4OH and CH_2O [63]. Therefore, HMT provides a slow, controlled supply of OH^- by NH_4OH base production upon decomposition. It gives rise to a pH increase in the solution and to the nucleation and growth of ZnO films by means of a growth-dissolution-recrystallization process [49]. The resulting films are generally made of arrayed ZnO nanorods, oriented along their *c*-axis perpendicular to the substrate [58]. The size of the rods can be controlled by the deposition time (between 1 h and 1 day). Their density depends on the substrate. For instance, high density ZnO nanorod arrays were obtained on a ZnO nanocrystallized SL [59, 60]. Conversely, growth on substrates which poorly nucleate ZnO (such as sapphire single crystals) yielded dispersed twinned multipod structures [49].Vayssieres [59] showed that the aspect ratio of the deposit could be easily monitored because of the decrease in reagent concentration while keeping the same 1:1 concentration ratio of Zn(II) and HMT yields more anisotropic ZnO crystals. Consequently, 1–2 μm-wide rods (up to 10 μm long) were obtained after several hours of growth at a concentration of 0.1 M zinc nitrate/HMT, while 100–200 nm wide nanorods and 10–20 nm nanowires were synthesized at 0.01 M and 0.001 M respectively.

Zinc precursor has been shown important for the control of the deposit shape. The crystallites become acicular if zinc nitrate is replaced by zinc chloride due to ZnO formation via a $Zn_5Cl_2(OH)_8$ intermediate [49].

Hollow microstructure could be prepared by aging an arrayed nanorod deposit in the deposition solution [17]. Using classical deposition conditions parameters (0.1 M $Zn(NO_3)_2$ and HMT, 90°C), arrays of ZnO microrods were obtained after 1 day of growth. After 2 days, the crystallite size was almost unchanged but hexagonal nanotubes were formed (Fig. 7). This was interpreted as due to the dissolution of the metastable (0001) face by aging. On x-ray diffraction (XRD) patterns, the (0002) reflection line was weak because the (000n) planes were poorly present in the hollow structure. One can note that the tube size was in the micrometer range but the tube walls were rather thin, with values in the 100–200 nm range (inset Fig. 7).

Tian et al. [61] showed that it is possible to control the growth direction of ZnO nanocrystal by varying the citrate concentration in the deposition bath. Increasing citrate concentration yields a decrease in the aspect ratio of the rods. Citrate ions slowed down the crystal growth along the *c*-direction, and at high citrate concentration, platelike ZnO crystals were produced. The researchers succeeded in growing complex nanostructures by preparing ZnO nanocolumns on a ZnO-seeded substrate in a citrate-free solution, and then by covering the nanorod by ZnO platelets through a secondary growth in the same deposition bath but in the presence of a high concentration of citrate ions.

4.3.2 Stimulated Emission of the Films

The luminescence of the as-grown ZnO nanorod array on Si surface was studied by Tam et al. [19]. The spectrum was dominated at room temperature by UV

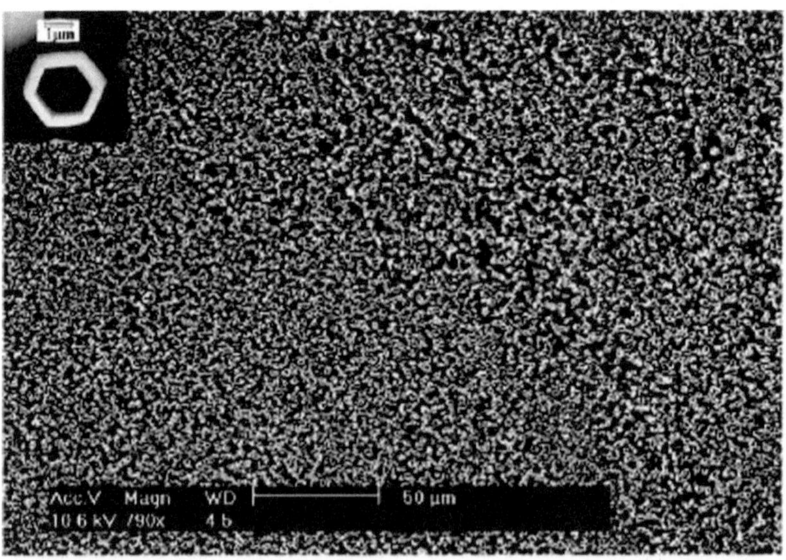

Fig. 7 FESEM micrographs of a chemically grown ZnO-oriented microtube array (after [17])

photoluminescence and presented a defect emission centered at about 550 nm. The films presented large defect concentration and the yellow emission was assigned to surface hydroxide groups. Annealing the sample at 200°C in air, forming gas or argon, markedly reduced the yellow defect emission. When the sample was heated at a higher temperature, green and red defect emissions appeared. The GL was assigned to defect complexes which contain V_{Zn}. Annealing the sample had a beneficial effect on the decay time of the UV emission. Its intensity was also considerably enhanced due to annealing of crystal defects.

Room-temperature lasing of the ZnO nanorod arrays was observed in the UV [62, 65] and in the whole visible wavelength range [64]. Microcavity UV lasing of ZnO was first reported in 2001 on nanorods epitaxially grown by vapor–liquid–solid mechanism on sapphire (110) substrate [11]. Excitonic lasing action with a threshold energy of 40 kW cm^{-2} (120 μJ cm^{-2}) was then observed and assigned to a stimulated emission from a resonator formed between the two {001} facets and lateral confinement of the column. Lasing ZnO nanorods have drawn much attention since then because they could be integrated in optoelectronic devices as ideal miniaturized laser light source. These short-wavelength nanolasers could have myriad applications, including optical computing, information storage and microanalysis ("lab-on-chip" spectroscopy). Govender et al. [64] showed that under short-pulse laser excitation at 337 nm, ZnO nanorods prepared from HMT on Au/Si substrate presented lasing spikes throughout the visible wavelength. The emission energy threshold was measured at pulse energy of ~2.4 mJ cm^{-2} (3 MW cm^{-2}). The spike spacing, $\Delta\lambda$, was 10 nm in the 400 nm region. In a cavity mode, for a linear resonator of length L, and at wavelength λ, the spacing is $\Delta\lambda = \lambda^2/2n\mathrm{L}$ and hence taking $n = 2.45$ for the refractive index of ZnO at 400 nm, the calculated resonator length was in good agreement with the rod length. UV lasing of ZnO nanorods prepared by the decomposition of HMT was first observed by Choy et al. [65]. Arrays of ZnO nanorod with a homogeneous diameter of 100 nm and a length of about 1.5 μm were grown on a 4 nm sized ZnO nanoparticle seed and buffer layer prepared by dip coating. The lasing power (energy) threshold was ~70 kW cm^{-2} (~140 μJ cm^{-2}). Djurišić et al. [62] studied smaller nanorods with cavity length of 300 nm. The lasing energy threshold was larger at ~460 μJ cm^{-2}. Time resolved experiments showed that the decay time of the nanorod lasing emission was very short in the picosecond range.

5 Hydrothermal Growth of ZnO Nanowhiskers on Zinc Foil

ZnO dense films are formed at a zinc foil surface by spontaneous corrosion or by applying a sufficient positive potential for anodizing metallic zinc. However, by using more drastic conditions, the growth of ZnO nanowhiskers has been described [67–69]. Typically, an ultrasonically washed metallic zinc plate is placed at the bottom of a cup filled with the solution. The solution is a mixture of water and ethylenediamine (*en*) with a pH adjusted to 14 with 1 M NaOH. The film growth is performed in an autoclave at 130–220°C. *en* is employed to regulate the ionic strength (mainly the pH value) of the reaction media and to modify the growth rate of ZnO whiskers. The whiskers grown in optimized condition are highly vertically oriented along their *c*-axis. They are single crystals with a hexagonal prismatic shape. They are also highly dispersed in size. The whiskers grown at 220°C exhibited single and/or a few modes of supernarrow spectral emission (~0.7 nm) at 378 nm [67]. Deep-level defect emissions were negligible. The peaks were due to the laser action with a pulse energy threshold of 70 $\mu J\,cm^{-2}$. This value is significantly lower than that reported for nanorods prepared by solid–vapor phase process [11]. The emission lifetime was below 30 ps. In the high-quality nanowhiskers, lasing was possible along the crystal *c*-axis since the two end facets act as end mirrors and the whisker forms a Fabry–Perot cavity (see Section 4.3.2). Pan et al. [68] examined the effect of growth temperatures between 130 and 200°C on the film photoluminescence and lasing. Increasing the growth temperature improved dramatically the crystal quality since the UV excitonic emission became intense and the green deep-level emission decreased. The deep-level PL emission was almost canceled in ZnO arrayed whisker films grown at 200°C. The PL intensity ratio of the NBE emission to the defect-related green emission is classically used to evaluate the concentration of structural defect and the quality of the ZnO crystal [70]. It was concluded that the whisker quality is gradually improved with increasing reaction temperature. Moreover, laser emission was observed only on the film grown at 200°C. The threshold excitation intensity was ~6 $kW\,cm^{-2}$. It was also found that the NBE emission peak red-shifted from 384 nm to 392 nm with increasing excitation intensity.

6 Electrochemical Preparation of Nanostructured ZnO Films

6.1 Electrodeposition of ZnO

The electrodeposition of ZnO requires conducting substrates. However, the method presents additional advantages compared to the chemical ones: the deposited layer presents a good electrical continuity with the conducting substrate and the amount of deposited material is directly accessible during the growth by the measure of the electrical charge exchanged during the process. The mechanism of ZnO film electrodeposition is based on the monitoring of the interfacial pH of an electrode by an electrochemical reaction which produces hydroxide ions. The flux of OH^- is controlled by the current density. Zinc ions present in solution react with hydroxides and ZnO is precipitated. Comparison of ZnO and $Zn(OH)_2$ solubility curves calculated between 25 and 90°C reveals that ZnO is thermodynamically more stable [28]. Figure 8a presents solubility curves of ZnO and $Zn(OH)_2$ calculated at 70°C in the presence of 0.1 M Cl^-. In Fig. 8a, S' denotes the total Zn(II) species concentration in solution. The solubility curves also show that supersaturation can be attained only if the initial pH of the solution is not too low; typically it was observed experimentally that it must range between 5 and 6.5. In solution, Zn^{2+} ions can be

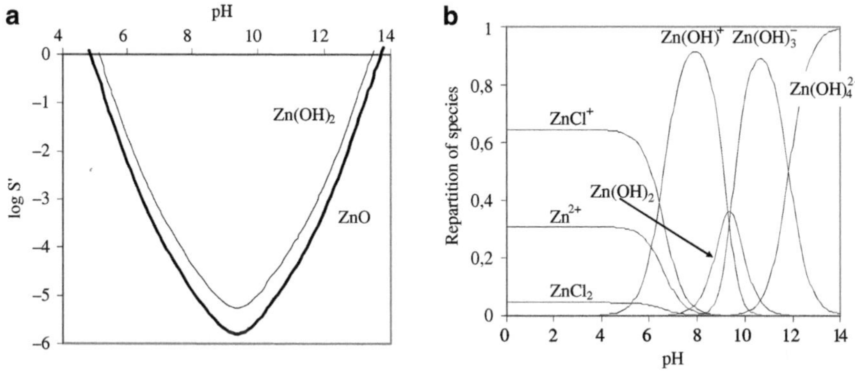

Fig. 8 (**a**) Solubility curve of ZnO and $Zn(OH)_2$ at 70°C. (**b**) Speciation diagram of Zn(II) in 0.1 M Cl^- at 70°C

complexed by OH^- and Cl^- ions and various Zn(II) species can be found depending on the species concentration and the solution pH. From the speciation diagrams, it can be concluded that the main reacting species at the ZnO deposition pH are Zn^{2+}, $Zn(OH)^+$ and $ZnCl^+$ in chloride solutions (Fig. 8b). Chloride ion is classically used as a supporting anion in ZnO electrodeposition experiments [28, 71–75].

6.1.1 Molecular Oxygen Precursor

The reduction of molecular oxygen for the preparation of ZnO was first proposed by Lincot et al. [71, 72]. The deposition reaction mechanism is

$$O_2 + 2H_2O + 4e^- \rightarrow 4OH^- \tag{7}$$

$$Zn^{2+} + 2OH^- \rightarrow ZnO + H_2O \tag{8}$$

The overall reaction process is

$$Zn^{2+} + {}^1\!/_2 O_2 + 2e^- \rightarrow ZnO \tag{9}$$

The mechanism was confirmed by in situ pH measurement with the help of a mini-electrode [72]. In the absence of Zn^{2+} in solution, the pH in the vicinity of the electrode increased from 6 to about 10 due to reaction (7). In the presence of zinc ions, the hydroxide ions produced were consumed by the oxide precipitation reaction and the interfacial pH was fixed at a value close to that calculated from the solubility value (6.7). In chloride medium, at Zn(II) concentration higher than 10 mM, the stable precipitated compound was no more ZnO but zinc hydroxychloride compounds [72].

The deposition rate is limited by the low solubility of molecular oxygen in aqueous solution (~0.8 mM at 70°C [28]). Pauporté et al. [76] have recently reported a kinetic study of oxygen reduction on ZnO electrode in conditions similar to those classically used in the electrodeposition process. A rotating disk electrode (RDE) was used and the rotation speed was varied in order to obtain the kinetic constants of the system by use of a Koutecky-Levich approach. In the absence of Zn^{2+} in solution, a Tafel slope of 139 mV dec^{-1} was obtained, a value close to that measured on bare platinum electrode (133 mV dec^{-1}) and ascribed to the limitation of the oxygen reduction reaction rate by the first electron transfer with the formation of an adsorbed intermediate. The main difference between the noble metal and the oxide electrode was a shift of the curves toward more negative potentials. In the presence of Zn^{2+} ions, the current density decreased significantly and the Tafel slope was measured at 282 mV dec^{-1}. The electrode was partially blocked by a reaction intermediate upon ZnO formation.

To increase ZnO film crystallographic quality, electrodeposition was carried out in nonaqueous dimethyl sulfoxide (DMSO) solution at 150°C by Gal et al. [77]. The films were highly transparent and crystallographically oriented with the *c*-axis perpendicular to the substrate. Oxygen concentration in the electrolyte was varied by mixing oxygen with dry nitrogen and controlling the ratio between them. As oxygen concentration in the electrolyte was decreased, the resistivity of the ZnO layer decreased. Electrodeposited ZnO film of high resistivity (prepared at high oxygen concentration) was deposited on a solar-quality $CuInSe_2$ layer and the cell was completed by an Al-doped ZnO top contact to form a CdS-free solar cell. The result was promising, giving an overall efficiency of 11.4%, compared to the 13.3% obtained when the electrodeposited ZnO is replaced by a classical CdS buffer layer.

High-quality ZnO was prepared electrochemically by epitaxial growth on GaN (0002) single crystal [73, 74, 78–80] and on gold single crystals with various orientations [81]. GaN and ZnO are two hexagonal crystals with close crystallographic parameters. The lattice mismatch between these two compounds is small ($a = 3.25$ Å and $c = 5.21$ Å for ZnO and $a = 3.19$ Å and $c = 5.18$ Å for GaN [82]). The in-plane crystallographic relationship was shown to be ZnO [100] || GaN [100]. Epitaxial ZnO was found to be of high quality, attested by the 0.07° full width at half maximum (FWHM) of the (0002) $\theta/2\theta$ peak and 0.74° FWHM of the $(10\bar{1}1)$ plane in the Ψ-scan diagram. Epitaxial dense and smooth ZnO single-crystalline films [73] as well as ZnO nanopillars [79] or nanorods [78] could be grown on GaN single crystals. ZnO nanopillars were also electrodeposited epitaxially onto Au(111), Au(110) and Au(100) single-crystal substrates [81]. The nanopillars grew with the same [0001] out-of-plane orientation on all three substrates but ZnO in-plane orientation varied with crystallographic orientation of the gold surface.

The bandgap of ZnO films deposited from oxygen precursor was measured between 3.4 and 3.55 eV at room temperature [74]. A significant energy shift was observed after annealing in air at 500°C with a bandgap of 3.27 eV. The PL of the as-electrodeposited layers was measured at 1.8 K [73, 74]. A broad excitonic emission was observed at 3.4 eV for the as-deposited films. The PL intensity was higher for films grown in perchlorate medium compared to chloride medium and it was concluded that perchlorate anion yields material of higher optical quality.

After film annealing at 500°C for 1 h, the emission became sharper and was centered at 3.37 eV. On the spectrum of a ZnO film electrodeposited on polycrystalline SnO_2, a shoulder was observed at 3.2–3.3 eV that could be assigned to two electron transition satellites. This spectral feature was less-marked on higher quality layers (perchlorate medium, single-crystal GaN substrate) [74].

6.1.2 Hydrogen Peroxide Precursor

On reduction of an aqueous hydrogen peroxide solution in the presence of zinc ions, the following two reactions occur:

$$H_2O_2 + 2e^- \rightarrow 2OH^- \tag{10}$$

$$Zn^{2+} + 2OH^- \rightarrow ZnO + H_2O \tag{11}$$

resulting in an overall deposition process of [75, 83–86]

$$Zn^{2+} + H_2O_2 + 2e^- \rightarrow ZnO + H_2O \tag{12}$$

Reduction of hydrogen peroxide on the working electrode leads to the production of hydroxide ions and hence an increase in the local pH. ZnO precipitates to form a layer at the working electrode surface. Compared to O_2, hydrogen peroxide is highly soluble in water and its use avoids the problem of gas-handling. Moreover H_2O_2 reduction, like O_2 reduction, is a "clean" reaction since the only reduction products are hydroxide ions which are consumed for ZnO formation. No undesirable byproducts that may pollute the deposition bath are formed. Films of ZnO were deposited at 70°C and –1 V vs saturated calomel electrode (SCE) (at standard conditions, the potential of the SCE is +0.244 V vs the SHE) at various H_2O_2 concentrations in the presence of Zn^{2+} in excess and using perchlorate as supporting anion to avoid the formation of basic zinc salts at large Zn^{2+} concentration [83]. As observed with molecular oxygen precursor, the films were highly textured and grew along the [0002] direction. Two growth regimes were observed. Up to 25 mM H_2O_2, the films were dense and the growth rate was proportional to H_2O_2 concentration in the deposition bath. At higher H_2O_2 concentration, the growth rate markedly increased and highly transparent porous ZnO was formed (see Section 5.2.4). The deposition mechanism was studied in a more detailed manner in chloride medium [75]. A parametric study was carried out by varying the substrates and peroxide concentration at constant zinc content. The film growth was under kinetic control since the catalytic activity of ZnO surfaces for the reduction of hydrogen peroxide species is low under these conditions. The nature of the substrates (tin oxide, gold) and their treatments prior to the deposition experiments had a marked influence on the electrochemical behavior of the system. It was possible to normalize the film growth behavior by taking the mean current density as the key parameter. The dependency of film growth rate on this parameter is shown in Fig. 9a. For values below the limiting diffusion current of zinc ions ~0.65 mA cm^{-2}, the faradaic efficiency (defined as the fraction

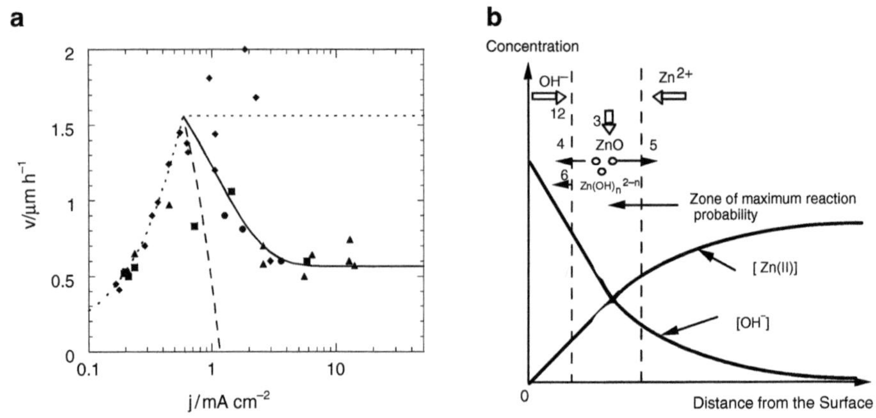

Fig. 9 (**a**) Variation of the growth rate of ZnO film formed from hydrogen peroxide precursor as a function of the mean current density. The dots are experimental data and the lines are fit curves with different models presented in [75]. (**b**) General scheme of the reaction steps occurring in bulk solution. Step 1 is the diffusion of electrogenerated OH^- from the electrode, step 2 is the diffusion of dissolved zinc, step 3 is the formation of ZnO particles in solution, step 4 their diffusion toward the electrode surface and step 5 toward bulk solution. Step 6 displays a potential change in the dissolved form of zinc in the vicinity of the electrode with the pH change (see Fig. 8b)

of charge consumed for the electrochemical formation of the film) was close to 1, whereas above this value the faradaic efficiency decreased markedly. A competition mechanism was suggested in which hydroxide ions produced in excess at the surface diffuse toward the solution and react partly with zinc ions diffusing toward the electrode surface (Fig. 9b). The reaction in solution reduces the availability of zinc ions for the heterogeneous deposition reaction at the surface and then decreases the deposition rate. Surprisingly, a plateau rate was found at high current density (Fig. 9a), which meant that a constant flux of Zn^{2+} could escape the precipitation reaction in solution and reach the surface or that a film was formed by aggregation on the surface of ZnO particles formed by homogeneous precipitation.

6.1.3 Nitrate Ion Precursor

Nitrate ions can be employed as the hydroxide precursor for ZnO electrodeposition [87–100]. On the reduction of nitrate in the presence of zinc ions, the following two reactions occur:

$$NO_3^- + H_2O + 2e^- \rightarrow NO_2^- + 2OH^- \quad (13)$$

$$Zn^{2+} + 2OH^- \rightarrow ZnO + H_2O \quad (14)$$

resulting in an overall deposition process of

$$Zn^{2+} + NO_3^- + 2e^- \rightarrow ZnO + NO_2^- \quad (15)$$

Employing nitrate ion is very convenient and simple as it does not require, for instance, gas-handling. However, the reduction reaction (13) produces NO_2^- by-product which progressively accumulates in the deposition bath. Moreover, it has been observed that the deposition solution cannot be stored because rapid decomposition occurs. ZnO thin films are classically prepared at 60–70°C and are well crystallized with the hexagonal wurtzite structure. Electrodeposition potential has a strong effect on film crystallographic orientation [88]. At low overvoltage, –0.7 V vs SCE, the film surface presents the hexagonal facets of the ZnO (0002) plane and XRD pattern is dominated by the (0002) plane reflection. However, at lower applied potential, the film crystallographic orientation changes and the *c*-axis is oriented parallel to the substrate. Crystallographic orientation changes were also observed with zinc nitrate concentration [87]. The mechanism of ZnO film growth with nitrate hydroxide nitrate precursor was studied by Yoshida et al. [90]. It was shown that the current density, and thus the film growth rate, was promoted by Zn^{2+} concentration. The improvement of the charge-transfer kinetics was interpreted as a "catalytic" role of Zn^{2+} adsorbed on the film surface to the reduction of nitrate ions. Zn^{2+} adsorption follows a Langmuir isotherm:

$$\theta = \frac{KC_{Zn^{2+}}}{1 + KC_{Zn^{2+}}} \tag{16}$$

where θ stands for zinc ion coverage. K is the equilibrium constant and $C_{Zn^{2+}}$ is the concentration of zinc ions in solution. However, the adsorbed Zn^{2+} ions cannot be catalysts in a true sense because they are spontaneously consumed to generate the ZnO film. Another specificity of nitrate precursor is that ZnO can be deposited at potential as low as –1.4 V, whereas with the other precursors metallic zinc is then formed in agreement with classical Pourbaix (potential-pH) diagrams [28]. Izaki et al. [91, 92] showed that the conductivity of ZnO films deposited from zinc nitrate solution could be controlled by adjusting the concentration of dimethylamine borane (DMAB) added to the bath (see also Section 4.2). Boron is included in the film and acts as an electron donor. Boron film content increases with increasing DMAB bath concentration and the film conductivity is significantly improved.

The as-deposited ZnO films present good structural and optical properties. Highly (0002) textured layers deposited on gold film substrates were luminescent at room temperature [94]. However, the PL signal was observed within a narrow deposition potential range. Films deposited at –0.6 V vs Ag/AgCl (at standard conditions, the potential of the Ag/AgCl reference is +0.235 V vs the SHE) had two emission peaks of similar intensity located at 3.30 eV and 2.28 eV. In the film deposited at –0.65 V, the defect emission was shifted to 2.8 eV. It is probable that the film quality was significantly improved by an epitaxial relationship between the gold layer and ZnO [81] (see also Section 6.1.1). PL measurements at 13 K confirmed that the films prepared at –0.6 V vs Ag/AgCl were of better quality than those prepared at lower potential [96]. It was observed that the UV NBE excitonic emission decreased with deposition overvoltage and the emission was canceled in films prepared at –1 V and –1.4 V. The annealing of films prepared between –0.6 and –1 V at 380°C for

1 h resulted in the enhancement and the sharpening of the excitonic PL band. They exhibited strong UV emission at 3.35 eV and negligible emission from defects. In annealing process, crystal quality of the films was improved because $Zn(OH)_2$ in ZnO film was changed to ZnO and defects such as oxygen vacancies were annealed by diffusion of oxygen from air. The annealing treatment also significantly red-shifted the optical bandgap of the films from 3.56 to 3.29 eV. The same behavior was observed for ZnO films deposited from molecular oxygen precursor (see Section 6.1.1) [73, 75].

We can note that low-temperature electrodeposition of ZnO was reported with nitrate precursor. Zhang et al. [97] worked at a bath temperature of 0°C and a potential of –1.30 V vs SCE. The ZnO films were highly transparent and presented a bandgap measured at 3.37 eV. They were made of nanocrystallites with grain sizes between 10 and 15 nm and a preferred growth orientation along the *c*-axis.

6.2 Electrochemical Growth of ZnO Nanorods and Nanowires

6.2.1 Nanorods

Electrochemical Preparation of Nanorod Arrays

ZnO nanorod is a nanostructure very commonly formed in aqueous solutions. This shape is greatly favored by the polar structure of ZnO with a positively Zn^{2+}-terminated (0001) and a negatively O^{2-}-terminated $(000\bar{1})$ polar surfaces which induce a net dipole moment along the *c*-axis. Thus, the surface energy of the polar $\{0001\}$ plane is higher than those of nonpolar $\{01\bar{1}0\}$ and $\{2\bar{1}\bar{1}0\}$ planes. Therefore, preferential growth along the *c*-axis is energetically favorable and the $\{0001\}$-oriented nuclei will grow faster [98]. The concentration of Zn(II) precursor has been shown to be an important parameter that governs the deposit morphology. At high Zn(II) concentration, dense and smooth films are obtained (Fig. 10a). Decreasing Zn(II) concentration leads to a decrease in the lateral growth rate of the crystallites. As a consequence, the aspect ratio of the ZnO crystallites increases and arrays of nanorods are formed (Fig. 10b). The phenomenon has been described with both oxygen [72, 78] and nitrate [100] hydroxide precursors. The substrate property is also important since it energetically governs the nucleation step and nanorods have been grown at high zinc concentration on nonactivated substrates with low density of nucleation center [71]. Combining high supersaturation and low density of surface nucleation site leads to the electrodeposition of spherulitic multipods of ZnO as illustrated in Fig. 10c with the example of a nonetched GaN(0002) single-crystal substrate.

The nanorods electrodeposited from an oxygen precursor are elongated along their *c*-axis and they are single crystals.

Arrays of free-standing ZnO nanorods with an aspect ratio value of about 8 were epitaxially grown at 80°C on GaN (0002) single crystals in a 0.1 mM $ZnCl_2$ solution bubbled by a mixture of O_2 and Ar (Fig. 11a) [78]. The top view of the layer clearly

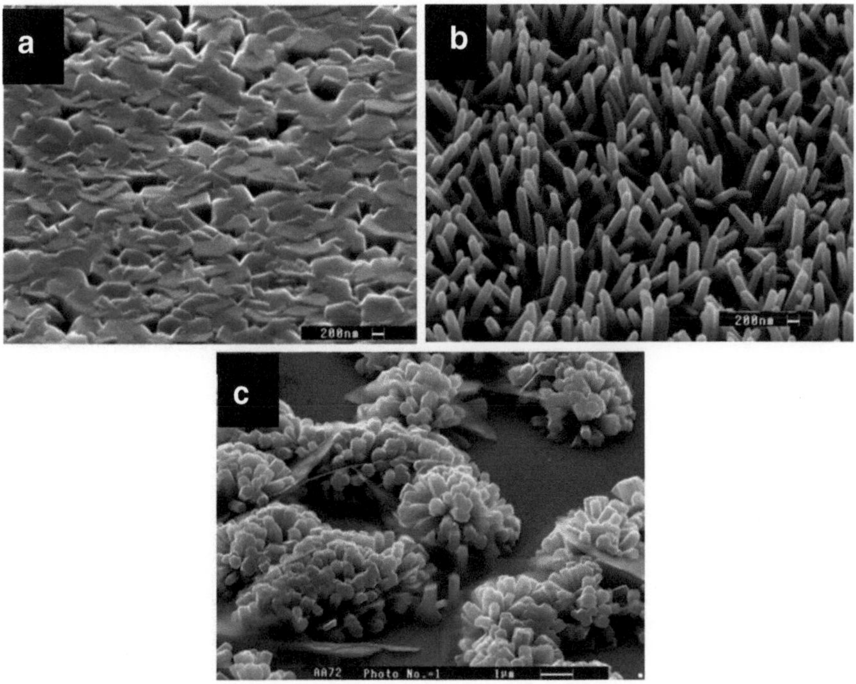

Fig. 10 (**a, b**) SEM views illustrating the effect of zinc concentration on film morphology. (**a**) Dense ZnO film electrodeposited on activated conducting glass surface from a solution containing 5 mM $ZnCl_2$ and molecular oxygen. (**b**) ZnO nanorod array electrodeposited on the same surface from a solution containing 0.2 mM $ZnCl_2$. (**c**) ZnO multipod structures grown electrochemically on nonetched GaN (0002) single crystal

shows the hexagonal cross section of the columns and that all the rods present the same in-plane orientation since the facets of the various crystallites are oriented in the same direction (Fig. 11b). Moreover, due to the epitaxial relationship between GaN and ZnO, the rods are perfectly oriented in a vertical direction. The average distance between the columns was 500 nm. The columns were illuminated at 266 nm with a pulsed laser at a low incident angle and the PL was measured in a direction perpendicular to the sample surface (Fig. 11c inset). At room temperature, the ZnO nanorods presented an UV-stimulated emission centered at 381 nm with an excitation threshold of 4.4 MW cm^{-2} (Fig. 11d). The two other emissions detected at 369 nm and 445 nm were due to a GaN substrate and were not amplified.

Application to Eta Solar Cells

Eta (extremely thin absorber) solar cells are based on a semiconducting extremely thin absorber layer sandwiched between two transparent highly interpenetrated nanostructured semiconductors. Figure 12a illustrates the basic design of an Eta solar cell. The cell is composed of a nano- or microstructured layer of a wide

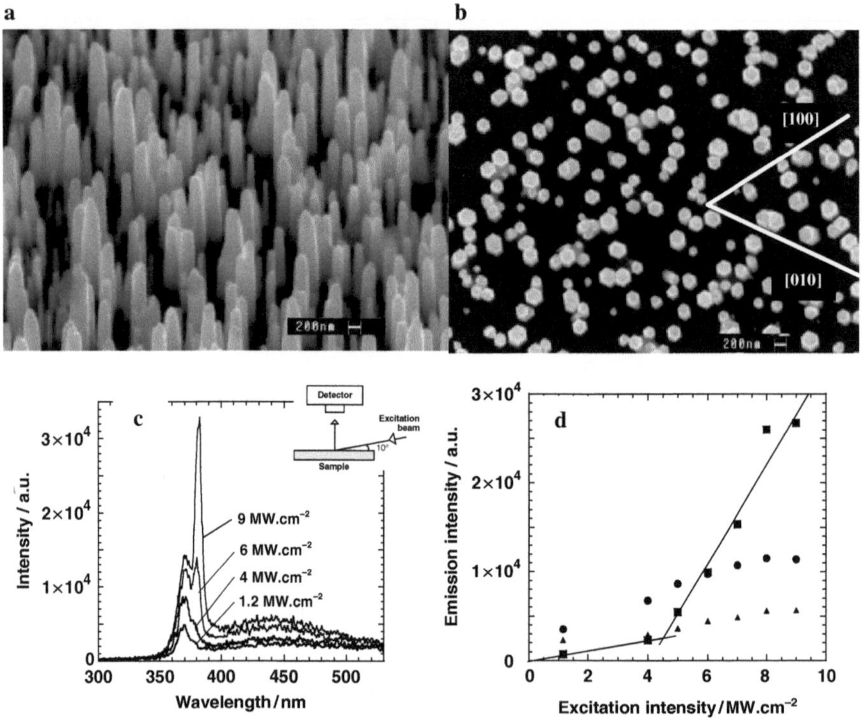

Fig. 11 (**a**) SEM view of ZnO nanorods epitaxially grown on GaN by electrodeposition. (**b**) Top view of sample (**a**). (**c**) PL spectra of the film as a function of the pumping laser energy intensity. **Inset**: experiment geometry. (**d**) Variation of ZnO (■ 381 nm) and GaN (● 369 nm and ▲ 445 nm) PL intensities with pumping laser power

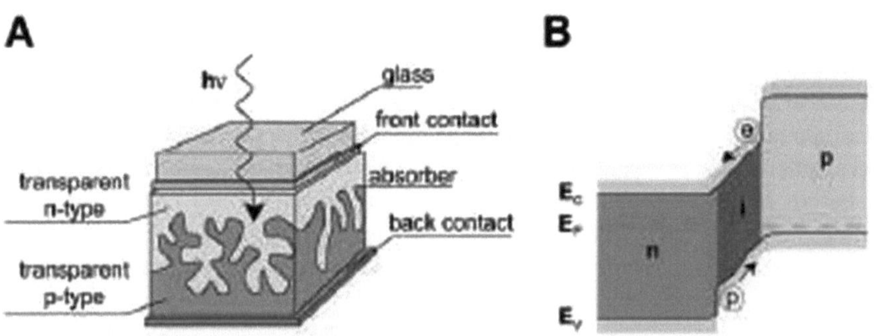

Fig. 12 (a) Schematic view of an extremely thin absorber (Eta) solar cell (after [104]). (b) Energy diagram of the heterojunction (after [106])

bandgap material ($E_g > 3$ eV) which also serves as an n-type window layer to the cell, an absorber (which absorbs solar light) with 1.1 eV $< E_g <$ 1.8 eV conformally coated on this window layer, and a void-filling p-type wide bandgap material ($E_g >$ 3 eV). When judiciously chosen in terms of their energy bands, the two transparent p- and n-type semiconductors develop a very high electrical field at their interface and facilitate the photogenerated charges separation (Fig. 12b). A highly structured interface can substantially reduce the transport path for excited charge carriers in the absorber, and at the same time increases the optical path length of photon absorption.

Electrodeposited arrays of ZnO free-standing nanocolumns were proposed as the nanostructured n-type window by Lévy-Clément et al. [101–104]. Since the columns were well separated and vertically oriented, it was shown that thin conformal layers of absorber could be deposited by chemical vapor deposition. Direct energy bandgap CdTe [102, 103] was preferred to indirect energy bandgap amorphous-Si [101]. The cell was completed by void filling with copper thiocyanate (CuSCN), a p-type semiconductor with a bandgap of 3.4 eV. The p-type window was deposited from a saturated solution of CuSCN in propylsulfide at 80°C [105]. However, better cell performances were obtained with a CdSe absorber layer in the 30–40 nm thickness range deposited by cathodic electrodeposition [104, 106]. The overall deposition reaction is written:

$$Cd^{2+} + Se^{\circ} + 2e^{-} \rightarrow CdSe \tag{17}$$

The gold electrical back contact was vacuum evaporated on the CuSCN layer. The best ZnO/CdSe/CuSCN Eta solar cell was obtained after CdSe layer annealing and showed at ⅓ sun an open circuit voltage of ~500 mV, a short-circuit current density close to 4 mA cm^{-2} and an overall efficiency of 2.3%.

Application to Light Emitting Diodes

Recently, LEDs have been prepared by low-temperature methods using solutions (Fig. 13a) [107, 108]. The active light emitting layer was made of a ZnO nanorod array electrodeposited at 80°C on conducting SnO_2-coated glass substrates using oxygen as a precursor. A good electronic contact between SnO_2 and ZnO was ensured by the electrosynthetic method. ZnO nanorods were robustly encapsulated in a thin insulating polystyrene film deposited from high molecular weight solutions. A p-type contact was taken by a 0.5–1.5 μm thick film of poly(3,4-ethylenedioxythiophene)/poly(styrenesulfonate) (PEDOT/PSS) deposited from a solution by spin coating. The thickness of the polystyrene layer between the rod tip and the PEDOT was found thin at 10–50 nm. The top contact on PEDOT was established by a 100 nm thick Au layer deposited by vacuum evaporation to provide a Schottky-barrier-type contact. In the device, hole charges were injected by the PEDOT layer and charge recombination occurred in ZnO nanorods. The diode emitted white light due to a broad defect-related emission centered at about 620 nm. Electroluminescence was observed from the device at a threshold voltage of 10 V with a

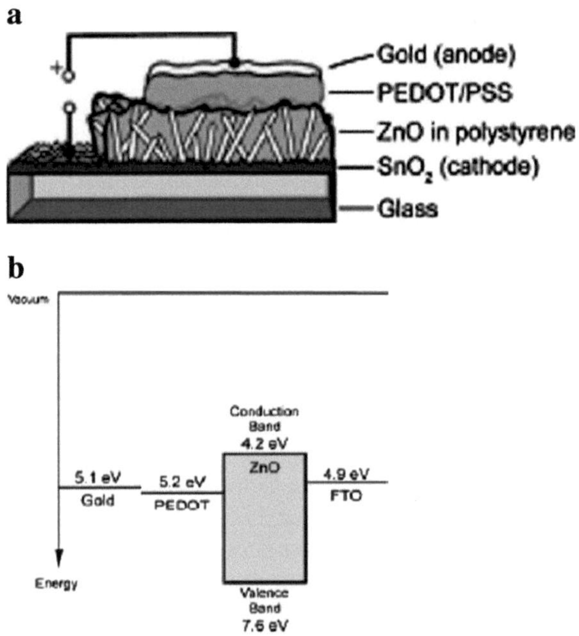

Fig. 13 (**a**) Schematic diagram of a nanorod LED arrangement with insulator filling the space between rods and a very thin film covering the rod tips (after [107]). (**b**) Energy diagram of the nanorod LED (after [108])

current density of ~100 mA cm^{-2}. First experiments showed stable emission for about 1 h. The device was subsequently improved [108]. The ZnO layer was heat-treated at 300°C to activate the UV emission and the top of the ZnO tips, covered by polystyrene, was plasma etched in order to improve the hole injection from PEDOT. The I–V characteristic of the diode showed excellent rectification and the luminescence onset voltage was typically about 5–7 V. Figure 13b shows the energy diagram of the system.

6.2.2 Nanowires

Seeded Growth of ZnO Nanowires

As observed in the case of films deposited by thermal decomposition of amino complexes (see Section 4.3.1), the presence of a ZnO SL induces morphological change in electrochemically grown ZnO [109, 110]. Lévy-Clément et al. [109] performed a systematic study of the preparation of ZnO nanowire arrays on an SL. The compact SL was prepared by electrodeposition at room temperature and was made of nanosized grains. After SL preparation, ZnO was grown at low Zn^{2+} concentration (0.1–1 mM) in a potassium chloride medium using molecular oxygen precursor. High density of free-standing ZnO nanowire array was obtained (Fig. 14a). Each nanowire was a single-crystal oriented along the *c*-axis. The aspect ratio of the ZnO wires grown on the SL was greatly improved at more than 20, whereas in its absence the value was about 8. Badre, Pauporté et al. [110] studied the wettability of these

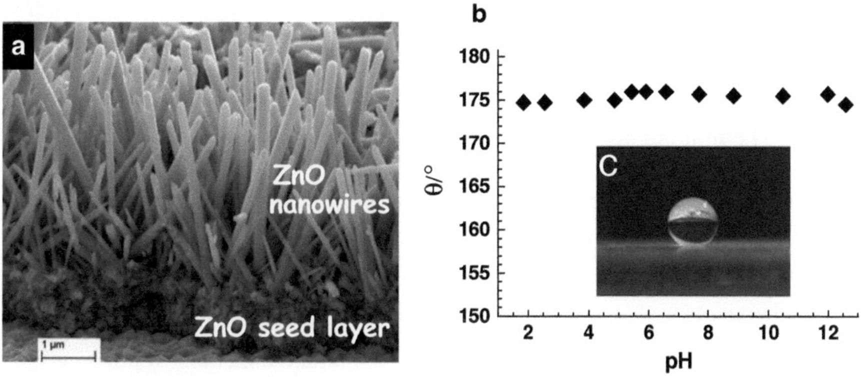

Fig. 14 (**a**) FESEM view of ZnO nanowire array electrodeposited on a ZnO seed layer. (**b**) Advancing contact angles at various buffered droplet pHs. (**c**) Quasi-spherical water droplet standing on the superhydrophobic ZnO layer

films. The as-deposited films were superhydrophilic with contact angles (CA) close to 0°. After surface modification by adsorption of stearic acid (SA) molecules, the free surface energy of the film was dramatically reduced and the surface became surperhydrophobic. Due to the very high roughness of the vertical nanowire array, the surface angles measured were remarkably high, indeed they found the highest value ever reported for an oxide surface with advancing/receding CA of 176°/175° (Fig. 14b and c). The low CA hysteresis observed on these films could be attributed to the low contact area between the substrate and the liquid droplet. Such structures entrap a large amount of air below the droplet, enhancing the surface hydrophobicity and inducing CAs of more than 160°. Basically, such a high CA and a very small advancing/receding CA difference means that the system is a composite surface consisting of ZnO and air where the Cassie–Baxter model applies. Moreover, the super repellent property of the surface was found remarkably stable over a wide pH range (1.8–12.5). The self-assembled monolayer of SA acted as an efficient barrier between the corrosive solution of the droplet and ZnO. The layer was also found stable after a thermal treatment at 250°C and after storage under ambient illumination condition.

Cao et al. [98] studied the electrochemical growth of ZnO on a (0001)-oriented SL. A zinc nitrate bath was used. In a first step, a 50 nm thick SL was deposited by sputtering on a silicon wafer of low resistivity. In a second step, ZnO was deposited on the SL at various constant applied current. The researchers observed a morphological transition with increase in the cathodic current density. At low current density, a flat and compact layer was grown. At higher current density, large-area well-aligned ZnO nanowire arrays in high density were obtained. The ZnO nanowires were all straight, smooth and relatively vertical to the substrate with uniform diameter of about 100 nm. Their length could be easily controlled by the growth time. They grew along the *c*-axis. At intermediate current density, the films, watched with the naked eyes, seemed to be composed of white ZnO particles. SEM view shows a singular morphology since the particles were connective microcalabashes, covered by densely arrayed nanospikes. Each spike was a single crystal and grew along the *c*-axis. At least at high current density, 2D ZnO nanosheets with tens of nanometers in thickness and several microns in dimension were formed by a fast nonequilibrium process. The film did not show any preferential orientation. All the nanostructured films were luminescent at room temperature with an UV NBE emission centered at 375 nm. The defect-related green emission was promoted by the deposition cathodic current density and dominated in the case of the nanosheets.

Electrochemical Growth of ZnO Nanowires in Membrane Templates

Electrodeposition in membrane template is an alternative method to produce ZnO nanowires. Either anodic alumina membranes (AAM) or nuclear-track-etched polymer membranes have been used in the literature as matrices for the preparation of nanowires [13, 84, 85, 93]. A continuous thin film of conducting gold is deposited on one side of the membrane by sputtering or by high-vacuum evaporation and electrodeposition is conducted starting from the bottom of the pore, at the gold electrode surface, and propagating continuously through the pores up to the membrane surface. The amount of solid deposited is theoretically proportional to the electrical charge passed. Therefore, the method should be straightforward for the preparation of nanorods and nanowires of perfectly defined shape and aspect ratio since their diameter is defined by the pore size and their length by the quantity of electricity passed through the electrode during the growth process.

Zheng et al. [93] used AAM templates and aqueous nitrate baths at 60°C. They obtained ZnO nanowires with a diameter equal to that of the pores. Selected area electron diffraction (SAED) pattern showed that the ZnO nanowires synthesized by the method were not single crystals. Under UV excitation, ZnO nanowires presented a broad orange luminescence centered at 608 nm. ZnO nanowire cathodic electrodeposition in AAM template was also achieved in nonaqueous DMSO solution [13] by the method reported by Gal et al. [77]. The solution contained chloride salts and dissolved molecular oxygen precursor. The deposit was polycrystalline. The main drawback of using AAM is the difficulty in dissolving the membrane after the synthesis and then to release the ZnO nanowires. The classical treatment consists in an etching in a concentrated NaOH solution of typically 0.5 M [93]. Unfortunately, ZnO is not stable in those conditions and partly dissolves.

The use of polymer membranes made of polycarbonate seems more promising since they can be easily dissolved in an organic solvent such as dichloromethane [84, 85]. However, the pores are randomly distributed on these templates whereas they are perfectly ordered in AAM templates. Polycrystalline ZnO nanowires were grown at room temperature in commercial nuclear-track-etched polycarbonate membranes using hydrogen peroxide precursor [85]. To achieve greater crystallinity, the researchers had to employ a higher growth temperature. ZnO wires displaying characteristics of single crystals were synthesized with NO_3^- precursor at 90°C [85]. The crystals were grown along the $[10\bar{1}1]$ direction which also corresponds to the preferential orientation of the nontemplated films [87]. It is of note that with both precursors the nanowire diameter was much larger than the nominal size of the membrane pore with values at 100–120 nm and 50 nm respectively. It was also shown that the nanowire length could not be directly calculated from the electrical charge exchange during the growth and that the deposition efficiency decreased with the wire length. Single-crystal ZnO nanowires were prepared in polycarbonate membranes of pore sizes ranging between 10 and 90 nm using an aqueous stirred solution maintained at 70°C containing 5 mM hydrogen peroxide precursor [84]. No preferential growth direction of the wires was detected. The wire surface roughness increased with pore size decrease and was related to the pore quality of the commercial membrane substrates.

6.3 Mesoporous ZnO Thin Film Grown by Electroposition

Mesoporous films refer to solid layers with pores greater than 2 nm and less than 50 nm in diameter. Several papers have described the preparation of mesoporous ZnO films by electrodeposition [83, 111–114]. The main advantages of the method are that the deposition is proceeded at a rather low temperature compatible with flexible plastic substrate and that the electrical continuity between the ZnO layer and the substrate is guaranteed by the deposition process in which charge exchange is involved. Two different strategies have been used: some researchers have manipulated deposition parameters (solvent, hydroxide precursor concentration, etc.) to induce the formation of pore, whereas some others have used organic additives which act as a self-assembled template on the electrode surface. In most case, electrodeposition of mesoporous ZnO was studied with the aim of developing cost-effective alternative preparation methods of photoanodes for dye-sensitized solar cells (or Grätzel cell) [115]. However, some other applications are emerging such as catalytic films for the photodegradation of pollutants [116].

6.3.1 Direct Growth

Electrodeposition of ZnO in aqueous solution at high H_2O_2 concentration ($c_{H2O2} \geq$ 40 mM) in the presence of an excess of zinc perchlorate salt (cH_2O_2/cZn^{2+} = 0.5) led to the formation of films with a very smooth surface [83]. It was shown that the measured film thickness was much larger than the equivalent dense film thickness calculated from the film's zinc content determined by titration. From these two film thickness values, it was concluded that the porosity of the film was about 33%. The presence of mesopores in the layer was observed by FESEM (Fig. 15a). O'Regan et al. [111] studied the preparation of smooth and homogeneously mesoporous ZnO films in an organic electrolytic solution of propylene carbonate. For 0.15 M $LiNO_3$ and 0.05 M $ZnCl_2$ oxygenated solutions at 70°C, the resulting mesoporous ZnO films consisted of vertically oriented semicontinuous columns between 10 and 30 nm in diameter. The presence of water trace at more than 0.03% in the deposition bath was necessary for the deposition of ZnO. Increasing water content above 0.05% had a detrimental effect on the porosity. At

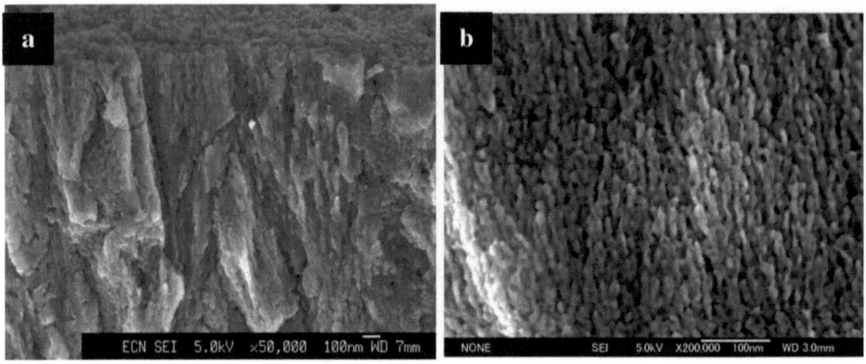

Fig. 15 Cross-sectional FESEM views of mesoporous ZnO films. (**a**) Film prepared from 40 mM H_2O_2 and 80 mM $ZnClO_4$. (**b**) Film prepared in the presence of 50 μM EY after dye desorption

low-deposition overvoltage, a higher density of pore and column was obtained and the pore diameter was smaller. However, it was shown that hot-nitrate-containing solutions have relatively short lifetime for ZnO deposition. The life of the solution could be significantly increased by slowly adding a low quantity of water in the bath upon deposition. The columnar mesoporous films were sensitized with phosphonated ruthenium polypyridyl dyes and the cell was completed by electrodepositing a p-type film of CuSCN [112]. The structure of the cell was similar to that of an Eta cell (Fig. 12a) except that the absorber layer was replaced by a monolayer of ruthenium complex dye. The ZnO/RuLL′NCS/CuSCN cells presented a maximum incident photon to current conversion efficiency (IPCE) of more than 50% at 500 nm. The efficiency of charge separation and collection was near 100% and the low IPCE was due to low overall light absorption. In simulated AM 1.5 (100 mW cm^{-2}) sunlight, the best performance reached was J_{SC} ~4.5 mA cm^{-2}, V_{OC} ~550 mV and a filling factor of 57% giving a full sun energy conversion at 1.5% for a cell which absorbs only 12% of the incident energy.

6.3.2 Dye-Assisted Growth

Electrodeposition of ZnO in the presence of organic dyes is of great importance in the preparation of nanostructured layers [80, 86, 113–123]. Among the various nanostructures reported, mesoporous films are of special interest due to their potential application in many fields such as solar cells [113, 114, 120], PL layers [117, 120] or photocatalysis [116]. The process was first discovered with eosin Y (EY), a red colored dye which absorbs light at 520 nm in its stable redox form (noted o-EY^{2-}) [113, 114, 119]. Due to the carboxylate function of the EY molecule which can complex zinc ions, hybrid ZnO/EY films are deposited electrochemically with oxygen precursor. At potentials below –0.9 V vs SCE, EY is included in the deposited film at low concentration (~50 mM) [113]. It is present in a closed structure, inside crystallites of ZnO. However, at lower deposition potential,

EY is electrochemically active upon the deposition process and the deposition mechanism of the hybrid compound is more complex since EY is electrochemically reduced and included in the film in an optically transparent form [113, 121]. EY concentration in the film is higher, reaching about 0.3 M. A voltametric analysis of the system at different zinc concentration has shown that Zn(II) ions were complexed by a radical reduced form of EY molecules resulting from the exchange of a first electron (noted $[\text{r-EY}^{3-}]^{\bullet}$). However, a recent in situ study of the Zn(II)-EY system by coupled ESR–visible spectrophotometry [121] has shown that upon the electrodeposition process, EY is further reduced with a total number of two electrons exchanged (noted r-EY^{4-}). The global mechanism is as follows:

$$\text{o-EY}^{2-} + e^- + Zn^{2+} \rightarrow [\text{r-EY}^{3-}]^{\bullet} \ldots Zn^{2+} \quad (18)$$

$$[\text{r-EY}^{3-}]^{\bullet} \ldots Zn^{2+} + e^- \rightarrow \text{r-EY}^{4-} \ldots Zn^{2+} \quad (19)$$

$$r - EY^{4-} \ldots Zn^{2+} + 1/2O_2 + 2e- \rightarrow r - EY^{4-} \ldots ZnO \quad (20)$$

Upon theelectrodeposition process, r-EY^{4-} is not only entrapped in the layer, but it is also self-assembled with the formation of continuous interconnected structures which go across the film. EY is reoxidized into o-EY^{2-} upon storage in air, and immediately after the deposition, the film, which is initially transparent, becomes red after several tens of minutes.

The films grown on conducting glass were highly textured along the *c*-axis. SAED technique revealed that they were made of large single crystals [80, 113]. Arrays of high-quality nanocolumns of ZnO/EY were epitaxially grown on GaN (0002) at –1.1 V vs SCE (Fig. 16a) [80]. The vertical cylindrical columns had spherical top and a diameter of about 200 nm. The high quality of the epitaxial ZnO was demonstrated by reflection high-energy electron diffraction (RHEED) (Fig. 16b and c) and five circle x-ray diffraction techniques (Ψ-scan). The RHEED spots and the

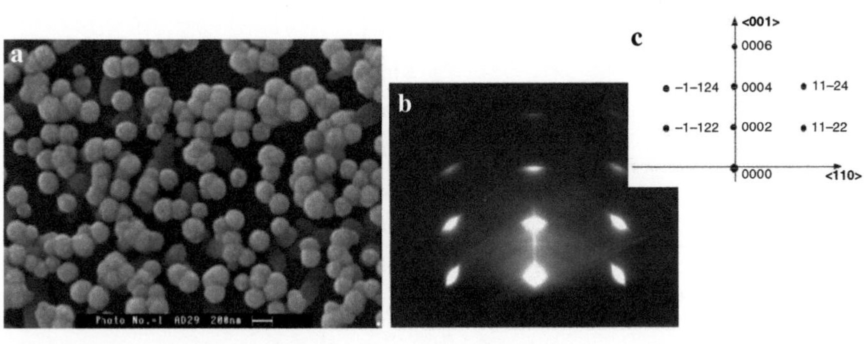

Fig. 16 Epitaxial ZnO/EY columns grown by electrodeposition at –1.1 V vs SCE on GaN (0002). (**a**) top SEM view. (**b**) RHEED pattern of the sample observed along the $< 11\bar{2}0 >$ azimuth. (**c**) Indexation diagram of the pattern

Ψ-scan peaks of ZnO were significantly broadened compared to pure epitaxial ZnO [79] showing a higher in-plane misorientation in the case of the highly dye loaded ZnO/EY columns.

Yoshida et al. showed that EY could be desorbed and removed from the film to leave an almost pure ZnO matrix after a dipping treatment of the film in a mildly alkaline (pH 10.5) solution of KOH [113]. The resulting film was mesoporous and developed a very large surface area (Fig. 15b). Their porosity and specific surface areas were measured by krypton BET (Brunauer Emmett Teller) technique [116]. Deposition parameters such as the deposition time and EY concentration in the electroplating baths were shown of great importance. The film porosity reached a maximum at about 60% above 40 μM EY in the deposition bath (Fig. 17a). The specific surface area increased continuously up to 60 μM EY (Fig. 17b), but above this value the films were found to be mechanically fragile. The deposition time also presented an optimized value corresponding to film thickness of 2–3 μm, since for long deposition times the layer was found to be fragile.

Interesting dye-assisted electrodeposition results were also obtained with Coumarine 343 (C343) dye in the same concentration range [120]. On SEM views, the films were formed of large single-crystalline platelets. XRD patterns were dominated by the reflection of the (1000) plane. The films were then textured in a direction different to that of ZnO/EY films since they were grown in the [1000] direction and the *c*-axis of ZnO was parallel to the surface. The dye could be removed by the same dipping treatment in mildly alkaline solution and the presence of mesopores was also found.

Mesoporous ZnO films prepared by the EY-assisted electrodeposition method were demonstrated efficient for dye pollutant photodegradation, with methylene blue and Congo red as model compounds [116]. Under UV illumination, holes photogenerated in the sample oxidize water and generate $OH^{\bullet}$. Hydroxide radicals are strong oxidizing agents that mineralize large dye molecules. The effect of EY

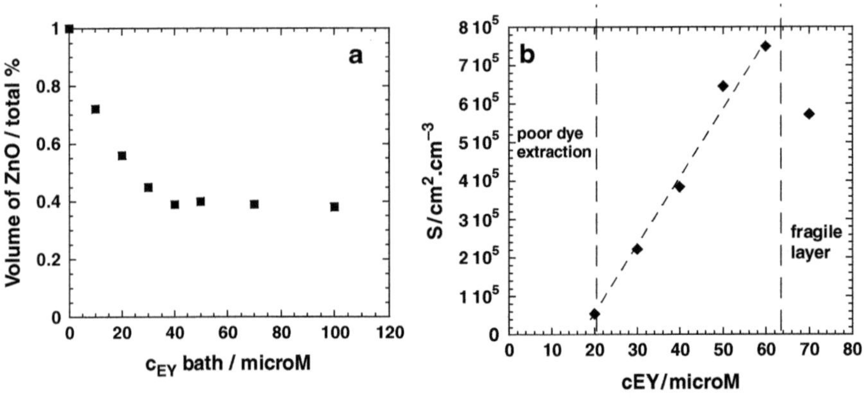

Fig. 17 (**a**) Variation of ZnO volume fraction as a function of EY bath concentration. (**b**) Specific surface area of the mesoporous films related to the film volume as a function of EY bath concentration

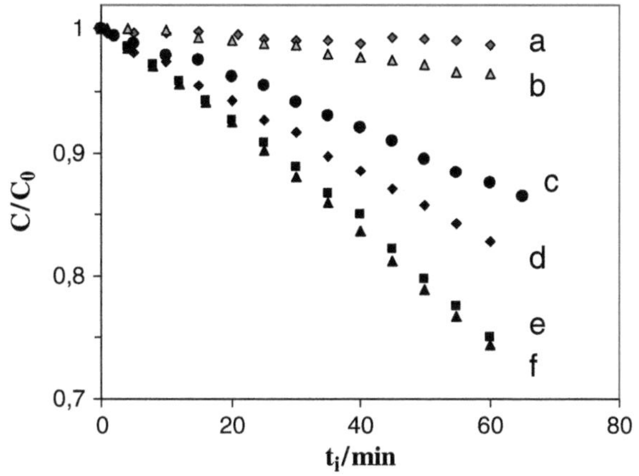

Fig. 18 Variation with time of the relative concentration of methylene blue. (**a**) 2 μm thick film prepared with 70 μM EY kept in dark. (**b–f**) Films irradiated with a xenon lamp. (**b**) Dense ZnO sample deposited on glass by sputtering. (**c**) 2 μm thick films prepared with 10 μM EY, (**d**) 20 μM EY, (**e**) 30 μM EY and (**f**) 70 μM EY

concentration in the electrodeposition bath of mesoporous ZnO is displayed in Fig. 18. Between 30 and 70 μM EY, the photodegradation rate was not significantly improved. BET study shows that increasing EY concentration increases the film surface area and that the pores become narrower. Photodegradation was found to occur preferentially in the larger pores of the films, whereas mesopores with pore diameters less than 8 nm did not contribute to a large extent to the process.

Mesoporous ZnO films prepared by dye-assisted electrodeposition were employed for the preparation of dye-sensitized solar cells (DSSC) [114]. After EY dye readsorption in mesoporous films, the incident-photon-to-current-conversion efficiency (IPCE) maximum was measured at 91%. The photocurrent density under AM 1.5 irradiation was 5.9 mA cm^{-2} and the overall energy conversion was 2.3%. In order to broaden the solar spectrum coverage of the photoanode, a mixture of EY and C343 dye was adsorbed in ZnO. A significant improvement of the photocurrent density was achieved with a value of J_{SC} ~6.9 mA cm^{-2}. Recently, an overall conversion efficiency of about 5.5% was achieved with readsorbed D149 dye under AM 1.5 simulated sunlight ($J_{SC} = 12.2$ mA cm^{-2}, $V_{OC} = 691$ mV and fill factor = 0.658) [120]. D149 is an indoline dye, free of precious metals that could become much cheaper than the Ru dyes classically used in DSSC when they are produced in a large volume in the future. This dye has been developed by Mitsubishi Paper Mills, Ltd.

Mesoporous ZnO thin films were also prepared by polyvinylpyrrolidone (PVP)-assisted deposition [99]. The films were grown by cathodic electrodeposition from aqueous zinc nitrate solutions containing various amount of PVP. They were subsequently annealed in air at 380°C to calcinate PVP. Mesoporous smooth films with grain size of 20–40 nm were obtained in the electrolyte containing 4 g L^{-1} PVP. They were sensitized with N3 dye from Solaronix S.A. and included in a sandwich-type DSSC. The best cell performance was found for 8 μm thick mesoporous ZnO films deposited on a 200 nm thick compact ZnO barrier layer. The overall efficiency was above 5% ($J_{SC} = 6.96$ mA cm^{-2}, $V_{OC} = 605$ mV and ff = 0.64 for 53 mW.cm^{-2}).

7 Polymer-Assisted ZnO Growth

Mixing ZnO or ZnO precursors with polymers in solution has been shown interesting to template and governs the growth of ZnO nanostructures, and polymer-assisted ZnO growth is a promising emerging field. Pauporté [124] used water-soluble polyvinyl alcohol (PVA) of high molecular weight as an additive for the electrodeposition of ZnO. PVA is able to complex zinc ions and under typical conditions for ZnO electrodeposition with molecular oxygen precursor ZnO was shown to be co-deposited with PVA to form a hybrid film. The PVA film content varied linearly with the initial PVA concentration in the deposition bath and a value of up to 50% (v/v) of the total film volume was attained. The layers were very smooth and highly transparent. They presented a good conductivity over a wide PVA content range. It was also shown that the film preferential orientation could be controlled by adjusting the PVA content in the deposition bath. At 0 $g\,L^{-1}$ PVA, the films were oriented along the [0002] direction, whereas at 2 $g\,L^{-1}$ and above 6 $g\,L^{-1}$ they were oriented along the $[10\bar{1}1]$ and $[10\bar{1}0]$ directions respectively. The films were luminescent at room temperature. The PL intensity ratio between NBE emission and defect green emission increased with PVA concentration showing that the density of defect in the crystals decreased with the polymer additive. Investigation of the electrochemical growth of nanocrystalline ZnO templated by PVP surfactant has also been reported (see Section 6.3.2) [99].

Electrospinning emerged recently as a promising method for the preparation of ZnO nanofibers (Fig. 19, route 1) [125]. In a first step, a sol precursor was prepared by mixing zinc acetate and PVA polymer. Then the mixture was electrospun by

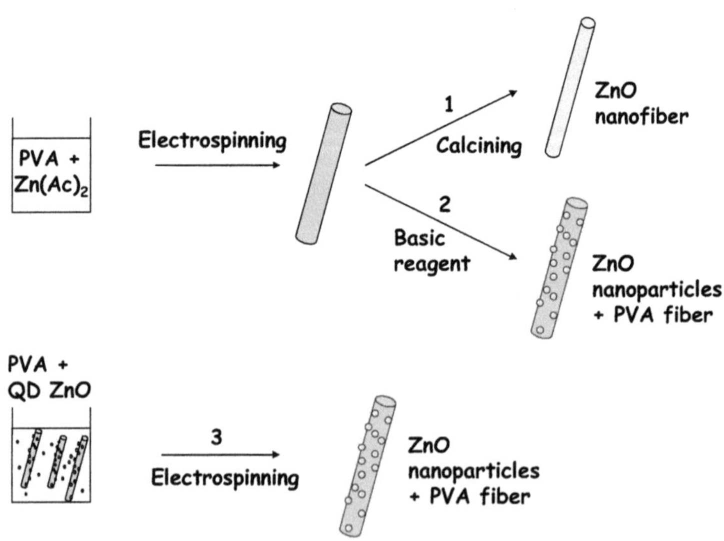

Fig. 19 Electrospinning routes for the preparation of ZnO (route 1) and ZnO/PVA (routes 1, 2) hybrid fibers

applying a high voltage (15–20 kV) between the needle of a syringe containing the sol and an aluminum foil used as a cathode. The distance between the tip of the needle and the aluminum foil was fixed at about 20 cm. A dense web of fibers was then collected on the aluminum foil. After drying and calcining the PVA/zinc acetate composite fibers at 700°C, pure ZnO fibers with smooth surface and diameters of 50–100 nm were produced.

Alternatively, ZnO nanoparticles/PVA hybrid fibers could be obtained by immersing the dried electrospun $Zn(Ac)_2$/PVA fibers in basic ethanolic solution of NaOH or LiOH at pH 10 (Fig. 19, route 2) [126]. SEM views showed that ZnO nanoparticles were homogeneously distributed at the fiber surface. Another approach for the preparation of PVA fiber coated with nanoparticles is presented in Fig. 19, route 3 [127]. Colloidal ZnO QDs with average radius of about 4 nm were first prepared by the method described by Spanhel and Anderson [29] (see Section 3.1) and mixed with PVA in solution. The mixture was electrospun and a dense mat of sprayed fibers was collected. The ZnO QDs were embedded in and aligned along the PVA matrix fiber. Thus, the interaction between ZnO QDs and PVA molecules was enhanced. PL results showed that the PVA/ZnO nanofibers had an intense white-light emission which originated from the simultaneous emission of three bands covering from the UV to the visible range: the UV and green bands were assigned to ZnO, whereas a blue band originated from the organic functional groups of PVA.

PVA templates were used to grow vertically well-aligned ZnO nanowires prepared on silicon substrates [128]. A mixed aqueous solution of zinc salt complexed by PVA was hydrolyzed by adjusting its pH to 8.5 and was spin coated onto the substrate. The substrate was subsequently heated up to 600°C in air. ZnO nucleation and growth occurred under the control of the polymer carbonized grid backbone that confined the ZnO diameter at about 20–80 nm. An array of ZnO nanorods oriented with the *c*-axis perpendicular to the surface and length up to about 1 μm was formed.

Duan et al. [51] investigated the effect of polyethylene glycol (PEG 2000) additive on the chemical growth of ZnO. Aligned ZnO nanotubes were grown on glass substrate by means of a two-step solution-phase procedure [51]. The tubular structures were grown with the *c*-axis perpendicular to the substrate. All the tubes had a hexagonal cross section. The PL spectra were dominated by the excitonic UV emission at 390 nm and a less intense defect emission at 530 nm.

8 Patterning of ZnO Nanostructures

The ability to precisely place nanomaterials at predetermined locations is required for realizing optoelectronic devices; therefore it is important to develop soft processes for patterning the ZnO deposits. Micropatterns of high-quality (0001)-oriented ZnO were prepared by electrodeposition in zinc nitrate solutions. Positive type photoresist and lithography techniques were employed [95]. ZnO was deposited at –0.45 V vs Ag/AgCl and low temperature (60°C) on 5 μm-wide gold microlines and 5 μm-diameter gold microdots. The ZnO micropatterns were UV luminescent at room temperature. However, using this method, high-resolution etching is difficult to achieve without damaging the substrate. Alternatively, direct patterning and selective growth process of ZnO were explored by several groups. For instance, phenylsilane SAMs were deposited on silicon substrate and photopatterned [56, 57]. Then Pd catalyst particles were selectively deposited on the phenyl surface regions of SAMs by immersing the substrate in a NaCl-based Pd/Sn colloidal solution. ZnO was subsequently deposited on the Pd catalyst particles by electroless growth in solution containing $Zn(NO_3)_2$ and DMAB (see Section 4.2). Cathodoluminescent images of visible light were successfully demonstrated for 1 μm wide lines. In more recent works, the use of Pd was avoided. In [63], patterned SAMs of 11-mercaptoundecanoic acid ($HSC_{10}H_{20}COOH$) were deposited on silver surfaces using polydimethalsiloxane stamps. Then ZnO was grown at 50–60°C by the HMT decomposition method (see Section 4.3.1). ZnO nanorods formation was observed on uncovered Ag substrate whereas the COOH-terminated regions covered by the SAMs remained free of ZnO. The deposition of ZnO was inhibited there by the formation of a COO^--HMT-H^+ complex. The deposition resolution was similar to the previous method. Masuda et al. [42] prepared micropatterned arrays of ZnO nanocrystals by using site-selective deposition of ZnO on self-assembled silane monolayers. The various silanes with hydrophobic chains were deposited on silicon wafers. Then the SAMs were patterned by illumination with UV light through a mask. Illuminated silanes were decomposed and patterned SAMs on the silicon surface were obtained (Fig. 20). The degradation of silane SAMs could be followed by CA measurements with value decreasing from 105° to 5° (Fig. 20). Then ZnO crystals were chemically deposited from alkaline solution in the presence of ammonia complexes on the hydrophobic parts of the substrate covered by a silane monolayer. The shape of the crystal could be changed (hexagonal rods, ellipses, multineedles) according to the ammonia/zinc acetate concentration ratio (see Section 3.2 and Fig. 6).

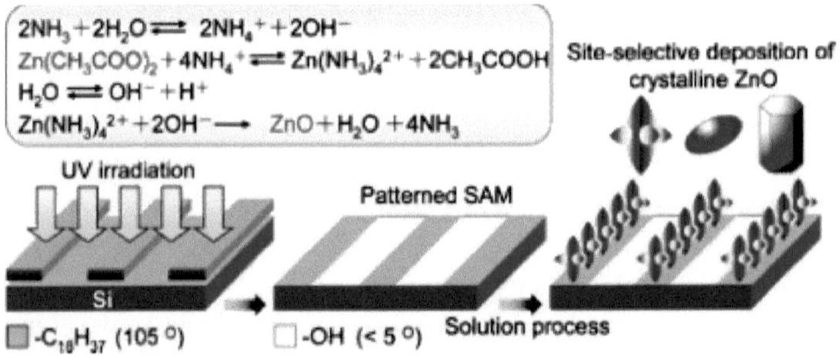

Fig. 20 Process for self-assembly patterning of light-emitting crystalline ZnO nanoparticles in aqueous solution (after [42])

9 ZnO Combined with Lanthanides for Visible Luminescence

We have illustrated that ZnO presents luminescence properties in the visible with broad emission bands centered between about 500 and 620 nm (see Section 2). However, for light emitting device applications such as field emission displays (FEDs), plasma display panels (PDPs) or electroluminescent displays, it would be greatly beneficial to better control and enlarge the palette of the colors emitted by ZnO with emitting centers presenting sharp and well-characterized emission peaks located in the three fundamental colors (450/520/610 nm). Incorporating these centers in ZnO matrices would be beneficial because of the outstanding stability of this oxide. One may take advantage of the energy transfer between ZnO and trivalent lanthanides (Ln^{3+}), which are well known to present sharp and intense emission peaks involving *4f*-*4f* transitions. The application requires the doping of ZnO matrix with different Ln^{3+}. Among lanthanides, Eu^{3+} is an interesting candidate with its strong red emission at about 610–620 nm. Tb^{3+} and Er^{3+} are two interesting luminescent centers for green emission, whereas Dy^{3+} emits in the blue wavelength range.

9.1 ZnO/Lanthanide Mixed Films

A first approach consists in introducing various Ln^{3+} luminescent centers by doping the oxide host. However, one of the difficulties of the approach is to mix ZnO and Ln at the molecular level in order to obtain an efficient energy transfer under UV illumination between ZnO and Ln^{3+}. In this context, the use of solution has been proposed for the preparation of the phosphor precursors.

Lima et al. prepared ZnO powders [129] and films [130] containing about 0.1–3% of Eu^{3+} by the Pechini method. Zinc and europium (III) citrate solutions were mixed, ethylene glycol was added and the solution was heated to obtain a polymeric mixture. The mixture was either directly heated at high temperature to obtain powders or was deposited on a glass substrate by dip coating before being heated for film preparation. Typical emission peaks of Eu^{3+} were observed under direct excitation at 465 nm. However, the structure of the doped compound was rather disordered and it was shown that at high europium concentration, Eu^{3+} occupied at least three low symmetry sites. Under UV light excitation, no energy transfer was found between ZnO and Eu^{3+} in the powders. The researchers suggested that the charge excess due to Eu(III) in these compounds was compensated by Zn(I) [129].

In order to improve the mixing of ZnO and europium, Gao et al. [131] proposed the use of a polymer-assisted preparation procedure. Eu-doped ZnO powders with urchinlike morphology, consisting of orderly aggregates of a large number of nanorods, were prepared by hydrothermal thermolysis at 100°C of zinc acetate in the presence of ammonia and a water soluble biopolymer, sodium alginate. The presence of the biopolymer was shown necessary for the formation of the mixed compound since in its absence a mixture of ZnO and Eu_2O_3 was obtained. The quantitative analysis of the product gave a molar ratio of Eu^{3+} to Zn^{2+} ions equal to about 0.07. Under UV excitation, the powder presented a white emission and the spectrum presented a broad emission ranging from 400 nm to above 630 nm. Two sharp peaks at 593 and 613 nm were assigned to 4*f*-4*f* intrashell transitions of $^5D_0 \rightarrow {}^7F_1$ and $^5D_0 \rightarrow {}^7F_3$ in Eu(III). The commission International de l'Eclairage (CIE) coordinate calculated from the spectrum fell within the white light region.

Pauporté et al. [132–135] explored the preparation of Ln/ZnO mixed films by electrodeposition. The researchers calculated the solubility curve of ZnO, $Eu(OH)_3$ and $Er(OH)_3$ at 70°C in the presence of chloride ions [133, 134]. The comparison of the different curves showed that co-precipitation of ZnO and Ln should be possible with a concentration ratio between Zn^{2+} and Ln^{3+} accurately adjusted. Experimentally, europium–ZnO films were prepared from a classical ZnO electrodeposition bath containing 5 mM $ZnCl_2$ and 0.6 mM or more $EuCl_3$ with oxygen precursor [134]. Fig. 21a shows a SEM view of an electrodeposited film. The ZnO crystallized rods contain a low amount of europium. On the bottom of the rods, a covering amorphous layer is formed. The layer contains about 50 at. % of europium and 50 at. % of zinc (Fig. 21b). The rods progressively disappeared with increasing europium ion concentration in the deposition bath.

In the case of the erbium–ZnO system, co-deposition of erbium and ZnO was observed in the presence of 5 mM $ZnCl_2$ and between 0.15 and 0.35 mM $ErCl_3$ [133]. A severe segregation phenomenon occurred with the formation of well-crystallized wurtzite ZnO cylindrical columns containing about 0.5 at.% Er. At their bottom, a thin layer rich in Er and ZnO could be observed (Fig. 21c and d).

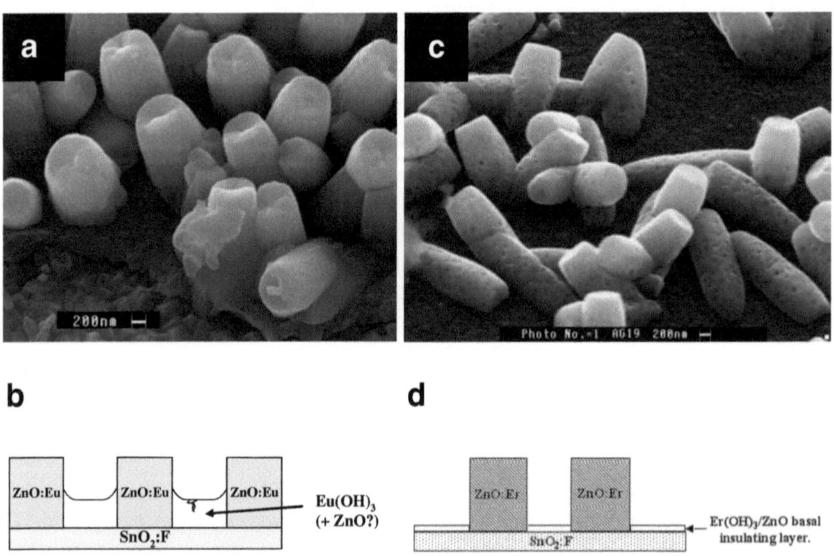

Fig. 21 SEM (**a**, **c**) and schematic (**b**, **d**) views of an europium–ZnO film (**a**, **b**) and an erbium–ZnO film (**c**, **d**) prepared by electrochemical precipitation

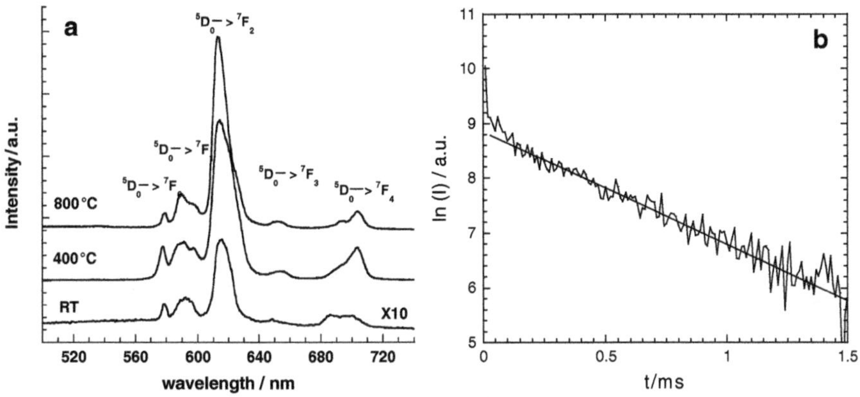

Fig. 22 (**a**) Effect of annealing temperature in air on the PL spectra of electrodeposited films containing 4.2 at. % Eu excited at 464 nm. (**b**) Variation in a logarithmic scale of the 612 nm emission intensity decay after illumination at 464 nm

The PL properties of as-electrodeposited and annealed europium–ZnO films were investigated [135]. The luminescence of europium included in the film was first probed at room temperature by exciting in the $^5D_2$ multiplet at 464 nm (Fig. 22a). The as-deposited films presented a weak PL signal, and emission peaks were observed at 570, 590, 612, 646 and 700 nm. They were assigned to the transitions from the $^5D_0$ (Eu^{3+}) excited state to the ^7F_J (with $J = 0, 1, 2, 3, 4$) multiplets respectively. The $^5D_0 \rightarrow {}^7F_2$ transition was the most intense as usually observed for Eu^{3+} embedded in materials and located in low symmetry sites. The films were subsequently annealed for 1 h at 400°C or 5 h at 800°C in different atmospheres, namely air, argon and forming gas (90% Arg/10% H_2). The best luminescence activation was found after annealing in air at 800°C. Increasing Eu content in the deposition bath was not significantly beneficial to the luminescence intensity. Therefore, it could be supposed that emission principally arises from the Eu-doped rods. The emission lifetime was measured at 420 μs (Fig. 22b).

The films were also studied under UV excitation at 266 nm with a 50 ns gate delay (Fig. 23) [135]. The curve of the as-deposited film was dominated by a broad yellow emission due to defects in ZnO. Due to the rather large gate delay, the UV emission of ZnO was not observed. There was no UV excitation of Eu(III) emission. After annealing at 400°C in air, the yellow emission became very weak and the film was red luminescent. The spectrum presented the different $^2D_0 \rightarrow {}^7F_J$ emission peaks of Eu^{3+}. After the treatment at 800°C in air, the spectrum was dominated by a broad green emission due to ZnO deep levels. It meant that defects, possibly O_i (Fig. 3), were formed upon heating at high temperature. Eu^{3+} electronic transitions were superimposed and the strong $^5D_0 \rightarrow {}^7F_2$ red emission at 612 nm was complemented by weak $^5D_0 \rightarrow {}^7F_1$ and $^2D_0 \rightarrow {}^7F_4$ emissions. The fact that Eu(III) emission under UV excitation was not observed in the as-deposited film but in annealed samples suggested that energy transfer occurs between ZnO and Eu(III) luminescent centers after annealing.

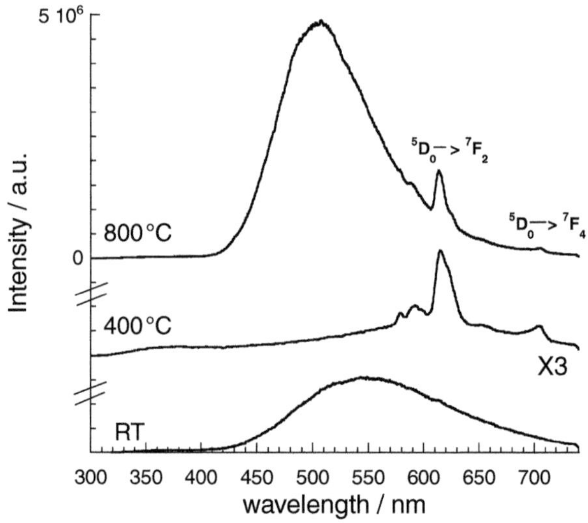

Fig. 23 PL spectra of electrodeposited films containing 4.2 at. % Eu under UV excitation at 266 nm. Effect of annealing in air

9.2 ZnO/Lanthanide Complexes Hybrid Films

Up to now, doping ZnO with Ln has not led to highly luminescent materials under UV excitation but another approach consisting in mixing ZnO and luminescent lanthanide complexes has been shown more successful [117, 118, 120]. A first proposed route was to add a ligand, 2,2′bipyridine 4,4′dicarboxylate (noted dcbpy) in the ZnO electrodeposition bath [118]. No current density decay was observed during the layer growth after the potential was applied, showing that the deposit was conducting. In the absence of dcbpy additive, the films were transparent and were made of dense hexagonal well-crystallized ZnO. Their *c*-axis was preferentially oriented perpendicular to the substrate as shown by XRD patterns. In the presence of 200 μM dcbpy in the deposition bath, the layers were highly light scattering and white colored (Fig. 24a). FESEM views show a fibrous aspect of the deposit, which presents large pores in the hundreds of nanometer size range (Fig. 24c). However, the presence of crystallized ZnO with the hexagonal wurtzite structure was demonstrated by XRD without preferential crystallographic orientation.

Fourier transform infra-red (FTIR) spectrum of the film showed its hybrid nature [118]. The comparison of FTIR spectra of free ligand and ligand included in the film showed the interaction between the dcbpy and the ZnO matrix by means of the bipyridine carboxylate groups and that dcbpy was not simply entrapped in the oxide. The as-deposited films were not luminescent but activation was successfully performed by refluxing the samples at 80°C during 1 h in a 1 mM ethanolic solution of $LnCl_3$ (Fig. 24b). This way the emission wavelength under near UV excitation could be easily tailored. In the film activated by $TbCl_3$ solution, the main emission line

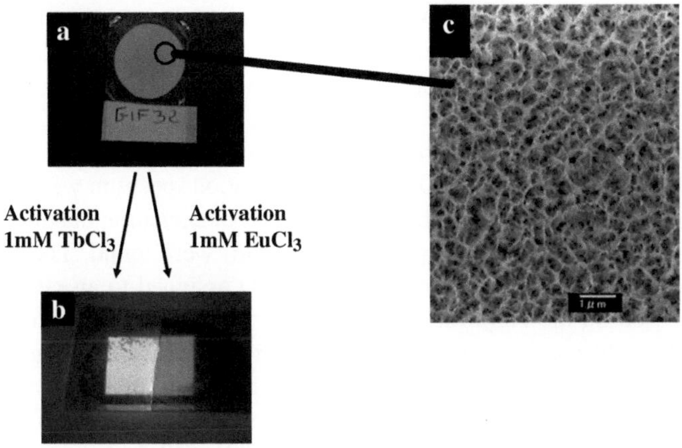

Fig. 24 (**a**) Global view of a ZnO/dcbpy film. (**b**) film illuminated at 320 nm with a xenon lamp. The *left-hand* part was activated with Tb (*green* luminescence), the right-hand part with Eu (*red* luminescence). (**c**) FESEM view of the ZnO/dcbpy hybrid films prepared with 200 μM dcbpy

was the Tb^{3+} $^5D_4 \rightarrow ^7F_5$ at 543 nm, and the film was green colored under UV light illumination. When $TbCl_3$ was replaced by $EuCl_3$, the main emission line under the same excitation conditions was the Eu^{3+} $^5D_0 \rightarrow ^7F_2$ transition at 613 nm, and the film emitted a red light. Under an excitation at 320 nm, the Ln^{3+} emissions are sensitized by the dcbpy ligand which absorbs light due to the $\pi \rightarrow \pi^*$ transition and an efficient energy transfer occurs from the ligands to the chelated Ln^{3+} ions. When compared to the free luminescent complex, additional excitation peaks were present at lower wavelength for the hybrid films which were assigned to an excitation by means of ZnO matrix. The room temperature lifetime of the Eu^{3+} emission in the hybrid film was measured at 0.23 ms. It is well-known that this parameter is strongly determined by the number of OH oscillators (and then water molecules) bounded to the lanthanide ions. Therefore, using Horrocks and Sudnick's formulae [136, 137], a total number of 3.9 water molecules bound to Eu^{3+} was found for a total coordination number of Eu^{3+} equal to 9. The lifetime of the Tb^{3+}-based hybrid film was measured at 0.35 ms, yielding an overestimated number of bound water molecules, close to the total coordination number of Tb^{3+} ion (9).

The second strategy was inspired by the preparation method of DSSC photoanode [117]. The electrodeposition of ZnO in the presence of EY in the deposition bath leads to the formation of highly crystallized ZnO developing very large surface area (see Section 6.3.2). The pores are filled with aggregated EY dye molecules, but almost pure ZnO matrix can be recovered by desorbing EY after soaking into dilute KOH solution (pH 10.5) for more than 3 h [117]. The internal surface area developed in these compounds is very high. It was measured by BET method at 150 cm^2/cm^2 in the case of 2.4 μm thick films prepared with 50 μM EY. FTIR study showed that refluxing the matrices at 80°C in an ethanolic solution of 200 μM dcbpy for 1 h resulted in hybrid ZnO/dcbpy films formation. During this treatment, dcbpy molecules were bound to the ZnO surface by means of their carboxylate functional groups. The last step was the luminescence activation by refluxing the sample in a 500 μM ethanolic solution of $LnCl_3$. The emission intensity was several times lower compared to the film in Fig. 24b.

The film luminescence properties were characterized by recording the excitation and the emission spectra. ZnO/Eu:dcbpy excitation spectrum was monitored by using the $^5D_0 \rightarrow {}^7F_2$ transition of Eu^{3+} at 613 nm. An excitation maximum at about 310 nm and another at lower wavelength (230 nm) were found. The former, also observed with free Eu^{3+}:dcbpy complex, could be assigned to an efficient $\pi \rightarrow \pi^*$ transition based on the conjugated double bond of dcbpy ligand. The other excitation band might involve the oxide matrix. After an efficient energy transfer to the chelated Eu^{3+}, the strong europium (III) red emission was observed. Under an excitation at 312 nm, the different classical emission lines of Eu^{3+} were recorded: $^5D_0 \rightarrow {}^7F_1$, $^5D_0 \rightarrow {}^7F_2$, $^5D_0 \rightarrow {}^7F_3$ and $^5D_0 \rightarrow {}^7F_4$ at 591, 613, 647 and 698 nm respectively. In a similar manner, the ZnO/dcbpy films were successfully doped with Tb^{3+}. The $^5D_4 \rightarrow {}^7F_j$ (j = 6, 5, 4, 3) transitions at 490, 543, 587 and 622 nm respectively were observed. The most intense emission line was the second at 543 nm, thus giving a green color to the film. The excitation spectrum presented a peak at 320 nm similar to that recorded with Eu^{3+} and confirmed that Ln excitation was not direct but occurred there by means of the ligand. Tb emission could also be excited at lower wavelength, below 260 nm, similar to Eu.

The two adsorption treatment steps were optimized [117]. The luminescence intensity reached a maximum with solutions above 200 μM of dcbpy and no improvement was observed with solutions of $LnCl_3$ above 500 μM and up to 5 mM. The emission intensities obtained from the ZnO matrix prepared with 50 μM of EY were more intense than those from films prepared with 30 μM EY. In fact, if the porosities of both films were close (60 % and 55 % respectively) (Fig. 17a), the total surface area or the surface accessibility of the former was larger (Fig. 17b) and a higher quantity of dcbpy was adsorbed.

10 Conclusions

The preparation of ZnO nanoparticles and nanostructured ZnO films is an exciting emerging field of research. In the present review we have shown the possibility of developing low-cost and low-temperature routes for the wet synthesis of a large variety of nanostructures with well-controlled shape and size. The crystallites can be prepared as individual nano-objects or deposited on a surface in a well-ordered manner and well-controlled crystallographic orientation. Solution-based preparation methods are easily scalable for mass production at low-cost. Therefore they present an interesting alternative to the physical methods that need high energy and expensive equipments to handle high vacuum and/or temperature.

The growth mechanism in solution is often complex and we have tried to explain the main physico-chemical processes that govern the ZnO nanostructure formation. A large variety of ZnO nanostructure has been described. 1D nanorods and nanowires can be prepared by chemical precipitation and electrodeposition. 1D nanowhiskers have been grown by hydrothermal corrosion of zinc foil whereas ZnO nanotubes were synthesized by controlled amino-complex decomposition and by growth in chemical solution in the presence of a polymer. Pumped micro-cavity lasing was observed on many of these 1D ZnO materials. 2D nanosheets and nanoplatelets were produced by chemical synthesis and by electrodeposition [138, 139]. Mesoporous films were also synthesized by electrodeposition. The mesopores can be obtained by controlling the growth conditions or by using a dye additive. By the latter method, a fine control of the porous structure is facilitated but the mechanism for mesopore formation and dye templating remains to be clarified. We have also shown that various nanostructures could be deposited in a patterned manner by soft methods. In the future the field will not only require experimental report on the formation of nanostructures but more importantly in-deep investigation of the physico-chemical processes that govern the formation of these objects. The issue of p-type ZnO formation from solution remains to be overcome. Epitaxial growth of ZnO has been successfully achieved and some optoelectronic devices have already been realized. It demonstrates that solution-grown ZnO nanostructures are of sufficient quality for being used in functional devices and that hole can be efficiently injected by means of n–p heterojunctions. Electrodeposited mesoporous ZnO films are efficient n-type matrices in dye-sensitized solar cells. Electrodeposited ZnO nanorods have been used as an n-type layer in Eta solar cells and in visible- and UV-light emitting diodes. The highly luminescent ZnO/lanthanide complex hybrid films recently discovered seem especially promising for this latter application, with the possibility of easily tuning the visible emission wavelength [118, 120].

References

1. Xiang B, Wang P, Zhang X, Dayeh S A, Aplin D P R, Soci C, Yu D, Wang D (2007) Rational synthesis of p-type zinc oxide nanowire arrays using simple chemical vapor deposition. Nano Lett. 7:323–328.
2. Kang H S, Kim G H, Kim D L, Chang H W, Ahn B D, Lee S Y (2006) Investigation on the p-type formation mechanism of arsenic doped p-type ZnO thin film. Appl. Phys. Lett. 89:181103
3. Sun J W, Lu Y M, Liu Y C, Shen D Z, Zhang Z Z, Li B H, Zhang J Y, Yao B, Zhao D X, Fan X W (2006) Hole transport in p-type ZnO films grown by plasma-assisted molecular beam epitaxy. Appl. Phys. Lett. 89:232101.
4. Hwang D K, Oh M S, Lim J H, Kang C H, Park S J (2007) Effect of annealing temperature and ambient gas on phosphorus doped p-type ZnO. Appl. Phys. Lett. 90:021106.
5. Mosnier J P, Chakrabarti S, Doggett B, McGlynn E, Henry M O, Meaney A (2007). P-type nitrogen- and phosphorus-doped ZnO thin films grown by pulsed laser deposition on sapphire substrates. Proc. SPIE 6474, 64740I.
6. Özgür Ü, Alivov Y I, Liu C, Teke A, Reshchikov M A, Dogan M A, Avrutin V, Cho S J, Morkoc H (2005) A comprehensive review of ZnO materials and devices. J. Appl. Phys. 98:041301_1–041301_105.
7. Wang Z L (2004) Nanostructures of zinc oxide. Mater. Today 7(6):26– 33.
8. Pan Z W, Dai Z R, Wang Z L (2001) Nanobelts of semiconducting oxides. Science

291:1947–1949.
9. Wang Z L (2004) Zinc oxide nanostructures: growth, properties and applications. J. Phys.:Condens. Matter 16:R829–R858.
10. Heo Y W, Norton D P, Tien L C, Kwon Y, Kang B S, Ren F, Pearton S J, LaRoche J R (2004) ZnO nanowire growth and devices. Mater. Sci. Eng. R 47:1–47.
11. Huang M H, Mao S, Feick H, Yan H, Wu Y, Kind H, Weber E, Russo R, Yang P (2001) Room-temperature ultraviolet nanowire lasers. Science 292:1897–1899.
12. Klingshirn C, Hauschild R, Priller H, Decker M, Zeller J, Kalt H (2005) ZnO rediscovered-once again!? Superlattices Microstructures 38:209–222.
13. Wang Q, Wang G, Xu B, Jie J, Han X, Li G, Li Q, Hou J G (2005) Non-aqueous cathodic electrodeposition of large-scale uniform ZnO nanowire arrays embedded in anodic alumina membrane. Mater. Lett. 59: 1378–1382.
14. Pan A P, Yu R, Xie S, Zhang Z, Jin C, Zou B (2005) ZnO flowers made up of thin nanosheets and their optical properties. J. Crystal Growth 282:165–172.
15. Ng H. T, Li J, Smith M, Nguyen P, Cassell A, Han J, Meyyappan M (2003) Growth of epitaxial nanowires at the junctions of nanowalls. Science 300:1249–1250
16. JCPDS-ICDD Card n° 36-1451.
17. Vayssieres L, Keis K, Hagfeldt A, Lindquist S E (2001) Three-dimensional array of highly oriented crystalline ZnO Microtubes. Chem. Mater. 12:4395–4397.
18. Park W I, Jun Y H, Jung S W, Yi G C (2003) Excitonic emissions observed in ZnO single crystal nanorods. Appl. Phys. Lett. 82: 964–966.
19. Tam K H, Cheung C K, Leung Y H, Djuristic A B, Ling C C, Beling C D, Fung S, Kwok W M, Chan W K, Phillips D L, Ding L, Ge W K (2006) Defects in ZnO nanorods prepared by hydrothermal method. J. Phys. Chem. B 110:20865–20871.
20. Lin B, Fu Z, Jia (2001) Green Luminescence in undoped zinc oxide films deposited on silicon. Appl. Phys. Lett. 79:943–945.
21. Lima S A M, Sigoli F A, Jafelicci M, Davolos M R (2001) Luminescent properties and lattice defects correlation on zinc oxide. Int. J. Inorg. Mater. 3:749–754.
22. Xu P S, Sun Y M, Shi C S, Xu F Q, Pan H B (2003) The electronic structure and spectral properties of ZnO and its defects. Nucl. Instr. Meth. Phys. Res. B 199:286–290.
23. Nikitenko V (2005) Optics and spectroscopy of point defects in ZnO. In *Zinc Oxide- A material for micro- and optoelectronic applications* Norberg N H, Terukov E, Eds. Springer, Dordrecht, The Netherland, pp. 69–81.
24. Vanheusden K, Seager C H, Warren L, Tallant D R, Voigt J A (1996) Correlation between photoluminescence and oxygen vacancies in ZnO phosphor. Appl. Phys. Lett. 68:403–405.
25. Vanheusden K, Seager C H, Warren L, Tallant D R, Voigt J A (1996) Mechanisms behind green luminescence in ZnO phosphor powders. J Appl. Phys. 79:7983–7990.
26. Leiter F, Alves H R, Hofstaetter A, Hofmann D M, Meyer B K (2001) The oxygen vacancy as the origin of a green emission in undoped ZnO. Phys. Stat. Sol. (b) 226:R4–R5.
27. Leiter F, Alves H R, Pfisterer D, Romanov N G, Hofmann D M, Meyer B K (2003) Oxygen vacancies in ZnO. Physica B 340–342:201–204.
28. Goux A, Pauporté T, Chivot J, Lincot D (2005) Temperature Effects on ZnO Electrodeposition. Electrochim. Acta, 50:2239–2248.
29. Spanhel L, Anderson M A (1991) Semiconductor clusters in sol-gel process: quantized aggregation, gelation and crystal growth in concentrated ZnO colloids. J. Am. Chem. Soc. 113:2826–2833.
30. Meulenkamp E A (1998) Synthesis and growth of ZnO nanoparticles. J. Phys. Chem. B 102:5566–5572.
31. Xiao Z Y, Liu Y C, Dong L, Shao C L, Zhang J Y, Lu Y M, Zhen D Z, Fan X W (2005) The effect of surface properties on visible luminescence of nanosized colloidal ZnO membranes. J. Colloid. Interface Science 282:403–407.
32. Peng W Q, Qu S C, Cong G W, Zhang Z G (2006) Structure and visible luminescence of ZnO nanoparticles. Mater. Sci. Semicond. Process. 9:156–159.

33. Pesika N S, Stebe K J, Searson P C (2003) Determination of the particle size distribution of quantum nanocrystals from absorbance spectra. Adv. Mater. 15:1289–1291.
34. Guan X H, Wang G S, Li C P, Lv Y Z, Guo L, Xu H B, Xu H B (2007) Synthesis and optical properties of ZnO multipod nanorods. J. Lumin. 122–123:770–772.
35. Lv Y Z, Zhang Y H, Li C P, Ren L R, Guo L, Xu H B, Ding L, Yang C L, Ge W K, Yang S H (2007) Temperature-dependent photoluminescence of ZnO nanorods prepared by a simple solution route. J. Lumin. 122–123:816–818.
36. Cheng B, Shi W, Russell-Tanner J M, Zhang L, Salmulski E T (2006) Synthesis of variable-aspect-ratio, single-crystalline ZnO. Inorg. Chem. 45:1208–1214.
37. Sekiguchi T, Miyashita S, Obara K, Shishido K, Sakagami N (2000) Hydrothermal growth of ZnO single crystals and their optical characterization. J. Crystal Growth 214–215:72–76.
38. Mass J, Avella M, Jimenez J, Callahan M, Grant E, Rakes K, Bliss D, Wang B (2005) Cathodoluminescence characterization of hydrothermal ZnO crystals. Superlattices Microstructures. 38:223–230.
39. Yoneta M, Yoshino K, Ohishi M, Saito H (2006) Photoluminescence studies of high-quality ZnO single crystals by hydrothermal method. Physica B 376–377:745–748.
40. Wang J, Gao L (2004) Synthesis of uniform rod-like, multipod-like ZnO whiskers and their photoluminescence properties. J. Crystal Growth 262:290–294.
41. Li W J, Shi E W, Zhong W Z, Yin Z W (1999) Growth mechanism and habit of oxide crystals. J. Crystal Growth 203:186–196.
42. Masuda Y, Kinoshita N, Sato F, Koumoto K (2006) Site selective deposition and morphology control of UV- and visible-light-Emitting ZnO crystals. Crystal Growth Design 6:75–78.
43. Wang J, Gao L (2003) Wet chemical synthesis of ultralong and straight single-crystalline ZnO nanowires and their excellent UV emission properties. J. Mater. Chem. 13:2551–2554.
44. Zou G., Yu D, Wang D, Zhang W, Xy L, Yu W, Qian Y (2004) Controlled synthesis of ZnO nanocrystals with column-, rosette- and fiber-like morphologies and their photoluminescence properties. Mater. Chem. Phys. 88:150–154.
45. Fan L, Song H, Li T, Yu L, Liu Z, Pan G, Lei Y, Bai X, Wang T, Zheng Z, Kong X (2007) Hydrothermal synthesis and photoluminescent properties of ZnO nanorods. J. Lumin. 122–123:819–821.
46. Kuo C L, Kuo T K, Huang M H (2005) Hydrothermal synthesis of ZnO microspheres and hexagonal microrods with sheetlike and platelike nanostructures. J. Phys. Chem. B 109:20115–20121.
47. Xie R, Li D, Zhang H, Yang D, Jiang M, Sekiguchi T, Liu B, Bando Y (2006) Low-temperature growth of uniform ZnO particles with controllable ellipsoidal morphologies and characteristic luminescence patterns. J. Phys. Chem. B 110:19147–19153.
48. Pal U, Santiago P (2005) Controlling the morphology of ZnO nanostructures in low-temperature hydrothermal process. J. Phys. Chem. B 109:15317–15321.
49. Govender K, Boyle D S, Kenway P B, O'Brien P (2004) Understanding the factors that govern the deposition and morphology of thin films of ZnO from aqueous solution. J. Mater. Chem. 14:2575–2591.
50. Yamabi S, Imai H (2002) Growth conditions for wurtzite zinc oxide films in aqueous solutions. J. Mater. Chem. 12:3773–3778.
51. Duan J, Huang X, Wang E (2006) PEG-assisted synthesis of ZnO nanotubes. Mater. Lett. 60:1918–1921.
52. Quang L H, Chua S J, Loh K P, Fitzgerald E (2006) The effect of post annealing treatment on photoluminescence of ZnO nanorods prepared by hydrothermal synthesis. J. Crystal Growth 287:157–161.
53. O'Brien P, Saeed T, Knowles J (1996) Speciation and the nature of ZnO thin films from chemical bath deposition. J. Mater. Chem. 6:1135–1139.
54. Izaki M, Omi T (1997) Transparent zinc oxide films chemically prepared from aqueous solution. J. Electrochem. Soc. 144:L3–L5.

55. Izaki M, Omi T (2000) Characterization of boron-incorporated zinc oxide film chemically prepared from aqueous solution. J. Electrochem. Soc. 147:210–213.
56. Saito N, Haneda H, Sekiguchi T, Ohashi N, Sakaguchi I, Koumoto K (2002) Low-temperature fabrication of light emitting zinc oxide micropatterns using self assembled monolayers. Adv. Mater. 14:418–421.
57. Saito N, Haneda H, Koumoto K (2004) Pattern-deposition of light-emitting ZnO particulate film through biomimetic process using self assembled monolayer template. Microelectronics J 35:349–352.
58. Vayssieres L, Keis K, Lindquist S E, Hagfeldt A (2001) Purpose-built anisotropic metal oxide materials: 3D highly oriented microrod array of ZnO. J. Phys. Chem. B 105: 3350–3352.
59. Vayssieres L (2003) Growth of arrayed nanorods and nanowires of ZnO from aqueous solutions. Adv. Mater. 15:464–466.
60. Boyle D S, Govender K, O'Brien P (2002) Novel low temperature solution deposition of perpendicularly oriented rods of ZnO: substrate effects and evidence of the importance of counter-ions in the control of crystallite growth. Chem. Commun. 80–81.
61. Tian Z R, Voigt J A, Liu J, Mckenzie B, Mcdermott M J, Rodriguez M A, Konishi H, Xu H (2003) Complex and oriented ZnO nanostructures. Nature Mater. 2:821–826.
62. Djuristic A B, Kwok W M, Leung Y H, Phillips D L, Chan W K (2005) Stimulated emission in ZnO nanostructures: a time resolved study. J. Phys. Chem. B 109:19228–19233.
63. Hsu J W P, Tian Z R, Simmons N C, Matzke C M, Voigt J A, Liu J (2005) Directed spatial organization of zinc oxide nanorods. Nano Lett. 5:83–86.
64. Govender K, Boyle D S, O'Brien P, Binks D, West D, Coleman D (2002) Room-temperature lasing observed from ZnO nanocolumns grown by aqueous solution deposition. Adv. Mater. 14:1221–1224.
65. Choy J H, Jang E S, Won J H, Chung J H, Jang D J, Kim Y W (2003) Soft solution route to directionally grown ZnO nanorod arrays on Si wafer; room-temperature ultraviolet laser. Adv. Mater. 15:1911–1914.
66. Ito K, Nakamura K (2006) Preparation of ZnO thin films using the flowing liquid film method. Thin Solid Film 286:35–36.
67. Qiu Z, Wong K S, Wu M W, Lin W, Xu H (2004) Microcavity lasing behaviour of oriented hexagonal nanowhiskers grown by hydrothermal oxidation. Appl. Phys. Lett. 84: 2739–2741.
68. Pan A L, Liu R B, Wang S Q, Wu Z Y, Cao L, Xie S S, Zou B S (2005) Controllable growth and optical properties of large scale ZnO arrays. J. Crystal Growth 282:125–130.
69. Dem'yanets L N, Li L E, Uvarova T G (2006) Hydrothermal synthesis and cathodoluminescence of ZnO crystalline powders and coatings. J. Crystal Growth 287:23–27.
70. Chen Y, Bagnall D M, Koh H J, Park K T, Hiraga K, Zhu Z, Yao T (1998) Plasma assisted molecular beam epitaxy of ZnO on c-plane of sapphire: growth and characterization. J. Appl. Phys. 84:3912–3918.
71. Peulon S, Lincot D (1996) Cathodic electrodeposition from aqueous solution of dense or open structured zinc oxide films. Adv. Mater. 8:166–170.
72. Peulon S, Lincot D (1998) Mechanistic study of cathodic electrodeposition of zinc oxide and zinc hydroxychloride films from oxygenated aqueous zinc chloride solutions. J. Electrochem. Soc. 145:864–874.
73. Pauporté T, Lincot D (1999) Heteroepitaxial electrodeposition of zinc oxide on gallium nitride. Appl. Phys. Lett. 75:3817–3819.
74. Pauporté T, Lincot D (2000) Electrodeposition of semiconductors for optoelectronic devices: results on zinc oxide. Electrochim. Acta 45:3345–3353.
75. Pauporté T, Lincot D (2001) Hydrogen peroxide oxygen precursor for zinc oxide electrodeposition. II- Mechanistic aspects. J. Electroanal. Chem. 517:54–62.
76. Goux A, Pauporté T, Lincot D (2006) Oxygen Reduction on Zinc Oxide Electrodes in KCl Aqueous solution at 70°C. Electrochim. Acta. 51:3168–3172.

77. Gal D, Hodes G, Lincot D, Schock H W (2000) Electrochemical deposition of zinc oxide films from non-aqueous solution: a new buffer/window process for thin film solar cells. Thin Solid Films 361–362:79–83.
78. Pauporté T., Lincot D, Viana B, Pellé F (2006) Toward laser emission of epitaxial nanorod array of ZnO grown by electrodeposition. Appl. Phys. Lett. 89:233112.
79. Pauporté T, Cortès R, Froment M, Beaumont B, Lincot D (2002) Electrocrystallization of epitaxial zinc oxide onto galium nitride. Chem. Mater. 14:4702–4708.
80. Pauporté T, Yoshida T, Cortès R, Froment M, Lincot D (2003) Electrochemical growth of epitaxial Eosin/ZnO hybrid films. J. Phys. Chem. B 107:10077–10082.
81. Liu R, Vertegel A A, Bohannan E W, Sorenson T A, Switzer J A (2001) Epitaxial electrodeposition of Zinc oxide nanopillars on single-crystal gold. Chem. Mater. 13:508–512.
82. Vispute R D, Talyansky V, Choopun S. Sharma R P, Venkatesan T, He M, Tang X, Halpern J B, Spencer M G, Li Y X, Salamanca-Riba L G, Iliadis A A, Jones K A (1998). Heteroepitaxy of ZnO on GaN and its implications for fabrication of hybrid optoelectronic devices. Appl. Phys. Lett. 73:348–350.
83. Pauporté T, Lincot D (2001) Hydrogen peroxide oxygen precursor for zinc oxide electrodeposition. I-Deposition in perchlorate medium. J. Electrochem. Soc. 148:C310–C314.
84. Leprince-Wang Y, Yacoubi-Ouslim A, Wang G Y (2005) Structure study of electrodeposited ZnO nanowires. Microelectronics J. 36:625–628.
85. Lai M, Riley J (2006) Templated electrosynthesis of Zinc oxide nanorods. Chem. Mater. 18:2233–2237.
86. Pauporté T, Yoshida T, Goux A, Lincot D (2002). One-step electrodeposition of ZnO/Eosin hybrid films from hydrogen peroxide oxygen precursor. J. Electroanal. Chem, 534:55–64.
87. Izaki M, Omi T (1996) Electrolyte optimization for cathodic growth of zinc oxide films. J. Electrochem. Soc. 143:L53–L55.
88. Izaki M, Omi T (1996) Transparent zinc oxide films prepared by electrochemical reaction. Appl. Phys. Lett. 68:2439–2440.
89. Izaki M, Omi T (1997) Characterization of transparent zinc oxide films prepared by electrochemical reaction. J. Electrochem. Soc. 144:1949–1952.
90. Yoshida T, Komatsu D, Shimokawa N, Minoura H (2004) Mechanism of cathodic electrodeposition of zinc oxide thin films from aqueous zinc nitrate baths. Thin Solid Films 451–452:166–169.
91. Ishizaki H, Izaki M, Ito T (2001) Influence of $(CH_3)_2NHBH_3$ concentration on electrical properties of electrochemically grown ZnO film. J. Electrochem. Soc. 148:C540–C543.
92. Ishizaki H, Imaizumi M, Matsuda S, Izaki M, Ito T (2002) Incorporation of boron in ZnO film from an aqueous solution containing zinc nitrate and dimethylamine-borane by electrochemical reaction. Thin Solid Films 411:65–68.
93. Zheng M J, Zhang L D, Li G H, Shen W Z (2002) Fabrication and optical properties of large-scale uniform zinc oxide nanowire arrays by one-step electrochemical deposition technique. Chem. Phys. Lett. 363:123–128.
94. Izaki M, Watase S, Takahashi H (2003) Low-temperature electrodeposition of room temperature ultraviolet-light-emitting zinc oxide. Adv. Mater. 15:2000–2002.
95. Izaki M, Watase S, Takahashi H (2003) Room-temperature ultraviolet light-emitting zinc oxide micropatterns prepared by low-temperature electrodeposition and photoresist. Appl. Phys. Lett. 83:4930–4932.
96. Wang Q, Wang G, Jie J, Han X, Xu B, Hou J G (2005) Annealing effect on optical properties of ZnO films fabricated by cathodic electrodeposition. Thin Solid Films 492:61–65.
97. Zhang L, Chen Z, Tang Y, Jia Z (2005) Low temperature cathodic electrodeposition of nanocrystalline zinc oxide thin films. Thin Solid Films 492:24–29.
98. Cao B, Teng X, Heo S H, Li Y, Cho S O, Li G, Cai W (2007) Different ZnO nanostructures fabricated by seed layer assisted electrochemical route and their photoluminescence and field emission properties. J. Phys. Chem C 111:2470–2476.

99. Chen Z, Tang Y, Zhang L, Luo L (2006) Electrodeposited nanoporous ZnO films exhibiting enhanced performance in dye sensitized solar cells. Electrochim. Acta 51:5870–5875.
100. Chen Q P, Xue M Z, Sheng Q R, Liu Y G, Ma Z F (2006) Electrochemical growth of nanopillar zinc oxide film by applying a low concentration of zinc nitrate precursor. Electrochem. Solid-State Lett. 9:C58–C61.
101. Könenkamp R, Boedecker K, Lux-Streiner M C, Poschenrieder M, Zenia F, Lévy-Clément C, Wagner S (2000) Thin film semiconductor deposition on free-standing ZnO columns. Appl. Phys. Lett. 77:2575–2577.
102. Lévy-Clément C, Katty A, Bastide S, Zenia F, Mora I, Munoz-Sanjose V (2002). A new CdTe/ZnO columnar composite film for Eta-solar cells. Physica E 14:229–232.
103. Tena-Zaera R, Katty A, Bastide S, Lévy-Clément C, O'Regan B, Munoz-Sanjosé (2005) ZnO/CdTe/CuSCN, a promising heterostructure to act as inorganic eta-solar cell. Thin Solid Films 483:372–377.
104. Lévy-Clément C, Tena-Zaera R, Ryan M A, Hodes G (2005) CdSe-sensitized p-CuSCN/Nanowire n-ZnO heterojunction. Adv. Mater. 17:1512–1515.
105. O'Regan B, Lenzmann F, Muis R, Wienke J (2002) A solid-state dye-sensitized solar cell fabricated with pressure-treated P25-TiO_2 and CuSCN: analysis of pore filling and I-V characteristics. Chem. Mater. 14:5023–5029.
106. Tena-Zaera R, Ryan MA, Katty A, Hodes G, Bastide S, Lévy-Clément C (2006) Fabrication and characterization of ZnO nanowires/CdSe/CuSCN eta-solar cell. C. R. Chimie 9: 717–729.
107. Könenkamp R, Word R C, Schlegel C (2004) Vertical nanowire light-emitting diode. Appl. Phys. Lett. 85:6004–6006.
108. Könenkamp R, Word R C, Godinez M (2005) Ultraviolet electroluminescence from ZnO/polymer heterojunction light emitting diodes. Nano Lett. 5:2005–2008.
109. Lévy-Clément C, Elias J, Tena-Zaera R (2007) Electrodeposition of arrays of ZnO nanostructures and application to photoelectrochemical devices. SPIE Proceedings 6340:63400R1–63400R12.
110. Badre C, Pauporté T, Turmine M, Lincot D (2007) ZnO nanowire array film with super water-repellent properties. Nanotechnology 18:365705.
111. O'Regan B, Sklover V, Grätzel M (2001) Electrochemical deposition of smooth and homogeneously mesoporous ZnO films from propylene carbonate electrolytes. J. Electrochem. Soc. 148:C498–C505.
112. O'Regan B, Schwartz D T, Zakeeruddin S M, Grätzel M (2000) Electrodeposited nanocomposite n-p heterojunctions for solid-state dye-sensitized photovoltaics. Adv. Mater. 12: 1263–1267.
113. Yoshida T, Pauporté T, Lincot D, Oekermann T, Minoura H (2003) Cathodic electrodeposition of ZnO/eosin Y hybrid thin films from oxygen-saturated aqueous solution of $ZnCl_2$ and eosin Y. J. Electrochem. Soc. 150:C608–C615.
114. Yoshida T, Iwaya M, Ando H, Oekermann T, Nonomura K, Schlettwein D, Wöhrle D, Minoura H (2004) Improved photoelectrochemical performance of electrodeposited ZnO/Eosin Y hybrid thin films by dye re-adsorption. Chem. Commun 400–401.
115. O'Regan B, Grätzel M (1991) A low-cost, high efficiency solar cell based on dye-sensitized colloidal TiO_2 films. Nature 353:737–740.
116. Rathousky J, Pauporté T (2007) Electrodeposited mesoporous ZnO thin films as efficient photocatalysts for the degradation of dye pollutants. J. Phys. Chem. C 111:7639–7644.
117. Pauporté T, Yoshida T, Komatsu D, Minoura H (2006) Highly porous electrodeposited zinc oxide films for red/green luminescence. Electrochem. Solid-State Lett. 9:H16–H18.
118. Pauporté T, Yoshida T (2006) Conducting hybrid layers of ZnO/lanthanide complexes with high visible luminescence. J. Mater. Chem. 16:4529–4534
119. Goux A, Pauporté T, Yoshida T, Lincot D (2006) Mechanistic study of the electrodeposition of nanoporous self-assembled ZnO/Eosin Y hybrid thin films. Effect of eosin concentration. Langmuir 22:10545–10553.

120. Yoshida T, Minoura H, Zhang J, Komatsu D, Sawatani S, Pauporté T, Lincot D, Oekermann T, Schlettwein D (2009) Electrodeposition of inorganic/organic hybrid thin films. Adv. Funct. Mater. 19:17–43.
121. Goux A, Pauporté T, Lincot D, Dunsch L (2007) In situ ESR and UV-visible spectroelectrochemical study of eosin Y upon reduction with and without Zn (II) ions. Chem. Phys. Chem. 8:926–931, DOI: 10.1002/cphc.200700009.
122. Aarão Reis F D A, Badiali J P, Pauporté T, Lincot D (2006). Statistical modelling of the electrodeposition of nanostructured hybrid films with ZnO- eosin Y as a case example. J. Electroanal. Chem. 598:27–35.
123. Pauporté T, Bedioui F, Lincot D (2005) Nanostructured zinc oxide- chromophore hybrid films with multicolored electrochromic properties. J. Mater. Chem. 15:1552–1559.
124. Pauporté T (2007) Highly transparent ZnO/Polyvinyl alcohol hybrid films with controlled crystallographic orientation growth. Crystal Growth Design 7:2310–2315.
125. Yang X, Shao C, Guan H, Li X, Gong J (2004) Preparation and characterisation of ZnO nanofibers by using electrospun PVA/zinc acetate composite fiber precursor. Inorg. Chem. Commun. 7:176–178.
126. Hong Y, Li D, Zheng J, Zou G (2006) In situ growth of ZnO nanocrystals from solid electrospun nanofiber matrixes. Langmuir 22:7331–7334.
127. Sui X M, Shao C L, Liu Y C (2005) White-light emission of polyvinyl alcohol/ ZnO hybrid nanofibers prepared by electrospinning. Appl. Phys. Lett. 87:113115.
128. He Y, Sang W, Wang J, Wu R, Min J (2005) Vertically well-aligned ZnO nanowires generated with self-assembling polymers. Mater. Chem. Phys. 94:29-33.
129. Lima S A M, Sigoli F A, Davolos M R, Jafelicci M (2002) Europium(III)-containing zinc oxide from pichini method. J. Alloys Comp. 344:280–284.
130. Lima S A M, Sigoli F A S, Davolos M R (2003) Pechini's solution precursor for Eu(III) –containing ZnO films. J. Solid State Chem. 171:287–290.
131. Gao S, Zhang H, Deng R, Wang X, Sun D, Zheng G (2006) Engineering white light-emitting Eu-doped ZnO urchins by biopolymer-assisted hydrothermal method. Appl. Phys. Lett. 89:123125.
132. Pauporté T, Goux A, Kahn-Harari A, de Tacconi N, Rajeshwar K, Lincot D (2003) Cathodic electrodeposition of mixed oxide thin films. J. Phys. Chem. Solids 64:1737–1742.
133. Goux A, Pauporté T, Lincot D (2006) Deposition of mixed zinc oxide/ lanthanide films by electrochemical precipitation. The ZnO/Er system. J. Electroanal. Chem. 587:193–202.
134. Goux A, Pauporté T, Lincot D (2007) Preparation of ZnO/Eu mixed films by electrochemical precipitation. Electrochim. 53:233–244.
135. Pauporté T, Pellé F, Viana B, Aschehoug P (2007) Luminescence of nanostructured ZnO/Eu mixed films prepared by electrodeposition. J. Phys. Chem. C 111:15427–15432.
136. Horrocks W D, Sudnick D R (1979) Lanthanide ion probes of structure in Biology. Laser-induced luminescence decay constants provide a direct measure of the number of metal-coordinated water molecules. J. Am. Chem. Soc. 101:334–340.
137. Lis S (2002) Luminescence spectroscopy of lanthanide(III) ions in solution. J. Alloys Compd. 341:45–50.
138. Xiang J H, ZhuP X, Masuda Y, Okuya M, Kaneko S, Koumoto K (2006) Flexible solar-cell from zinc oxide nanocrystalline sheets self-assembled by an in-situ electrodeposition process. J. Nanosci. Nanotechnol. 6:1797–1801.
139. Cao B Q, Cai W P, Zeng H B, Duan G T (2006) Morphology evolution and photoluminescence properties of ZnO films electrochemically deposited on conductive glass substrates. J. Appl. Phys. 99:073516_145–073516_150.

Self-Assembled Metal Nanostructures in Semiconductor Structures

Francesco Ruffino, Filippo Giannazzo, Fabrizio Roccaforte, Vito Raineri, and Maria Grazia Grimaldi

Abstract Understanding the effects of downscaling the devices' dimensions to the nanometer size is one of the most important topics in the modern material science applied to microelectronics. In fact, the confinement of electrons in dimensions typical of atoms and molecules obliges to consider their quantum behavior. Therefore, a new class of effects is characterization of ultra-scaled devices. In the last years, these ideas led to the birth of the "nanotechnology and nanoelectronic revolution" with the aim to understand the effects of downscaling the matter in the atomic range and to develop innovative nanostructured materials and quantum effects–based devices following a bottom-up procedure with respect to the traditional top-down scaling scheme.

In particular, the nanometric level knowledge of the structural characteristics of such innovative materials and the nanometric control and manipulation of these characteristics acquired a fundamental importance in the design and realization of innovative electrical nanodevices. In fact, it is well known that the local electrical characteristics of such devices are dramatically dependent on the local structural characteristics. Hence, a precise control and manipulation (at atomic level) of the structural characteristics allow the precise control and manipulation of the electrical ones that are always innovative properties with respect to the traditional devices.

A promising topic of nanotechnology research is, surely, the study of the structural and electrical properties of nanometric metal clusters deposited on or embedded in semiconductor/insulating substrates in view of the realization of nanostructured materials with electrical properties dependent on and tunable by the structural ones (clusters size, density, etc.).

In the first part of this chapter, we illustrate some methods to fabricate nanostructured materials using metallic nanoclusters in connection with insulator and semiconductor substrates and matrices. The methodologies to control and manipulate

F. Ruffino (✉)
Dipartimento di Fisica ed Astronomia and MATIS CNR-INFM, Università di Catania, via Santa Sofia 64, 95123 Catania, Italy; Consiglio Nazionale delle Ricerche-Instituto per la Microelettronica e Microsistemi (CNR-IMM), Stradale Primosole 50, 95121, Catania, Italy
e-mail: francesco.ruffino@ct.infn.it

Z.M. Wang (ed.), *Toward Functional Nanomaterials*, Lecture Notes in Nanoscale Science and Technology 5, DOI 10.1007/978-0-387-77717-7_3,

the clusters structural properties based on the self-organization mechanism of the Au nanoclusters on the SiC and SiO_2 surfaces induced by thermal and ion beam processes are also discussed. In particular, the thermal-induced self-organization kinetic mechanism of Au nanoclusters on SiC hexagonal and SiO_2 surfaces is shown to be a ripening process of three-dimensional structures controlled by surface diffusion, and the application of the ripening theory enabled us to derive the surface diffusion coefficient and all other parameters necessary to describe the entire process. Then a detailed study of the morphological modifications occurring during the irradiation, by an Ar beam, of Au nanoclusters on SiO_2 surface is presented showing how it is a suitable methodology to tailor the clusters size and surface density. These characterizations allow us to clarify the evolution of the Au clusters during ion bombardment: at low beam current, the cluster mean size increases with ions fluence, and at high beam current, the clusters mean size decreases with ions fluence. This behavior has been described in terms of a conventional ripening process (in which small clusters shrink promoting the growth of the large clusters) at low beam current and of an inverse ripening process (in which the large clusters shrink) at high beam current. The kinetics of clusters self-organization has been modeled using the theory of cluster ripening and inverse ripening under ion beam irradiation that we integrated, for our experimental case, introducing the effect of the sputtering process of Au atoms from the surface by the ion beam irradiation.

Therefore, we suggest to apply the self-organization of Au nanoparticles as a nanotechnology step to control the metal–semiconductor (MS) and the metal–insulator interface at atomic level to fabricate innovative nanostructured devices: the fabricated 6H-SiC/Au cluster-based nanostructured materials were used to probe, by local conductive atomic force microscopy (C-AFM), the electrical properties of nano-Schottky-contact Au nanocluster/SiC. The main observation was the Schottky barrier height dependence of the nano-Schottky contact on the cluster size. Such a behavior was interpreted considering the physics of few electron quantum dots merged with the concepts of ballistic transport and thermoionic emission finding a satisfying agreement between the theoretical prediction and the experimental data. The local nanoscale electrical properties of the SiO_2/Au cluster-based nanostructured material revealed a rectifying behavior characterized by a threshold voltage tunable by the clusters size. This behavior is interpreted by comparing physical considerations on metal-oxide-semiconductor structure and on double barrier tunnel junction (DBTJ) device.

Finally, the longitudinal electronic collective transport properties in a disordered array of $TiSi_2$ nanocrystals (with surface density of 10^{12} cm^{-2}) embedded in Si polycrystalline matrix as a function of temperature are described. The system is characterized by a high degree of disorder compared to the standard disordered nanocrystals array usually studied in the literature. Despite this fundamental difference, we demonstrate that the theoretical models used to describe the collective electronic transport in standard systems are adequate to describe the electrical behavior of such a "non-standard" system.

Contents

1 Self-Assembling of Metal Nanostructures

Nowadays, metal nanoclusters have attracted increasing attention due to their novel properties and potential applications. In particular, with the advent of the "nanotechnology revolution" [1, 2], the study of their self-organization acquired a fundamental importance [3] as a fundamental step toward the fabrication of complex nanodevices.

In this first part of the chapter, we expose some recent studies concerning the fabrication of nanostructured materials formed by metal nanoclusters, following a bottom-up approach (self-organization procedures). Traditionally, the top-down scaling scheme has been reported considering patterning by lithography. Here, a self-organization procedure is demonstrated (Section 1.1), some novel structural investigations on the metal nanoclusters are reported (Section 1.2), and, in particular, some methods to induce an atomic-level controlled self-organization of the nanoclusters are illustrated (Sections 1.3, 1.4, and 1.5). Aim of the first part of the chapter is connected with this last topic. In the second part of the chapter, the atomic-level modification and control of the structural properties will be considered to show how to modify and control the material electrical properties, which originate innovative behaviors.

1.1 Nanoclusters Formation

The experimental methods for producing nanoclusters have been reviewed by de Herr [4], Milani and Iannotta [5], and Binns [6], and such studies acquired an increasing importance in view of the fabrication of innovative nanostructured materials. In this section we present a brief summary of the production of nanoclusters on surfaces, focusing on those aspects closely related to the subject treated in this chapter.

It is well known that, generally, thin films deposited on a surface can growth in three different ways [7]: Volmer–Weber mode (growth by islands), Frank–van der Merwe mode (layer-by-layer growth), and Stranski–Krastanov mode (mixed growth by layers and islands). The occurrence of one of these growth modes is determined by the relationship between γ_{sv} (substrate-surface energy), γ_{fv} (film-surface energy), and γ_{fs} (film–substrate interface energy):

$$\gamma_{sv} < \gamma_{fs} + \gamma_{fv} \quad \text{Volmer–Weber} \tag{1}$$

$$\gamma_{sv} > \gamma_{fs} + \gamma_{fv} \quad \text{Frank–van der Merwe} \tag{2}$$

$$\gamma_{sv} \approx \gamma_{fs} + \gamma_{fv} \quad \text{Stranski–Krastanov.} \tag{3}$$

The Volmer–Weber growth mode occurs due to the "non-wetting nature" of the film on the substrate, while the Frank–van der Merwe growth occurs due to the "wetting nature" of the film on the substrate. The concepts of "wetting" and "non-wetting" are correlated with the concept of adhesion energy E_{adh}: it is defined as the energy needed, in vacuum, to separate the film/substrate interface. It is given by [8] the following equation:

$$E_{adh} = \gamma_{fv} + \gamma_{sv} - \gamma_{fs}. \tag{4}$$

Therefore, another way to specify the thermodynamic criterion for wetting is simply that if wetting occurs at equilibrium, then $E_{adh} = 2\gamma_{fv}$. If $E_{adh} < 2\gamma_{fv}$, the layer is not wetting the surface and the Volmer–Weber mode is expected to describe the mode. Furthermore, in this last case, the islands grow on the substrates as three-dimensional (3D) or two-dimensional (2D) structures on the basis of the strength of the non-wetting nature: "highly" non-wetting nature ($E_{adh} << 2\gamma_{fv}$) determines the 3D structure of the islands (high contact angle), while "slightly" non-wetting nature ($E_{adh} < 2\gamma_{fv}$) determines the 2D structure of the islands (low contact angle). The dimensionality of the growing islands is a basic feature in the determination of the particular kinetic mechanisms involved in the growth process [9].

Typically, metal films grow in the Volmer–Weber mode as 3D structures on insulator substrates and as 3D or 2D structures on semiconductor substrates [7]. As an example, Au grows on SiO_2 and on SiC in the Volmer–Weber mode as 3D structures; the adhesion energy is given by the following equations [10, 11, 12]: $E_{adh}(Au/SiO_2) = 227 \div 246\,mJ/m^2$, $E_{adh}(Au/SiC) = 445\ mJ/m^2$, and $\gamma_{fv} = \gamma_{Au} = 1500\,mJ/m^2$ (and experimentally shown by the data in Section 1.3).

Recently, several experimental methods have been developed to deposit metal nanoclusters on insulator and semiconductor substrates. These methods allow to control the features of the deposited nanoclusters (size, density, fraction of covered area, etc.) by simply controlling the process parameters [13]: plasma-enhanced chemical vapor deposition (PE-CVD), sol-gel "wet" chemical routes, layer-by-layer self-assembly, ion beam–assisted techniques, evaporation, sputtering, radio frequency (RF) magnetron sputtering, and cosputtering.

In particular, for the preparation of the nanostructured materials, Au nanoclusters/SiO_2 and Au nanoclusters/SiC, analyzed in the following, a RF sputtering process of Au on the substrates was adopted. In fact, among the various synthetic approaches, RF sputtering is one of the most feasible, thanks to its inherent versatility and the capability of obtaining a homogeneous surface coverage under controlled processing conditions and at low temperatures. Other key advantages over other preparation routes are the possibility to control the metal content and the possibility to tailor the nanocluster size and distribution by exploiting the ubiquitous competition between ablation and deposition processes occurring in glow discharges. Such a methodology gives the possibility of tailoring the Au/insulator or Au/semiconductor morphology from cluster-like to island-like systems or continuous films, with consequent modification of the system properties, by a proper choice of the total pressure, RF power, deposition time, and growth temperature. As a general rule, a careful control of Au nanoparticle size, shape, and composition might result in systems with well-tailored functional properties.

Specifically, three major categories of Au/substrate nanocomposites have been recognized as a function of morphology and overall metal content [10]:

(1) Cluster-like systems, where gold nanocrystals (*guest*), mostly single domain, are either embedded in an insulator/semiconductor *host* (inside clusters) or dispersed on its surface (outside clusters). In the latter case, gold-substrate chemical interactions cover a variety of situations, ranging from simple physisorption to the formation of chemical bonds between the shell atoms of the clusters and the matrix.
(2) Island-like systems, where Au crystallites are partially interconnected between each other and their geometry undergoes progressive variations from the typical spherical-like shape of isolated clusters to oblate or prolate particles resulting from cluster agglomeration. In this case, the average aggregate size is typically greater than in cluster-like systems, because multidomain particles are predominant.
(3) Continuous films, where particles are all interconnected to form a coating of variable thickness on the substrate surface.

The occurrence of one of these cases is determined by the deposition rate, time of deposition, chamber pressure, and substrate temperature during the deposition. Each of these cases occurs, progressively, for increasing amount of deposited Au.

For the samples analyzed in this chapter, the deposition of Au on SiO_2 and SiC substrates was performed by a K550x RF Sputter Coater using Ar plasma. In particular, it was found [14] that while performing depositions at room temperature, optimum conditions to obtain cluster-like systems are pressure of 10^{-4} Pa, emission current of 10 mA, and 30 s of deposition time for the SiO_2 substrates and 15 s for the SiC ones. The total amount of the deposited Au was evaluated by Rutherford backscattering spectrometry (RBS), and the result was as follows: $Q \approx 9 \times 10^{15}\,\text{cm}^{-2}$ on the SiO_2 substrates and $Q \approx 4.5 \times 10^{15}\,\text{cm}^{-2}$ on the SiC substrates (within a statistical error of 5%). As an example, Fig. 1 shows a plan-view

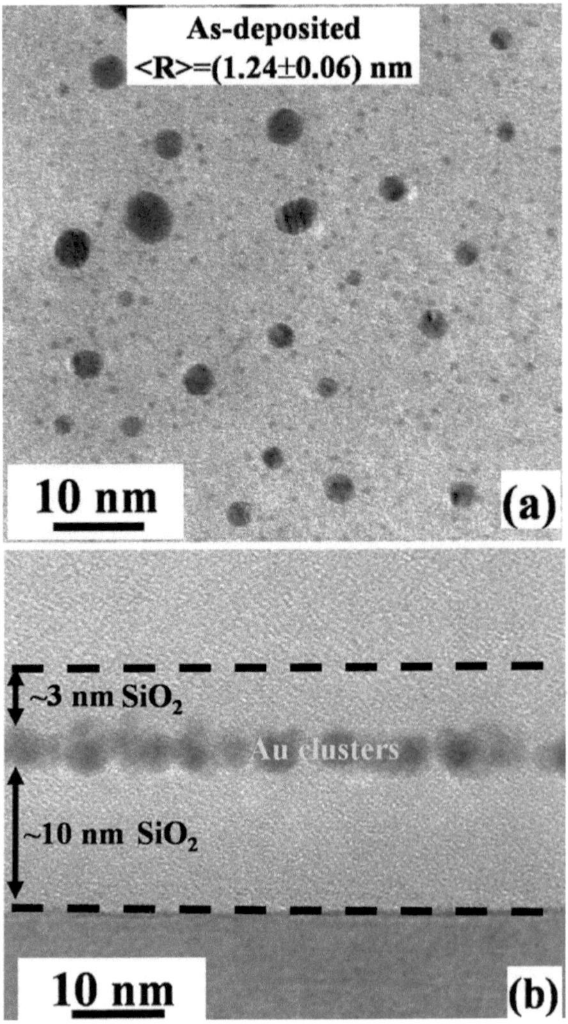

Fig. 1 (**a**) Plan-view TEM image of the as-deposited sample with the SiO_2-substrate. (**b**) Cross-view TEM of the same sample

(a) and a cross-view (b) transmission electron microscopy (TEM) image of the Au nanoclusters deposited as said on the SiO_2 substrate (the second 3-nm SiO_2 layer over the nanoclusters was subsequently deposited by sputtering in order to protect the nanoclusters during preparation for TEM); the mean nanoclusters radius was evaluated by TEM as $< R > \approx (1.24 \pm 0.06)$ nm.

The mean radius of the as-deposited Au nanoclusters on SiC was evaluated as $< R > \approx (0.73 \pm 0.02)$ nm. The evaluation of the mean radius of the Au nanoclusters is explained in Section 1.3. The difference between the Au nanoclusters deposited on SiC and SiO_2 substrates is due to the higher activation energy for the diffusion process of Au on SiO_2 with respect to Au diffusion on SiC.

Table 1 Summary of the sputtering conditions in which the Au nanoclusters were grown on SiC and SiO_2 substrates, with the relative measured amount of deposited Au and mean clusters radius of the formed nanoclusters

Substrate	Pressure during deposition (Pa)	Substrate temperature (K)	Emission current (mA)	Deposition time (s)	Amount of Au deposited (atoms/cm^2)	Mean radius of the obtained clusters (nm)
SiC	10^{-4}	300	10	15	4.5×10^{15}	0.73±0.02
SiO_2	10^{-4}	300	10	30	9×10^{15}	1.24±0.06

Table 1 summarizes the condition in which the nanoclusters were deposited for the analyzed substrates and the main results obtained for the as-deposited samples.

The obtained results confirm the strong non-wetting nature of Au on SiO_2 and SiC, allowing to consider the Au nanoclusters on SiC and SiO_2 substrates as 3D structures with a spherical shape, which will be deeply discussed in Section 1.3.

1.2 Effect of the Anisotropy on the Atomic Structure and Equilibrium Shape of Nanocrystals

In this section, we present some recent experimental data concerning the effect of surrounding environment on atomic structure and equilibrium shape of growing nanocrystals, focusing, in particular, on the case of growing Au nanocrystals on and in amorphous SiO_2 during their self-organization induced by thermal processes (whose details are exposed in the next section).

Starting from samples prepared as described in Section 1.1 (with the SiO_2 substrate), performing a systematic high-resolution TEM and extracting quantitative data on the clusters' surface free energy (Wulff plot) [15, 16], it is possible to correlate the environment spatial isotropy (Au nanocrystals in SiO_2) or anisotropy (Au nanocrystals on SiO_2) with the structural characteristics (internal atomic structure and equilibrium shape) of such small systems.

On the other hand, these experimental data acquire a more general importance in view of the comprehension of the chemical and physical nature of the interaction between nanocrystals and the surrounding environment in order to control the structural and all physical properties of complex nanostructured materials [17]. Furthermore, the obtained experimental data can improve the capacity of simulation models [18–20] of the atomic structure and equilibrium shape of nanocrystals (in order to understand the correlation between micro- and macroscopic matter [21]).

In the described experiment, to avoid any possible difference in the nanocrystals structures and shape caused by different sample preparation, we used the same procedure to prepare the samples with nanocrystals in or on amorphous SiO_2. The cluster-like Au film was deposited by sputtering on the SiO_2 substrate. In some samples, the Au layer was covered by a 3-nm-thick SiO_2 layer deposited (at room temperature) by sputtering using an AJA RF Magnetron sputtering apparatus (Ar plasma, 1.3×10^{-8} Pa pressure). The Au/SiO_2- and SiO_2/Au/SiO_2-fabricated samples were made to undergo contemporary annealing in Ar ambient at different temperatures (873 K $\div$ 1073 K) for several times (5 min $\div$ 60 min) to obtain the nanocrystals' self-organization and growth. Before the cross-sectional TEM analyses of the samples with Au on SiO_2, a thin SiO_2 cap-layer ($\sim$3 nm) was deposited (by sputtering) on the samples in order to protect the nanocrystals during the sample preparation. It is worth noting that this deposition was performed at room temperature, so that we assume that the supported Au nanocrystals do not change their atomic structure and equilibrium shape in consequence of the SiO_2 protective-layer deposition, being quite improbable that they release energy during such a process.

The nanocrystal growth on and in the SiO_2 substrate was described by a ripening process limited by diffusion [13, 22] (and Sections 1.3 and 1.4 of the present chapter). The size distributions were determined for each sample by TEM performed on significant statistical population of nanocrystals (100 for each sample). In particular, the mean radius $< R >$ increases from (1.24 ± 0.16) nm to (2.95 ± 0.09) nm for the Au nanocrystals in SiO_2 and to (3.15 ± 0.06) nm for those supported on SiO_2 (in the same range of annealing temperature and time).

Nanocrystals' internal structures and morphologies were examined using Fresnel contrast and HR-TEM (high-resolution TEM) analyses. These images show a good resemblance with the proposed model for Au nanocrystals structure viewed along (110) [23]. The spacing between the crossing (111) planes is 0.235 nm in this projection and the Au–Au distance is 0.25 nm.

The HR-TEMs indicate that, at every fixed annealing temperature and time, i.e. every thermodynamic equilibrium situation, the internal structure of each nanocrystal in the distribution is determined only by its size and not by the annealing temperature.

For the Au nanocrystals embedded in SiO_2, the HR-TEM images indicate that, in the range of radius 1 nm $< R <$ 3.5 nm the nanocrystals always exhibit single icosahedral crystal shape (close-packing of atoms structure) [24] free of internal defects, as indicated by the representative HR-TEM images in Fig. 2(a–c). This means that the equilibrium shape symmetry is conserved by increasing the nanocrystals size, without introduction of internal defects.

Differently, for the self-assembled Au nanocrystals on SiO_2, the data seem to delineate three large groups of structures, determined by the nanocrystals size:

- Group 1 (representative nanocrystals in Fig. 2d): it is formed by nanocrystals with a radius $R <$ 1.5 nm; they are single icosahedral crystals (close-packing of atoms structure) free of internal defects [24];
- Group 2 (representative nanocrystal in Fig. 2e): it is formed by nanocrystals with a radius 1.5 nm $< R <$ 2 nm; they have a twinned icosahedral structure [24] (twins of (111) planes);
- Group 3 (representative nanocrystal in Fig. 2f): it is formed by the nanocrystals with a radius 2 nm $< R <$ 3.5 nm; they have a complicate decahedral multi-twinned (and lamellar) structure [24].

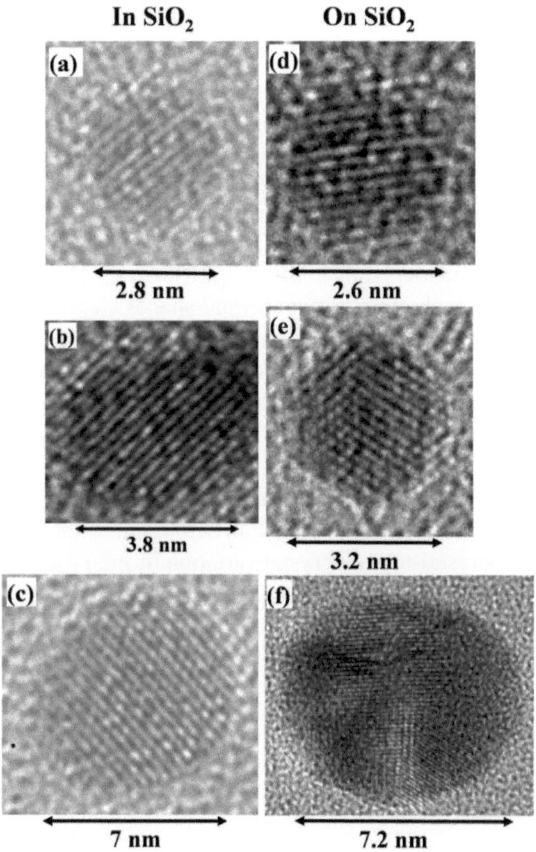

Fig. 2 (**a–c**) HR-TEM images of Au nanocrystals in SiO_2 with similar sizes chosen of the nanocrystals showed in (**d–f**) that are the Au nanocrystals grown on SiO_2 and representative of the three determined main groups

From this classification it is clear that the evolution in size of the nanocrystals on SiO_2 corresponds to an evolution of the equilibrium shape with a progressive loss of symmetry and a corresponding introduction of internal defects (twins).

The extraction of quantitative information on the surface free energy from these nanocrystals shapes was carried out via the inverse Wulff construction [15]: first, the "Wulff point" of the nanocrystal was identified and then the application of the Wulff construction in reverse order allowed to obtain the surface energy plot to within a constant scale factor. The Wulff point (center of mass) of a nanocrystal was defined as the intersection point between the perpendicular bisectors of two (111) facets. This point was taken as the center of a polar plot, with angular and radial coordinates θ and r. For each θ_i (with a 5° spacing), the distance r_i between the Wulff point and the nanocrystal surface (determined by the accurate analyses of the HR-TEM images) was measured. According to the Wulff relation $\gamma_i/r_i = \lambda = const$, the value of the surface free energy $\gamma_i \equiv \gamma(\theta_i)$ for that orientation was obtained. In this way the $\gamma(\theta)$ plot was determined. In particular, to make a direct comparison between the nanocrystals in and on SiO_2, the nanocrystal surface was distinguished in two parts. The line passing through the Wulff point and parallel to the substrate plane was drawn. For the nanocrystals on SiO_2, the nanocrystal surface above this line was defined as the surface exposed to the Ar (the gas environment in which the annealing processes were performed), while the below one as the surface exposed to SiO_2. The same procedure was adopted for the nanocrystals in SiO_2. According to this consideration, the surface energy relative to the upper surface part was named $\gamma_{\text{up}}(\theta)$ and the surface energy relative to the bottom part was named $\gamma_{\text{down}}(\theta)$, so that for the nanocrystals on SiO_2, $\gamma_{\text{up}}(\theta) \equiv \gamma_{\text{Au/Ar}}(\theta)$ and $\gamma_{\text{down}}(\theta) \equiv \gamma_{\text{Au/SiO}_2}(\theta)$, while for the nanocrystals in SiO_2, $\gamma_{\text{up}}(\theta) \equiv \gamma_{\text{Au/SiO}_2}(\theta)$ and $\gamma_{\text{down}}(\theta) \equiv \gamma_{\text{Au/SiO}_2}(\theta)$.

For each class of nanocrystal shapes (representative images reported in Fig. 2), an averaged $\gamma(\theta)$ plot was calculated where the relative energies were normalized to the surface free energy for the (111) plane [12]. In particular, for the three groups of nanocrystals grown on SiO_2 (representative images in Fig. 2d–f), the calculated averaged $\gamma(\theta)$ plot is shown in Fig. 3.

In these plots, the zero in the angle scale is taken as the direction of one (111) plane for all the nanocrystals. However, the r_i values for each θ_i are averaged over 10 different nanocrystals belonging to the same group (i.e., similar size). The nanocrystal-to-nanocrystal variation in $\gamma(\theta)$ for a given orientation was indicated as the error bars reported in Fig. 3.

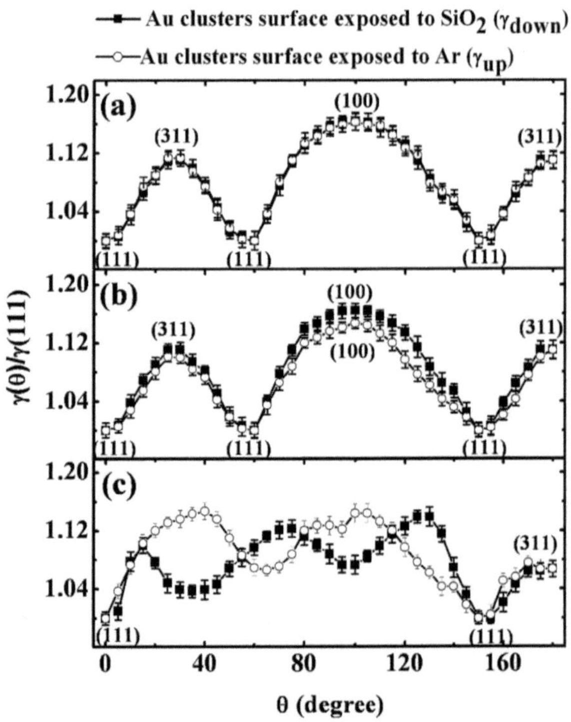

Fig. 3 Wulff plot ($\gamma_{\text{up}}(\theta) = \gamma_{\text{Au/Ar}}(\theta)$) and $\gamma_{\text{down}}(\theta) = \gamma_{\text{Au/SiO}_2}$) for the three groups of nanocrystals on SiO_2

In the case of Au nanocrystals in SiO_2, the peaks and valleys in the $\gamma_{up}(\theta)$ and $\gamma_{down}(\theta)$ lie in identical angular position (consequence of nanocrystal angular symmetry) and assume identical values (the $\gamma(\theta)$ are completely analogous to those showed in Fig. 3a). The latter point means that the facets with the same Miller indexes in the upper and lower parts of the nanocrystal, respectively, have identical radial coordinates (radial symmetry). Both the angular and radial symmetries are conserved by increasing the cluster size in the cases of nanocrystals in SiO_2. This fact is reflected in the symmetrical single crystal icosahedral structure for the equilibrium shape of nanocrystals in SiO_2. These nanocrystals grow always subjected only to the thermodynamic process determined by the Au/SiO_2 interface in any direction, i.e., in an "isotropic spatial situation."

Now, if the case of Au clusters on SiO_2 is considered, it is clear, from Fig. 3(a), that for the smaller sizes (group 1), a situation similar to the case of nanocrystals in SiO_2, exists: $\gamma_{up}(\theta)$ and $\gamma_{down}(\theta)$ lying in identical angular position and assuming identical values ($\gamma_{up}(\theta) \equiv \gamma_{Au/Ar}(\theta) \approx \gamma_{down}(\theta) \equiv \gamma_{Au/SiO_2}$): both angular and radial symmetries are present. However, differently from the nanocrystals in SiO_2, for cluster sizes in the group 2, the angular symmetry is still conserved, but the radial one begins to be lost ($\gamma_{up}(\theta) \equiv \gamma_{Au/Ar}(\theta) \neq \gamma_{down}(\theta) \equiv \gamma_{Au/SiO_2}$). In fact, the nanocrystals growing on SiO_2 are subjected to two thermodynamic competitive processes, due to the different surface energies of the Au/Ar interface ($\gamma_{up}(\theta) = \gamma_{Au/Ar}(\theta)$) and of the Au/$SiO_2$ interface ($\gamma_{down}(\theta) = \gamma_{Au/SiO_2}$). This is reflected in a progressive loss of symmetry in the nanocrystals' equilibrium shape with increasing the cluster size beyond a "critical size." In particular, both the angular and radial symmetries are lost for the nanocrystals of group 3 (Fig. 3c), where any correspondence between $\gamma_{up}(\theta)$ and $\gamma_{down}(\theta)$ is not present. Furthermore, the different surface energies at Au/Air and Au/SiO_2 interfaces determine an internal strain accumulation during the nanocrystals growth that is released in the formation of internal defects for larger sizes.

The exposed experimental data allow to conclude that the main effect of the surrounding environment on atomic structure and equilibrium shape of growing nanocrystals is related to its spatial "isotropy" or "anisotropy": nanocrystals surrounded by an "isotropic" environment exhibit an angular and radial symmetry $\gamma(\theta)$, determining a symmetrical equilibrium shape, while nanocrystals in a "non-isotropic" environment exhibit a $\gamma(\theta)$ that lost its angular and radial symmetry for sufficiently large sizes, determining a loss of symmetry in the equilibrium shape of the nanocrystal and the formation of internal defects.

1.3 Growth of Au Nanoclusters on Surfaces

In this section, we expose some recent data concerning the self-organization and growth of Au nanoclusters on hexagonal SiC and SiO_2 surfaces induced by annealing processes, in the 873 K $\div$ 1073 K temperature range, focusing our attention on the fundamental physics phenomena at the basis of the processes.

Starting from samples prepared following the methodology exposed in the Section 1.1, it is possible to reach a predetermined cluster distribution on several surface, such as SiC or SiO_2, controlling their growth during subsequent thermal processes. Annealing processes performed in Ar ambient, in the 873 K ÷ 1073 K temperature range for times ranging from 300 ÷3600 s, have been reported, allowing to determine the evolution of clusters sizes and their distribution, of surface–surface clusters distances and their distribution, of the clusters surface density, and of the fraction of area covered by the clusters. As an example, some cross-view TEM images of the Au clusters deposited on SiO_2 and annealed at a fixed temperature (973 K) for some increasing annealing times (0 s, 2400 s, 3600 s) are reported in Fig. 4 (the second 3-nm SiO_2 layer over the nanoclusters was subsequently deposited by sputtering in order to protect the nanoclusters during preparation for TEM). RBS analyses allowed to demonstrate that no Au loss occurs during thermal treatments (out diffusion, evaporation, reaction with C and/or Si) [14], i.e., the amount of Au remains the same of those measured in the as-deposited samples for all the annealing processes performed.

Microscopic analyses, such as TEM, scanning electron microscopy (SEM), and AFM, indicated that the clusters are spherical (plan and cross-sectional views are circular), feature obviously deriving from the strong non-wetting nature [13, 25] of Au both on SiC and SiO_2 and from the limited cluster size (1 ÷ 10 nm), and, in particular, that their mean radius, mean surface–surface distance, and fraction of area covered, for each fixed temperature, change as a function of annealing time as indicated in Fig. 5, for the case of SiC (similar data were found in the case of SiO_2) [17].

These data allowed to determine the mechanism at the basis of Au nanoclusters growth, for both SiC and SiO_2 surfaces, in a surface diffusion-limited ripening of 3D structures, with different diffusion coefficient for SiC and SiO_2.

At any stage during a general ripening process of small particles, there is a so-called critical particle radius R^* being in equilibrium with the mean matrix composition; particles with $R > R^*$ will grow and particles with $R < R^*$ will shrink [26]. In particular, the theoretical study of Allmang–Feldman–Grabov [9] and Shorlin–Krylov–Allmang [27] based on the Lifshitz–Slyozov–Wagner [28, 29] ideas led to the following relation between R and the time t (for sufficiently long time, i.e., in stationary state):

$$R^n(t) - R_0^n = K^* t \tag{5}$$

with R_0 the radius of the particle at time $t = 0$, and K^* an appropriate constant depending on the diffusion coefficient. The scaling exponent n is the parameter that takes account of the particular physical case in which the ripening process takes place, with $n = 2$ (2D/2D case) for the growth of 2D particles on a surface (2D), $n = 3$ (3D/3D case) for the growth of 3D particles embedded in a bulk matrix (3D), and $n = 4$ (3D/2D case) for the growth of 3D particles on a surface (2D).

In order to determine the most adequate mechanism for experimental cases examined, in Fig. 6 (a) the experimental $< R >^4 - < R_0 >^4$ are compared, for example

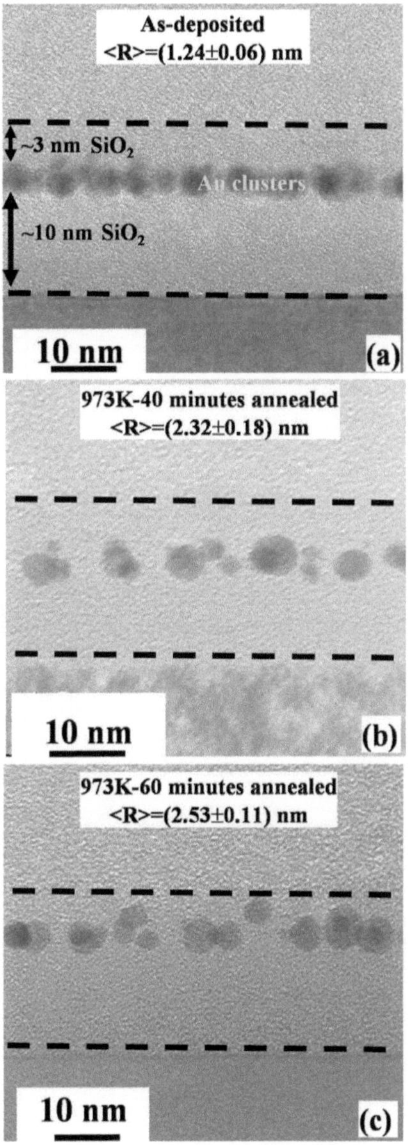

Fig. 4 Cross-view TEM images of Au nanoclusters deposited on SiO_2 and subjected to thermal treatments at 973 K for different time. It is evident the change of mean cluster size as a function of annealing temperature. The 3-nm-thick SiO_2 layer over the Au film was deposited by sputtering to protect it during the samples preparation for the TEM analyses

in the samples with the SiC substrate and annealed at 1073 K, with the best fit calculated for $n = 4$, $n = 3$, and $n = 2$. It is evident that the best agreement is reached for $n = 4$ (as expected). Similar results [14] were obtained for the samples with the SiO_2.

Therefore, in the assessed growth the cluster mean radius increases with time as follows:

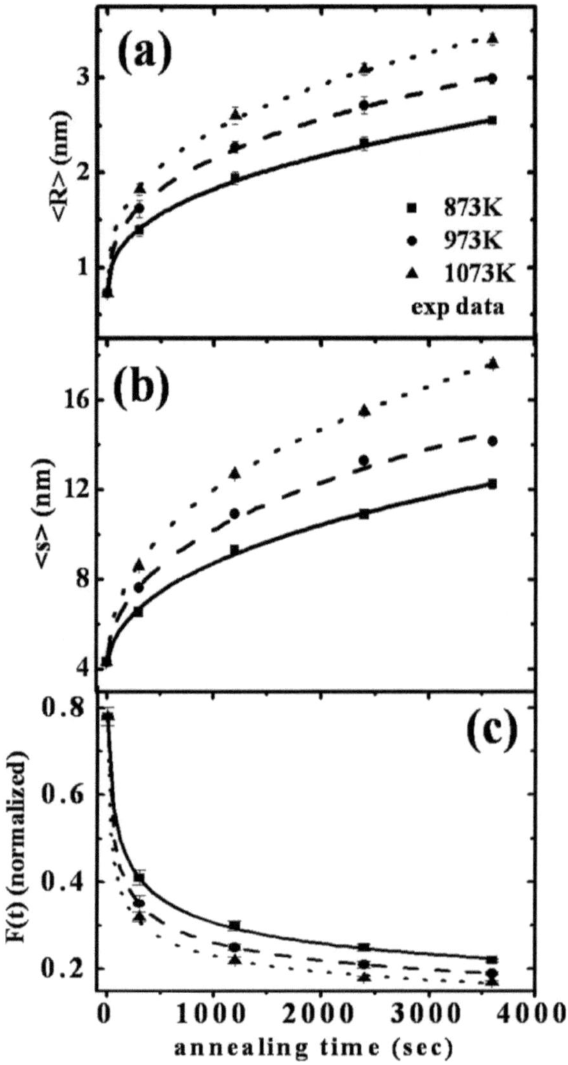

Fig. 5 Au on SiC substrate: for each fixed temperature (**a**) experimental (*dots*) mean clusters radius $< R >$ as a function of annealing time; (**b**) experimental (*dots*) mean center-to-center distance between nearest clusters $< s >$ as a function of annealing time; (**c**) experimental (*dots*) fraction F (normalized) of surface area covered by clusters as a function of annealing time. All the curves indicate the theoretical simulations according to the model exposed in the text

$$< R >^4 - < R_0 >^4 = K^* t \tag{6}$$

with K^* defined by [30]

$$K^* = \frac{8 N_0 D_s \gamma \Omega^2}{45 k_B T \ln(L)} \tag{7}$$

where D_s is the surface diffusion coefficient of the Au atoms on the surface, $L \approx 3$ a characteristic length (in units of R), N_0 the density of nucleation sites,

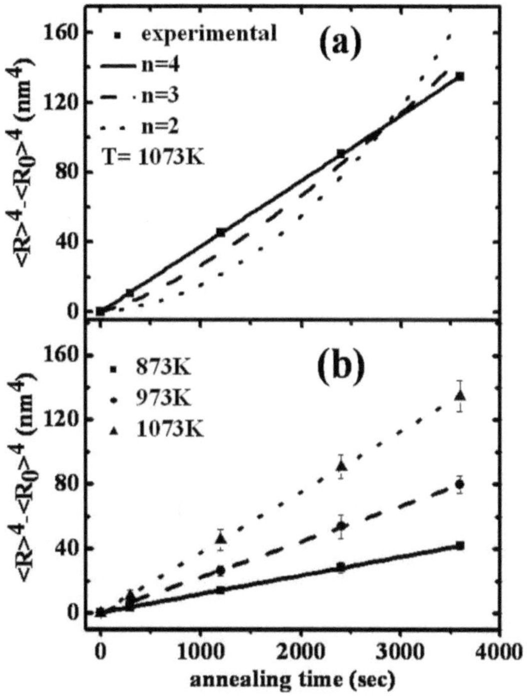

Fig. 6 (**a**) Comparison between the theoretical (*continuous*, *dashed*, and *dotted curves*) mean clusters radius $< R >^4 - < R_0 >^4$, at $T = 1073$ K, plotted, as a function of annealing time, for the cases $n = 4$, $n = 3$, $n = 2$, and the experimental data (*square dots*), for the case of Au nanoclusters on SiC substrate. (**b**) For the same substrate, experimental $< R >^4 - < R_0 >^4$ (*black dots*) as a function of annealing time for each fixed temperature and the relative linear fits (*continuous lines*), being K^* the fits parameter

Ω the Au atomic volume, γ the Au/air interface energy (for gold [31, 32, 12], $N_0 = (a_{Au}^4/8\Omega^2)\,cm^{-2}$, where $a_{Au} = 0.409$ nm is the Au lattice constant at room temperature, $\Omega = 1.69 \times 10^{-29}\,m^3$, and $\gamma = 1.5\,J/m^2$ for Au/Air). The fit of experimental data by Equation (6), as showed in Fig. 6(b), for example in the case of SiC, allowed to determine $K^*(T)$. Moreover, inversion of Equation (7) allowed to determine the surface diffusion coefficient of Au on SiC and SiO_2. The Arrhenius plots of the resulting $D_s(T)$, showed in Fig. 7 for both cases of SiC and SiO_2 substrates, indicate the occurrence of the thermally activated diffusion process [33] described by $D_s(T) = D_0 e^{-E_a/k_B T}$ with $D_0 = (2.59 \times 10^{-12} \pm 1.6 \times 10^{-13})\,cm^2/s$ and $E_a = (0.55 \pm 0.01)\,eV/atom$ for Au on the SiC surface, and $D_0 = (7.5 \times 10^{-11} \pm 1.1 \times 10^{-12})\,cm^2/s$ and $E_a = (0.90 \pm 0.01)\,eV/atom$ for Au on the SiO_2 surface.

Furthermore, the simulation of temporal variation of the clusters' surface–surface distance and fraction of covered area as imposed by the mechanism described by Equations (6) and (7), and indicated, in Fig. 5, by the lines, is in good agreement with the observed experimental behavior. These theoretical simulations are obtained considering that according to Equation (6) the average cluster section is $< S >^2 - < S_0 >^2 = \pi^2 K^* t$ (with $S_0 = \pi < R_0 >^2$) and therefore the average number of atoms per cluster is $< n > = < V > /\Omega = \left[4\pi \left(K^*t + < R_0 >^4\right)\right] / \left(3\Omega\sqrt[4]{K^*t + < R_0 >^4}\right)$ and therefore the clusters surface density is $N_s(t) = Q\Omega / < V > = [(3Q\Omega)/4\pi]$

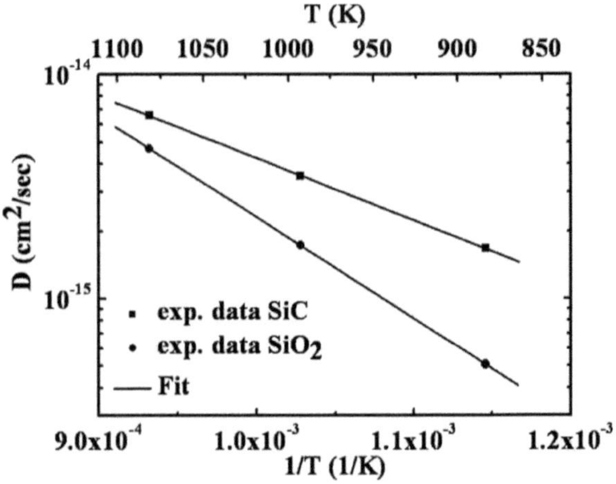

Fig. 7 Experimental value of $D_s(T)$ (*square dots* for the SiC substrate, *triangular dot* for the SiO_2 substrate) as a function of T^{-1} with the relative Arrhenius fit (*continuous line*). The error bar for each point is reported, but in this scale it is so small that it is not visible

$\left\lfloor\left(\sqrt[4]{K^*t+<R_0>^4}\right)/\left(K^*t+<R_0>^4\right)\right\rfloor$. As a consequence, the fraction of the total surface area covered by the clusters is [14]:

$$F(t)=<S(t)>N_s(t)=\left(\frac{3Q\Omega}{4\pi}\frac{\sqrt[4]{K^*t+<R_0>^4}}{K^*t+<R_0>^4}\right)\left(\sqrt{\pi^2 K^*t+<S_0>^2}\right). \tag{8}$$

Finally, the average center-to-center $<s(t)>$ distance between nearest clusters can be estimated supposing that the clusters are placed in an ordered squared matrix of side l, $\sqrt{N_s(t)}$ is the number of clusters per side, and $l\left(1-2\sqrt{N_s(t)}<R>\right)$ the linear space occupied by clusters per side. So [14]:

$$\begin{aligned} s(t) &= \frac{l(1-2\sqrt{N(t)}<R>)}{l\sqrt{N(t)}}+2<R> \\ &= \frac{1-2\sqrt{\frac{3Q\Omega}{4\pi}\frac{\sqrt[4]{K^*t+<R_0>^4}}{K^*t+<R_0>^4}}\sqrt[4]{K^*t+<R_0>^4}}{\sqrt{\frac{3Q\Omega}{4\pi}\frac{\sqrt[4]{K^*t+<R_0>^4}}{K^*t+<R_0>^4}}}+ \\ &\quad +2\sqrt[4]{K^*t+<R_0>^4} \end{aligned} \tag{9}$$

with the boundary condition $s(t=0)=<s_0>$.

1.4 Growth of Au Nanoclusters in SiO_2

The systematic study of the growth kinetics of Au nanoclusters in SiO_2 was performed by A. Miotello et al. [34] and G. De Marchi et al. [22]. In such works, the authors fabricated Au nanoclusters in fused SiO_2 matrix by implantation. They implanted Au^+ at 200 keV energy and at a fluence of $4 \times 10^{16}\,cm^{-2}$. Because such a concentration exceeds the Au solubility threshold in SiO_2 ($10^{16}\,cm^{-2}$), the Au implantation results in a first stage of precipitation process with the formation of small Au spherical (3D) nanoclusters. Then, the authors performed sample annealing in air (oxidizing atmosphere) and other different atmosphere (Ar, H_2). By structural (TEM) and optical (absorption spectra) measurements, the cluster size evolution as function of annealing temperature, time, and atmosphere was studied.

As an example, Fig. 8, shows some TEM images and related clusters size distributions, for sample annealed at same temperature (1173 K) for different times (in air).

The experimental data, i.e., the mean cluster radius $< R >$ as a function of the annealing time, suggests that the kinetics of cluster growth changes with annealing times. To expose this point, Fig. 9(a) reports, as a function of annealing time,

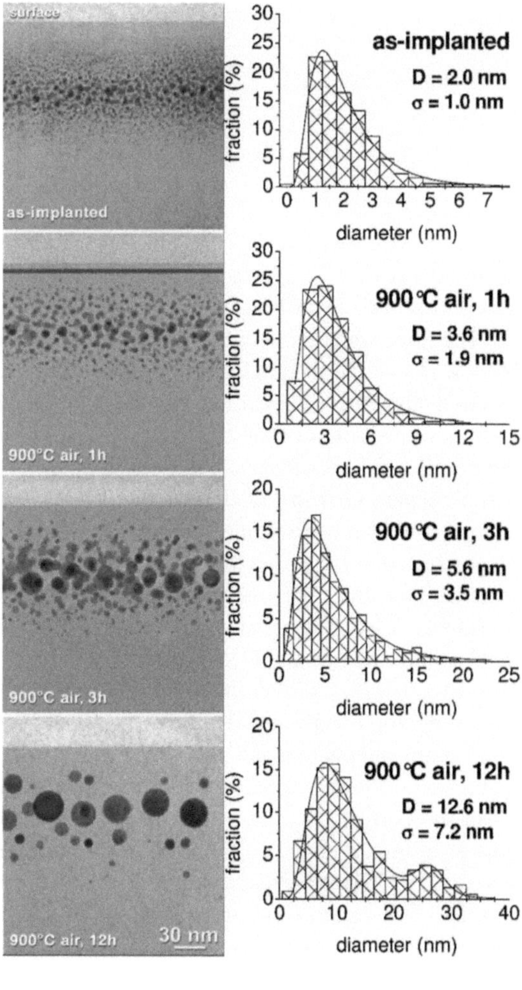

Fig. 8 Cross-sectional TEM results on Au-implanted silica sample (**a**) and annealed in air at 1173 K for 1 (**b**), 3 (**c**), and 12 (**d**) hours; (**e, f, g, h**) relative clusters size distributions derived from the analyses of the previous TEM images. (Reprinted with permission from De Marchi [22]. © (2002) American Institute of Physics)

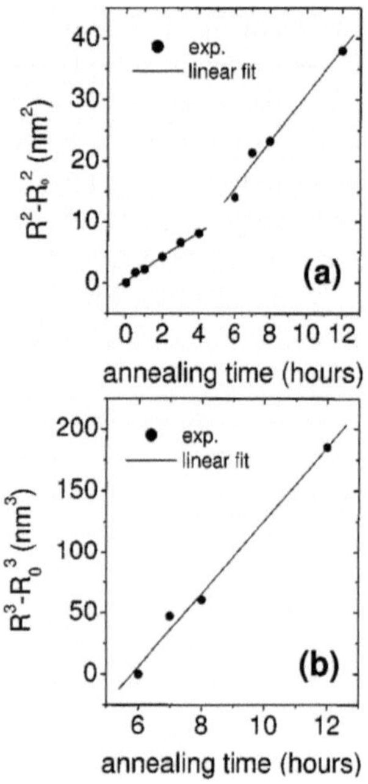

Fig. 9 (**a**) $< R >^2 - < R_0 >^2$ and (**b**) $< R >^3 - < R_0 >^3$ evolution in Au-implanted silica samples annealed in air at 1173 K for different time intervals. Solid lines are the linear fits to the experimental data (*filled circles*). (Reprinted with permission from De Marchi [22]. © (2002) American Institute of Physics)

the square of the average radius of the growing particles: there is a linear relation between the two quantities with a change of the slope in the range between 4 and 5 h. The linear relation between $< R >^2$ and t, of type $< R >^2 - < R_0 >^2 = K^{**}t$ (with $< R_0 >$ the mean cluster radius at $t = 0$ and K^{**} a constant depending on the Au diffusion coefficient in SiO_2, D), is indicative of a cluster growth due to the precipitation process of the supersaturated solution. The change of the slope in the plot ($< R >^2 - < R_0 >^2$)) vs t (evident in Fig. 9(a)) suggests the modification of the clusters growth kinetics. Without considering a possible change in D, the discontinuity can be explained by considering that the ripening regime occurs, after the precipitation stage due to supersaturation, when clustering is allowed over long times. In fact, the experimental and calculated values, corresponding to annealing times of 6, 7, 8, 12 h, agree well with a $t^{1/3}$ law of growth (Fig. 9b), as expected in the Ostwald ripening of 3D structures in a matrix.

In Fig. 10, it is shown the Arrhenius plot $\ln\left(< R >^2\right)$ as a function of $1/kT$ comparing the experimental data of the post-implanted samples annealed in different atmospheres (air and Ar). Also, from such a figure it is clear the existence of the two different stages in the cluster growth in SiO_2 when the annealing processes are performed in air (in the analyzed temperature range). Furthermore, this graphic allows to evaluate the activation energy for the diffusion process. Below 973 K ÷ 1073 K, the cluster radius increases very slowly with the annealing temperature, at constant time, independently from the atmosphere composition. This very small temperature dependence suggests a diffusion mechanism controlled by

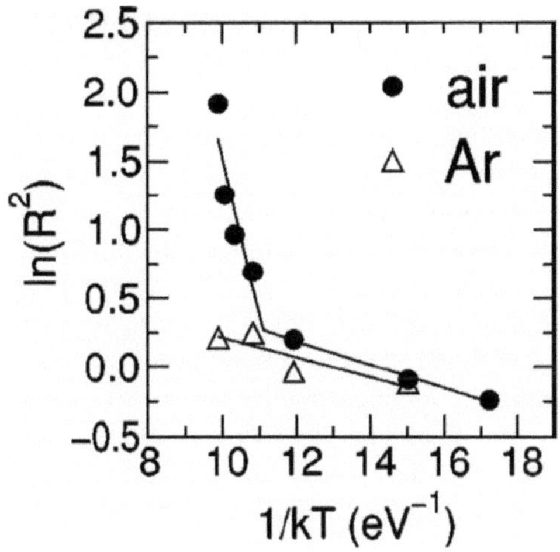

Fig. 10 Arrhenius plot of the squared average clusters radius $< R >^2$ after 1-h annealing in air (*filled circles*) or argon (*empty triangles*). Solid lines are linear fits to the experimental data. (Reprinted figures with permission from Miotello [34] © (2001) American Physical Society)

radiation damage. The measured activation energy of 1.17 eV/atom for gold diffusion in SiO_2, during annealing in air, in the temperature range 1023 K $\div$ 1173 K, is very different from the literature data [35] of 2.14 eV/atom. This is consistent with the fact that the activation energy for the molecular oxygen diffusion mechanism through an interstitial mechanism is in the range from [36] 1.1 to 1.3 eV/atom. So, it is reasonable to suppose that the heating in air favours, through oxygen migration coming from the ambient atmosphere, the annealing of radiation-induced defects in SiO_2, mainly connected to the Si–O bond breaking with oxygen atoms displaced from their equilibrium lattice position.

1.5 Growth of Nanoclusters During ion Irradiation

Ion beam–based techniques are a powerful tool to synthesize nanostructured materials and to modify their structural properties [37–39].

In this section, a methodology to induce the self-organization of metal nanoclusters deposited on a surface is described and some recent data are presented. In particular, the original topic concerns the study of the morphological modifications of Au nanoclusters on SiO_2 surface occurring during the irradiation by an Ar beam. As already seen for the case of the Au nanocluster self-organization induced by thermal processes on surfaces (in particular SiO_2), the knowledge of the related kinetics mechanism is fundamental toward an atomic-level control of the structural properties.

Samples were prepared on SiO_2 as described in Section 1.1 and annealed to reach an Au nanocluster mean radius of (3.15 ± 0.11) nm. These samples were irradiated at room temperature using 200 keV-Ar^+ beam at several fluences in the $5 \times 10^{13} \div 10^{16}\,cm^{-2}$ range. The beam current was kept constant at $I = 1.4\,\mu A$ for fluences of $5 \times 10^{13} cm^{-2}$, $10^{14} cm^{-2}$, and $5 \times 10^{14} cm^{-2}$, at $I = 2.5\,\mu A$ for the samples irradiated with fluences of $10^{15} cm^{-2}$, $5 \times 10^{15} cm^{-2}$, and $10^{16} cm^{-2}$.

To follow the morphology evolution, we applied several microscopic methods. In particular, the merging of the AFM and TEM analyses allows to determine the clusters size distribution and the variation of the mean clusters size as a function of the beam fluence. As an example, in Fig. 11 the cross-sectional TEM analyses are reported for the samples: (a) un-irradiated, (b) irradiated with 1×10^{14} cm^{-2} Ar (at beam current $I = 1.4\,\mu A$), (c) irradiated with $5 \times 10^{15} cm^{-2}$ Ar, and (d) irradiated with $10^{16} cm^{-2}$ Ar (at beam current $I = 2.5\,\mu A$), respectively. In the same figure, the respective cluster size distributions derived are drawn (for each sample a statistical population of 100 clusters was considered).

The TEM images of Fig. 11 clearly indicate a decrease of the clusters density after irradiation due to the sputtering of Au atoms by the incoming ions. This was confirmed by the RBS analyses that allowed to quantify the efficiency of this process

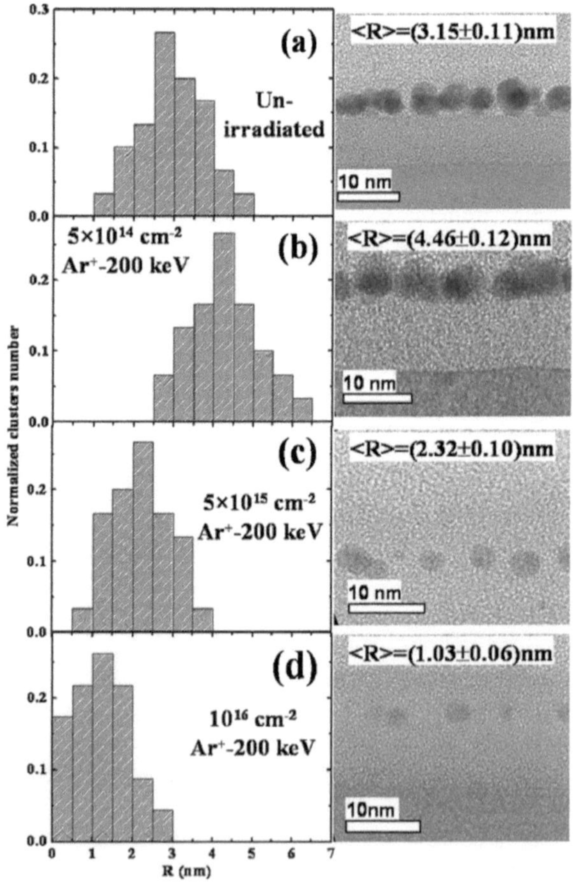

Fig. 11 Cross-TEM analyses for the samples with Au nanoclusters on SiO_2 substrate: (**a**) un-irradiated, (**b**) irradiated (room temperature) with $5 \times 10^{14} cm^{-2}$ Ar^+-200 keV ($I = 1.4\,\mu A$), (**c**) irradiated (room temperature) with $5 \times 10^{15} cm^{-2}$ Ar^+-200 keV ($I = 2.5\,\mu A$), and (**d**) irradiated (room temperature) with $10^{16} cm^{-2}$ Ar^+-200 keV ($I = 2.5\,\mu A$). To each TEM images is associated the relative clusters size distribution

[40]: a continuous decrease of Au atoms surface density down to a minimum of $4 \times 10^{13}\,\text{cm}^{-2}$ with the increasing fluence was observed (the statistical error on the RBS measurements is below 5%), while the change of the current has only a slight effect on the sputtering yield. Because the deposited Au film thickness is less than the projected range of ions in Au (~59 nm, from TRIM calculation), a considerable fraction of ions interacts and penetrates the interface between Au and SiO_2, giving rise to some direct recoil implantation of Au in the substrate. In this regime, the available Au atoms for sputtering will be comparatively less due to preferential sputtering and so the sputtering yields of Au [41]. When the sputtering is carried out at fluences greater than the fluence needed to completely remove the Au film, only residual Au atoms, remaining out of recoiled implantation, will contribute to the yield. This consideration explains the continue decrease of the residual Au density on the SiO_2 surface when the ion fluence is increased.

In Fig. 12(a, b) the experimental mean cluster radius $< R >$ is reported as a function of the Ar fluence (dots), for the two beam currents, respectively ($I = 1.4\,\mu\text{A}$ and $I = 2.5\,\mu\text{A}$). The beam current has a relevant role on the cluster average size evolution: in fact, a conventional clusters growth driven by enhanced mobility during ion bombardment occurs at low current ($I = 1.4\,\mu\text{A}$) for any fluence below $5 \times 10^{14}\,\text{cm}^{-2}$; irradiation with a beam current of $I = 2.5\,\mu\text{A}$ produces, instead, a reduction of the cluster size for any of the used fluences above $10^{15}\,\text{cm}^{-2}$.

These experimental data can be modeled in terms of ripening of nanoclusters in the far from equilibrium condition generated by ion beam irradiation taking into account the loss of Au atoms due to the sputtering process during ion irradiation (as determined by RBS analyses).

It is known [26] that, in a solid at equilibrium, the growth rate of individual clusters can be predicted solving a continuity equation using the Gibbs–Thompson relation $C_R = C_e e^{-(R_C/R)} \approx C_e\,[1 + (R_C/R)]$, where C_e is, in the present case, the Au concentration in the gas phase in equilibrium with a solid Au film, C_R is the equilibrium Au concentration at the surface of a spherical Au cluster with radius R, R_C is a parameter that is usually referred to as the capillary length and at equilibrium $R_C = 2\gamma\Omega/k_B T > 0$, where γ the surface energy Au/air, Ω is the Au atomic volume, k_B is the Boltzmann constant, and T is the absolute temperature. The difference between C_R and C_e induces a diffusive flux of atoms from the smaller to the larger particles. Thus, the average particle radius increases and the total number of particles decreases with time. As already said in Section 1.3, in a conservative system in which 3D cluster growth is limited by surface diffusion, the mean cluster radius increases according to $< R >^4 - < R_0 >^4 = K^* t$, with K^* a constant depending on D_s and R_C, where D_s is the atomic surface diffusion coefficient and $< R_0 >$ the radius of the particle at time $t = 0$. Such a process was demonstrated, in Section 1.3, to regulate the self-organization of Au nanoclusters on SiO_2 induced by annealing processes.

During the ion beam irradiation, the ripening process takes place in far-from-equilibrium condition. This process can be modeled for a conservative system, in steady-state condition, as a ripening process characterized by Gibbs–Thompson relation similar to the standard relation but with parameters that take into account

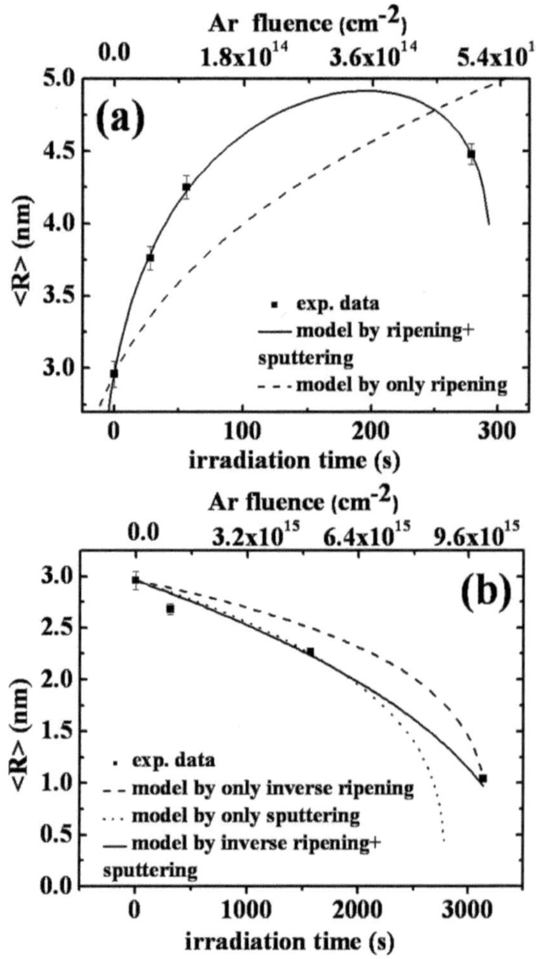

Fig. 12 (**a**) Experimental mean clusters size (*dots*) as a function of irradiation time (fluence) for the fixed ion flux $1.8 \times 10^{12} \mathrm{cm}^{-2}\mathrm{s}^{-1}$ (1.4 μA of beam current) and fit by ripening+sputtering model as indicated by Equation (10) (*continues curve*) and by only ripening theory (*dashed curve*). (**b**) Experimental mean clusters size (*dots*) as a function of irradiation time for the fixed ion flux $3.2 \times 10^{12} \mathrm{cm}^{-2}\mathrm{s}^{-1}$ (2.5 μA of beam current) and fit by inverse ripening+sputtering model as indicated by Equation (11) (*continues curve*) and by only inverse ripening model (*dashed curve*) and by only sputtering model (*dotted curve*)

the ions irradiation effects [42]: $C^I(R) = C_e^I\left[1 + \left(R_C^I/R\right)\right]$, $C_e^I = C_e(1+\Delta)$, $R_C^I = \left[R_C - (5\lambda\Delta/4)\right]/(1+\Delta)$, $\Delta = \left(q\Phi\lambda^2/D_s^I C_e\right)$, where Φ is the ion flux, $q\,(\mathrm{nm}^{-1})$ the mean mixing rate ($q\Phi$ is the damage rate), λ (nm) the mean displacement distance, and D_s^I the atomic surface diffusion under ion irradiation. Therefore, under ion beam irradiation the capillary length R_C^I can be positive or negative according to the boundary conditions (temperature and ion flux). These two different possibilities lead two different behaviors in the particles' growth: at $R_C^I > 0$ the growth process is a conventional ripening process in which the particles with a radius smaller than a critical radius dissolve themselves to increase the particles size with a radius larger than the critical one; at $R_C^I < 0$ the process is the opposite. In particular, considering as variable only the ion flux (because the substrate temperature is fixed in these experiments), it is clear that a critical flux $\Phi^* = \left(4R_C D_s^I C_e/5q\lambda^3\right)$ exists: at $\Phi < \Phi^*$, it is $(< R >) > (< R_0 >)$ $< R >^4 - < R_0 >^4 = K_1^* t$, with $K_1^* > 0$ depending also on Φ, and $< R >$ decreases when Φ increases ($\Phi \rightarrow \Phi^*$); for Φ approaching Φ^* the mean radius $< R >$ tends to $< R_0 >$, until it reaches the condition $< R >=< R_0 >$ at $\Phi = \Phi^*$. At $\Phi > \Phi^*$ it is $(< R >) < (< R_0 >)$ $< R >^4 - < R_0 >^4 = -K_2^* t$, $K_2^* > 0$ depending also on Φ, and $< R >$ decreases when Φ increases. In the present case, irradiation processes fixing two different beam currents, $I = 1.4\,\mu A$ and $I = 2.5\,\mu A$ were performed, corresponding, respectively, to the two different ion fluxes $\Phi_1 = 1.8 \times 10^{12}\,cm^{-2}s^{-1}$ and $\Phi_2 = 3.2 \times 10^{12}\,cm^{-2}s^{-1}$. It is not possible to predict Φ^* because no reliable value of D_s^I under irradiation and of C_e exist for Au on SiO_2. However, based on the present experimental data, it is possible to argue it that $\Phi_1 = 1.8 \times 10^{12}\,cm^{-2}s^{-1} < \Phi^*$ and $\Phi^2 = 3.2 \times 10^{12}\,cm^{-2}\,s^{-1} > \Phi^*$ so that must be $2.9 \times 10^5 cm^{-1}\,s^{-1} < D_s^I C_e < 5.1 \times 10^5 cm^{-1}\,s^{-1}$.

Furthermore, we need to consider that the ion beam irradiation of the clusters produces a decrease of the cluster mean radius, as a consequence of a sputtering process, that is a competing process with ripening and inverse ripening. To describe this process, we can consider that the number of sputtered atoms is proportional to the time of sputtering: $\Delta n \propto t$; so supposing that at $t = 0$ the mean cluster radius is $< R_0 >$ and the average number of atoms forming the cluster is $< n_0 >= \left[(4/3)\pi R_0^3/\Omega\right]$ and after a time t of irradiation it reaches a mean radius $< R_t >$ with a number of atoms $< n_t >= \left[(4/3)\pi R^3/\Omega\right]$, then the variation $\Delta(< R >^3)$ is $\Delta(< R >^3) =< R_0 >^3 - < R_t >^3 \propto t$. Hence, to model the clusters growth process, we used the following equations:

$$< R(t) >= \left(< R_0 >^4 + K_1^* t\right)^{\frac{1}{4}} - \left[< R_0 > - \left(< R_0 >^3 - a_1 t\right)^{\frac{1}{3}}\right] \quad (10)$$

at $\Phi < \Phi^*$

$$< R(t) >= \left(< R_0 >^4 - K_2^* t\right)^{\frac{1}{4}} - \left[< R_0 > - \left(< R_0 >^3 - a_2 t\right)^{\frac{1}{3}}\right] \quad (11)$$

at $\Phi < \Phi^*$

In Equation (10), the first term describes the mean cluster radius increasing due to ripening process and the second term the mean cluster decreasing due to the competing sputtering process. In Equation (11), the first term describes the mean cluster radius decreasing due to inverse ripening process and the second term the mean cluster decreasing due to the added sputtering process. The constants K_1^* and K_2^* depend on T (fixed in this experiments), γ, Ω, and R_C^I (and therefore, respectively, on Φ_1, Φ_2 and D_{s1}^I, D_{s2}^I). The constants a_1 and a_2 depend on Φ_1 and Φ_2, respectively (but also from ion energy, which is fixed in this experiments). In Fig. 12(a) and (b), the experimental mean cluster radius (dots), respectively, for the fixed ion flux $\Phi_1 = 1.8 \times 10^{12}\,cm^{-2}\,s^{-1}$ and $\Phi_2 = 3.2 \times 10^{12}\,cm^{-2}\,s^{-1}$, as a function of the irradiation time (i.e., Ar fluence) are reported, and the fit of the experimental data was performed using Equations (10) and (11), respectively, with K_1^*, a_1 and K_2^*, a_2 as the fit parameters. The results are showed as the continue lines with

$K_1^* = (5.5\pm0.2)\,\mathrm{nm}^4/\mathrm{s}$, $a_1 = (0.090\pm0.001)\,\mathrm{nm}^3/\mathrm{s}$, $K_2^* = (0.010\pm0.001)\,\mathrm{nm}^4/\mathrm{s}$, and $a_2 = (0.010 \pm 0.005)\,\mathrm{nm}^3/\mathrm{s}$. As a comparison, the prediction considering the pure ripening or inverse ripening process (dashed lines) and the sputtering process (dotted line) is reported in the same figure. According to this interpretation, it is interesting to note, for example, that in Fig. 12(a) until about 180 s the mean cluster radius increases because the ripening process is dominant on the sputtering process while for higher irradiation time the sputtering process is dominant on the ripening process so that the mean cluster size decrease until the total amount of Au disappears for irradiation time higher than 300 s.

So, the data presented in Section 1.3 and those presented in this section show that thermal processes and ion beam irradiation processes are powerful methods to modify and control the structural properties of Au nanoclusters deposited on SiO_2. On the other hand, the data exposed in Section 1.4 show that thermal processes are a powerful tool to modify and control the structural properties of Au nanoclusters embedded in SiO_2, too.

As a difference, ion beam irradiation allows to modify nanoclusters embedded in $SiO_2$with some peculiarities [42–44]. It is worth to compare the data concerning the self-organization of Au nanoclusters induced by ion beam irradiation with those concerning the self-organization of Au nanoclusters induced by ion beam irradiation when they are embedded in SiO_2.

Typically, the Au nanoclusters are formed in SiO_2 by ion implantation (Section 1.4) or chemical synthesis and irradiated with ion beams of opportune energy and fluence to promote their self-organization. Recently [43], G. Rizza et al. showed as the irradiation by energetic Au ions (~MeV) of large Au nanoclusters embedded in SiO_2 led to the nanoclusters dissolution by the loss of Au atoms, consequence of the ion collisions, and such atoms, remaining trapped in the matrix, nucleates other small nanoclusters around the original one. The nanoclusters size reduction is accompanied by a characteristic narrowing of the size distribution: a typical example carried out by Rizza et al. is reported in Fig. 13. Au nanoclusters formed in SiO_2 matrix by chemical synthesis present an unimodal size distribution with mean size of ~16 nm. The sample is then irradiated with 4 MeV Au^{2+} at room temperature in the irradiation fluence range of $0 \div 8\times10^{16}$ cm^{-2}. The nanocluster evolution is illustrated in Fig. 13 as a function of the irradiation fluence: Figure 13(a) shows a plan-view TEM image of the as-prepared nanoclusters. They consist of almost spherical and isolated particles. The corresponding size distribution is nearly symmetric with a centroid at 15.9 nm and a small dispersion of 2.1 nm (Fig. 13f). The nanocluster evolution vs the irradiation fluence is shown in Fig. 13(b–e). The associated size distributions are reported in Fig. 13(f–j). The irradiation promotes the formation of a halo of satellite precipitates around the original nanocluster, as shown in Fig. 13(b). This corresponds to the emergence of a new peak in the size distribution graph (black to indicate the original nanoclusters distribution and gray to indicate the satellites distribution). For irradiation fluences up to 0.5×10^{16} cm^{-2}, the two peaks are well separated, Fig. 13(g). The centroid of the nanoclusters is at (14.5 ± 2.4) nm, whereas that of the satellites is at (2.1 ± 1.2) nm. Increasing the irradiation fluence, the continuous dissolution of the original nanocluster corresponds to a shift of the

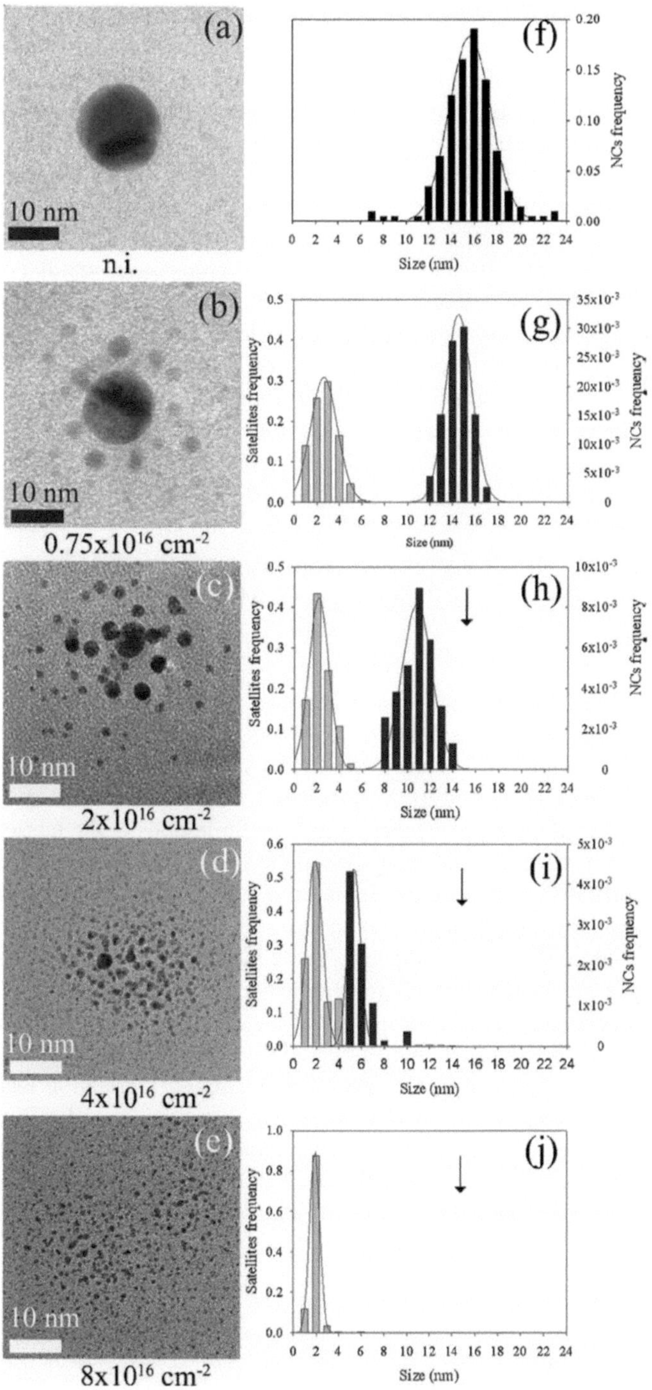

Fig. 13 (continued)

mean diameter toward the lower values of the cluster dimension (Fig. 13 g–j). The arrow indicates the average size of the un-irradiated particles. It is worth noting that this shift does not correspond to the broadening of the size distribution, which remains limited to 2–3 nm. This means that all the original nanoclusters were dissolved by the irradiation. On the other hand, the nucleation and growth of a first generation of satellites close to the nanoclusters surface are followed by the precipitation of new satellites at increasing distances. For irradiation fluences higher than about $4 \times 10^{16}\,cm^{-2}$, the two size distributions merge (Fig. 13i). This corresponds to the complete dissolution of the original nanocluster and the distribution is of the satellites considered as a whole. For irradiation fluences ranging from 4 to $8 \times 10^{16}\,cm^{-2}$, a partial dissolution of the larger satellites is observed. This is associated with an increase of the precipitate density. Finally, for the maximum fluence of $8 \times 10^{16}\,cm^{-2}$, a very narrow distribution is found (Fig. 13f): 0.4 nm (much lower than that of the original nanoclusters distribution, 2.1 nm).

It is worth noting that these experimental data observed by Rizza et al. can be interpreted in terms of the inverse ripening model [44].

2 Electronic Transport Properties of Metal Nanoclusters–Based Materials

The main reason that led to the "nanotechnology revolution" is, surely, connected with the decennial effort toward the scaling-down of the electronic devices and with the potential end of the "Moore law" [45]. In fact, for the past 40 years, the electronic devices have grown more powerful, as their basic units, the transistor, has shrunk. However, the law of quantum mechanics and the limitations of the fabrication techniques may soon prevent further reduction in the size of today's conventional diode, transistor, field-effect transistor, etc. In order to continue the miniaturization of circuitry elements down to the nanometer scale, several alternatives to the traditional electronic technology were investigated. These innovative nanometer-scale devices are the basis of the nanotechnology revolution.

Metal nanoparticle–based nanostructured materials and devices are one of the most potential and prolific alternative, already investigated since some years ago. Whereas in the first part of this chapter, we illustrated fabrication techniques to fabricate metal nanoclusters–based nanostructured materials and to control their structural properties, in this second part we show how such materials can realize prototypes of nanodevices that could find application in the nanoelectronic circuitry of next generation. In particular, strong efforts are dedicated to correlate the novel

Fig. 13 (continued) (**a, e**) TEM images of the time sequence of Au nanocluster evolution in SiO_2 under 4 MeV Au irradiation at 300 K. The samples were irradiated, increasing fluences up 8×10^{16} cm^{-2}. (**f, j**) The corresponding size distributions of nanoclusters and satellites. (Reprinted with permission from Rizza [43] © (2007) American Institute of Physics)

observed electrical properties of such devices with the structural properties. In fact, the capability to tailor at atomic level the structural properties of the fabricated nanostructured materials gives the possibility to control, in a wide range, their electrical functionalities.

2.1 Schottky Barriers in Metal Nanoclusters/Semiconductors Contacts

This section presents the first main example concerning the application of the metal nanoparticles–based nanostructured materials studied in the Section 1, toward the demonstration of possible nanodevices.

In particular, here the Au nanoclusters/SiC system is considered and its electrical properties are examined by local nanoscale measurements by C-AFM, a powerful technique to probe atomic-level properties. These studies show how the single Au nanocluster/SiC contact forms a "nano-Schottky contact" [46]: its *I-V* characteristics are similar to those of a well-known macroscopic Schottky contact, but the quantum effects concerning the electron confinement in small spatial regions and the energy discreteness effects (effects clearly shown at these nanometric level) concur in the observation of a cluster size dependence of the MS nanocontact.

It represents a typical example of the effects coming in evidence due to the scaling-down of the traditional devices in the nanoscale regime [45], in particular the first example of the nanoscale scaling-down of Schottky diodes. On the other hand, it is also a typical example of how, inevitably, the quantum effects, typical of atomic and molecular level, modify the properties (in this particular case the electrical) of a device when it is scaled in nanoscale spatial range and how those properties, typical of the macroscopic matter, derive from the microscopic ones.

Finally, from a technological point of view, the proposed nanostructured material could be the base for the development of complex nanoelectronic circuits (e.g., in power applications, in which the SiC is one of the candidate to replace Si in the next devices generation) characterized by novel functionalities [45] and with electrical properties tunable in a critical manner simply controlling the structural features (i.e., the cluster size and size distribution, the relative distance distribution, and the covered area fraction by the nanoparticles).

The Au nanoclusters on the 6H-SiC surface were formed as described in Section 1.1, and their self-organization was promoted by annealing process as described in Section 1.3. As illustrated in that section, once that the kinetics mechanisms of the Au nanoclusters self-organization on the SiC surface were studied, it is possible, by opportune annealing processes, to form on this surface Au nanoclusters of predetermined size, density, surface–surface distance, and fraction of area covered. So, the as-deposited and the 873 K-5 min, the 973 K-20 min, the 973 K-60 min, and the 1073 K-60 min annealed samples were considered. They showed a mean cluster diameter of (1.46 ± 0.03) nm, (2.78 ± 0.12) nm, (4.54 ± 0.11) nm, (5.98 ± 0.09) nm, and (6.82 ± 0.12) nm; a mean center-to-center distance between nearest clusters of (4.31 ± 0.18) nm, (6.53 ± 0.18) nm, (10.9 ± 0.13) nm, (14.15 ± 0.15) nm, and

(17.6 ± 0.19) nm; and a fraction of area covered of 78%, 41%, 35%, 19%, and 17%, respectively (Fig. 5a–c). The local transversal current–voltage (I-V_{tip}) analyses of this fabricated nanostructured material were carried out at room temperature using the C-AFM, one of the more powerful scanning probe microscopy techniques for the analyses of the atomic-level properties [47]. For these analyses ultra-sharpened diamond-coated Si tips were used. The conductive diamond coating is polycrystalline, and the effective tip diameter is ultimately set by the very small diamond grain (a few nanometers) placed at the apex of the tip. For each sample, 400 I-V_{tip} acquisitions in 400 different positions were performed in a matrix of 20 × 20 points with a step of 500 nm.

According to Giannazzo et al. [48], a biased C-AFM tip in contact with a continuous ultra-thin metal film on a semiconductor forms a nano-Schottky diode due to the nanometric localization of the current across the MS interface. In the present case of a discontinuous film, for each tip position on the sample surface, the typical rectifying Schottky-contact I-V_{tip} characteristics were found, with the threshold voltage (correlated with the Schottky Barrier Height (SBH)) depending on the tip position. As an example, in Fig. 14 the characteristics recorded in bare SiC (Fig. 14a) and in SiC covered with Au clusters of different size are compared (Fig. 14b, c). Each I-V_{tip} curve is typical of thermoionic emission [49], and in the reference sample the I-V_{tip} characteristics belong to a unique family that can be associated with the Schottky contact between the diamond tip and the 6H-SiC substrate. In Au-covered samples, the I-V_{tip} curves are split in two families: one corresponds to the diamond/6H-SiC Schottky contact (area not covered by Au) and the second one corresponds to the Au-nanocluster/SiC Schottky contact. The second family shift toward higher voltage when the mean nanoclusters size increases. It is possible to model this behavior assuming that in each point the AFM tip probes a single Au nanocluster. In fact,

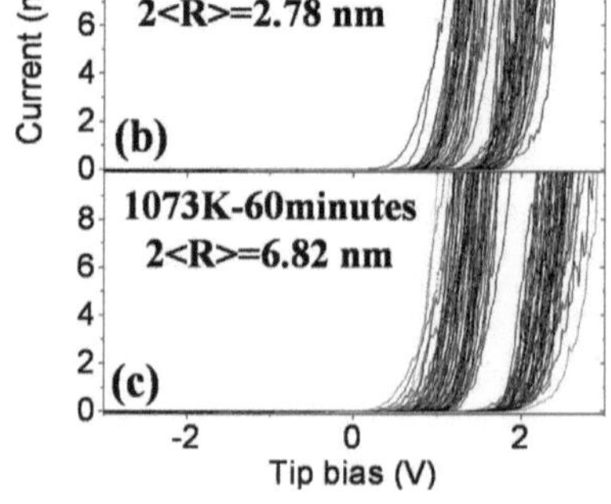

Fig. 14 I-V_{tip} curves measured (by C-AFM) in SiC covered with Au nanoclusters with different mean size

according to the used tip shape and to the average cluster-distance/cluster-dimension ratio (and the step of 500 nm between each point), it is quite unlikely that more than one cluster could be simultaneously contacted by the tip. Hence, when the tip is in contact with a single cluster, the nearest clusters can contribute to the total current only by a tunnel component through the air that is negligible, being, according to realistic estimations, at least two orders of magnitude smaller than the current due to the direct tip-cluster-substrate contact. On the other hand, the fraction of I-V_{tip} curves belonging to the family in which the tip is in contact with Au corresponds to the fraction of area covered by Au for that sample. Therefore, it is assumed that each curve is representative of a nanometric Au/SiC Schottky contact and the barrier height depends on the cluster size.

To determine the SBH Φ_{B} for each nanometric MS contact, we fitted the current onset region of the relative I-V_{tip} curve with a parabolic function and determined the SBH value as the parabola vertex derived by the fits. In fact, considering the thermoionic emission law [49], $I = I_0 e^{-(q\Phi_{\mathrm{B}}/k_{\mathrm{B}}T)}\left(e^{qV_{\mathrm{tip}}/nk_{\mathrm{B}}T} - 1\right)$, for opportunely small values of $q(n\phi_{\mathrm{B}} - V_{\mathrm{tip}})/nkT$, the second-order Taylor expansion gives $I \approx C\left(V_{\mathrm{tip}} - V_{\mathrm{th}}\right)^2$ with $\phi_{\mathrm{B}} = (V_{\mathrm{th}}/n) + (kT/q)$ (at room temperature $\phi_{\mathrm{B}} \approx V_{\mathrm{th}}$). Therefore, for each I-V curve, the V_{th} (i.e., ϕ_{B}) was evaluated from the parabolic fit.

So the SBH spatial distribution for each sample was obtained and reported in the normalized distributions (Fig. 15). For the reference sample (sample without Au clusters), the SBH distribution is peaked at (1.24 ± 0.02) eV, with a broadening due to statistical fluctuations (see Fig. 14a). This measured value of 1.24 eV is associated with the SBH of the diamond-tip/6H-SiC Schottky contact. The measured value is in reasonable agreement with the difference between the heavily doped diamond work function and the SiC electron affinity, considering the peculiar surface electronic properties of diamond crystals [50]. For the samples with Au clusters on the surface, by increasing the mean cluster sizes, the SBH distributions exhibit

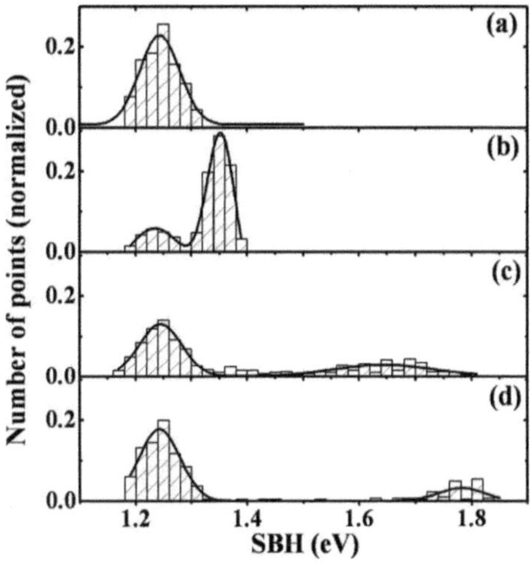

Fig. 15 Schottky barrier height distributions: (**a**) Reference sample (sample without Au cluster); sample with Au cluster mean diameter of (**b**) (1.46 ± 0.03) nm; (**c**) (2.78 ± 0.12) nm; (**d**) (6.82 ± 0.12) nm

a bimodal shape (see Fig. 14b–d), with two broad peaks fitted by two Gaussian curves. For all the samples, the first peak is centred at 1.24 eV, i.e., the value of the tip-diamond/6H-SiC SBH. Hence, the presence of this first peak can be associated to the surface regions in which the diamond tip is directly in contact with the 6H-SiC substrate. The position of the second peak changes with the nanocluster average dimension. As already said, the SBH values in the histograms around the second peak can be associated with the direct contact with a single Au cluster, i.e., to a single Au cluster/6H-SiC nano-Schottky diode. The data in Fig. 15 demonstrate the dependence of the SBH on the cluster size. Accordingly, to each sample a unique SBH corresponding to the cluster mean size (the peak at higher SBH in Fig. 15) is associated, and the error bar on each SBH value was evaluated from the σ (full width at half maximum in the Gaussian distribution) of each histogram in Fig. 15.

In Fig. 16, the evaluated SBH (dots) is reported as a function of the mean cluster size. It increases with increasing average cluster size, tending asymptotically to the ideal SBH value of a continuous Au film/SiC contact (~1.9 eV) [51]. This latter indicates that the larger Au clusters (> 7 nm) on SiC approach the behavior of the Schottky barriers formed by continuous Au films.

The interpretation of the observed SBH dependence on the clusters size [45] was based on merging the thermoionic transport theory through the MS barrier coupled with the concept of ballistic transport and the Constant Interaction (CI) model for the electron transport in few electrons quantum dots, which takes into account the charging energy of small systems [52].

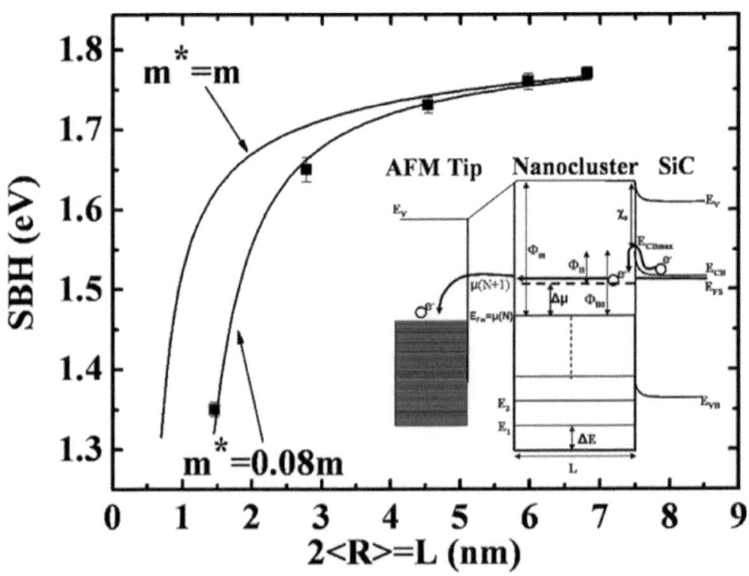

Fig. 16 Experimental values (*dots*) of the SBH as a function of mean cluster size and theoretical prediction for $\phi_B = 1.85\,\text{eV}$, $m^* = 0.08m$, and $m^* = m$. The inset shows the considered band diagram of the system (AFM) tip-cluster-SiC substrate

For macroscopic MS contact, the Schottky barrier height Φ_{B0} is defined as $\Phi_{B0} = \Phi_m - \chi = E_v - \mu - \chi$, with Φ_m the work function of the metal, χ the electron affinity of the semiconductor, E_v the energy vacuum level, and μ the electrochemical potential, usually referred as the Fermi level E_F of the metal (see inset of Fig. 16). In the present case of Au/SiC nano-barrier, each Au nanocluster is described as a 3D box with energy levels equally spaced by [53] $\Delta E = \left(3\pi^2 N\right)^{-1/3} \left(\hbar^2\pi^2/m^* L^2\right)$ and occupied until $E_F = \left(\hbar^2/2m^*\right)\left(3\pi^2 N/V\right)^{2/3}$, where m^* is the electron effective mass, N the number of electrons in the system, and V system volume (in the present case the mean cluster volume); $L = 2 < R >$ the cluster diameter. Assuming N equal to the number of atoms in the gold cluster, i.e., one valence electron for each atom, and 100% packing, E_F and ΔE can be expressed as $E_F = \left(\hbar^2/2m^*\right)\left(3\pi^2/\Omega\right)^{2/3}$ and $\Delta E = \left(3\pi^2\right)^{-1/3} (\pi/6\Omega)^{-1/3} \left(\hbar^2\pi^2/m^* L^3\right)$, with Ω the atomic volume.

The process was schematized as indicated in the inset of Fig. 16. For a forward (positively) biased tip, an additional electron from the substrate overcomes the SBH by thermoionic emission and falls on the lowest unoccupied energy level $\mu(N+1)$ within the 3D box containing N electrons. As the electron mean free path λ_e for the considered electron kinetic energy in the Au nanoclusters is ranges from [54] 10 to 20 nm (i.e., λ_e larger than the average cluster dimension), the electron moves ballistically within the Au dot and it is collected to the tip in ohmic contact with the Au grain. Hence, the Au grain/SiC SBH is given by $\Phi_B(N) = \Phi_{B0} - \Delta\mu(N)$, with $\Delta\mu(N)$ the energetic distance between E_F and $\mu(N+1)$ (inset of Fig. 16); $\Delta\mu(N)$ accounts for the SBH dependence on the clusters size for the present system. According to the CI model of few electron quantum dots [52], $\Delta\mu(N) = \left(e^2/C\right) + \Delta E$, where $E_c = e^2/C$ is the electrostatic energy (charging energy) necessary to add or subtract one electron to the dot, taking into account the Coulomb interactions of that electron with all other electrons, in and outside the dot. This "charging energy" contribution, which is negligible for macroscopic systems, is instead fundamental in ultra-scaled systems. Within the approximation of the electrostatic interaction among the nearest neighbor dots, the capacitance C is derived as $C = C_0 + n_s C_c$. $C_0 = 2\pi\varepsilon_r\varepsilon_0 L$ is the self-capacitance of the dot, i.e., the capacitance of a sphere of diameter L embedded in a dielectric of constant ε_r, $C_c \approx \left(\pi^3\varepsilon_r\varepsilon_0 L^2\right)/4(< s > +L)$ is an approximated expression for the coupling capacitance between two nearest clusters, described as two spheres with the same diameter L and sited at center-to-center distance $< s > +L (< s >$ is the surface–surface distance between the two clusters) [55]. Finally, $n_s + 2\pi(< s > +L)\sqrt{N_s}$ is the number of nearest neighbor clusters (in the random distribution of the clusters on the surface, a circular symmetry around each fixed cluster can be assumed), where $N_s(L) = 3Q\Omega/4\pi L^3$ is the cluster surface density, as indicated in Section 1.3. So

$$\Phi_B(L) = \Phi_{B0} - \frac{e^2}{2\pi\varepsilon_r\varepsilon_0 L + \sqrt{\frac{3Q\Omega}{4\pi}\frac{\pi^3\varepsilon_r\varepsilon_0}{2}} L^{\frac{1}{2}}} - \left(\frac{1}{3\pi^2}\right)^{\frac{1}{3}} \left(\frac{\pi}{6\Omega}\right)^{-\frac{1}{3}} \frac{\hbar^2\pi^2}{m^* L^3} \quad (12)$$

Clearly, $\Phi_B(L)$ tends to saturate to Φ_{B0} for sufficiently large L values, i.e., the behavior of large clusters approaches that of the bulk material. In Equation (12), the second and third terms on the right hand side, i.e., $E_c(L)$ and $\Delta E(L)$, play a competitive role. Because by increasing L the first term $E_c(L)$ decreases more slowly with respect to $\Delta E(L)$, for sufficiently high L (in our case $L \geq 4$ nm), $E_c(L) > \Delta E(L)$ and the electrical behavior is described by a "metallic Coulomb blockade regime" [56]. For sufficiently small L ($L < 4$ nm), the term $\Delta E(L)$ begins to be significant with respect to $E_c(L)$ and the effects due to the energy discreteness in the dot are relevant. From Equation (12), it is evident that, decreasing L, the barrier height Φ_B becomes zero for a certain L^* value, which in our case is $L^* = 0.85$ nm. For $L < L^*$, Φ_B would be negative. Because $\Phi_B(L) = \Phi_m - \chi - \Delta\mu(L) = \Phi_{meff}(L) - \chi$, a negative Φ_B value is equivalent to an effective metal work function Φ_{meff} (L) lower than the electron affinity χ in the semiconductor. In such a situation the Au cluster/6H-SiC contact would be not any more a Schottky contact, but an ohmic one and the transport mechanism at the interface would change in consequence [49].

Although the developed model is valid for $T = 0$, it gives a good approximation even at room temperature (our experimental case) for which $\left(e^2/C\right) >> kT$.

It is worth noting that the proposed model was considered only in the thermoionic process as the main process of electron transport, and was neglected in the electron transport by tunneling through the Schottky barrier. This hypothesis is very good in the particular experimental case analyzed of doping concentration of the substrate $N_D \approx 5 \times 10^{17}$ cm^{-3} (in such a case the rate of emission of electrons by tunneling process through the Schottky barrier is much lower than the rate of emission by thermoionic process). For higher doping concentration also the tunneling process could begin significant and would be considered in the model.

Comparing the theoretical prediction by Equation (12) with the experimental data of Fig. 16 using the values $\Phi_{B0} = 1.85$ eV, $\varepsilon_r = 1$, $Q = 4.5 \times 10^{15}$ cm^{-2} (by RBS analyses), and $\Omega = 1.69 \times 10^{-29}$ m^3 the continuous curve reported in Fig. 3 is obtained. The only free fitting parameter was m^*. The best agreement was obtained for $m^* = 0.08m$ (m is the electron free mass). This significant difference from the electron mass in Au bulk (where $m^* \approx m$) can be attributed to the greater coupling of electron with phonons and plasmons in nanostructures respect to the bulk [57] and to the effect of electrons exchange interactions that are more significant when the system dimensionality is scaled. For cluster size larger than 10 nm, $\Phi_B(L) \approx \Phi_{B0}$, thus suggesting that these clusters acquired bulk properties. This is confirmed by other experimental evidences as Au cluster melting point and structural properties dependence on size [58].

2.2 *Rectifying Behavior of Au Nanoclusters Embedded in SiO_2*

Nanoclusters embedded in dielectrics can be used to control the vertical current flow considering the charge blocking effects and tunneling trough the dielectrics. In particular, a possible nanodevice is illustrated here. It is based on Au nanoclusters embedded in SiO_2. A nanometric sandwich SiO_2/Au nanoclusters/SiO_2 was realized whose electrical properties were probed by C-AFM. Such a system clearly exhibits an electrical rectifying behavior characterized by a threshold voltage for positive bias and an asymmetric shape with respect to the bias inversion. Furthermore, as already seen for the system Au nanocluster/SiC, the electrical properties of the system here studied are strongly dependent on and controllable by the structural ones (cluster size).

On the other hand, another general feature of the fabricated nanostructured based rectifier is the integration of the traditional Si based metal-oxide-semiconductor (MOS) devices with quantum effects typical of a single electron transistor (SET). In fact, the system can be seen as a nanometric MOS embedding Au nanoclusters in the SiO_2 dielectric layer. This is another example of the efforts toward the scaling-down of the devices in the nanometric regime (using a bottom-up approach) with the introduction of nanoscale physics concepts.

The samples were prepared following the sputtering process of Au nanoclusters on SiO_2 as described in Section 1. Then, by annealing processes clusters were formed of predetermined sizes, density, surface–surface distance, and fraction of area covered, as described in Section 1.3. In particular, samples with Au nanoclusters of mean radius $< R >$ of (1.24 ± 0.06) nm, (1.37 ± 0.11) nm, and (1.98 ± 0.26) nm, with mean surface–surface distance between the nearest clusters (1.29 ± 0.28) nm, (1.36 ± 0.15) nm, and (2.17 ± 0.20) nm, are considered. Finally, a second layer of 3 nm was deposited on the Au nanoclusters by a next step of sputtering. The local transversal I-V_{tip} analyses were performed at room temperature by C-AFM analyses using Si-tips covered by a thin ($\sim$2 nm) Au film.

The characterizations were so performed: first, a conductive scan of the sample surface (1 μm $\times$ 1μ m) was performed to locate the more conductive nanometric isolated regions (i.e., the regions in which an Au cluster is present); then, the I-V_{tip} acquisitions were performed fixing the tip on one of these positions and then acquiring multiple I-V_{tip} curves; the same for a large array of positions in each sample. In each position the I-V_{tip} curves are stable in time (up to $\sim$15 min), while they are strongly correlated with the tip position. To each sample a unique I-V_{tip} curve was associated, obtained by averaging 10 different curves corresponding to different positions (Fig. 17). For each sample, the I-V_{tip} is strongly asymmetric with respect to the potential sign, and for V_{tip}>0, it presents a threshold voltage V_{th} that decreases with increasing the Au cluster mean size.

To exclude the possibility that the observed I-V_{tip} characteristics are due to anodic oxidation when the tip is negatively biased, we performed AFM topographical images of the samples surface before and after the electrical measurements: no morphological change has been observed. On the other hands, the current measured during anodic oxidation is in the range of μA, while the current measured in the described experiments never exceeds 10^{-10} A, too low to cause any significant oxidation.

For the theoretical interpretation of the experimental data, a model was implemented considering the tunneling transport of single electrons through an asymmetric DBTJ junction with the gate of the MOS structure represented by the metallic

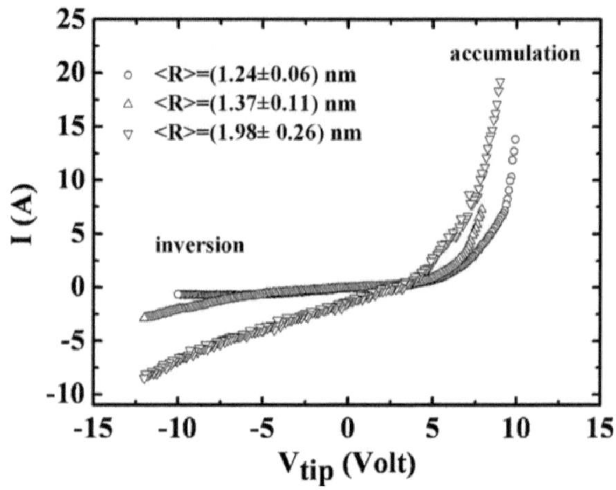

Fig. 17 *I*-V_{tip} characteristics (on average on multiple curves) for all the analyzed samples

AFM tip. In fact, a sample with a metallic cluster embedded between two dielectric thin films can be modeled as a double barrier tunnel junction (DBTJ [59, 60]. If the cluster is opportunely small, a DBTJ presents, also at room temperature, a conduction characterized by a Coulomb blockade with a threshold voltage [61] V_{th}. For a standard DBTJ, the current–voltage characteristic is symmetric with respect to the potential sign change and it is characterized by a V_{th} tunable by the cluster dimensions and the oxide thickness [59]. In the present case, considering the energy band diagram in Fig. 18, the first tunnel barrier is represented by the δ_1 thick oxide between the right side metallic contact (AFM tip) and the cluster. The second barrier is represented by the δ_2 thick oxide between the cluster and the Si substrate for a positively biased tip, i.e., for the MOS in the accumulation regime (Fig. 18a), while it is formed by the series of the δ_2 thick oxide and the semiconductor depletion region, of thickness x, for a negatively biased tip (Fig. 18b).

In this adopted configuration, the electrons flow from the Si substrate to the tip for $V_{\text{tip}} > 0$, while they flow from the tip to the Si substrate for $V_{\text{tip}} < 0$. The main physical parameters involved in the description are the coupling capacitances C_1 and C_2 and the tunneling resistance R_{t1} and R_{t2} associated with the two tunnel junctions. Coherent quantum processes, consisting of several simultaneous tunneling events (co-tunneling), are ignored (this assumption is valid if the tunneling resistance R_{tj} of all the tunnel barriers of the system is $R_{tj} >> R_q = h/4q^2 \approx 6.5\,\text{k}\Omega$, with R_q the quantum unit of resistance and $q > 0$ the electron charge) [56].

Under these assumptions, the tunneling of a single electron through a tunnel barrier is always a random event, with a certain rate which depends solely on the reduction of the free (electrostatic) energy of the system as a result of this tunneling event [56]. The current I_{tot} flowing through a DBTJ can be expressed as $I_{\text{tot}} = q\Gamma_{\text{tot}}$, where the net tunneling rate Γ_{tot} is the sum of four terms: the tunneling rate from tip to grain $\Gamma_{(\text{TG})}$, the rate from substrate to grain $\Gamma_{(\text{SG})}$, the rate from grain to tip $\Gamma_{(\text{GT})}$, and the rate from grain to substrate $\Gamma_{(\text{GS})}$. In particular, if the DBTJ is asymmetric

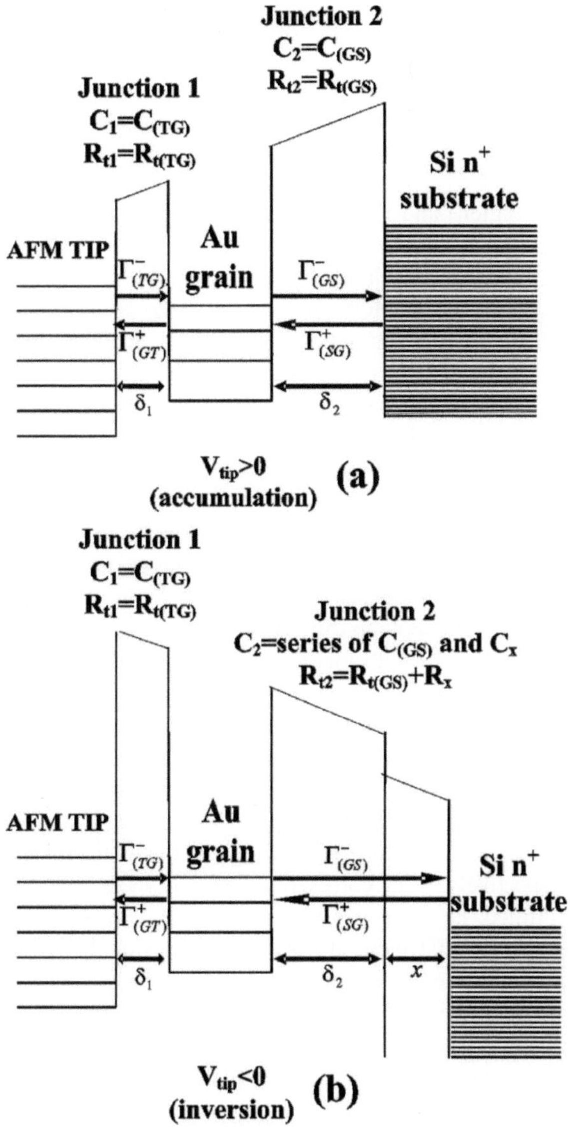

Fig. 18 (**a**) Qualitative energy band model of the DBTJ for $V_{\text{tip}} > 0$: in the nanometric size AFM tip and Au grain the electronic states are supposed to be quantized, and in the Si(n$^+$) substrate the electronic state are supposed to be a quasi-continuous. The second tunnel barrier is represented by the oxide of thickness δ_2 only; (**b**) qualitative energy band model of DBTJ for $V_{\text{tip}} < 0$: in this case the second tunnel barrier is represented by the oxide of thickness and by the Si depletion region of thickness x

($R_{t1} << R_{t2}$ and it is surely true if $\delta_1 < \delta_2$), the total rate can be reduced to the net rate for the less "transparent" barrier [61], i.e., $\Gamma_{\text{tot}} \approx \Gamma_{(\text{SG})} + \Gamma_{(\text{GS})}$. Therefore, the tunneling event from a quantized state of metallic grain to a state of a quasi-continuum electronic spectrum of Si substrate can be expressed by the following equation, according to D. V. Averin and A. N. Korotkov [56, 59, 62]:

$$\Gamma_{\text{tot}} \approx \Gamma_{(\text{GS})} + \Gamma_{(\text{SG})} = -\Gamma_{0(\text{GS})}\left[1 + e^{-\frac{\Delta W_{(\text{GS})}}{k_B T}}\right]^{-1} + \Gamma_{0(\text{SG})}\left[1 + e^{-\frac{\Delta W_{(\text{SG})}}{k_B T}}\right]^{-1} \quad (13)$$

where the first term describes the electron tunnel rate for $V_{tip} < 0$ and the second term for $V_{tip} > 0$.

Considering the circuital representation of the DBTJ [59, 60], the different terms presented in Equation (13) can be expressed in terms of the tunneling resistance R_{t2} and of the capacitances C_1 and C_2 (see Fig. 18). In particular [56], the rates $\Gamma_{0(GS)} \approx V_{tip}/q(R_{t(GS)} + R_{tx})$ (where R_{tx} is the tunneling resistance of the depletion region) and $\Gamma_{0(SG)} \approx V_{tip}/qR_{t(SG)}$, respectively, while the energies $\Delta W_{(GS)} = -(q^2 + 2C_2qV_{tip})/[2(C_1 + C_2)]$ and $\Delta W_{(SG)} = (-q^2 + 2C_2qV_{tip})/[2(C_1 + C_2)]$ considering the clusters initially neutral. In the case of a positively biased tip ($V_{tip} > 0$), the MOS structure tip-SiO_2-Si is in accumulation regime (assuming the flat band potential equal to zero). Hence, the tunneling resistance R_{t2} is associated only with the oxide layer with thickness δ_2 between the grain and the substrate ($R_{t(SG)}$), and the capacitances C_1 and C_2 are the coupling capacitances between tip–grain $C_{(GT)}$ and the coupling capacitance grain–substrate $C_{(SG)}$, respectively. Moreover, if $V_{tip} > 0$, Equation (13) assumes a more simplified form, because the first term of the sum is negligible with respect to the second one. Instead, for a negatively biased tip ($V_{tip} < 0$), the second tunnel barrier is characterized by the tunneling resistance $R_{t2} = R_{t(GS)} + R_{tx}$ and the capacitance $C_2 = C_{(GS)}C_x/(C_{(GS)} + C_x)$, where C_x is the capacitance associated with the depletion region in the semiconductor. This depletion region increases with the applied negative bias and reaches its maximum width for the surface inversion potential of the MOS structure [49], $\Phi_{inv} = -2(k_BT/q)\ln(N_D/n_i) = -1.08$ V. Both the first and the second terms in Equation (13) exhibit a threshold behavior. In particular, the first term, describing the electron tunnel rate for $V_{tip} < 0$, is negligible for $\Delta W_{(GS)} < 0$ and increases rapidly for $\Delta W_{(GS)} \geq 0$. Similarly, the second term, describing the electron tunnel rate for $V_{tip} > 0$, increases for $\Delta W_{(SG)} \geq 0$. Therefore, two threshold voltages V_{th1} and V_{th2} can be defined for the conduction in the inversion and accumulation regime, respectively. They correspond to the V_{tip} values for which $\Delta W_{(GS)}$ and $\Delta W_{(SG)}$ are nullified, i.e., $V_{th1} = -q(C_{(GS)} + C_x)/2C_{(GS)}C_x$ and $V_{th2} = q/2C_{(SG)}$. It is worth observing that these are the expressions of the Coulomb blockade thresholds for conduction in an asymmetric DBTJ. Clearly, such a system must be characterized by electrical rectifying properties. In fact, for $V_{tip} \geq 0$ we expect a conduction characterized by a threshold voltage $V_{th2} << |V_{th1}|$ and by a tunneling resistance $R_{t(SG)} << R_{(GS)} + R_{tx}$. Moreover, the rectifying properties of this DBTJ would be tunable by the cluster size. In fact, both $C_{(SG)}$ and $C_{(GS)}$ can be modeled [63] as the capacitance of a sphere of fixed radius $< R >$ with an infinite plane sited at a fixed distance δ_2, i.e., $C_{(SG)} = C_{(GS)} \approx 2\pi\varepsilon_0\varepsilon_{r(SiO_2)} < R > (2 + < R >/\delta_2)$. Consequently, the threshold voltage in accumulation (V_{th2}) is expected to strongly depend on the cluster size, while the threshold voltage in inversion (V_{th1}) is expected to have a lower dependence. In agreement with these considerations, in the experimental I-V_{tip} curves in Fig. 17 the observed experimental threshold voltages in the accumulation (V_{th2}) and inversion (V_{th1}) regimes decrease when the cluster size increases. Clearly, V_{th2} is the threshold voltage value with stronger dependence on the cluster size. In the following, V_{th2} will be simply indicated as V_{th}.

In the proposed model, three fitting parameters are required: the tip–grain coupling capacitance $C_{(GT)}$, the grain–substrate coupling capacitance $C_{(SG)}$, and the tunneling resistance between grain and substrate $R_{t(SG)}$, and the fits well reproduce the experimental data over four orders of magnitude current range when [65] $R_{t(SG)}$, $C_{(GT)}$, and $C_{(SG)}$ are $(3.7\times10^{11} \pm 4\times10^{9})$ Ω, $(7\times10^{-21} \pm 1\times10^{-22})$ F, $(5.7\times10^{-19} \pm 1.3\times10^{-20})$ F; $(3.3\times10^{11} \pm 8\times10^{9})$ Ω, $(8\times10^{-21} \pm 9\times10^{-23})$ F, $(6.3\times10^{-19} \pm 1.8\times10^{-20})$ F; $(3\times10^{11} \pm 1.4\times10^{9})$ Ω, $(9\times10^{-21} \pm 1\times10^{-22})$ F, $(9\times10^{-19} \pm 2.8\times10^{-20})$ F, respectively for the three samples examined. The tunneling resistance so estimated results much greater than quantum unit of resistance as required by the model.

To verify the consistence of the fitting model, we correlated the values derived by the fitting parameters with structural information (in particular with the grain size). In fact, using $C_{(SG)} \approx 2\pi\varepsilon_0\varepsilon_{r(SiO_2)} < R > (2+ < R > /\delta_2)$, where $< R >$ is the cluster radius and $\varepsilon_{r(SiO_2)} = 11.9$, $\delta_2 = 10\,\text{nm}$, we obtained the Au grain radius: $< R >= (1.3\pm0.1)$ nm for the as-deposited, $< R >= (1.4\pm0.1)$ nm for the 873 K-5 min annealed sample, and $< R >= (1.9\pm0.4)$ nm for the 873 K-60 min annealed sample. For the three cases, the average values are obtained fitting the average I-V_{tip} curves, while the errors on these values derive from fitting the I-V_{tip} curves with the largest deviation from the average ones. The obtained average $< R >$ values and the errors are in good agreement with the values obtained by the AFM–TEM analyses. This indicates that the dispersion in the measured I-V_{tip} curves is related to the clusters size distribution.

2.3 Electronic Collective Effects in Disordered Array of Nanocrystals

In Sections 2.1 and 2.2, some local nanometric electrical properties were described concerning metal nanoclusters interacting with semiconductor or insulator materials, trying to focus the attention on the electrical properties of the single nanocluster and its correlation with the structural ones. In view of the technological applications, a fundamental step toward the design and production of complex nanoelectronic circuits is the implementation of several interacting single components (nanocluster) in a complex system [65]. As a consequence, in the nanoelectronic research a fundamental importance was acquired by the study of the electrical properties of array of nanocrystals and in particular what are the effects of factors such as the nature of cluster interaction, the system's structural disorder, and the temperature operation on such properties [65–69].

In this section, some recent experimental data are exposed concerning the effect of disorder and temperature on the collective electron transport in an array of metal nanoclusters ($TiSi_2$) embedded in a semiconductor matrix (Si).

First, a self-organization methodology, based on the annealing processes of a silicon/titanium multilayer (each layer of opportune thickness) prepared by e-beam deposition, was performed [67]. A total of 20 alternate layers of 1.7 nm Ti and 25 nm Si were deposited (in vacuum of 4×10^{-6} Pa). The Ti and Si thickness were adjusted in order to obtain $C_{Ti}/C_{Si} = 6\%$, where C_{Ti} and C_{Si} are the average Ti and Si atomic concentrations in the 250-nm-thick multilayer structure, respectively (as checked by Rutherford backscattering analyses). Rapid Thermal Annealing in Ar (873 K-3 min) was used to induce the reaction between the Si and the Ti to form the metallic $TiSi_2$ nanoclusters embedded in Si polycrystalline matrix. The energy-filtered TEM (EFTEM) analyses in plan view and cross-section, of the annealed sample, are reported in Fig. 19(a) and (b), respectively. EFTEM allows the mapping of the elemental composition of the analyzed sample with spatial resolution typical of TEM analyses, by selecting electrons with a well-defined energy from the transmitted beam. In particular, the images in Fig. 19(a, b) were obtained selecting those electrons which lost 45 eV (edge M of Ti) with respect to the transmitted beam so that the bright areas correspond to Ti-rich regions while the Ti is below the detection limit in the dark regions. Therefore, the bright spots give the distribution of the $TiSi_2$ clusters in the Si matrix. It is interesting to note that, although the plan-view image gives evidence of separated silicide grains, the cross-sectional view indicates that the in-depth grain distribution maintains a "memory" of the initial multilayer structure. Furthermore, X-ray diffraction (XRD) measurements indicated the presence of C49 phase $TiSi_2$ (and the absence of the C54 phase) and of the polycrystalline Si.

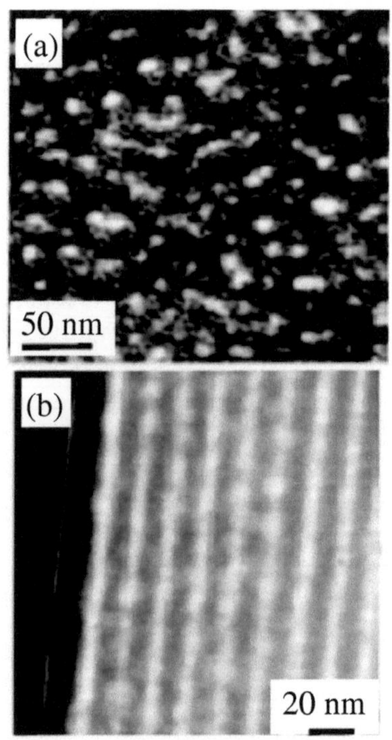

Fig. 19 Plan-view (**a**) and cross-view (**b**) EFTEM of the sample after the annealing

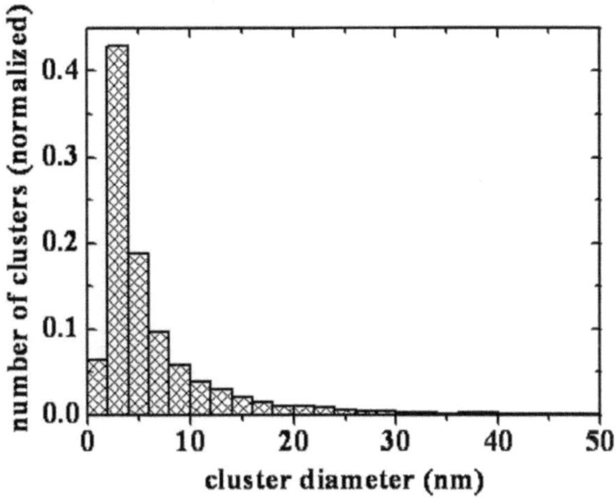

Fig. 20 Distribution of grains dimensions; the mean radius is $< r > = (3.6 \pm 0.2)$ nm

The analyses of several plan-view and cross-view TEM images by the Gatan digital micrograph software allowed to determine the distribution of the grains dimension (Fig. 20) and of the surface–surface distance between nearest grains. From those, the mean cluster size and mean surface–surface distance were determined to be $< R >= (3.6 \pm 0.2)\,\text{nm}$ and $< s >= (10.3 \pm 0.5)\,\text{nm}$, respectively. The surface density Q of $TiSi_2$ grains can be calculated using the formula $Q = \dfrac{(Nt)_{\text{Ti}}}{\rho_{\text{Ti}}^{\text{TiSi}_2} \times < V >}$ $Q = (Nt)_{\text{Ti}} / \left(\rho_{\text{Ti}}^{\text{TiSi}_2} \times < V >\right)$, where $(Nt)_{\text{Ti}}$ is the surface density of Ti, $\rho_{\text{Ti}}^{\text{TiSi}_2}$ is the density of Ti in $TiSi_2$ $(2.2 \times 10^{22}\,\text{atoms/cm}^3)$, and $< V >= (4/3)\pi(< R >^3)$ the mean volume of the grain and it results in $Q \approx 10^{12}\,\text{cm}^{-2}$.

I-V measurements were performed using a Keithley 2410 1100 V system equipped with a cryostat SpectrostatCF Oxford utilizing liquid N_2. The applied voltage was ranging from –5÷5 V. Samples were cut in $7 \times 7\ \text{mm}^2$ square piece for the *I-V* measurements. Metallic tips gave a good electrical ohmic contact in the as-deposited sample. Contacts in the annealed sample were made by deposition of a 100-nm-thick Ti film (subsequently annealed at 773 K for 10 min) over a 0.5-mm^2 area at the corners of the square using a mask.

The *I-V* characteristics of the as-deposited and annealed samples were measured at several temperatures in the 70÷300 K range. While the as-deposited sample shows an ohmic behavior (as expected [70]), the *I-V* curves of the annealed sample are highly symmetric respect to voltage inversion although they do not exhibit an ohmic behavior (Fig. 21 shows some of the *I-V* at different temperatures). They seem to be characterized by a clear voltage threshold V_t that decreases when T increases. The differential resistance R_t can be estimated to be in the $10^4\ \Omega$ (at $T \sim 200$ K)÷$10^6\ \Omega$ (at $T \sim 70$ K) range.

The discussion of such behavior is based on a model developed by Middleton and Wingreen [71] (MW approach). The main idea of the model is that the nanocrystals array can be seen as system of capacitively coupled conductors with electrons

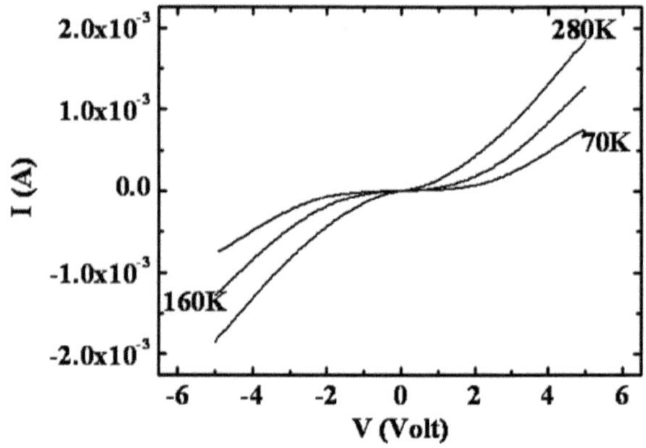

Fig. 21 *I-V* measurements of the annealed sample for three different temperatures (70 K, 160 K, 280 K) in the –5÷5 Volt range

allowed to tunnel between neighboring dots (according to a "metallic Coulomb blockade regime" [56]) and the total electrostatic energy can be expressed as

$$E = \frac{1}{2}\sum_{i,j}(Q_i + q_i)C_{ij}^{-1}(Q_j + q_j) + V_L Q_L + V_R Q_R + \sum_i V_i^{ext} Q_i \qquad (14)$$

C_{ij} is the matrix of coupling capacitance between the dots, Q_i the charge in each dots, q_i an offset charge that account for the quenched disorder (deriving from background charges present in the substrates or matrix surrounding the particles), $Q_{L,R}$ is the charges on the leads which are at voltages $V_{L,R}$. In the present case, average quantities are used: C_0 is the mean self-capacitance of the grain and C is the mean coupling capacitance between nearest grains. Based on these assumptions, the MW model interprets the electrical transport through the system in the temperature limit $T << \frac{e^2}{2k_B \max(C, C_0)}$ as collective electrons tunneling between the dots characterized by a second-order transition (with associated critical phenomena) that separates a static non-conducting state from a dynamic conducting state. The background quenched disorder induces random shifts in the local nanoparticles potentials and leads to a distribution of electrostatic energy costs experienced by electrons as they tunnel from grain to grain. Under these assumptions, the *I-V* characteristics present a clear non-linear behavior characterized by a threshold voltage V_t: at $V \leq V_t$ non-conducting state ($I = 0$), at $V > V_t$ conducting state ($I \neq 0$). In the conducting state the *I-V* characteristics exhibit a scaling behavior:

$$I = A(V - V_t)^{\zeta} \qquad (15)$$

where A is a proportionality constant and ς is a parameter that depends on the system dimensionality. In particular for a 2D conduction $\zeta \approx 2$ is predicted.

By increasing the temperature a smear of the local threshold occurs (R. Parthasarathy [65, 66] and K. Elteto [72]) and a certain fraction $p(T)$ of the tunnel junctions becomes truly ohmic. At the critic temperature T^*, the fraction

$p(T^*)$ will become large enough for a continuity of junctions to form a percolating ohmic network that bridges the array. So, Parthasarathy's calculations [65] predict the existence of critic temperature T^* defined by $V_t(T^*) = 0$, which separates two conduction regimes: at $T < T^*$ the MW model is valid and the *I-V* curves are described by the scaling law of Equation (15) with V_t depending on T. At $T \geq T^*$ the differential conductance of the system $G = dI/dV$, at low field, acquires a thermally activated form ("over barrier" transport):

$$G \propto \exp\left(-\frac{U}{k_B T}\right) \tag{16}$$

with activation energy $U = bk_B T^*$, with b predicted in $b \approx 2.4$.

Furthermore, V_t is predicted to decrease linearly as a function of T according to

$$V_t(T) = a_1 + a_2 T = V_t(0) + V_{\text{shift}}(T) \tag{17}$$

where $V_{\text{shift}}(T)$ is the voltage shift, according to the scaling behavior, needed to collapse all the *I-V* curves (at $T < T^*$) on the master curve characterized by the threshold voltage $V_t(0)$. Obviously, Equation (17), in this picture, is valid at $T < T^*$ (in the "MW regime") and not at $T > T^*$ because in this regime a threshold voltage for conduction no longer exists. However, it has been observed [72] that at $T > T^*$, for opportune high bias, the *I-V* curves should still be described by Equation (15) with an effective negative threshold voltage that experimentally is the shift required to collapse the *I-V* characteristics on the master curve. This prediction derives from the consideration that above T^*, when the global threshold disappears along the optimal percolating path, there are still many other paths that have finite thresholds and are accessible at higher bias. These thresholds will keep decreasing linearly with temperature as more and more steps are linearized by thermal fluctuations. As a consequence, the MW non-linear *I-V* curve of a 2D array again emerges when summing the contributions of all the accessible paths.

While the MW model is well established and experimentally confirmed, only recently the Parthasarathy's model was coherently formulated [65, 72] and experimentally confirmed by Parthasarathy's group in systems formed by a low number (100÷10000) of metallic nanoclusters (Au) on/in insulator substrates (Si_3N_4) [65, 66]. A limited amount of disorder was introduced in the array in order to determine the validity limit of the theory when applied to disordered systems. However, the experimental data exposed in this section show that the same picture adequately describes the collective electronic transport in the very disordered system analyzed here: in fact, such models, well settled for ordered array, well fit the observed experimental data concerning the electrical properties of the disordered array if average quantities are used in the calculations.

The fundamental quantities involved in this analysis are the mean self-capacity of a grain $C_0 = 4\pi\varepsilon\varepsilon_0 < R > \approx 5 \times 10^{-18}$ F (with $\varepsilon = 11.9$ for Si-poly), the mean coupling capacity for nearest grains $C = \varepsilon\varepsilon_0 (2 < R >)^2 / < s > \approx 1 \times 10^{-18}$ F, the mean charging energy of the grain $E_C = e^2/2C_0 \approx 17\,\text{meV}$, and the tunneling resistance $R_t \approx 10^4 \div 10^6\Omega >> 6.5\,k\Omega$.

The *I-V* curves of the annealed sample were interpreted using the scaling function described by Equation (15) in the 70 K÷190 K temperature range. The threshold voltage V_t was preliminarily determined by plotting [68] $I\,(dV/dI)$ as a function of V. The resulting V_t is plotted in Fig. 22 as a function of the temperature: it decreases linearly in agreement with Equation (17). The linear fit is

$$V_t(T) = (0.98 \pm 0.01) - (5 \times 10^{-3} \pm 1 \times 10^{-4})T = V_t(0) + V_{\text{shift}}(T). \quad (18)$$

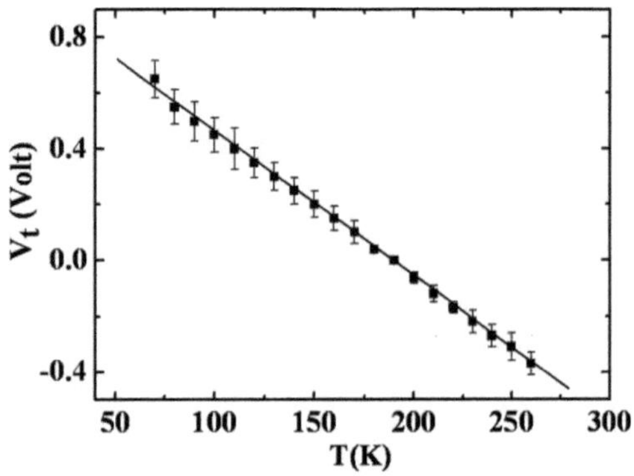

Fig. 22 Threshold voltage V_t as a function of T: black dots indicate the experimental data and the line the linear fit

From this, the threshold voltage extrapolated at $T = 0$ in given by $V_t(0) = (0.98 \pm 0.01)$ Volt and $V_{\text{shift}}(T)$ is the translating curves voltage. The extrapolation of the linear fit allows to evaluate the critical temperature $T^* \approx 196\,K$, at which $V_t(T^*) = 0$. The *I-V* curves collapse into a single master curve if they are translated by the appropriate voltage $V_{\text{shift}}(T)$ confirming the scaling behavior (apart a small statistical spread) [70]. Once the threshold voltage has been determined, the *I-V* characteristics were fitted using A and ς as fitting parameters. For the *I-V* characteristics in the 70 K÷190 K, $\zeta \approx 2$, independent from the temperature, consistent with a 2D conduction mechanism. This implies that the longitudinal electronic transport is mainly confined in the planes visible in the cross-view TEM in Fig. 19(b). It is reasonable because the mean surface–surface distance between clusters in the plane is close to 10 nm while the mean distance between adjacent planes is about 25 nm so that the tunneling between clusters lying in adjacent planes is unlike with respect to the tunneling between adjacent clusters. At $T > 196\,\text{K}$ and low field, a thermally activated conduction is expected (Equation 2.3.3). To verify such behavior, we calculated the differential conductance $G = dI/dV$ as a function of temperature. Effectively, the linearity of $\log G$ versus T^{-1} was verified, for each fixed applied voltage until 1 V, with activation energy of (38.5±0.5) meV obtained by the fit of the experimental data [70]. According to Parthasarathy's model, the activation energy of the conduction process is expected to be $U = bk_BT^*$, where b is a parameter that should be equal to 2.4. Coupling our determination of T^* and of the activation energy above the critical temperature, we evaluated $b = 2.35 \pm 0.03$ in excellent agreement with the theoretical expectation. The activation energy $U = (38.5 \pm 0.5)\,\text{meV}$

indicates a mean value of the barrier height between two adjacent nanograins in this "over barrier" transport regime. For applied voltage above 1 V, the activated form of the differential conductance is not more respected. So, in agreement with the theoretical expectation of a scaling law at high field, we translate the *I*-*V* characteristics measured at $T > 196\,\text{K}$ on the master curves (to obtain their coincidence at least for voltage above 1 V) in order to determine $V_{\text{shift}}(T)$. In this way, we obtained the effective negative threshold voltages evaluated by the Equation (18) and reported in Fig. 22. The *I*-*V* characteristics at $T > 196\,\text{K}$ and $V > 1$ Volt were again fitted by MW scaling function described by Equation (15) using the previously determined effective threshold voltage using A and ζ as fitting parameters, where $\zeta \approx 2$, independent from the temperature, in the 200 K÷300 K range, confirming the Elteto's prediction [70] in examined "non-standard system."

A final remark on the consistence between the structural and electrical characterizations is that the structural analyses furnished a mean cluster radius $< R > \approx 3.6\,\text{nm}$ and from this the mean charging energy $E_C = e^2/2C_0 \approx 17\,\text{meV}$, which allows to evaluate a critical temperature (under which the Coulomb blockade phenomenon is visible) $T^* = E_C/k_B \approx 196\,\text{K}$, which is coincident with the critical temperature determined by the analyses of the *I*-*V* characteristics using the Parthasarathy's model.

References

1. P. Moriarty, Rep. Prog. Phys. 64, 297, 2001.
2. D. Goldhaber-Gordon, M. S. Montemerlo, J. C. Love, G. J. Opiteck, J. C. Ellenbogen, Proc. IEEE 85, 521, 1997.
3. Nanoparticles, edited by G. Schmid, Wiley-VCH, 2004.
4. W. A. de Herr, Rev. Mod. Phys. 65, 611, 1993.
5. P. Milani, S. Iannotta, Cluster Beam Synthesis of Nanostructured Materials, Sprinter, Berlin, 1999.
6. C. Binns, Surf. Sci. Rep. 44, 1, 2001.
7. J. A. Venables, G. D. T. Spiller, M. Hanbücken, Rep. Prog. Phys. 47, 399, 1984.
8. A. W. Adamson, Physical Chemistry of Surfaces, Wyley, New York, 1990.
9. M. Zinke-Allmang, L. C. Feldman, M. H. Grabov, Surf. Sci. Rep. 16, 377, 1992.
10. R. Sangiorgi, M. L. Muolo, D. Chatain, N. Eustathopoulos, J. Am. Ceram. Soc. 71, 742, 1988.
11. Z. Wang, P. Wynblatt, Mat. Sci. Eng. A 259, 287, 1999.
12. P. Buffat, J. P. Borel, Phys. Rev. A 13, 2287, 1976.
13. D. Barreca, A. Gasparotto, E. Tondello, G. Bruno, M. Losurdo, J. Appl. Phys. 96, 1655, 2004.
14. F. Ruffino, A. Canino, C. Bongiorno, F. Giannazzo, F. Roccaforte, V. Raineri, M. G. Grimaldi, J. Appl. Phys. 101, 064306, 2007.
15. C. Herring, in Structures and Properties of Solid Surface, edited by R. G. Gomer and C. S. Smith, University of Chicago Press, Chicago, 1953.
16. R. F. Sekerka, J. Cryst. Grow. 275, 77, 2005.
17. A. S. Barnard, P. Zapol, J. Chem. Phys. 121, 4276, 2004.
18. M. Brack. Rev. Mod. Phys. 65, 677, 1993.
19. T. P. Martin, Phys. Rep. 273, 199, 1996.
20. R. N. Barnett, C. L. Cleveland, H. Häkkinen, W. D. Luedtke, C. Yannoules, U. Landman, Eur. Phys. J. D 9, 95, 1999.
21. R. L. Johnston, Atomic and Molecular Clusters, ed. Taylor and Francis, 2002.

22. G. De Marchi, G. Mattei, P. Mazzoldi, C. Sada, A. Miotello, J. Appl. Phys. 92, 4249, 2002.
23. L. R. Wallenberg, J. O. Bovin, G. Schmidt, Surf. Sci. 156, 256, 1985.
24. L. D. Marks, Rep. Prog. Phys. 57, 603, 1994.
25. Z. Wang, P. Wynblatt, Surf. Sci. 398, 259, 1998.
26. A. Baldan, J. Mat. Sci. 37, 2171, 2002.
27. K. Shorlin, S. Krylov, M. Zinke-Allmang, Phys. A 261, 248, 1998.
28. I. M. Lifshitz, V. V. Slyozov, J. Phys. Chem. Solids 19, 35, 1961.
29. C. Wagner, Z. Electrochem. 65, 581, 1961.
30. B. K. Chakraverty, J. Phys. Chem. Solids 28, 2401, 1967.
31. Á. W. Imre, E. G. Gontier-Moya, D. L. Beke, I. A. Szabo, G. Erdélyi, Surf. Sci. 441, 133, 1999.
32. Beszeda, E. G. Gontier-Moya, Á. W. Imre, Appl. Phys. A 81, 673, 2005.
33. K. N. Tu, J. W. Mayer, L. C. Feldman, Electronic Thin Film Science, Macmilian Publishing Company, New York, 1992.
34. A. Miotello, G. De Marchi, G. Mattei, P. Mazzoldi, C. Sada, Phys. Rev. B 63, 075409, 2001.
35. D. R. Collins, D. K. Schroder, C. T. Sah, Appl. Phys. Lett. 8, 323, 1966.
36. M. A. Lamkin, F. L. Riley, R. J. Fordham, J. Eur. Ceram. Soc. 10, 347, 1992.
37. G. Battaglin, in Modifications Induced by Ion Beam Irradiation in Glasses, edited by P. Mazzoldi, Elsevier, Amsterdam, 1992.
38. L. Thomé, G. Rizza, F. Garrido, M. Gusso, L. Tapfer, A. Quaranta, Appl. Phys. A 67, 241, 1998.
39. F. Gonnella, Nucl. Instr. and Meth. B 166–167, 831, 2000.
40. F. Ruffino, R. De Bastiani, C. Bongiorno, F. Giannazzo, F. Roccaforte, C. Spinella, V. Raineri, M. G. Grimaldi, Nucl. Instr. and Meth. B 257, 810, 2007.
41. D. Datta, S. R. Bhattacharyya, Nucl. Instr. and Meth. B 212, 201, 2003.
42. K. H. Heinig, T. Müller, B. Schmidt, M. Strobel, W. Möller, Appl. Phys. A 77, 17, 2003.
43. G. Rizza, H. Cheverry, T. Gacoin, A. Lamasson, S. Henry, J. Appl. Phys. 101, 014321, 2007.
44. G. Rizza, M. Strobel, K. H. Heinig, H. Bernas, Nucl. Instr. Meth. B 178, 78, 2001.
45. A. Ionescu, Ecole Polytechnique Fédérale de Lausanne, "An Outlook of technology scaling beyond Moore's law", 2005, http: si.epfl.ch.
46. F. Ruffino, F. Giannazzo, F. Roccaforte, V. Raineri, M. G. Grimaldi, Appl. Phys. Lett. 89, 243113, 2006.
47. R. Wiesendanger, Scanning Probe Microscopy and Spectroscopy: Methods and Applications, Cambridge University Press, New York, 1994.
48. F. Giannazzo, F. Roccaforte, V. Raineri, S. F. Liotta, Europhys. Lett. 74, 686, 2006.
49. A. S. Groove, Physics and Technology of semiconductor Devices, John Wiley and Sons, Inc., New York, 1967.
50. W. Zhu, G. P. Kochanski, S. Jin, Science 282, 1471, 1998.
51. L. A. Kosyachenko, V. M. Sklyarchuk, Ye. F. Sklyarchuk, Solid-State Electronics 42, 145, 1998.
52. L. P. Kouwenhoven, D. G. Austing, S. Tarucha, Rep. Prog. Phys. 64, 701, 2001.
53. C. Kittel, H. Kroemer, Thermal Physics, Freeman and Company, San Francisco, 1980.
54. A. Zangwill, `Physics at Surfaces', Cambridge University Press, Cambridge, 1988.
55. C. Wasshuber, About single-electron devices and circuits, Ph. D. Dissertation, 1997.
56. K. K. Likharev, Proc. IEEE, 87, 606, 1999.
57. Y. W. Tan, J. Zhu, H. L. Stormer, L. N. Pfeiffer, K. W. Baldwin, K. W. West, Phys. Rev. Lett. 94, 016405, 2005.
58. M. B. Cortie, E. van der Lingen, Mat. Forum 26, 1, 2002.
59. D. K. Ferry, S. M. Goodnick, Transport in Nanostructures, Cambridge University Press, Cambridge, 1997.
60. A. E. Hanna, M Tinkham, Phys. Rev. B, 44, 5919, 1991.
61. L. I. Glazman, J. Low Temp. Phys. 118, 247, 2000.
62. D. V. Averin, A. N. Korotkov, J. Low Temp. Phys. 80, 173, 1990.

63. D. Sarid, Exploring Scanning Probe Microscopy with Mathematica, John Wiley and Sons, Inc., New York, 1997.
64. F. Ruffino, F. Giannazzo, F. Roccaforte, V. Raineri, M. G. Grimaldi, Appl. Phys. Lett. 89, 263108, 2006.
65. R. Parthasarathy, X.-M. Lin, K. Elteto, T. F. Rosenbaum, H. M. Jaeger, Phys. Rev. Lett. 92, 076801, 2004.
66. R. Parthasarathy, X.-M. Lin, H. M. Jaeger, Phys. Rev. Lett. 87, 186807, 2001.
67. A. S. Cordan, Y. Leroy, A. Goltzené, A. Pépin, C. Vieu, M. Mejias, H. Launois, J. Appl. Phys. 87, 345, 1999.
68. M. G. Ancona, W. Kruppa, R. W. Rendell, A. W. Snow, D. Park, J. B. Boos, Phys. Rev. B 64, 033408, 2001.
69. K. C. Beverly, J. F. Sampaio, J. R. Heat, J. Phys. Chem. B 106, 2131, 2002.
70. F. Ruffino, A. M. Piro, G. Piccitto, M. G. Grimaldi, J. Appl. Phys. 101, 024316, 2007.
71. A. A. Middleton, N. S. Wingreen, Phys. Rev. Lett. 71, 3198, 1993.
72. K. Elteto, E. G. Antonyan, T. T. Nguyen, H. M. Jaeger, Phys. Rev. B 71, 064206, 2005.

Nanocrystal-Based Polymer Composites as Novel Functional Materials

M. Striccoli, M.L. Curri, and R. Comparelli

Abstract This chapter provides an overall picture of nanocrystal-polymer based composites and describes the key properties of these original functional materials, particularly suited for advanced applications in photonic, optoelectronic as well as in sensing. Here, we aim at pointing out the relevance of the incorporation of inorganic colloidal nanocrystals with size-dependent properties in highly processable polymers. Due to the countless different combination of material types and, accordingly, the large extent of the topic, this contribution will focalize mainly on luminescent semiconductor nanocrystals embedded in plastic structurable matrices.

First, an overview on the complex and various scenarios of the nanocomposite preparation strategies will be provided. Next, the original properties of the prepared nanocomposites will be illustrated, paying particular attention to their fabrication by means of conventional and emerging micro- and nanoscale processing techniques. Finally, recent examples of applications of nanocomposite materials in photonic, optoelectronic and sensing devices will be reviewed.

Contents

M. Striccoli (✉)
CNR-IPCF Institute for Physical Chemistry Processes c/o Department of Chemistry, University of Bari, Italy
e-mail: m.striccoli@ba.ipcf.cnr.it

Z.M. Wang (ed.), *Toward Functional Nanomaterials*, Lecture Notes in Nanoscale Science and Technology 5, DOI 10.1007/978-0-387-77717-7_4,

1 Introduction

Significant research efforts in the last years have been dedicated to the design, preparation and characterization of inorganic particles-polymer based nanocomposite materials. The term "nanocomposite" refers to compound materials in which at least one of the components has characteristic dimensions in the nanometer range. The enormous interest in nanocomposite materials relies mainly on the extraordinary range of properties deriving from the synergy between the properties of the single components, such as the original physical and chemical characteristics of the nanoparticles and the high processability of the polymers. When different materials are combined to form heterogeneous structures, the properties of the resulting material depend on the properties of the constituent materials, as well as on the chemical and morphological details of the dispersion. In addition, nanometer-sized particles present an inherently large surface-to-volume ratio, and the diminished size of the particles induces the modification of certain properties, preventing the alteration of others, e.g., mechanical resistance enhancement while maintaining optical transparency, etc. Also, at high concentration of nanoparticles in the polymer matrix, when the inter-particle distance is reduced, proximity effects and cooperative phenomena can arise in nanocomposite systems. For instance, in closely spaced particle arrangements, dipolar interactions can arise, resulting in new collective excitation modes that dramatically impact the materials' optical or magnetic properties.

Special attention must also be paid to the occurrence of structural synergism between the nanocomposite components, with the concomitant "emergence" of physical properties which are not inherent in the single component but a consequence of the micro-structural arrangements of the nanosized objects in the polymer [2]. In addition, the huge potential for applications arises not only from the enhancement of static properties, but even from the dynamic response of nanocomposite materials to external stimuli [50].

The incorporation of fluorescent semiconductor nanocrystals in polymers has recently interested the scientific community, thanks to the high technological impact of these composite materials in optical display and LED [9, 17], nonlinear optical devices [45, 46] and biolabelling. Indeed, the incorporation of emitting nanoparticles in suitable polymers allows adding new functionalities to polymer materials that can be suitable for the fabrication of all-polymer devices. The interest in semiconductor nanocrystals is mainly due to their peculiar size-dependent optical properties exhibited in quantum confinement regime, when their dimensions are comparable with the de Broglie wavelength associated with the charge carriers. Their narrow emission, high photoluminescence quantum yield, large absorption spectra as well as their spectral tunability [4, 31] (Fig. 1) and fast relaxation dynamics make the nanocrystals potential candidates for the development of devices in several fields ranging from photonic to optoelectronic and sensing [39, 54].

Among the several strategies for the preparation of nanocrystals, the chemical colloidal routes offer numerous advantages. In particular, the decomposition of defined precursors in hot solvents in the presence of coordinating agents accesses a good control on size and shape of nanocrystals and narrow size distribution. The

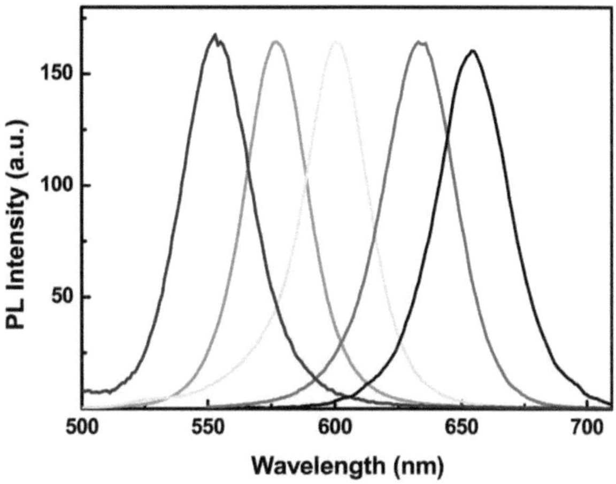

Fig. 1 Emission spectra of different sized CdSe/ZnS nanocrystals in $CHCl_3$

nanocrystal dimensions and shapes can be, thus, finely controlled and suitably tailored by changing reaction parameters such as temperature, time, surfactants and ratio between the precursors, as proved by the large number of papers on the subject [6]. Such synthetic routes provide nanocrystals presenting a surface coated with organic agents, which can play a twofold role, since they allow dispersion in organic solvents and at the same time improve the surface passivation. II–VI semiconductor nanocrystals such as CdS, CdSe and CdTe emitting in the visible range and InAs, PbS and PbSe emitting in the near infrared can be easily synthesized by the colloidal method. Other notable benefit of the synthetic chemical procedures relies on the consideration that the synthesized nanocrystals can be easily functionalized by changing the capping ligands, in order to make them compatible with several host environments.

While nanocrystals can be easily dissolved in various solvents, they need to be homogeneously dispersed in an inert or functional matrix in order to effectively exploit their specific functionality in practical applications. The basic requirement is then the good dispersibility of the nanocrystals in such organic materials; several strategies have been followed in order to produce stable and highly luminescent composites.

Here, we will focus on the incorporation of luminescent nanocrystals in plastic and structurable polymers with defined properties, i.e., transparent in the visible range, in order to obtain composite materials that can be flexibly fabricated for integration in optoelectronic, photonic and sensing devices.

2 Strategy of Nanocomposites Preparation

Several routes have been accounted for the preparation of hybrid nanocrystal–polymer nanocomposites, and diverse classifications can be made.

A broad distinction can be made considering "in situ" nanoparticle formation methods, in which nanoparticles are directly synthesized in the polymeric environment and "ex situ" type of strategy, in which pre-made nanoparticles can be incorporated in monomers before the polymerization process or directly in the organic matrix (Fig. 2).

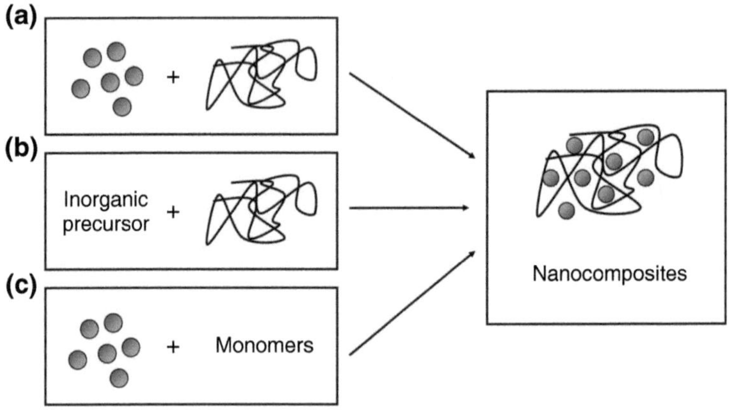

Fig. 2 Strategies of nanocomposites preparation: (**a**) mixing of pre-formed nanoparticles and polymer; (**b**) nanoparticle precursors incorporated in the polymer matrix; (**c**) pre-synthesized nanoparticles incorporated in the monomers before the polymerization process

"In situ" nanocrystal growth of CdS nanoparticles has been successfully performed within a polystyrene network [20]. The morphology, the surface properties and, hence, the fluorescence properties of the obtained nanoparticles have been found to be dependent on the Cd^{2+} feed ratio. The resulting hybrid composites have been fabricated into films, used as coatings, or even in further polymer blending, showing a high value of the effective nonlinear refractive index n_2 variable with the input laser energy and the concentration of CdS nanoparticles. Several experiments have been carried out on nanoparticle formation in block copolymer. In particular, nanoparticles have been synthesized in lamellar, cylindrical and spherical domains of a functionalized block copolymer in which the monomers were preloaded with metal salts. Semiconductor nanoparticles were prepared within the domains of a norbornene-derived block copolymer preloaded with lead salt [69]. Recently, however, the enhanced ability to synthesize block copolymer with suitable characteristics enabled access to another fabrication approach based on the nanoreactor scheme. In this case, amphiphilic block copolymer which contains polar groups is first microphase separated and subsequently immersed in a salt solution; due to the chemical affinity, the salt selectively infiltrates the hydrophilic copolymer domain. Nanocrystals are then formed upon reduction within the precursor-loaded domains. Following this approach, Boontongkong [3] demonstrated that block copolymer nanoreactor scheme can be applied to a variety of metals as well as semiconductor nanocrystals. Although the mechanism of the nanoparticle formation within the polymer domains is still disputed, recent studies suggest a reaction diffusion controlled nucleation and growth process, which is initiated by the formation of small clusters that act as nucleation sites and subsequently grow, depleting the reactant from the surrounding polymer matrix.

The ability to control both the length scales and the spatial and orientational organization of block copolymer morphologies makes these nanocomposite fabrication routes particularly attractive. The induced long-range order of the nanoscopic inclusions in the polymer scaffold results particularly interesting for applications that require specific composite microstructure geometry. On the other hand, the "in situ" nanocomposite preparation approaches generally reveal a limited control on the synthetic parameters together with relatively low interaction between nanocrystals and surrounding host medium, which weaken the surface passivation ability of the polymer component at nanocrystal surface [56, 22].

From this perspective, among the different approaches for accessing inorganic nanoparticle-based polymer nanocomposites, handling nanoparticles and polymer moiety as separate entities seems to be particularly profitable, with the nanocomposite preparation decoupled from the nanocrystal synthesis, which is carried out "ex situ."

Nanocomposites based on pre-synthesized nanosized inorganic particles and crystals have gained increasing interest due to the fact that such a preparation scheme allows to directly convey the unique size-dependent nanocrystal properties into the host matrix and to effectively combine them with the other defined properties of the materials. An additional advantage of such hybrid composites of particles in organic polymer matrix is the possible protection of systems that are sensitive towards environmental conditions (oxygen, humidity, etc.) by coverage with a passivating layer in the polymer matrix. Such an effect can be suitably tailored by tuning the interaction between the specific polymer chemistry and the nanocrystal surface. Furthermore, this class of approach allows a fine control of the inorganic moiety content, which can be thus gradually varied in the polymer matrix. Finally, a fine tune of the ordering capability is also achievable, thus making it possible to range from highly ordered superlattices, i.e., in suitable block copolymer host systems, to the so-called stochastic mixtures, where the particles are randomly dispersed.

One of the major drawbacks of this approach is that, due to the high surface energy of the nano-objects, physical mixture of organic polymers and preformed inorganic nanoparticles usually lead to separation in discrete phases, resulting, for instance, in a deterioration of mechanical, optical and electrical properties. In addition, particle agglomeration tends to weaken mechanical or optical properties of the resulting composite material. Unmodified particles tend to aggregate in the polymer matrix independent of the material size and composition.

There are several possibilities to overcome the phase separation that generally occurs in hybrid particulate systems (Fig. 3), either acting on the organic component by using functional polymers, which interact with the surface of nanoparticles, or by surface modifying the particles, to tailor their specific chemistry. Alternative procedures are sometimes viable, especially when the chemical modification of one of the two moieties could be detrimental for the overall properties and processability of the system. In such cases, the nanocrystal incorporation could rely on

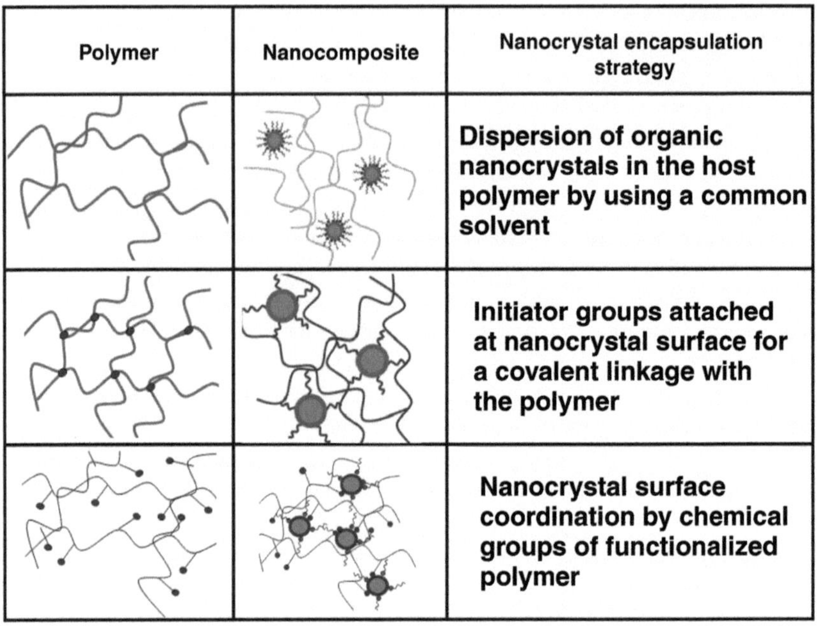

Fig. 3 Strategies for homogeneous dispersion of nanocrystals in polymer matrix

the selection of a common solvent that is able to safely disperse pre-synthesized organic-capped nanoparticles in a host polymer matrix. Such a common medium should then be selected as a function of the systems characteristics (i.e., polarity, processability) and be compatible with polymer properties and processing requirements, as well as able to preserve the distinctive properties of the functional nano-sized components [32].

The "ex situ" composite preparation methods transfer the peculiar size-dependent features of the inorganic nanocrystalline components in the nanocomposiite property profile. Indeed the optimization of the nanocrystal properties is carried out prior to the incorporation in the host matrix. Accordingly, a fundamental step in such a nanocomposite design is given by the synthesis of the suitable nanocrystalline building block.

The electronic and optical properties of semiconductor nanocrystals are strongly dependent on their size, shape and surface chemistry; thus the preparation of nearly monodisperse (with size distribution $\sigma \leq 5\%$) nanocrystals with fine control over size and shape is essential to fabricate composite materials with tailored characteristics [10].

The key point to synthesize semiconductor nanocrystals with controlled dispersion in size and shape is the careful control on the separation between nucleation and growth stages. This goal can be achieved in the presence of one or more organic molecules in the reactor, generally termed as "surfactants." Such amphiphilic compounds are molecules composed of one hydrophilic moiety (a polar or a charged functional group) and one hydrophobic moiety (usually one or several hydrocarbon chains). They can act in two main different ways: (i) they can assemble in solution yielding to *organic templates* or (ii) they can act as *terminating agents*.

(a) *Organic Templates*: In certain experimental conditions, depending on surfactant concentration, temperature, solvent polarity, additives and ionic strength, surfactants self-assemble into soluble micellar templates [62]. The shape of the amphiphilic compound, its concentration and the chemical composition of the dispersing agent (i.e., the water/oil ratio) play a fundamental role in controlling the shape of such micellar templates, which can assume spherical shapes or evolve into rod-like or cylindrical aggregates, flexible bilayers, and planar bilayers [62]. Therefore, the liquid phase can contain hydrophilic or hydrophobic compartments of nanometer size prior to the synthesis, which act as nanocrystal templates. The size and shape of the nanocrystal are predetermined by the micellar size and shape, while the nanocrystal growth is naturally terminated once a nanocrystal fills the compartment volume [38, 43, 62].

(b) *Terminating Agents*: Amphiphilic molecules carrying functional moieties having good affinity for the nanocrystal surface can control the growth acting as "terminating" agents. Such molecules direct the growth of nanostructures by dynamically coordinating their surface under the reaction conditions. In general, suitable surfactants are composed of one or more alkyl chains (usually $C_8 \div C_{18}$) and a functional group with electron–donor atoms such as carboxylic, alkylthiols, phosphines, phosphine oxides, phosphates, phosphonates, amides or amines and nitrogen-containing aromatics [43, 16, 48, 40, 6, 60]. In a typical synthesis, each of the atomic species that will form the nanocrystals is introduced into a reactor at high temperature (250–300°C) in the form of a precursor. A precursor is a molecule or a complex containing one or more atomic species representing the nanocrystal building blocks. Once the precursors are introduced into the reaction flask, they react or decompose, generating the reactive monomers that will cause the nucleation and growth of the nanocrystals [52]. The size and shape control during the growth is achieved by the presence of one or more terminating agents in the reactor. The choice of the proper terminating agent varies from case to case: a molecule that binds too strongly to the surface of the nanocrystal is not useful, as it would hinder the growth. On the other hand, a weakly coordinating molecule would yield large particles or aggregates [59]. The surfactant molecules must be chemically stable at the reaction temperature in order to be suitable candidates for controlling the growth. The coordinating bond between the terminating agent and the nanocrystal surface can be temperature controlled. At high temperature, the terminating agent adsorbs and desorbs quickly from the cluster surface, thus preventing aggregation, but, at the same time, allowing for the continuous supply of new monomers for the growth. The growth can be stopped by simply cooling the reaction vessel, as at room temperature the surfactants are strongly bound to the surface of the synthesized nano-object and provide solubility in a wide range of organic solvents. This organic coating allows for great synthetic flexibility, since it can be exchanged with organic molecules having different functional groups or polarity. More importantly, surface defects (i.e., unsaturated valences) that act as traps for electron and holes, in turn degrading the nanocrystal optoelectronic properties, can be effectively passivated through coordination bonds with the surface ligands. In addition, the surfactants can be temporarily removed and an epitaxial layer of another material with different electronic, optical or magnetic properties can be grown on the initial nanocrystal (core/shell structures) [18, 58]. Delicate control of parameters, such as strain, surface reactivity, interfacial energy and crystal solubility at the nanoscale [13, 14, 90, 49, 73, 51, 37], opens access to various complex nanocrystal morphologies and even hybrid architectures, ranging from simple core/shell systems to individual nanostructures with linear and/or branched topology [49, 73], with site-specific deposits of a different material [37, 51], and heterogroups made of magnetic, metal or fluorescent spherical nanocrystals [28, 80].

3 Nanocrystal Functionalization

Due to the typically unfavorable interactions between the particle surface and the host polymer, most particles will need to suitably modify their surface. The ligand binding the particle surface plays a critical role in determining the overall interaction. Selective chemical compatibilization of the nanoparticles into a block copolymer can be achieved by attaching an oligomer or polymer to the particle surface that will favorably interact with the target copolymer domain [2]. On the other hand, the capping agent used in the synthesis can be replaced by a second one having higher compatibility with the polymer matrix by post-synthesis treatments. Conversely, the mixing of isolated and naked inorganic nanocrystal powders with polymers can usually induce the formation of strongly connected aggregates, due to the high specific surface energy of nanocrystals [7]. In fact, only inorganic colloidal nanocrystals with a strongly coordinated surface layer of organic molecules can prevent the formation of aggregates within the polymer matrix, due to the encapsulation of organic-capped nanocrystals within the polymer through ionic or hydrophobic interaction.

The capping exchange procedure can be effectively carried out in particularly complex cases when (i) ligand conformation, (ii) ligand chain packing on a faceted surface and (iii) polymer chain diffusion into the ligand layer are critical parameters interplaying in a delicate equilibrium [12]. Therefore, a careful choice of a suitable ligand for nanocrystals can ensure a good dispersion and maximize the interaction between nanocrystals and host matrix.

The capping exchange can occur either by mass action or by using functional groups with an affinity for the nanocrystal surface stronger than the pristine ligands.

For instance, oleic acid (OLEA)-capped CdS nanocrystals can be functionalized with alkylamine differing in the alkyl chain length (C_8–C_{16}) by refluxing the as-prepared nanocrystals in alkylamine for 24 h under mild heating, exploiting the mass action principle [11, 75, 92]. Likewise, the original OLEA coating on TiO_2 nanorods can be easily exchanged by capping with an alkylphosphonic acid by removing the excess of OLEA by repeatedly washing with a suitable precipitating agent. Subsequently, an alkylphosphonic acid solution in $CHCl_3$ can be added dropwise under stirring at room temperature until a clear solution is obtained [14].

On the other hand, electron-donor groups (thiols [33], phosphines [36], amines [5], and nitrogen-containing aromatics like pyridine and its derivative [94]) can be used to displace the original ligand exploiting their strong affinity for the nanocrystal surface.

A plethora of different ligands have been used in literature to promote the embedding of pre-synthesized colloidal nanocrystals in polymer matrices. For example, the surface of trioctylphosphine oxide (TOPO)-capped CdSe and CdSe/ZnS nanocrystals has been modified by Wang et al. with poly(*N*,*N*-dimethylaminoethyl methacrylate) (PDMA) and p-(dimethylamino)pyridine (DMAP) [87].

A polymer ligand that can be attached to semiconductor nanocrystals via a phase transfer reaction has been recently proposed by Potapova et al. The ligand consists of a chain of reactive esters, which can be substituted with different molecules containing amino functionalities. Such a polymer ligand has been successfully used to functionalize TOPO-capped CdSe nanocrystals [63]. Alkylammonium-functional polymers (AFCPs) have been used to coat fluorescent CdTe nanocrystal modified with a polymerizable surfactant allowing to fabricate nanocrystal–polymer complexes. The resultant complexes have been processed to form both macroscopic and microscopic fluorescent structures and materials [94].

Another concept for nanocomposite preparation is based on the coated nanoparticles which can act as starting material for improved nanocomposites. The peculiar and original properties of nanocomposites rely on the properties of isolated nanoparticles and can be lost after combining them with a polymer matrix. Therefore, it is necessary to avoid or at least reduce the interaction of the particles. This can be achieved by coating each individual particle with a shell of another material, i.e., a polymer layer.

The presence of an initial hydrophobic layer on the surface of the particles seems to be crucial for the formation of the polymer shell. This can be achieved by either using a suitable synthetic route or introducing nonionic surfactants, amphiphilic block copolymers, etc., at nanoparticle surface. The concentration of such compatibilizing molecules in solution needs to be tuned carefully because a latex formation in free micelles in the emulsion can be observed at higher concentrations. Additionally, these molecules are usually only weakly bonded to the surface and can therefore be desorbed easily. Hence, the covalent attachment of organic groups, which potentially can interact in the polymerization reaction, is used for surface modification. Encapsulation of inorganic nanoparticles can be carried out by applying an emulsion polymerization process. Polymerization occurs primarily at the surface of unmodified particles due to the adsorption of the monomer on the surface, followed by the polymerization of the adsorbed layer [1, 82]. Another experiment has been reported on the "in situ" encapsulation of nanocrystals into polymer microspheres by using functionalized oligomeric phosphine ligands [71]. However, in spite of the retained quantum efficiency and the uniform distribution of the nanocrystals in the polymer matrix of the spheres, it has been found that at higher nanocrystal concentrations the formation of self-polymerized nanoparticle aggregates limits the loading fractions possible and the addition of nanocrystals and ligands perturbs the nucleation and growth processes of polymer beads, resulting in a widening of the size distributions. In fact, such a strategy of chemically incorporating nanocrystals into a polymer matrix by polymerizable ligands hinders the formation of uniform microspheres.

The surface of nanocrystals can also be effectively modified by encapsulation with their initial ligand shell, using organic dendrons [29], amphiphilic polymers [57, 85–87], polyelectrolytes [61] and oligomeric phosphine [36] as multidentate ligands. In particular, the role of poly(acrylic acid) and PDMA derivatives in stabilizing the CdSe/ZnS nanocrystal surface has been successfully investigated by several authors, in fields ranging from molecular recognition to supramolecular organization of nanocrystals [53, 42, 88, 95]. Interactions based on multivalent ligands result extremely efficient, due to the amplifying effect of multiple bonding sites, which can ultimately increase the coordination degree [36] and avoid, at the same time, nanoparticle aggregation while preserving their optical properties. Following the above considerations, two distinct transparent copolymers of poly(methyl methacrylate) (PMMA), namely poly(methyl methacrylate-co-acrylic acid) (PMMA-co-MA) and poly(methyl methacrylate-co-dimethylaminoethyl methacrylate) (PMMA-co-DMAEMA), carrying carboxylic acid and amine groups respectively on the repeating polymeric unit, have been used for incorporating luminescent colloidal TOPO-capped CdSe/ZnS nanocrystals. Amines and carboxylic acids on the polymer chains can interact directly with the nanocrystal surface, replacing or intercalating the original capping ligands, resulting in an homogeneous dispersion of nanoparticles in the organic matrix and in the enhancement of nanocrystal luminescence quantum yield [78].

4 Nanocomposite Engineering

The design of new functional materials, characterized by novel and tailored physicochemical properties and, at the same time, an enhanced processing capability, is one of the most crucial challenges in modern material science. New composite materials, based on polymers functionalized with inorganic nanoparticles, are particularly attractive, thanks to their peculiar properties [34, 68].

Original mechanical and optical properties arise when the size-dependent characteristics of nanosized particles are effectively integrated into the high processability of polymers, providing novel advanced materials [27, 72]. Examples of nanocomposites have been demonstrated to exhibit extraordinary optical [77, 84] and magnetic [25] properties, as well as conductivity [24], permeability [15], catalytic activity [21] and mechanical strength [67].

At the same time, these functional materials are suitable for flexible and innovative fabrication processes, allowing to achieve micro- and nanometric patterning, which are required for the manufacturing of original devices in photonic and optoelectronic applications [19].

The potential for applications of nanocomposite materials relies on the ability to design and fabricate these materials at the nanoscale. Such functional materials need to be structured in order to exploit their original properties for integrated devices and fabrication processes. Different techniques are currently used for nanocomposite patterning and fabrication, ranging from conventional microstructuring approaches, such as UV photolithography and E-beam lithography, to unconventional patterning methods, such as nanoimprint lithography (NIL), inkjet printing, etc.

The synergetic combination of colloidal nanostructures with lithographic patterning can enable the precise position control necessary to produce highly integrated nanostructured assemblies on all length scales. Conventional and emerging methodologies of deposition and patterning techniques have been used to process nanocomposite materials in order to fabricate original devices both at micro and nano scales. Thick structures realized by the well-known procedure of casting have been demonstrated by Yang and coworkers by using polymerizable surfactants such as AFCP on CdTe nanocrystals, allowing to produce highly luminescent and transparent solids or thick films [93, 94]. The solid nanocomposite materials appear transparent in the visible range, highly fluorescent and photostable against UV radiation, and can be easily processed in a large variety of shapes and forms. Also, the solutions of AFCP–CdTe nanocrystals can be used as a polymer coating and as an ink for the fabrication of ordered structures by micro contact printing and soft lithography techniques [74, 89].

In addition, the preparation of fluorescent microspheres, simply dispersing emulsion droplets of the nanocomposites in a water solution of poly(vinylalcohol) (PVA), followed by evaporation of the solvents has been obtained. Multicolor materials can also be prepared by using distinct sized CdTe nanocrystals, with emission of different colors, by adjusting the relative intensities and the concentration of the nanocrystals in solution. Full-color emission was demonstrated also for CdSe/ZnS nanocrystals in polymers from Lee [39] by using a poly(lauryl methacrylate) (PLMA) matrix in the presence of extra trioctylphosphine (TOP) organic ligands. Randomly mixed nanocrystal blends suffer from the energy transfer phenomena from larger to smaller nanocrystals, with a relevant fluorescence quenching result. In this perspective, spin-coating deposition is a choice technique for fabrication of layered geometries in luminescent nanocomposite materials. In particular, white light generation has been demonstrated in layered structures of blue-yellow emitting nanocrystal–polymer composites.

Among the innovative deposition techniques, inkjet printing is an attractive tool for microscale patterning since it allows the local positioning of tiny droplets of functional materials onto addressable sites of a substrate.

The used additive approach with low waste of materials, the large choice of available substrates and the possibility of multiple depositions are the main advantages of this technique and it is recently used for the fabrication of multicolor LED, solar cells and transistors [70, 91]. CdTe nanocrystals in PVA were recently used as ink in inkjet printing to produce combinatorial libraries, by changing nanocrystal size (emission color) and loading (intensity) into the polymer matrix [79]. The homogeneity of the droplet thickness on the substrate was improved by the addition of ethylene glycol to the nanocomposite solution, avoiding the accumulation of functional materials at the edge of the deposited structure.

CdSe/ZnS nanocrystals have been incorporated also in epoxy resin-based negative photoresist, particularly suitable for high aspect-ratio surface micromachining (MEMS and NEMS) permanent applications [32]. The proper selection of a common solvent was the critical point that allowed the embedding of the nanocrystals, even in the presence of partial nanoparticle aggregation. The nanocrystal-modified resist was patterned by standard UV lithography down to micrometer resolution and high aspect-ratio structures were successfully fabricated on a 100-mm wafer-scale, showing the emission features of the incorporated nanocrystals without deterioration of the lithographic performances of the resist.

A very homogeneous dispersion of the nanoparticles in the polymer has been recently demonstrated by the incorporation of CdSe/ZnS nanocrystals into properly functionalized PMMA-based copolymers, associated with an increase in the luminescence quantum yield (Fig. 4a,b) [76, 78]. The excellent processability of the thermoplastic PMMA-based polymers and the easiness of such nanocrystal–polymer

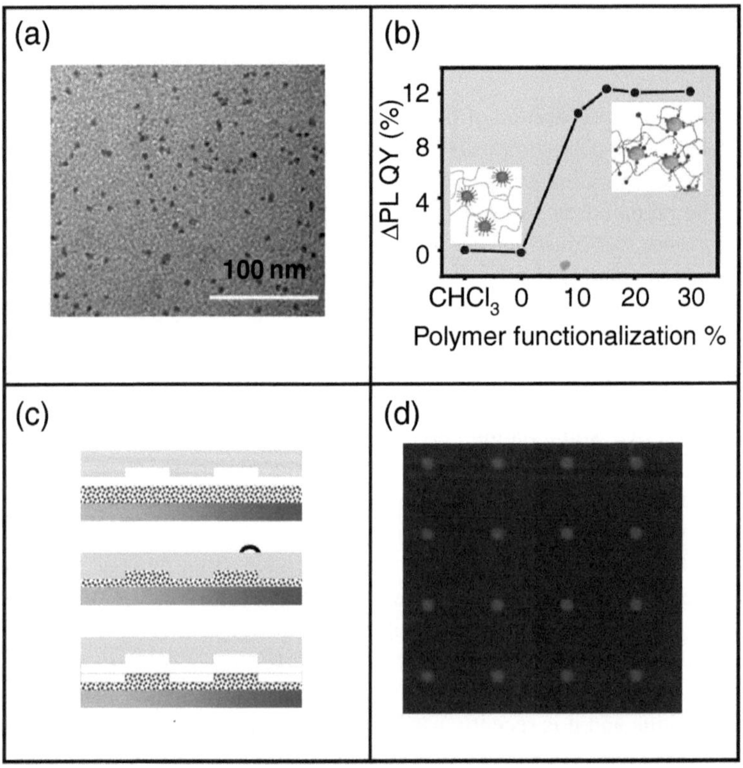

Fig. 4 Schematic diagram illustrating the results of the incorporation of CdSe/ZnS nanocrystals in PMMA-based functionalized copolymers: (**a**) TEM picture shows the homogeneous dispersion of the nanocrystals in the organic matrix; (**b**) enhancement of the luminescence quantum yield with the percentage of polymer functionalization; (**c**) principle of Nanoimprint lithography; (**d**) fluoroscent microscopy image of imprinted structures

composite preparation route strongly encouraged the use of the obtained luminescent nanocomposite materials for the fabrication of micro/nano devices in optoelectronic and photonic applications. To demonstrate the viability of such approach, the emerging technique of NIL [30] has been used to fabricate luminescent and permanent structures as waveguides and photonic bandgap slabs [65, 66, 78].

Nanoimprint is an innovative lithographic technology that allows to obtain high-throughput patterning of nanostructures with excellent resolution (down to 10 nm), beyond the limitations set by light diffractions or beam scatterings in other conventional technique. Based on the mechanical embossing principle, nanoimprint technology can not only create resist patterns as in lithography but also imprint functional device structures directly in polymers.

A hard mould obtained by electron beam or UV lithography with micro- and nanoscale features is embossed into the polymer film spun on the substrate under controlled pressure and temperature above the glass transition temperature of the polymer (Fig. 4c). Then, after the reduction of the temperature and the pressure release, demolding occurs, creating a thickness contrast in the polymer material. Residual polymer layers can be removed by a subsequent plasma-based anisotropic etching treatment.

The NIL has been successfully applied to nanocrystal-based polymer composites in presence of PMMA-based copolymers [78], demonstrating the fabrication of structures with micro and nano resolutions (Fig. 4d), but also in commercial thermoplastic and UV curable resist [65, 66] without significant alterations in the optical properties of the composites. Indeed, the composite materials after the NIL still retain the original luminescence, in spite of the hard conditions used during the imprinting process (temperature up to 170°C and pressure up to 60 bars).

In the last years, electrospinning has emerged as a versatile and effective method for the fabrication of fibers with diameters in the range of nanometers to a few microns on a variety of materials, such as polymers and inorganic and hybrid compounds. The obtained fibers have a large surface area per unit mass so that, collected on a screen, they can be used for example for filtration of submicron particles in separation industries and in biomedical applications, such as tissue engineering scaffolds, optical sensors and biosensors and artificial blood vessels. In the electrospinning process, a high voltage is applied to an electrode attached to a capillary tube containing the spinning solution, while the other electrode is attached to a collector tube to create an electrically charged jet of polymer solution, which dries or solidifies in a polymer fiber [23, 35]. The electrospinning technique has been used for the production of well-dispersed PbS and CdS nanocrystals in poly(vinyl pyrrolidone) (PVP) fibers by exposing the pre-made PVP electrospun fibers loaded with lead and cadmium ions to H_2S gas at room temperature to grow the nanocrystals "in situ" [47]. Experiments have also been reported on electrospun polymer nanofibers incorporating CdSe/ZnS nanocrystals in water [44]. Suspended inorganic–organic heterostructure cylindrical waveguides have been fabricated by a one-step electrospinning approach and the use of luminescent nanocrystals as an internal light source has been demonstrated.

5 Nanocomposites as Functional Materials for Applications

The possible applications of the semiconductor nanocrystal-based nanocomposite materials are manifold and the future perspectives do really span over a range of technological fields, ranging from optoelectronic and photonic devices, as well as sensors and bio-sensors. In particular, the huge number of possibilities of purposely joining inorganic nanocrystals and organic moieties offers a variety of potential combinations which cannot be obtained by conventional materials, with significant advantages in nanocomposite-based devices such as low cost of device production, large-area devices, new material morphologies and original device geometries [26].

In addition, the improved long-term stability of nanocomposite-based devices with respect to the all-polymer equivalent represents a unique benefit for technological purposes. The typical nanoparticle dimensions are well below the visible wavelength, so the composite media does not suffer from Rayleigh scattering phenomena and the refractive index of the materials can be easily modified by changing the nanocrystal concentration in the organic matrix.

Here we report only on a restricted number of peculiar applications of nanocomposites in photonic, optoelectronic and sensing devices, limiting our interest to colloidal nanocrystals embedded in a nonconductive polymer matrix.

Recent efforts have been devoted towards the development of integrated photonic devices based on the optical properties of the nanocrystals. For example, by trench filling method, working waveguides based on InAs and PbSe nanocrystals in perfluorocyclobutane (PFCB) on silicon substrate have been fabricated [55]. The PFCB was chosen for the very low optical loss in the NIR, the low dielectric constant and the high thermal stability. The experimental results indicate that the coupling losses are larger with respect to propagation losses and the optical properties of the nanocrystals are preserved during the fabrication procedure.

The same authors demonstrate optical-induced population inversion and gain at 1.55 μm in InAs nanocrystals–PFCB composite, opening the way to fabricate planar photonic devices on silicon [8]. Triangular and honeycomb photonic crystal structures have been successfully fabricated both in properly functionalized PMMA-based copolymers and in commercial UV curable thermoplastic polymer, after the incorporation of CdSe/ZnS nanocrystals, by NIL (Fig. 5) [65, 66].

The preservation of the emission properties of the nanocrystals incorporated in polymers and after the imprinting process testifies the high stability of the nanocomposite materials. A significant enhancement of nanocrystal spontaneous emission light extraction has been obtained, up to 240%, due to the slowing in the photon propagation speed inside the two-dimensional lattice, which increases the coupling with the out-of-plane radiative modes.

By using ultrafast femtosecond laser, a direct writing of woodpile photonic crystals has been demonstrated in PbS nanocrystal-UV sensitive photo-polymerizable resin, and an increase in refractive index has been measured [81]. Similar experiments have been performed by using a two-photon polymerization process [41].

The physical dimensions and the optical properties of nanocrystals naturally lend them to sensing on small length scales. At the same time, the size-

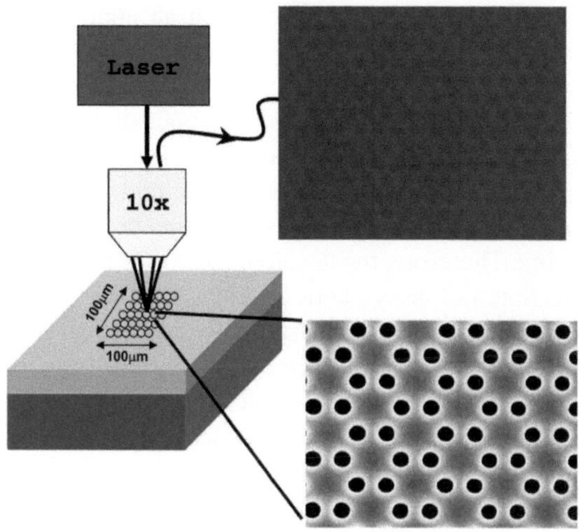

Fig. 5 SEM and fluorescent microscopy picture of honeycomb two-dimensional photonic crystals fabricated on functionalized PMMA copolymer incorporating CdSe/ZnS colloidal nanocrystals (in collaboration with the group of C. Sotomayor Torres, Tyndall National Institute, Cork, Ireland)

tunable absorption and emission wavelength, the high photostability and the large luminescence quantum yield are extremely attractive properties in optical thermometry applications. A practical temperature sensor based on the luminescence modification of CdSe/ZnS nanocrystals embedded in PLMA has been demonstrated by the M. G. Bawendi group [83]. The changes in photoluminescence intensity with respect to the temperature have been calculated to be linear and reversible and the sensor is un-sensitive to the oxygen. In the absence of protective inorganic shells on the CdSe nanocrystals, a chemical selective sensor for polar and nonpolar vapors in air at atmospheric pressure has been recently set up by using thin film with different sizes of nanocrystals in PMMA [64]. After the incorporation in polymers, nanocrystals of different size demonstrated a distinct luminescence response after exposure to toluene and methanol vapors, ascribed to the effects of dielectric media on the size-dependent dipole moment of the nanocrystals.

6 Conclusion

We have discussed here the latest results in the field of inorganic nanoparticle-based polymer nanocomposites as novel functional materials with original optical and electronic properties. Polymers incorporating pre-synthesized nano-objects present an expanding field with immense potential on technological applications, due to the large availability of functional nanocrystals enabling unique and exclusive properties within polymer materials. The high quality of the nanocrystals which can be afforded by using the more advanced synthetic routes to carefully control size, size dispersion and structural parameters, combined with the wide choice of the polymer moiety, offers an extremely versatile nanocomposite preparation route, which can be largely customized according to a precise and defined material design.

Taking into consideration such an approach, the availability of monodisperse inorganic nanoparticles, with tailored shape, size and composition, and a clear understanding of the most crucial preparative issues become critical.

Moreover, also in view of the application potential of such class of materials, the question of mass production of nanoparticles represents an hot issue, since only the availability of suitable inorganic components on a large scale for the nanocomposite fabrication would ensure the spreading of the developed materials towards manufacturing. Therefore, the development and establishment of versatile and reliable but at one time simple synthetic schemes together with a better understanding of the guiding principles of nanocrystal growth would allow further progress for fine tuning of the size, shape and surface of nanocrystals for nanocomposite preparation.

In this perspective, the interaction between the nanocomposite developing laboratories and the manufacturers can represent a unique opportunity to advance the technology issues in such an emerging field, opening new venues and perspectives for original functional materials and devices in optoelectronics.

References

1. Al-Ghamdi GH, Sudol ED, Dimonie VL, El-Aasser MS (2006) Encapsulation of titanium dioxide in styrene/n-butyl acrylate copolymer by miniemulsion polymerization. J. Appl. Polym. Sci 101:3479–3486
2. Bockstaller MR, Mickiewicz RA, Thomas EL (2005) Block copolymer nanocomposites: perspectives for tailored functional materials. Advanced Mater. 17:1331–1349
3. Boontongkong Y, Cohen RE, (2002) Cavitated block copolymer micellar thin films: lateral arrays of open nanoreactors. Macromolecules 35:3647–3652
4. Brushez M, Moronne M, Gin P, Weiss S, Alivisatos AP (1998) Semiconductor nanocrystals as fluorescent biological labels. Science 281:2013–2016
5. Bullen C, Mulvaney P, (2006) The effects of chemisorption on the luminescence of cdse quantum dots. Langmuir 22:3007–3013
6. Burda C, Chen X, Narayanan R, El-Sayed MA (2005) Chemistry and properties of nanocrystals of different shapes. Chem. Rev. 105:1025–1102
7. Caseri W (2000) Nanocomposites of polymers and metals or semiconductors: historical background and optical properties. Macromol. Rapid Commun. 21:705–722
8. Chen G, Rapaport R, Fuchs DT, Lucas L, Lovinger A, Vilan S, Aharoni A, Banin U, (2005) Optical gain from InAs nanocrystal quantum dots in a polymer matrix. Appl. Phys. Lett. 87:251108
9. Colvin VL, Schlamp MC, Alivisatos AP (1994) Light-emitting diodes made from cadmium selenide nanocrystals and a semiconducting polymer. Nature 370:354–357
10. Comparelli R, Curri ML, Cozzoli PD, Striccoli M (2007) Optical biosensing based on metal and semiconductor colloidal nanocrystals. In: Kumar C (ed) Nanomaterials for Biosensors. Wiley-VCH, Weinheim, pp. 123–151
11. Comparelli R, Zezza F, Striccoli M, Curri ML, Tommasi R, Agostiano A (2003) Improved optical properties of CdS quantum dots by ligand exchange. Mat. Sci. Eng. C 23:1083–1086
12. Corbierre MK, Cameron NS, Sutton M, Laaziri K, Lennox RB (2005) Gold nanoparticle/polymer nanocomposites: dispersion of nanoparticles as a function of capping agent molecular weight and grafting density. Langmuir 21:6063–6072
13. Cozzoli PD, Fanizza E, Curri ML, Laub D, Agostiano A (2005) Low-dimensional chain like assemblies of TiO_2 nanorod-stabilized Au nanoparticles. Chem. Commun. 942–944

14. Cozzoli PD, Kornowski A, Weller H (2003) Low-temperature synthesis of soluble and processable organic-capped anatase. TiO_2 nanorods J. Am. Chem. Soc. 125:14539–14548
15. Csetneki I, Filipcsei G, Zrinyi M (2006) Smart nanocomposite polymer membranes with on/off switching control. Macromolecules 39:1939–1942
16. Curri ML, Agostiano A, Manna L, Della Monica M, Catalano M, Chiavarone L, Spagnolo V, Lugara M (2000) Synthesis and characterization of CdS nanoclusters in a quaternary microemulsion: The role of the cosurfactant. J. Phys. Chem. B 104:8391–8397
17. Dabbousi BO, Bawendi MG, Onitsuka O, Rubner MF (1995) Electroluminescence from CdSe quantum-dot/polymer composites. Appl. Phys. Lett. 66:1316–1318
18. Dabbousi BO, Rodriguez-Viejo J, Heine JR, Mattoussi H, Ober R, Jensen KF, Bawendi MG (1997) (CdSe)ZnS core-shell quantum dots: synthesis and characterization of a size series of highly luminescent nanocrystallites. J. Phys. Chem. B 101:9463–9475
19. Ding L, Li Y, Chu H, Li X, Liu J (2005) Creation of cadmium sulfide nanostructures using afm dip-pen nanolithography. J. Phys. Chem. B 109:22337–22340
20. Du H, Xu GQ, and Chin WS, Huang L, and Ji W (2002) Synthesis, characterization, and nonlinear optical properties of hybridized cds-polystyrene nanocomposites. Chem. Mater., 14:4473–4479
21. Esumi K, Houdatsu H, Yoshimura T (2004) Antioxidant action by gold-pamam dendrimer nanocomposites. Langmuir 20:2536–2538
22. Fan X, Xia C, Advincula RC (2005) On the formation of narrowly polydispersed pmma by surface initiated polymerization (sip) from aibn-coated/intercalated clay nanoparticle platelets. Langmuir, 21:2537–2544
23. Fang X, Reneker DH (1997) DNA fibers by electrospinning. J. Macromol. Sci-Phys. B36:169–173
24. Gangopadhyay R, De A (2000) Conducting polymer nanocomposites: a brief overview. Chem. Mater. 12:608–622
25. Gass J, Poddar P, Almand J, Srinath S, Srikanth H, (2006) Superparamagnetic polymer nanocomposites with uniform Fe_3O_4 nanoparticle dispersions. Adv. Funct. Mater. 16:71–75
26. Godovsky DY (2000) Device applications of polymer-nanocomposites. Adv. Polym. Sci. 153:163–205
27. Gomez-Romero P (2001) Hybrid organic-inorganic materials – in search of synergic activity. Adv. Mater.13:163–174
28. Gu H, Zheng R, Zhang X, Xu B (2004) Facile one-pot synthesis of bifunctional heterodimers of nanoparticles: A conjugate of quantum dot and magnetic nanoparticles. J. Am. Chem. Soc. 126:5664–5665
29. Guo W, Li J, Wang YA, Peng X (2003) Luminescent CdSe/CdS Core/Shell nanocrystals in dendron boxes: superior chemical, photochemical and thermal stability. J. Am. Chem. Soc. 125:3901–3909
30. Guo LJ (2004) Recent progress in nanoimprint technology and its applications. J. Phys. D: Appl. Phys. 37:R123–R141
31. Hung ND, Meyer YH (1991) Simple generation of 400–700 nm picosecond dye laser pulses with nanosecond laser pumping. Appl. Phys. B-Lasers Opt. 53:226–230
32. Ingrosso C, Fakhfouri V, Striccoli M, Agostiano A, Voigt A, Gruetzner G, Curri ML, Brugger J (2007) Luminescent nanocrystal modified epoxy photoresist for the fabrication of 3-D high aspect-ratio microstructures. Adv. Funct. Mater in press
33. Jeong S, Achermann M, Nanda J, Ivanov S, Klimov VI, Hollingsworth JA (2005) Effect of the thiol-thiolate equilibrium on the photophysical properties of aqueous CdSe/ZnS nanocrystal quantum dots. J. Am. Chem. Soc. 127:10126–10127
34. Kickelbick G (2003) Concepts for the incorporation of inorganic building blocks into organic polymers on a nanoscale. Prog. Polym. Sci. 28:83–114
35. Kim JS, Reneker DH (1999) Polybenzimidazole nanofiber produced by electrospinning. Polymer Eng. Sci. 39:849

36. Kim S, Bawendi MG (2003) Oligomeric ligands for luminescent and stable nanocrystal quantum dots. J. Am. Chem. Soc. 125:14652–14653
37. Kudera S, Carbone L, Casula MF, Cingolani R, Falqui A, Snoeck E, Parak WJ, Manna L (2005) Selective growth of PbSe on one tips of colloidal semiconductor nanorods. Nano Lett. 5:445–449
38. Lade M, Mays H, Schmidt J, Willumeit R, Schomaker R (2000) On the nanoparticle synthesis in microemulsions: Detailed characterization of an applied reaction mixture. Colloids Surf. A 163:3–15
39. Lee J, Sundar V, Heine JR, Bawendi MG, Jensen KF (2000) Full color emission from II-VI semiconductor quantum dot-polymer composites. Adv. Mater. 12:1102–1105
40. Lee SM, Cho SN, Cheon J (2003) Anisotropic shape control of colloidal inorganic nanocrystals. Adv. Mater. 15:441–444
41. Li J, Jia B, Zhou G, Gu M (2006) Fabrication of three-dimensional woodpile photonic crystals in a PbSe quantum dot composite material. Opt. Express 14:10740–10745
42. Li W, Gao C, Qian H, Ren J, Yan D (2006) Multiamino-functionalized carbon nanotubes and their applications in loading quantum dots and magnetic nanoparticles. J. Mater. Chem.16:1852–1859
43. Lisiecki I (2005) Size, shape and structural controls of metallic nanocrystals. J. Phys. Chem. B 109:12231–12244
44. Liu H, Edel JB, Bellan LM, Craighead HG (2006) Electrospun polymer nanofibers as subwavelength optical waveguides incorporating quantum dots. Small 2:495–499
45. Lu C, Guan C, Liu Y, Cheng Y, Yang B (2005) PbS/Polymer nanocomposite optical materials with high refractive index. Chem. Mater. 17:2448–2454
46. Lu X, Zhao Y, Wang C (2005) Fabrication of PbS nanoparticles in polymer-fiber matrices by electrospinning. Adv. Mater. 17:2485–2488
47. Lu X, Zhao Y, Wang C, Wei Y, (2005) Fabrication of CdS nanorods in pvp fiber matrices by electrospinning Macromol. Rapid Commun. 26:1325–1329
48. Manna L, Scher EC, Alivisatos AP (2002) Shape control of colloidal semiconductor nanocrystals. J. Clusters Sci. 13:521–532
49. Milliron DJ, Hughes SM, Cui Y, Manna L, Li J, Wang LW, Alivisatos, AP (2004) Colloidal nanocrystal heterostructures with linear and branched topology. Nature 430:190–195
50. Milton GW (2002) The Theory of Composites. Cambridge University Press, Cambridge
51. Mokari T, Rothenberg E, Popov I, Costi R, Banin U (2004) Selective growth of metal tips onto semiconductor quantum rods and tetrapods. Science 304:1787–1790
52. Murray CB, Norris DJ, and Bawendi MG (1993) Synthesis and characterization of nearly monodisperse CdE (E = S, Se, Te) semiconductor nanocrystallites. J. Am. Chem. Soc. 115:8706–8715
53. Nagasaki Y, Ishii T, Sunaga Y, Watanabe Y, Otsuka H, Kataoka K, (2004) Novel molecular recognition via fluorescent resonance energy transfer using a biotin-PEG/Polyamine stabilized CdS quantum dot. Langmuir 20:6396–6400
54. Nazzal AY, Qu L, Peng X, Xiao M (2003) Photoactivated CdSe nanocrystals as nanosensors for gases. Nano Lett. 3:819–822
55. Olsson YK, Chen G, Rapaport R, Fuchs DT, Sundar VC, Steckel JS, Bawendi MG, Aharoni A, Banin U (2004) Fabrication and optical properties of polymeric waveguides containing nanocrystalline quantum dots. Appl. Phys. Lett. 85:4469–4471
56. Pang L, Shen Y, Tetz K, Fainman Y (2005) PMMA quantum dots composites fabricated via use of pre-polymerization. Optic Express, 13:44–49
57. Pellegrino T, Manna L, Kudera S, Liedl T, Koktysh D, Rogach AL, Keller S, Radler J, Natile G, Parak WJ, (2004) Hydrophobic nanocrystals coated with an amphiphilic polymer shell: a general route to water soluble nanocrystals. Nano Lett. 4:703–707
58. Peng XG, Schlamp MC, Kadavanich AV, Alivisatos AP (1997) Epitaxial-Growth of highly luminescent CdSe/CdS Core/Shell nanocrystals with photostability and electronic accessibility. J. Am. Chem. Soc. 119:7019–7029

59. Peng X, Wickham J, and Alivisatos AP (1998) Kinetics of II-VI and III-V colloidal semiconductor nanocrystal growth: "focusing" of size distributions. J. Am. Chem. Soc. 120: 5343–5344
60. Peng X (2002) Green chemical approaches toward high-quality semiconductor nanocrystals. Chem. Eur. J. 8:334–339
61. Petruska MA, Bartko AP, Klimov VI, (2004) An amphiphilic approach to nanocrystal quantum dot-titania nanocomposites. J. Am. Chem. Soc. 126:714–715
62. Pileni MP (2003) The role of soft colloidal templates in controlling the size and shape of inorganic nanocrystals. Nature Mater. 2:145–150
63. Potapova I, Mruk R, Prehl S, Zentel R, Basché T, Mews A (2003) Semiconductor nanocrystals with multifunctional polymer ligands. J. Am. Chem. Soc. 125:320–321
64. Potyrailo RA, Leach AM (2006) Selective gas nanosensors with multisize CdSe nanocrystal/polymer composite films and dynamic pattern recognition. Appl. Phys. Lett. 88:134110
65. Reboud V, Kehagias N, Sotomayor Torres CM, Zelsmann M, Striccoli M, Curri ML, Agostiano A, Tamborra M, Fink M, Reuther F, Gruetzner G (2007) Spontaneous emission control of colloidal nanocrystals using nanoimprinted photonic crystals. Appl. Phys. Lett. 90:011115
66. Reboud V, Kehagias N, Zelsmann M, Striccoli M, Tamborra M, Curri ML, Agostiano A, Mecerreyes D, Alduncın JA, Sotomayor Torres CM (2007) Nanoimprinted photonic crystals for the modification of the (CdSe)ZnS nanocrystals light emission. Microelectron. Eng. 84:1574–1577
67. Russo GM, Simon GP, Incarnato L (2006) Correlation between rheological, mechanical, and barrier properties in new copolyamide-based nanocomposite films. Macromolecules 39: 3855–3864
68. Sanchez C, Soler-Illia CG, Ribot F, Lalot T, Mayer CR, Cabuil V (2001) Designed hybrid organic-inorganic nanocomposites from functional nanobuilding blocks. Chem. Mater. 13:3061–3083
69. Sankaran V, Cummins CC, Schrock RR, Cohen RE, Silbey RJ, (1990) Small lead sulfide (PbS) clusters prepared via ROMP block copolymer technology. J. Am. Chem. Soc., 112:6858–6859
70. Service RF (2004) Printable electronics that stick around. Science 304:675
71. Sheng W, Kim S, Lee J, Kim S-W, Jensen K, and Bawendi MG (2006) In-Situ encapsulation of quantum dots into polymer microspheres. Langmuir 22:3782–3790;
72. Shenhar R, Norsten TB, Rotello VM (2005) Polymer-mediated nanoparticle assembly: structural control and applications Advanced Mater. 17:657–669
73. Shieh F, Saunders AE, Korgel BA (2005) General shape control of colloidal CdS, CdSe, CdTe quantum rods and quantum rod heterostructures. J. Phys. Chem. B 109:8538–8542
74. Suh KY, Kim YS, Lee HH, (2001) Capillary force lithography. Adv. Materials 13:1386–1389
75. Talapin DV, Rogach AL, Kornowski A, Haase M, Weller H (2001) Highly luminescent monodisperse CdSe and CdSe/ZnS nanocrystals synthesized in a hexadecylamine-trioctylphosphine oxide-trioctylphospine mixture. Nano Lett. 1:207–211
76. Tamborra M, Comparelli R, Curri ML, Striccoli M, Petrella A, Agostiano A (2005) Hybrid nanocomposites based on cds and cdse colloidal nanocrystals in organic polymers. in Nanotechnology II Proc. SPIE vol. 5838 Paolo Lugli (ed), pp. 236–244
77. Tamborra M, Striccoli M, Comparelli R, Curri ML, Petrella A, Agostiano A (2004) Optical properties of hybrid composites based on highly luminescent CdS nanocrystals in polymer. Nanotechnology 15:S240–S244
78. Tamborra M, Striccoli M, Curri ML, Alducin JA, Mecerreyes D, Pomposo JA, Kehagias N, Reboud V, Sotomayor Torres CM, Agostiano A (2007) Nanocrystal based highly luminescent composites for nano imprinting lithography. Small 5:822–828
79. Tekin E, Smith PJ, Hoeppener S, van der Berg AMJ, Susha AS, Rogach AL, Feldmann J, Schubert US (2007) Inkjet printing of luminescent cdte nanocrystal-polymer composites. Adv. Funct. Mater 17:23–28
80. Teranishi T, Inoue Y, Nakaya M, Oumi Y, Sano T (2004) Nanoacorns: anisotropically phase-segregated CoPd sulfide nanoparticles. J. Am. Chem. Soc. 126:9914–9915

81. Ventura MJ, Bullen C, Gu M (2007) Direct laser writing of three-dimensional photonic crystal lattices within a PbS quantum-dot-doped polymer material. Opt. Express 15:1817–1820
82. Vollath D, Szabò DV (1999) Coated Nanoparticles. A new way to improved nanocomposites. J. Nanopart. Res. 1:235–242
83. Walker GV, Sundar VC, Rudzinski CM, Wun AV, Bawendi MG, Nocera DG (2003) Quantum-dot optical temperature probes. Appl. Phys. Lett. 83:3555–3557
84. Wang Y, Tang Z, Correa-Duarte MA, Liz-Marzan LM, Kotov NA (2003) Multicolor luminescence patterning by photoactivation of semiconductor nanoparticle films. J. Am. Chem. Soc.125:2830–2831
85. Wang X-S, Dykstra TE, Salvador MR, Manners I, Scholes GD, Winnik MA (2004) Surface passivation of luminescent colloidal quantum dots with Poly(Dimethylaminoethyl methacrylate) through a ligand exchange process. J. Am. Chem. Soc. 126:7784–7785
86. Wang M, Dykstra TE, Lou X, Salvador MR, Scholes GD, Winnik MA (2006) Colloidal CdSe nanocrystals passivated by a dye-labeled multidentate polymer: quantitative analysis by size-exclusion chromatography. J. Am. Chem. Soc. 45:2221–2224
87. Wang M, Kwon-Oh J, Dykstra TE, Lou X, Scholes GD, Winnik MA (2006) Surface modification of CdSe and CdSe/ZnS semiconductor nanocrystals with poly(dimethylaminoethyl methacrylate). Macromolecules 39:3664–3672
88. Wu X, Liu H, Liu J, Haley KN, Treadway JA, Larson JP, Ge N, Peale F, Bruchez MP (2003) Immunofluorescent labeling of cancer marker Her2 and other cellular targets with semiconductor quantum dots. Nature Biotechnol. 21:41–46
89. Xia Y, Kim E, Zhao XM, Rogers JA, Prentiss M, Whitesides GM (1996) Complex optical surfaces formed by replica molding against elastomeric. Masters Sci. 273:347–349
90. Yin Y, Alivisatos AP (2005) Colloidal nanocrystal synthesis and the organic–inorganic interface. Nature 437:664–670
91. Yoshioka Y, Calvert PD, Jabbour GE (2005) Simple modification of sheet resistivity of conducting polymeric anodes via combinatorial ink-jet printing techniques. Macromol. Rapid Commun. 26:238–246
92. Zezza F, Comparelli R, Striccoli M, Curri ML, Tommasi R, Agostiano A, Della Monica M (2003) High quality CdS nanocrystals: surface effects. Synthetic Met 139:597–600
93. Zhang H, Cui Z, Wang Y, Zhang K, Ji X, Lu C, Yang B, Gao M (2003) From water-soluble CdTe nanocrystals to fluorescent nanocrystal-polymer transparent composites using polymerizable surfactants. Adv. Mater. 15:777–780
94. Zhang H, Wang C, Li M, Zhang J, Lu G, Yang B (2005) Fluorescent nanocrystal-polymer composites from aqueous nanocrystals: methods without ligand exchange. Chem. Mater. 17:853–857
95. Zucolotto V, Gattás-Asfura KM, Tumolo T, Perinotto AC, Antunes PA, Constantino CJL, Baptista MS, Leblanc RM, Oliveira Jr. ON (2005) Nanoscale manipulation of CdSe quantum dots in layer-by-layer films: influence of the host polyelectrolyte on the luminescent properties. Appl. Surf. Sci. 246:397–402

Large-Scale *Ab Initio* Study of Size, Shape, and Doping Effects on Electronic Structure of Nanocrystals

Jingbo Li and Su-Huai Wei

Abstract Semiconductor nanocrystals, such as quantum dots (QDs) and wires (QWs), often contain from a few thousands to more than 10^6 atoms. It has been a great challenge to calculate the electronic structure of these large nanosystems using self-consistent first-principles method. In this chapter, recent development of calculations of nanocrystal physical properties using the large-scale *ab initio* pseudopotential or charge-patching methods is reviewed. The calculated size-dependent exciton energies and absorption spectra of QDs and QWs are in good agreement with experiments. The calculated ratios of bandgap increases between QWs and QDs are found to be material-dependent, and for most direct bandgap materials, this ratio is close to 0.586, as predicted by simple effective-mass approximation. We show that the electronic structure of a nanocrystal can be tuned not only by its size, but also by its shape. Therefore, the shape can be used as an efficient way to control the electronic structure of the nanocrystals. Changing the shape is expected to be more flexible and provides more variety of the electronic states than simply changing the size of the system. The special features of the electronic states obtained in different shapes of the nanocrystals can be used in various device applications. We also show that defect properties in QDs could be significantly different from those in bulk semiconductors. For example, although negatively charged DX^- center is unstable in bulk GaAs:Si with respect to the tetrahedral coordinated $Si_{Ga}{}^-$, when the dot size is small enough, it becomes stable.

Contents

J. Li (✉)
State Key Laboratory for Superlattices and Microstructures, Institute of Semiconductors, Chinese Academy of Sciences, P. O. Box 912, Beijing 100083, China
e-mail: jbli@semi.ac.cn

Z.M. Wang (ed.), *Toward Functional Nanomaterials*, Lecture Notes in Nanoscale Science and Technology 5, DOI 10.1007/978-0-387-77717-7_5,

1 Introduction

During the past 20 years, semiconductor nanocrystals such as quantum dots (QDs) and quantum wires (QWs) [1–8] have attracted much attention because the physical properties of the nanocrystals such as the band gap and optical transitions can be tailored by size or shape. This feature opens up a great potential for novel device applications, from lasers [9] to solar cells [10] to single-electron transistors [11]. Many of these applications require the quantitative knowledge of the size or shape dependence of the nanocrystal's optical properties, which, in a semiconductor, is related to the transitions near the electronic band gaps. Therefore, studying the size and shape dependence of the electronic band gaps and the related exciton transition energies is one of the most important topics in semiconductor nanocrystals research. Moreover, nanocrystals are usually grown in various media, such as polymers, cavities of zeoliths, glasses, solutions, and organic molecules or biomolecules [4]. In most of these cases, the surface dangling-bond states of the nanocrystal are removed by the medium through passivation. In these cases, the electronic and optical properties of the nanocrystal are close to the intrinsic properties of the nanosystem, i.e., independent of the enclosure medium and surface passivation. Therefore, the first step in the study of nanostructures is to understand these intrinsic properties of nanocrystals.

Most of the previous studies are focused on the size dependence of physical properties of QDs. In recent years, development in chemical synthesis has shown that, in addition to the control of size, the shape of a nanocrystal can also be controlled during growth, such as nanorods, nanodrops, nanoarrows, nanotetrapods, and nanoribbons [12–15]. Recent theoretical study [16] has shown that change in shape can lead to drastically different electronic states and energy band gaps in nanocrystals. This implies that the physical properties of semiconductor nanocrystals can be tuned by both size and shape. For example, by changing the shape, it is possible to change the location of the holes and electrons; therefore, there is either an increase or decrease in the recombination rate, which is very useful in light emission and solar cell applications or other optoelectronic devices.

The application of semiconductor nanocrystals as novel electronic devices also depends critically on doping properties. Most of the semiconductors will not be very useful if insufficient charge carriers are generated by the dopants at normal working temperature. Defect properties have been extensively studied in the past for bulk semiconductors, and various approaches have been proposed to overcome the doping limit in semiconductors [17]. However, very few studies have been carried out to understand how the formation of QDs affects the defect properties in these systems.

For example, it is not clear how the size of QDs affects the defect formation energies and ionization energy levels, as well as the relative stability between different defects [18–21].

In this chapter, we first describe our recent development on the theoretical study of the electronic structure of semiconductor nanocrystals by large-scale numerical calculations in Section 2. In Section 3 we discuss the size dependence of exciton transition energies and absorption spectra of nanocrystals. Section 4 presents the calculated ratios of bandgap increases between QWs and QDs. Section 5 discusses the shape effects on electronic states of nanocrystals. Section 6 demonstrates the unique features of defect physics in small QDs. A brief summary is given in Section 7.

2 Method of Calculations

Several methods have been used in our study of electronic structure of nanocrystals. The first method is the first-principles band-structure method within the density functional theory (DFT) as implemented in the pseudopotential codes such as VASP [22] and PEtot [23]. This is the most accurate approach and is used as a benchmark to test all other approaches. However, despite recent rapid growth of computing power, the system that can be handled using this approach is still limited to a few hundred atoms per cell. It also suffers, in some cases, from the bandgap error associated with the DFT method. To overcome this disadvantage, in the last few years, several new approaches that can give approximate, but reliable, results of the electronic structure of nanocrystals have been developed. One of the approaches is the semi-empirical pseudopotential method (SEPM). In the SEPM approach, the electron wave functions are expanded by plane-wave basis functions as in the *ab initio* pseudopotential methods. The screened SEPM atomic potentials are fitted to the DFT potentials and experimental band structures, and are tested extensively for transferability and reliability. Using the SEPM pseudopotential, the single particle's Schrödinger equation is solved non-self-consistently using the linear scaling folded spectrum method [23] for several states near the band gap. This approach solves two problems simultaneously: the bandgap error in the DFT and the inability to solve large systems using the current self-consistent DFT method. Combined with modern large-scale supercomputers, this method can be used to calculate systems containing tens of thousands of atoms.

Although quite successful, the SEPM approach suffers from the difficulty of finding a suitable surface-passivation potential because the potentials are usually fitted to bulk semiconductor properties. To overcome this difficulty, the "charge patching method" (CPM) was developed [23], which has the quality of first-principles DFT accuracy, while also handling large systems. In the CPM method, the *ab initio* quality electron charge density $\rho_{patch}(r)$ is first constructed using charge motifs generated from prototype systems with similar atomic environments. After that, the total local-density-approximation (LDA) potential *V(r)* is obtained by solving the Poisson's equation and standard LDA formula. The single-particle Schrödinger equation can then be solved using the linear scaling folded spectrum method for the band edge states. The surface of the nanocrystal is passivated by pseudo-hydrogen atoms (e.g., 1.25-charge H to passivate the surface Ga-atom dangling bonds, and 0.75-charge H to passivate the surface As-atom dangling bonds in GaAs QDs). Furthermore, to describe the quantum confinement effects accurately, the effective mass, which is also underestimated in DFT, is corrected. This method is denoted as "LDA+C". The details of this entire procedure are described in [24].

3 Size Dependence of Exciton Energies and Absorption Spectra

The exciton transition energies of QDs and QWs have been systematically calculated for III–V (GaAs, InAs, InP, GaN, AlN, and InN) and II–VI (CdSe, CdS, CdTe, ZnSe, ZnS, ZnTe, and ZnO) systems [24] as a function of the diameter. The electron-hole Coulomb binding energies are included in these calculations for QDs. The exciton transition energy is calculated as

$$E_{i,j} = \varepsilon_{j,c} - \varepsilon_{i,v} - J_{ij} \tag{1}$$

where J_{ij} is the Coulomb energy between the electron and hole. In this approach, the small exchange and correlation interactions are neglected. J_{ij} is calculated as

$$J_{i,j} = \int\int \frac{|\psi_{i,v}(x_1)|^2 |\psi_{j,c}(x_2)|^2}{\varepsilon(r_1 - r_2)|r_1 - r_2|} d^3x_1 d^3x_2 \tag{2}$$

Unlike the case in bulk exciton, here the electron and hole are spatially close to each other, so the distance-dependent dielectric function is needed to obtain an accurate result. The screening dielectric function $\varepsilon(\mathbf{r}_1\text{-}\mathbf{r}_2)$ includes both the electronic and the ionic contributions, which are described using the modified Penn model and the Haken formula, as discussed in [24].

The wave function charge densities of conduction band minimum (CBM and valence band maximum (VBM states (here, we use the convention of the bulk states) are plotted in Fig. 1 for diameter $d = 4.33$ nm CdSe QDs, and Fig. 2 for diameter $d = 4.37$ nm CdSe QWs. First, no surface state exists in the gap region, indicating that the pseudo-hydrogen passivation works very well. The wave functions in both QD and QW extend all the way to the surfaces. As pointed out in previous studies, the pseudopotential calculated wave functions are less confined than what has been predicted by the simple effective-mass model that sets the wave function to be zero at the boundary of the QD. In comparison to the QD and the QW CBM states, although they look very different due to different viewing perspectives, the QD CBM and QW CBM have the same atomic characteristics. However, for the VBM state, due to the crystal field splitting, the QW VBM has z direction polarization, whereas the QD VBM has predominant xy polarization.

CdSe QDs [5] have been one of the most extensively studied nanosystems during the past decades, and many works have reported the size dependence of exciton

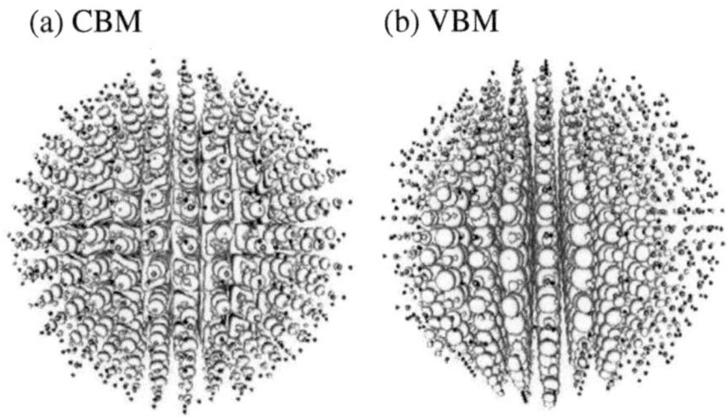

Fig. 1 Charge density isosurfaces of wurtzite CdSe QDs with diameter of 4.33 nm for (**a**) CBM and (**b**) VBM states. The isosurfaces are drawn at 20% of maximum

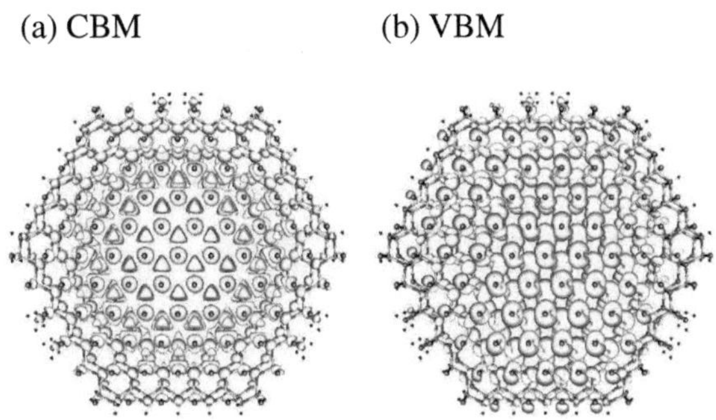

Fig. 2 Same as Fig. 1, but for CdSe QWs with diameter of 4.37 nm

energy of CdSe QDs. However, only in recent years, the size dependence of exciton transition energy of CdSe QWs has been reported experimentally [25]. The calculated results for exciton transition energy shift of CdSe QDs and QWs are plotted in Figs. 3 and 4, respectively. The exciton transition energies of QDs take into account the electron-hole Coulomb interaction, whereas for the QWs systems, we assume

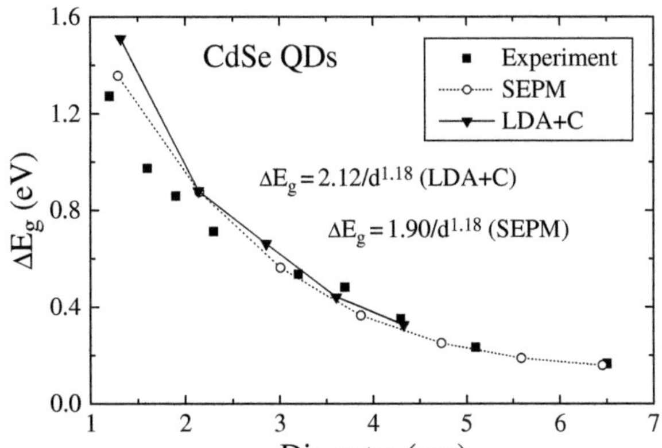

Fig. 3 Comparison of the relative exciton transition energy shift (with respect to bulk) of CdSe QDs between experiment, "LDA+C", and SEPM calculations. Coulomb energies are considered in this calculation. Experimental data are from Ref. [5]

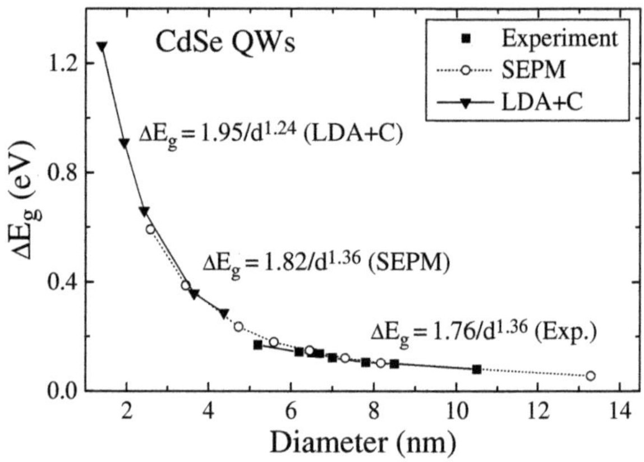

Fig. 4 Comparison of the relative bandgap energy of CdSe QWs as a function of QW size between experiment, "LDA+C", and SEPM calculations. Experimental data are from Ref. [25]

the Coulomb interaction is small due to the one-dimensional (1D) periodicity. The calculations are performed using both the "SEPM" and "LDA+C" approaches. We find that in both cases, the calculated results are generally in good agreement with each other and with experimental measurements [5, 25]. The calculated results are fitted to the formula $\Delta E_g = \beta | d^{\alpha}$, where *a* is close to 1.2 for QDs and 1.36 for QWs, respectively, which are significantly smaller than the simple effective-mass value of 2.

The other well-studied nanosystems are InP QDs and QWs [8, 26, 27, 28]. The calculated results for InP QDs and QWs using the "LDA+C" and the "SEPM" methods are shown in Figs. 5 and 6, respectively, together with experimental measurements [8, 26, 27, 28]. The agreement with experimental observation is also very good, except that for the 2 nm QD, the calculated results are higher in energy than the experimental values.

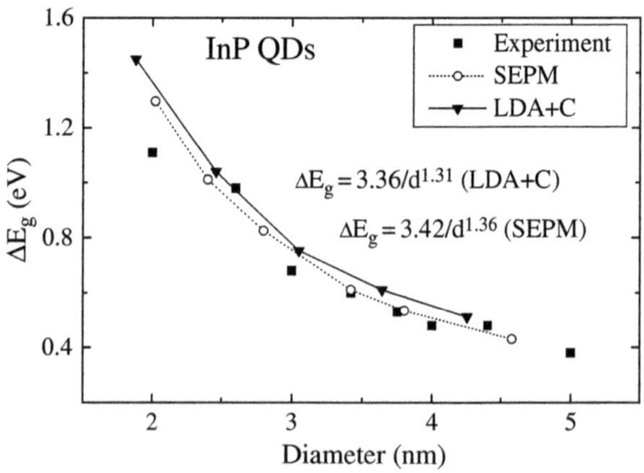

Fig. 5 Comparison of the exciton transition energy shift of InP QDs between experiment, "LDA+C", and SEPM calculations. Experimental data are from Ref. [26, 27]

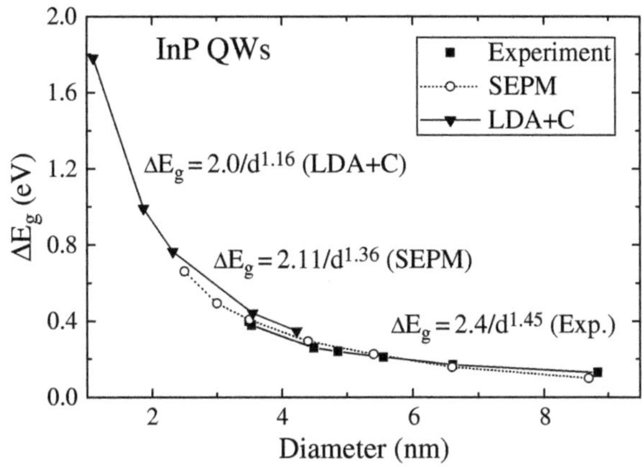

Fig. 6 Comparison of the bandgap energy shift of InP QWs between experiment, "LDA+C", and SEPM calculations. Experimental data are from Ref. [8, 28]

The optical absorption intensity is obtained by summing over the dipole matrix elements' coupling hole state (i,v) and electron state (j,c), i.e.,

$$I_\alpha(E) = \sum_{i,j} \frac{4e^2}{3m^2c^2} |\langle \psi_{i,v} | P_\alpha | \psi_{j,c} \rangle|^2 f(E - E_{ij}) \tag{3}$$

Here, $f(E\text{–}E_{ij})$ is a Gaussian broadening function, and P_α is the momentum operator with the subscript α denoting polarization directions.

The calculated optical absorption spectra of CdSe quantum rods with 2-nm diameter are shown in Fig. 7(a) and (b) for the aspect ratio of 1.0 (spherical QD) and 3.1 (quantum rod), respectively. It identifies the transition peaks as A, B, C, D, E, F, and G in Fig. 7. Solid curves show the z-direction polarized transitions, while the dotted curves show the xy-plane polarized transitions. The conduction band (CB) states is denoted as CB_n in increasing energy order from the conduction band minimum using $n = 1, 2, \ldots$, while the valence band (VB) states are denoted as VB_n in decreasing energy order from the valance band maximum using $n = 1,2,\ldots$ The line shapes and intensities of transition peaks in Fig. 7(a) are consistent with previous calculations for QD systems. Group A peaks originate from $VB_1 \rightarrow CB_1$ transition (A_1) and $VB_2 \rightarrow CB_1$ transition (A_2). The splitting between A_1 and A_2 peaks corresponds to the crystal-field splitting of the wurtzite CdSe QDs. B and C peaks originate from $VB_5 \rightarrow CB_1$ and $VB_7 \rightarrow CB_1$ transitions, respectively. The strong D and E peaks involve high energy level transitions. For example, E peak originates from $(VB_5,VB_6,VB_7) \rightarrow (CB_3,CB_4)$ transitions. The xy-plane polarized and z-direction polarized curves are similar in Fig. 7(a). Thus, there is no strong polarization effect in a spherical QD although wurtzite structure is intrinsically asymmetric. In contrast, a high polarization of the absorption spectra is found in the quantum rod, as shown in Fig. 7(b). It shows strong polarization dependence up to 0.8 eV above the absorption edge; beyond that there are many peaks and the polarization effects are averaged out. The polarization dependence of the absorption spectra was confirmed by recent

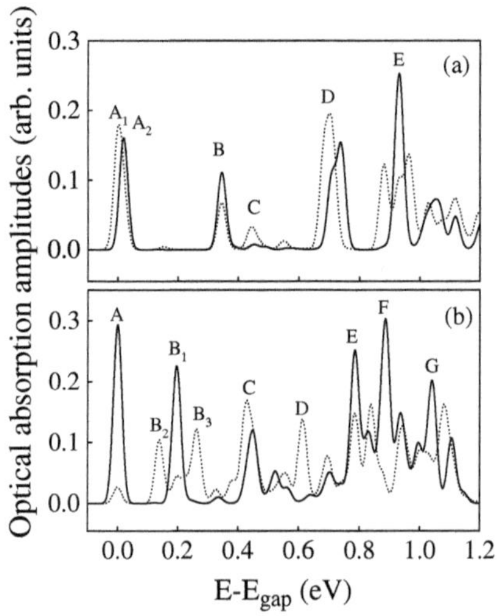

Fig. 7 Theoretical optical transition spectra of CdSe quantum rods calculated using the screened pseudopotential Hamiltonian. Coulomb interaction is taken into account in the calculation. E_{gap} is the ground exciton energy. Solid curve shows the z-direction polarized transition and dotted curve shows xy-plane polarized transition. (**a**) the aspect ratio of the rod is 1, and (**b**) the aspect ratio of the rod is 3.1

photoluminescence excitation (PLE) experiment of single CdSe quantum rod [29]. In this experiment, it was found that the absorption coefficient depends on the exciting photon polarization at a given photon energy. However, based on the calculated results in Fig. 7(b), one expects to see more rich experimental results by scanning the PLE energy from the absorption edge. The polarization direction should change with the change of excitation energy. This is waiting for experimental confirmation. In Fig. 7(b), A and B_1 peaks originate from $VB_1 \rightarrow CB_1$ and $VB_2 \rightarrow CB_2$ transitions, respectively. Their transitions are from the z-direction polarization modes. But, B_2 and B_3 peaks involve xy-plane polarized transitions. Their intensities are weaker than that of B_1 peak. They originate from $VB_3 \rightarrow CB_1$ transition and $VB_4 \rightarrow CB_2$ transition, respectively. In Fig. 7(b), E, F, and G peaks also involve high-energy transitions: E peak corresponds to $(VB_6,VB_{15}) \rightarrow$ (CB_6,CB_4) transitions, F peak corresponds to $(VB_7,VB_{11}) \rightarrow$ (CB_7,CB_8) transitions, and G peak corresponds to $(VB_{14},VB_{19}) \rightarrow (CB_9,CB_6)$ transitions.

Recent experiments of PLE spectra for CdSe quantum rods with very large aspect ratio are reported by Katz et al. [30]. To compare with their experiment, the optical absorption spectra of CdSe QWs (used to mimic quantum rods with large aspect ratio) are calculated with diameters of 2.0, 2.5, 3.0, and 3.8 nm. The band gaps E_g are 2.52 eV, 2.3 eV, 2.22 eV, and 2.09 eV respectively. This is the first theoretical calculation of optical absorption spectra of CdSe QWs. The results are shown in Fig. 8. The shape of the main peaks in Fig. 8 is similar to the idealized density of states of one-dimensional QWs. In this calculation, 10 k_z points are used for each QW to get the eigen-energies and optical transition matrix elements, then interpolation is done to generate the results of 100 k_z points. This procedure gives well-converged results for both band structure and absorption spectra. The major transition peaks are assigned the letters a–e. The connection of the peaks across

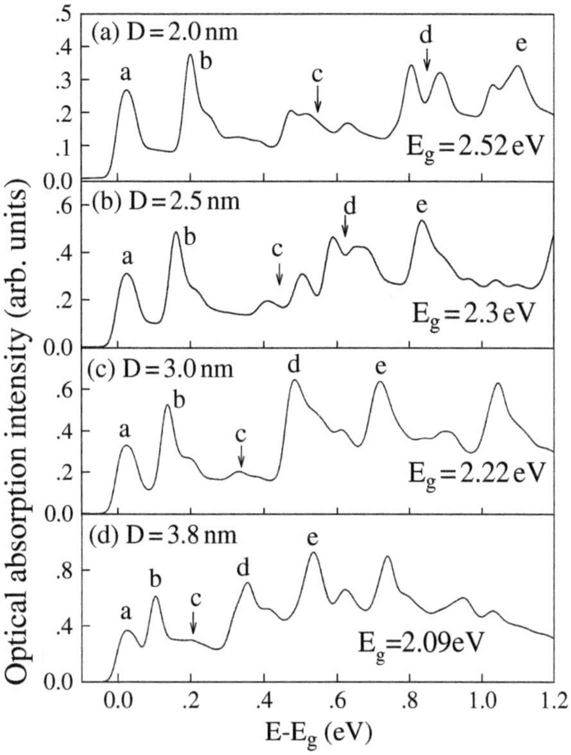

Fig. 8 Theoretical excitonic transition spectra for four CdSe QWS with diameters of 2.0, 2.5, 3.0 and 3.8 nm from the top to bottom curves, respectively. E_g is the ground exciton energy. The spectra are generated with a Gaussian broadening of 20 meV. The major peaks are labeled from letters *a–e*

different QW sizes is done by examining the symmetries of the initial and final single-particle states related to each peak. Some peaks might contain a few closely packed small peaks due to the use of the Gaussian broadening function to fit these peaks. For example, in Fig. 8(a) and (b), peak *c* and peak *d* contain two or three peaks around the position. Thus, this procedure assigns only a few large peaks (*a–e*) in the low-energy region and uses them to compare with the experiment. Except peak *c*, all other peaks are originated from the Γ–point ($k_z = 0$) transition. At Γ–point transition, peak *a* and peak *b* are related to $VB_1 \rightarrow CB_1$ and $(VB_2,VB_3) \rightarrow CB_1$ transitions, respectively. The peak *d* and peak *e* include high-energy transitions. The main contribution to peak *d* and peak *e* include transitions from $(VB_3,VB_4) \rightarrow$ (CB_2,CB_3) and $(VB_5,VB_6,VB_7,VB_8) \rightarrow$ (CB_2,CB_3), respectively. Unlike the other peaks, the peak *c* originates from transitions at the shoulder of the band structure: $0.15(2\pi/c) < k_z < 0.2(2\pi/c)$.

Figure 9 plots the position of the above peaks (filled dots) and experimental PLE and tunneling spectroscopy results (diamonds) as functions of the ground exciton transition energy. The pseudopotential calculations show reasonable agreement with experimental results without any adjustable parameters. Especially, the overall positions of *b*, *c*, and *d* peaks are in good agreement with the experiment [30]. At this

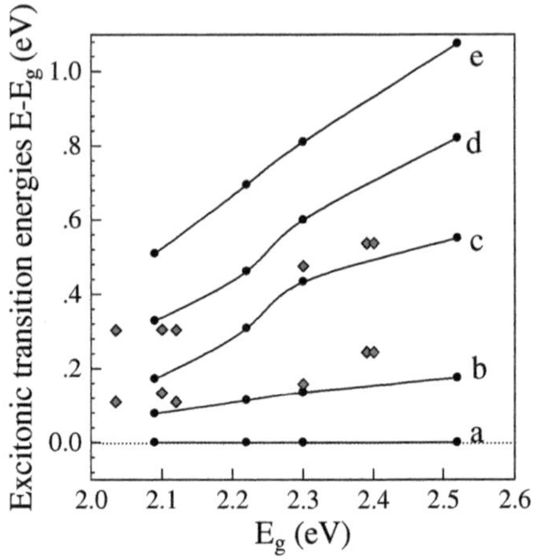

Fig. 9 Transition energies (relative to the ground exciton state) vs. the energy of the first ground exciton of CdSe QWs. Diamonds correspond to experimental data from Ref. [22] by PLE and scanning-tunneling spectroscopy. Theoretically predicted transition peaks are shown as filled dots and the solid curves are guides for eyes

point, the experimental results are rather scattered, especially at the higher energy region. Better experimental results are needed for more precise comparison.

4 Ratios of BandGap Increases Between QWs and QDs

According to an overly simplified particle-in-a-box effective-mass model [31, 32], the bandgap increases of QDs and QWs from the bulk value are given by $\Delta E_g = \frac{2\hbar^2\zeta^2}{m^*d^2}$ where $\frac{1}{m^*} = \frac{1}{m_e^*} + \frac{1}{m_h^*}$ (m_e^* and m_h^* are electron and hole effective masses, respectively) and d is the diameter. For spherical QDs, $\zeta = \pi$ is the zero point of the spherical Bessel function, whereas for cylindrical QWs, $\zeta = 2.4048$ is the zero point of the cylindrical Bessel function. Thus, the ratio of bandgap increases between QWs and QDs should be $\Delta E_g^{wire}/\Delta E_g^{dot} = 0.586$ in this simple model.

To test this model, the electronic structures of surface-passivated QDs and QWs have been systematically calculated for a wide variety of II–VI and III–V semiconductor compounds [33]. It is found that both the calculated QW and QD band gaps can be fitted well by the formula $\Delta E_g = \beta | d^\alpha$ with material-dependent parameters α and β. The fitted results of α and β are listed in Table III of [33] for all the studied systems. By comparing the QD and QW values, the following are observed: (1) The difference for α between QD and QW for the same material is very small. Typically, these differences are within 4%, except in the case of CdS and CdTe, where the difference is ∼ 6–8%. (2) For these small differences, there is no systematic trend. For example, one cannot say the α value from QW is larger or smaller than that from QD. (3) Given the small differences and the lack of a trend, one can assume that the difference for α between QW and QD is probably due to fluctuation of the fitting; thus, they can be set as the same. Note that the differences for α between different materials are more significant. Roughly, the IV–IV material of Si has α ≈ 1.6, while the III–V materials have α ≈ 1.0, and the II–V materials have α ≈ 1.2.

By assuming α_{dot} to be the same as α_{wire}, one can use α_{dot} to fit ΔE_g of the QWs. This gives us a new β'_{wire}, which is also listed in Table III of [33]. As a result, the $\Delta E_g^{wire}(d)/\Delta E_g^{dot}(d)$ ratio will just be the ratio of $\beta'_{wire}/\beta_{dot}$. The calculated ratios range from 0.408 (for Si) to 0.595 (for InAs). Except for Si and CdTe, these ratios are close to the simple effective-mass value of 0.586. The small ratio for Si is probably due to the indirect nature of its band gap, where its effective mass near the X point is highly asymmetric. For InP and CdSe, the LDA ratios of 0.566 and 0.572 are very close to their EPM values of 0.570 and 0.561, respectively [25, 28]. This indicates that, despite the LDA bandgap errors, the ratios between QWs and QDs are captured accurately in the LDA calculations. Notice that these ratios are calculated before the electron and hole Coulomb interactions are included to get the exciton transition energies. If the Coulomb interaction energy is added, then these ratios will be slightly larger.

5 Shape Effects on Electronic States of Nanocrystals

Using wet chemical methods, CdSe quantum nanostructures have been synthesized into spherical dots, rod, arrow, teardrop, and tetrapods [12–15]. However, despite the variety of these shapes, the difference of their electronic states, and whether these differences can be used towards device applications, is not known. Through pseudopotential calculations, it is found that the electronic states in different nanocrystal shapes are dramatically different, and indeed these differences have profound implication in their potential device applications. Theoretically calculated electron and hole states in wurtzite (WZ) structure CdSe spherical dot, rod, arrow, teardrop, and tetrapod are shown in Figs. 1, 2, 3, 4, and 5 of [16]. The spherical QD contains 1115 atoms, while the tetrapod contains 2817 atoms. The dot, rod, arrow, and teardrop are in similar sizes in order to focus on their shape effects. While the dot, rod, arrow, and teardrop have a principle *c* axis and a C_{3v} symmetry along this axis, the tetrapod has an overall T_d symmetry. By analyzing calculated results, one can get the following observations.

(1) *The band gaps*: The calculated band gaps agree well with the experiments; for example, the band gaps of CdSe quantum rods listed in Table I of [16] are in good agreement with that of the experiments [34]. This again demonstrates the accuracy and reliability of the calculations. The shape dependence of the band gap is shown in Table II of [16]. Though the dots, rods, arrows, and teardrops have similar volume, they have very different band gaps. The band gap of the rod is about 270 meV larger than that of the dot due to stronger quantum confinement in the lateral direction. The most compact shape is the spherical dot, which has the smallest band gap, and hence the smallest quantum confinement effect.

(2) *Hole state polarization:* The atomic parts of all the electron states have an *s*-like character. For the hole states, the atomic parts mainly come from two bulk Bloch states at the Γ point. One is the Γ_9 state, which has a p_{xy} character, and another is the Γ_7 state, which has a majority p_z character; here z is the *c*-axis direction. These

atomic characters and polarizations can be seen directly in the wave function cross-section plots as shown in Figs. 1, 2, and 3 of [16]. They also demonstrate themselves in the optical matrix elements $Q_{xy(z)} = | < VB_1|P_{xy(z)}|CB_1 > |^2$ as listed in Table II of [16]. It is found that the polarization of the first valence band state is along the *xy* direction in the spherical shape, while it is along the *z* direction for the other prolonged shapes. This difference causes the state crossing as the shape changes from sphere to rod as reported in [35]. More interestingly, these two polarizations could be mixed in the lower energy hole states. For example, as shown in Fig. 3(k) of [16], the VB_5 state has a p_z character in the handle region, while a p_{xy} character is in the arrow prism. A similar situation is found for the teardrop shape. In general, the p_{xy} character is in the wider part of the shape, while p_z character is in the elongated part.

(3) *Overall shapes of electron and hole states*: It is found that the electron states can be qualitatively described by simple effective-mass theory, while the valence band states are more complicated due to the mixing of the two states as discussed above. As a result of this, the overall shapes of the electron states and the hole states are qualitatively different; there is no one-to-one correspondence between them. Nevertheless, for the quantum rod, the first few electron and hole states look similar. The hole states in this case consist of only the p_z atomic orbital, thus they can be described by a single band effective-mass model like the electron states. For the more fat shapes (spherical dot, arrow, and teardrop), the higher excited electron states tend to develop nodes in the envelope function along the principle *c* axis (WZ (0001) direction), while the nodes of the first few hole states are in the *xy* plane. As a result, they have a hollow center in the *xy*-plane cross section as shown in Figs. 1(i) and 3(i) of [16].

(4) *Effects of the overall shape of the nanostructure*: Though the wave function difference between the dot, the rod, and the arrow are dramatic, the difference between the arrow shape and the teardrop is rather small. The arrow shape has a sudden change in its diameter along the *c*-axis, whereas the change in the teardrop shape is much smoother. However, these detailed features of the shape do not affect the wave function strongly. What does matter is the overall shape of the nanostructure. In our case, beside the tetrapod, one can classify three overall shapes: (i) sphere, the most compact shape; (ii) rod, the symmetrically elongated and slim shape; and (iii) the fat asymmetric shape, with one side large and another side small. Interestingly, for the third shape, it is the higher level electron states (CB_2, CB_3) which tend to occupy the smaller end of the asymmetric side, while the hole states tend to avoid it.

(5) *Wave functions of the tetrapod*: The most dramatic finding of the calculation is the wave functions of the tetrapod. The tetrapod has a T_d symmetry and a tetrahedron with zinc-blende (ZB) crystal structure at its center. This central tetrahedron is joined by four WZ structure nanorods at the (111), $(\bar{1}\bar{1}1)$, $(1\bar{1}\bar{1})$, and $(\bar{1}1\bar{1})$ surfaces. The tetrapod shown in Fig. 5 of [16] has 2817 atoms and the four arms of the nanorod are 2.2 nm in diameter and 4.6 nm in length. The most interesting feature is that the lowest electron state CB_1 is completely localized in the central tetrahedron, while the first hole state VB_1 enters the arms of the tetrapod. Due to the symmetry of the tetrapod shape, the singlet CB_1 state has Γ_6 symmetry, while the doublet VB_1 state has Γ_8 symmetry. The localization of the electron state has

significant implications on the device application of the tetrapod. With its four arms, it is tempting to use the tetrapod for electronic devices like transistors. Imagine two arms are used as the two electrodes of a transistor and the others arms are used as the electric field control gate. If the conduction band is used for the current channel, then the existence of the localized CB $_1$ state at the center of the current pathway will have a strong effect on the magnitude of the current. If the application of an electric field from the control gate can drive the CB $_1$ state away from the tetrahedron center, then current control should be possible. Recently, a shape with two tetrapods sitting at the two ends of a nanorod has been synthesized [12, 13]. If an electric field is applied along this nanorod, then the localized CB $_1$ state can be switched between the two possible locations of the two tetrapod centers, providing a current control mechanism. The long range movement of the CB $_1$ state due to the applied electric field can also produce large nonlinear polarizations.

(6) *Electronic states in WZ/ZB structure*: To understand the reason why the electron state is localized in the central tetrahedron of the tetrapod, a nanorod with half of it in the WZ structure and the other half in the ZB structure has been studied [16]. The *c*-axis of the rod is in the [0001] direction of the WZ and [111] direction of the ZB, the rod diameter is 3.0 nm, and its total length is 8.4 nm. This is an interesting structure by its own right, since it has been synthesized experimentally [12, 13]. As shown in Fig. 10, the electron state is localized in the ZB region

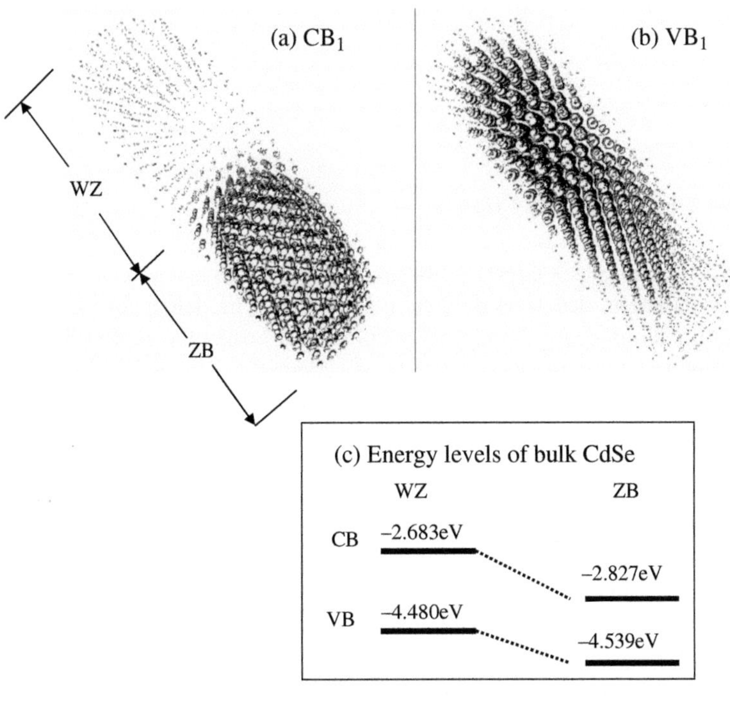

Fig. 10 The isosurface plots of the wave-function squares of (**a**) CBM and (**b**) VBM states, for the CdSe ZB/WZ quantum rod. The total length of the rod is 10.6 nm and the diameter is 2 nm

while the hole state is more localized in the WZ region, this is similar to the tetrapod case. This separation of the electron and hole charge is caused by a band alignment shown in Fig. 10(c), which is the same as in a recent *ab initio* calculation [36]. In semiconductors, such band alignment and the consequent electron-hole charge separation are often achieved by two different materials. The significance here is that it is achieved using the same material with two different crystal structures. The charge separation in this WZ/ZB rod structure makes it a good candidate for solar cell applications.

Figure 11(a) and (b) show branched tetrapod with CdSe central tetrahedron and terminal CdTe branches. This system is interesting because the type-II band alignment between CdSe and CdTe leads to unusual charge-separating properties. The sharp reduction in spatial overlap between the electrons and the holes, apparent in Fig. 11(a) and (b), effectively quenches the bandgap photoluminescence [15].

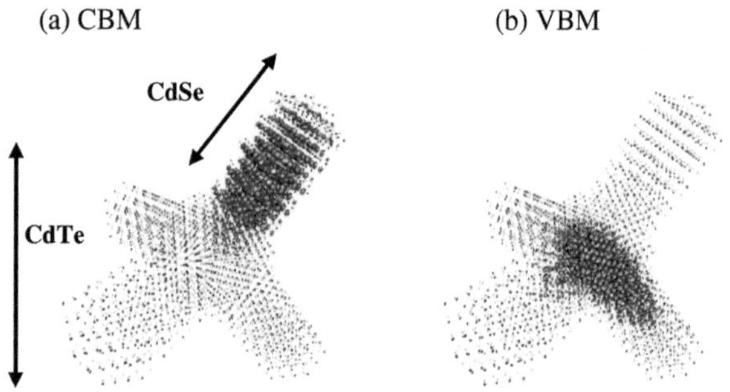

Fig. 11 The isosurface plots of the wave-function squares of (**a**) CBM and (**b**) VBM states, for the type II CdSe/CdTe tetrapod, which show the electron and hole are spatially separated

6 Defect Properties of QDs

It is very interesting to study how the quantum confinement affects the stability and transition energy levels of defects in QDs. [18] The defect formation-energy and defect transition-energy levels are calculated [18] using the supercell approach, where a defect is put at the center of a large supercell and periodic boundary conditions are applied. All the internal structural parameters are relaxed by minimizing the quantum mechanical force and total energy until the changes in the total energy are less than 0.1 meV/atom.

Figure 12(a) shows the formation energy of neutral $Si_{Ga}{}^{0}$ in GaAs QDs as a function of the QDs diameter d. The diameter $d = \infty$ corresponds to the bulk system. As the size of the QDs decreases, the formation energy of $Si_{Ga}{}^{0}$ increases from 1.55 eV for bulk GaAs:Si to 2.99 eV for QD with $d = 1.55$ nm. This increase is because $Si_{Ga}{}^{0}$ creates a singly occupied level near the CBM. This level has a strong CBM *s* character and moves up in energy with the CBM as the QDs' size decreases, thus increasing the formation energy. However, because this defect level is not a pure CBM state, as the CBM moves up in energy with the decreasing QDs' size, the energy differences between the defect level and the CBM, thus the (0/+) transition

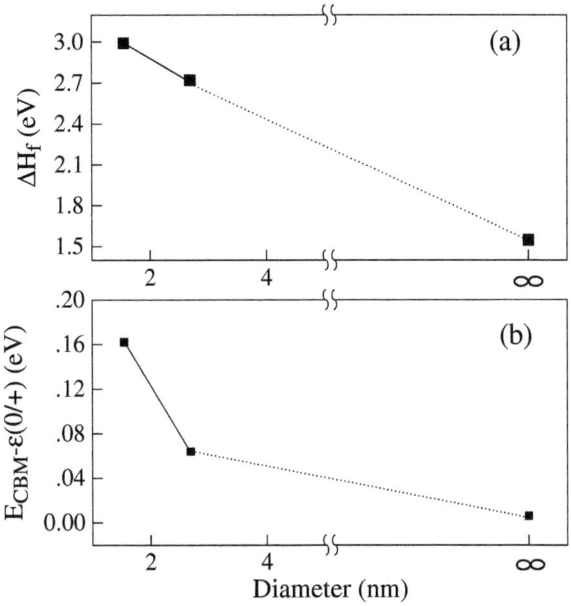

Fig. 12 (**a**) The formation energy of neutral $Si_{Ga}{}^0$ in GaAs:Si and (**b**) the transition energy ε(0/+) (referenced to CBM) of Si_{Ga} in GaAs:Si as a function of the QDs diameter

energy level from the CBM, also increases. The calculated transition energy levels as a function of QDs' size are shown in Fig. 12(b). It is found that the calculated ε(0/+) level of Si_{Ga} is very shallow at 6 meV below the CBM in the bulk system, in good agreement with experimental results. It increases to 162 meV for the smallest QD studied in this work. These results indicate that *n*-type doping using Si as a dopant will be much more difficult in small QDs than in bulk GaAs.

The LDA-calculated Ga–As bond length in pure GaAs is $d_0 = 2.42$ Å. For bulk GaAs:Si, the Si–As bond lengths of $Si_{Ga}{}^+$, $Si_{Ga}{}^0$, and $Si_{Ga}{}^-$ are 2.375 Å, 2.377 Å, and 2.378 Å, respectively, which are slightly smaller than d_0. The small increase of the bond length as q decreases from +1, 0 to −1 is consistent with the fact that the As atom in GaAs is negatively charged; therefore, when Si_{Ga} becomes more negatively charged, Coulomb repulsion will push As away, increasing the Si–As bond length. However, because the Si_{Ga} level is very shallow, that is, the defect charge is delocalized (Fig. 13a), the variation of the bond length as a function of q is rather small. However, in small QDs with diameter $d = 1.55$ nm, the Si–As bond length increases significantly for $Si_{Ga}{}^+$, $Si_{Ga}{}^0$, and $Si_{Ga}{}^-$, changing from 2.387 Å, to 2.438 Å, to 2.489 Å. This is consistent with the fact that the Si_{Ga} level is deep and more localized around Si in the QDs (Fig. 13c). Thus, changing the charge state at an Si site will have a large effect on the Coulomb interaction and will lead to a large variation of the Si–As bond lengths.

The DX-like defect center, which converts shallow hydrogenic donor into a deep level [19], is one of the major "killer" defects that limits n-type doping in II–VI and III–V semiconductors [20]. The DX centers in bulk semiconductors have

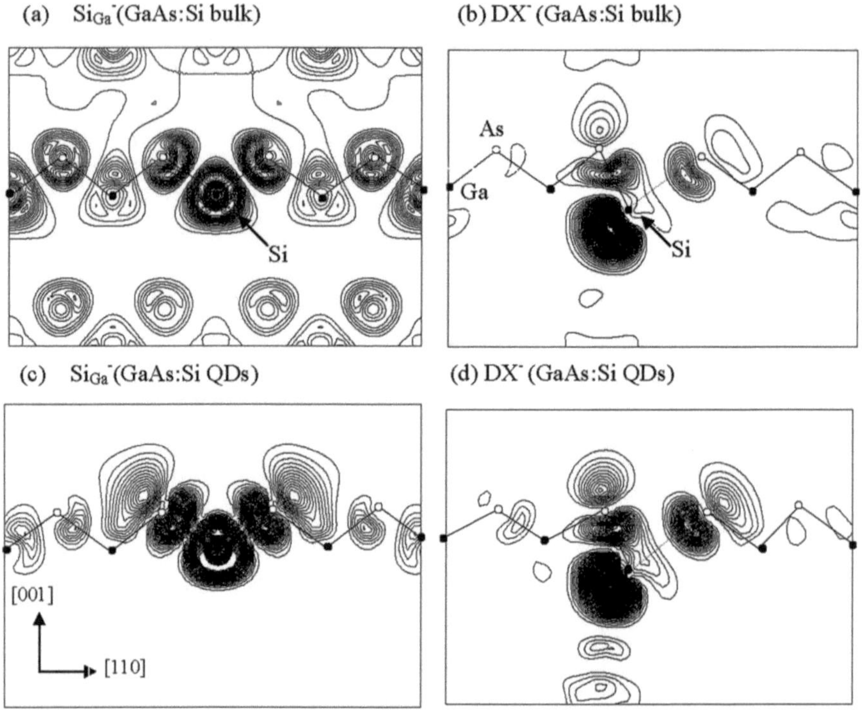

Fig. 13 Contour plot of the charge density distribution of (**a**) Si_{Ga}^- and (**b**) DX^- of bulk GaAs:Si; (**c**) Si_{Ga}^- and (**d**) DX^- of GaAs:Si QDs (diameter d = 2.67 nm)

been intensively studied for many years [19–21]. The original single-bond-breaking model, proposed by Chadi and Chang [21] for Si-doped GaAs, suggests that the DX center forms by displacing the substitutional Si defect along the <111> direction, breaking one bond and changing the local symmetry from T_d to C_{3v} (see Fig. 1 of [18]). Accompanied by the displacement, the t_{2c} state in the conduction band splits into an a_1 and a doubly degenerated e state. The resulting $a_1(t_{2c})$ state couples with the original $a_1(a_{1c})$ state, pushing one of them down. If the a_1 state is occupied (preferentially by two electrons, such as in Si_{Ga}^-), such an atomic-displacement-induced level repulsion can lead to electronic energy gain, although breaking the bond in the <111> direction also costs energy. When the DX center becomes stable, it converts a shallow donor (e.g., Si_{Ga}^+) into a deep acceptor (e.g., DX $(Si_{Ga})^-$), thus limiting the doping process [18].

Recently, we studied the relative stability of the DX center in GaAs:Si QDs as a function of the dot size [18]. The DX formation energy is defined as the energy difference: $\Delta E(DX)=E(DX^-)-E(Si_{Ga}^-)$, where $E(DX^-)$ is the total energy of the negatively charged DX center and $E(Si_{Ga}^-)$ is the total energy of the corresponding tetrahedral-coordinated defect Si_{Ga} at the same charge state. A negative $\Delta E(DX)$ will indicate that the DX center is more stable than Si sitting at the substitutional

Fig. 14 DX formation energy as a function of the calculated band gap of GaAs QDs. Arrow indicates the band gap $E_g = 1.78$ eV, at which the DX^- is stabilized. The corresponding QD diameter is about 14.5 nm

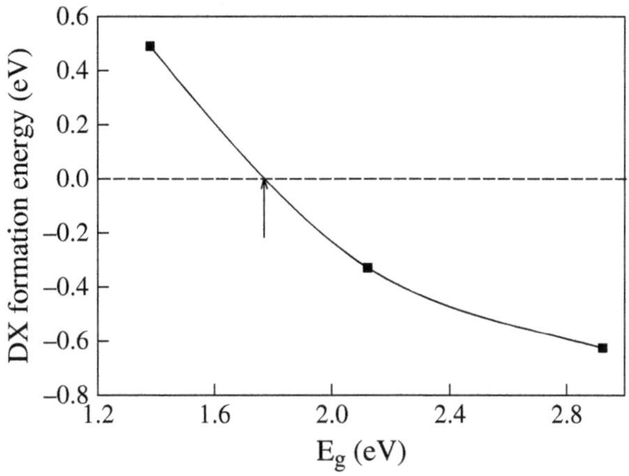

Ga site. The calculated results are shown in Fig. 14. In bulk GaAs, the Si-doped DX formation energy is positive, indicating that the formation of the DX center is not favored in bulk GaAs:Si. However, as the size of the QDs decreases and the corresponding band gap E_g increases, the DX formation energy becomes less positive and changes sign when the band gap is close to $E_g = 1.78$ eV, which is about 0.26 eV larger than the experimental band gap. Previous calculations show [24] that the bandgap increases of GaAs QDs due to quantum confinement can be expressed as $\Delta E_g = 3.88/d^{1.01}$. Using this expression, it can therefore be estimated that the Si DX center in GaAs will become stable when the diameter of the QD is less than 14.5 nm.

The origin of the enhanced stability of the DX center due to the quantum confinement can be understood as follows: The quantum confinement increases the CBM energy. For the negative-charged Si_{Ga}^- at the T_d site, the shallow defect level has mostly the CBM *s* wave function character. Thus, the energy level of Si_{Ga}^- is expected to follow closely with the CBM. But for the DX center, the Si impurity undergoes a large Jahn–Teller distortion along the <111> directions. Consequently, the level repulsion between the $a_1(a_{1c})$ with $a_1(t_{2c})$ states mixes a significant amount of atomic *p* orbital into the wave function, so the DX^- level does not follow closely to the CBM. Therefore, in QDs, when the band gap increases, the energy difference between the occupied DX^- and Si_{Ga}^- levels also increases, thus stabilizing the DX center.

7 Conclusion

In summary, recent development of calculations of nanocrystal physical properties using the large-scale *ab initio* pseudopotential or charge-patch methods is reviewed. The calculated size-dependent exciton energies and absorption spectra of QDs and

QWs are in good agreement with experiments. The calculated ratios of bandgap increases between QWs and QDs are found to be material-dependent, and for most direct bandgap materials, this ratio is close to 0.586, as predicted by simple effective-mass approximation. We show that the electronic structure of a nanocrystal can be tuned not only by its size, but also by its shape. Therefore, the shape can be used as an efficient way to control the electronic structure of the nanocrystals. Changing the shape is expected to be more flexible and provides more variety of the electronic states than simply changing the size of the system. We also show that defect properties in QDs could be significantly different from those in bulk semiconductors. For example, although negatively charged DX^- center is unstable in bulk GaAs:Si with respect to the tetrahedral coordinated $Si_{Ga}{}^-$, when the dot size is small enough, it becomes stable.

Acknowledgments J. Li gratefully acknowledges financial support from "One-hundred Talents Plan" of the Chinese Academy of Science. We would like to thank Dr. L. W. Wang for his contribution in this work and helpful discussions. The work is partially supported by the National Natural Science Foundation of China and by the Foundation of the Chinese Academy of Science. The work at NREL is supported by the U.S. DOE under contract No. DE-AC36-99GO10337. The use of computer resources of the NERSC is greatly appreciated.

References

1. Alivisatos, A. P. *Science* **1996**, 271, 933.
2. Alivisatos, A. P. *J. Phys. Chem.* **1996**, 100, 13226.
3. Brus, L. E. *J. Chem.. Phys.* **1984**, 80, 4403.
4. Woggon, U. *Optical Properties of Semiconductor Quantum Dost,* Springer-Verlag, Berlin, **1996**.
5. Murray, C. B., Norris, D. J., Bawendi, M. G. *J. Am. Chem. Soc.* **1993**, 115, 8706.
6. Ozin, G. A. *Adv. Mater.* **1994**, 4, 612.
7. Yoffe, A. D. *Adv. Phys.* **1993**, 42, 173.
8. Duan, X., Lieber, C. M. *Adv. Mater.* **2000**, 12, 298.
9. Kazes, M., Lewis, D. Y., Ebenstein Y., Mokari, T., Banin, U. *Adv. Mater.* **2002**, 14, 317.
10. Huynh, W. U., Dittmer, J. J., Alivisatos, A. P. *Science* **2002**, 295, 2425.
11. Bruchez, M., Moronne, M., Gin, P., Weiss, S., Alivisatos, A. P. *Science* **1998**, 281, 2013.
12. Manna, L., Scher, E., Alivisatos, A. P. *J. Am. Chem. Soc.* **2000**, 122, 12700.
13. Manna, L., Milliron, D., Meisel, A., Scher, E., Alivisatos, A. P. *Nat. Mater.* **2003**, 2, 382.
14. Wang, Z. L., Pan, Z. W. *Adv. Mater.***2002**, 14,1029.
15. Milliron, D., Hughes, S. M., Cui, Y., Manna, L., Li, J., Wang, L. W., Alivisatos, A. P. *Nature* **2004**, 430, 190.
16. Li, J., Wang, L. W. *Nano Lett.* **2003**, 3, 1357.
17. Wei, S. -H. *Comp. Mater. Sci.* **2004**, 30, 337.
18. Li, J., Wei, S. -H., Wang, L. W. *Phys. Rev. Lett.* **2005**, 94, 185501.
19. Lang, D. V., Logan, R. A. *Phys. Rev. Lett.* **1977**, 39, 635.
20. Zhang, S. B., Wei, S. -H., Zunger, A. *Phys. Rev. Lett.* **2000**, 84, 1232.
21. Chadi, D. J., Chang, K. J. *Phys. Rev. Lett.* **1988**, 39, 10063.
22. Kresse, G., Hafner, J. *Phys. Rev. B* **1993**, 47, 558.
23. Wang, L. W.*Phys. Rev. Lett.* **2002**, 88, 256402.
24. Li, J., Wang, L. W. *Phys. Rev. B* **2005**, 72, 125325 (and references therein).
25. Yu, H., Li, J., Loomis, R. A., Gibbons, P. C., Wang, L. W., Buhro, W. E. *J. Am. Chem. Soc.* **2003**, 125, 16168.

26. Guzelian, A., Katari, J. E. B., Kadavanich A. V., Banin, U., Hamad, K., Juban, E., Alivisatos, A. P., Wolters, R. H., Arnold, C. C., Health, J. R. *J. Phys. Chem.* **1996**, 100, 7212.
27. Micic, O. I., Curtis, C. J., Jones, K. M., Sprague, J. R., Nozic, A. J. *J. Phys. Chem.* **1994**, 98, 4966.
28. Yu, H., Li, J., Loomis, R. A., Wang, L. W., Buhro, W. E. *Nat. Mater.* **2003**, 2, 517.
29. Chen, X., Nazzal, A., Goorskey, D., Xiao, M., Peng, Z. A., Peng, X. *Phys. Rev. B*, **2001**, 64, 245304.
30. Katz, D., Wizansky, T., Millo, O., Rothenberg, E., Mokari, T., Banin, U. *Phys. Rev. Lett.* **2002**, 89, 86801.
31. Li, J., Xia, J. B. *Phys. Rev. B,* **2000**, 61, 15880.
32. Xia, J. B., Li, J. *Phys. Rev. B,*. **1999**, 60, 11540.
33. Li, J., Wang, L. W. *Chem. Mater.* **2004**, 16, 4012
34. Li, L. S., Hu, J., Yang, W., Alivisatos, A. P. *Nano Lett.* **2001**, 1, 349.
35. Hu, J., Li, L. S., Yang, W., Gin, P., Weiss, P., Alivisatos, A. P. *Science* **2001**, 292, 2060.
36. Wei, S. -H., Zhang, S. B. *Phys.Rev B* **2000**, 62, 6944.

Chaotic Behavior Appearing in Dynamic Motions of Nanoscale Particles

K. Miura, R. Harada, M. Kato, M. Ishikawa, and N. Sasaki

Abstract The case of one-directional motion, under which graphite and mica flakes are driven on an octamethylcyclotetrasiloxane (OMCTS) liquid surface, is presented. The dynamical forces needed to move these bodies increase linearly with the logarithm of scanning velocity, which is a typical energy dissipation process. A transition from quasi-periodic to chaotic motions occurs in the dynamics of a graphite flake when its velocity is increased. The dynamics of graphite flakes pulled by the nanotip on an OMCTS liquid surface can be treated as that of a nanobody on a liquid. On the other hand, there do not appear chaotic motions in the dynamics of a mica flake because the contact area between a mica flake and an OMCTS liquid surface is larger than that between a graphite flake and an OMCTS liquid surface.

Contents

K. Miura (✉)
Department of Physics, Aichi University of Education, Hirosawa 1, Igaya-cho, Kariya 448-8542, Japan
e-mail: kmiura@auecc.aichi-edu.ac.jp

Z.M. Wang (ed.), *Toward Functional Nanomaterials*, Lecture Notes in Nanoscale Science and Technology 5, DOI 10.1007/978-0-387-77717-7_6,

1 Introduction

At present, an understanding of the dynamics of a small body in a liquid is of great importance because it will provide fundamental knowledge of lubrication, molecular electronics and microfluid. This leads to the problem that is presently facing nanoelectrical mechanical systems (NEMS) and microelectrical mechanical systems (MEMS) because it becomes difficult to make the nanometer-scale body to move on a surface. In addition, when the size of the body on the liquid is reduced to the nano- and micrometers, it becomes the problem of a small body in one direction under Brownian motion. However, there has been little study on this topic, because there are no applicable methods for investigating the dynamics of a small body in a liquid. We present the case of one-directional motion, under which a flake on a liquid surface is driven by a nanotip. This method has been applied for the studies on atomic friction of solid on solid [8, 9, 10]. Here, we report the results of the one-directional motion of a flake on a liquid examined by this method. The force needed to drive a flake on a liquid surface is investigated as functions of loading force and scanning velocity. Friction forces and energy dissipation, which appear in the dynamics of a flake propulsive on a liquid, are discussed.

2 Experiment

Octamethylcyclotetrasiloxane (OMCTS), a molecule which consists of forty atoms in an ellipsoidal shape on average, as shown in the upper part of Fig.1, was used as the liquid material. Graphite and mica substrates were prepared by cleaving highly oriented pyrolytic graphite (HOPG) and muscovite mica blocks, respectively. An OMCTS droplet was dropped onto each of the graphite and mica substrates. Cleaved graphite and mica flakes with an area of 1 mm square and a thickness of several micrometers were placed on the OMCTS droplet and then an atomic force microscope (AFM) tip was brought into contact with them. At this time, using a CCD camera, it was confirmed that the graphite and mica flakes are not immersed in the OMCTS droplet but float on it. Within 10 minutes, lateral forces were measured at room temperature using a commercially available instrument (Seiko Instruments Inc., SPI-3700) because the OMCTS droplet quickly vaporizes and disappears. A rectangular silicon cantilever with a normal spring constant of 0.75 N/m was used. The zero normal force is defined as the force at the position at which the cantilever is not bent. The friction forces were calibrated by the method reported in our previous papers [8, 9, 10].

3 Load and Velocity Dependence of Friction Force

3.1 Graphite Flake Case

Figure 2 shows the lateral force loops needed to drive the graphite flake on the OMCTS droplet as functions of loading force and scanning velocity. All

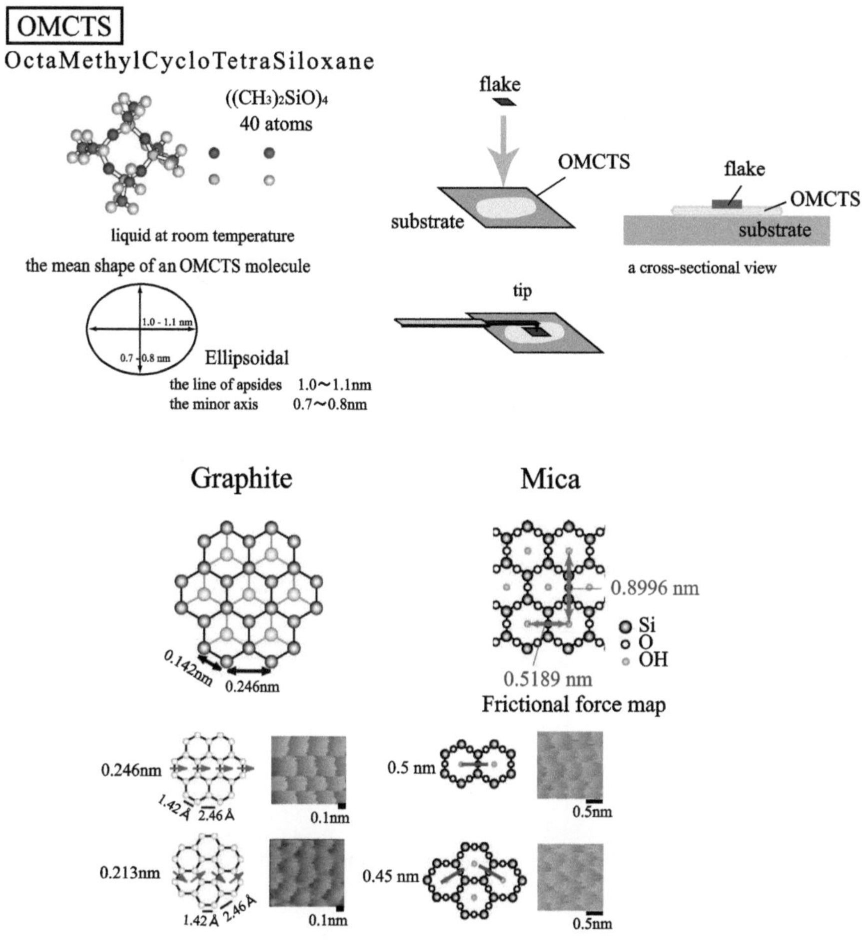

Fig. 1 Octamethylcyclotetrasiloxane (OMCTS) consisting of forty atoms in an ellipsoidal shape, and sample preparation. An OMCTS droplet was dropped onto graphite and mica substrates. Cleaved graphite and mica flakes with an area of 1 mm square and a thickness of several micrometers were placed on the OMCTS droplet. The lateral force maps and tip movements on the mica flake and the graphite flake are shown at the *bottom*

experiments were done within 10 minutes of introducing the OMCTS droplet onto the graphite substrate because the OMCTS droplet rapidly volatilizes and disappears. The lateral force loops versus scan velocity exhibit fluctuating oscillations at a low scan velocity for each loading force; namely, the behaviors of fluctuating oscillations seem to depend strongly on scan velocity whereas they are insensitive to loading force. The mean friction force at a loading force of 0 nN increases only slightly linearly with the logarithm of the scan velocity, as seen at the bottom of Fig. 2. The mean friction forces in relation to loading force at a scan velocity of

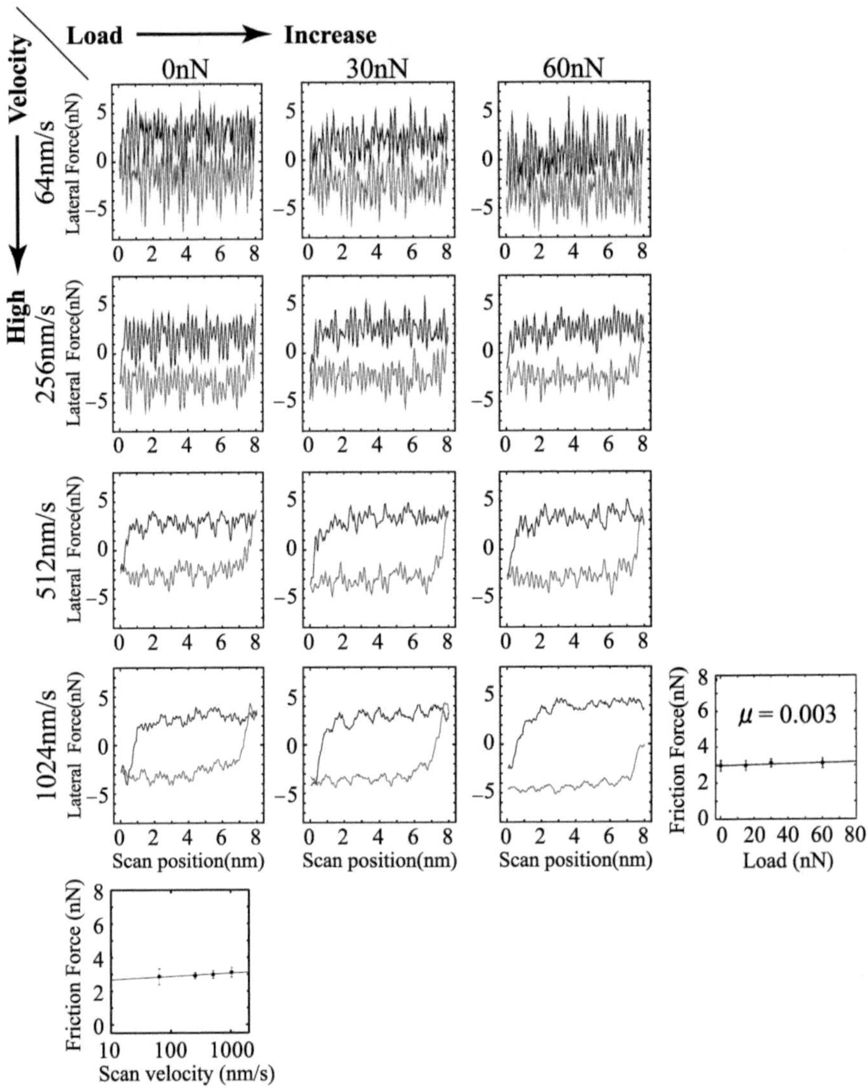

Fig. 2 Lateral forces needed to drive a graphite flake on an OMCTS liquid as functions of loading force and scanning velocity, where the *black* and *red lines* indicate one direction and the opposite direction, respectively. The mean friction forces at a loading force of 0 nN are shown on the scale of the logarithm of the scan velocity at the *bottom*. The mean friction forces in relation to loading force at a scan velocity of 1024 nm/s are shown on the *right-hand* side

1024 nm/s are almost constant (friction coefficient is 0.003), as shown on the right-hand side of Fig. 2.

In a friction experiment of a graphite flake on a graphite substrate where an AFM tip pushes the graphite flake, it was found that the actual contact area between the graphite flake and graphite surface becomes very small and is on the order of a nanometer square, compared to the total area of the flake, 1 mm square, reflecting the prominent surface deformations of graphite, as shown in our previous paper [10], Guo et al.'s study [6] and Dienwiebel et al.'s study [2]. In a similar manner to the experiment of a graphite flake on the graphite substrate, even in this experiment the contact size between the flake and the OMCTS droplet is thought to be of nanometer order, because the mean friction forces on the right-hand side of Fig. 2, relative to loading force, is almost constant (friction coefficient is 0.003 and comparable to that (0.001) of a graphite flake on a graphite substrate). Furthermore, the emergence of fluctuations of the graphite flake at a low scan velocity supports the contact size between the flake and the OMCTS droplet being of nanometer order.

The Fourier spectra of Fig. 2 are shown in Fig. 3, where the vertical axis of the inset on the right-hand side is in *decibels*. Peaks of these spectra consist of quasi-periodic ones of $\mathrm{m} f_1+\mathrm{n} f_2$(m, n: integers) where main peaks f_1 and f_2 are produced. The f_2 peak is the strongest at a low scanning velocity which gives high-frequency oscillations; however, as the scanning velocity increases, the f_1 peak becomes the strongest and the noise level increases, as shown by the arrows in the background of the inset on the right-hand side. This is the typical pattern of a transition from a quasi-periodic state to a nonperiodic (or chaotic) state [11]. Thus, the dynamics of the graphite flake changes from quasi-periodic motion to chaotic motion as the scanning velocity increases.

Here the important point is that the change of the scanning velocity of the graphite flake reveals the different time-scaled dynamics of an OMCTS molecule. If the low scanning velocity of the graphite flake is comparable to the effective velocity of dynamics of the internal degrees of freedom of the OMCTS molecule, the graphite flake could detect mainly structural fluctuations derived from rotations, oscillations and stretches of the OMCTS molecule, which results in the f_2 peak corresponding to a scale of about 0.2 nm (a scan velocity of 64 nm/s in Fig. 3). Although it is interesting to confirm exactly which internal degree of freedom of the OMCTS molecule contributes to the f_2 peak, it is difficult to determine it under the present experimental setup. On the other hand, as the scanning velocity increases, the graphite flake could mainly detect the stationary arrangement of OMCTS molecules. Compared to the scanning velocity of the graphite flake, the OMCTS molecule appears to be fixed, which results in the f_1 peak corresponding to a scale of about 1 nm (a scan velocity of 1024 nm/s in Fig. 3). Thus, the interactions between the graphite flake and the OMCTS liquid consist of the combined ones of the two modes, f_1 and f_2. The competition between these modes, as a result, keeps viscosity (or friction) against one-directional motion of the graphite flake almost constant. It is interesting to note that at a higher scanning velocity, this system is not yet in the adiabatic state but in the nonequilibrium stationary state. Thus, as the scanning velocity increases, a transition to the chaotic state is necessary to lower the energy dissipation of this system in the nonequilibrium stationary state, which is derived from the minimum entropy production principle of Gransdorff and Prigogine [5].

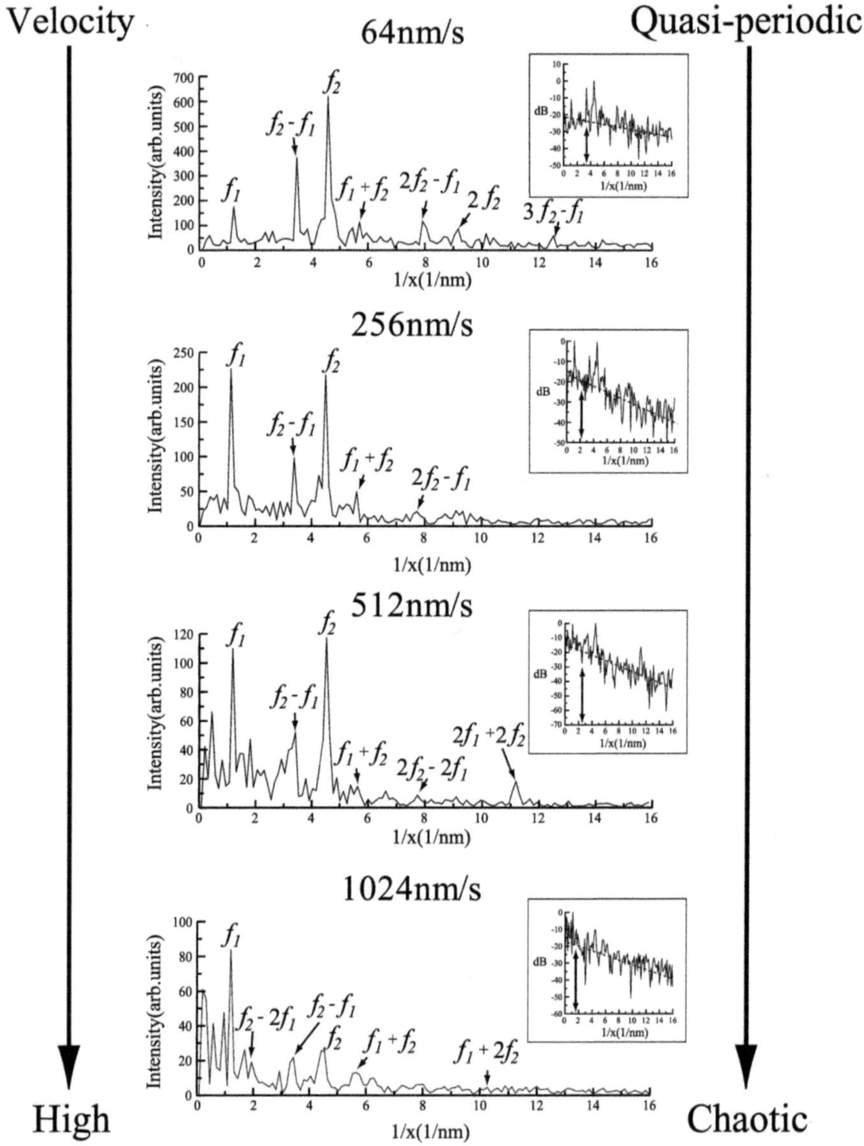

Fig. 3 Fourier spectra of the samples on the case of Fig. 2 are shown, where the vertical axis of the inset on the *right-hand* side is in *decibels*

3.2 Mica Flake Case

Figure 4 shows the lateral force loops needed to drive the mica flake on the OMCTS droplet as functions of loading force and scanning velocity. In a similar manner to the graphite flake experiments, all experiments were done within 10 minutes

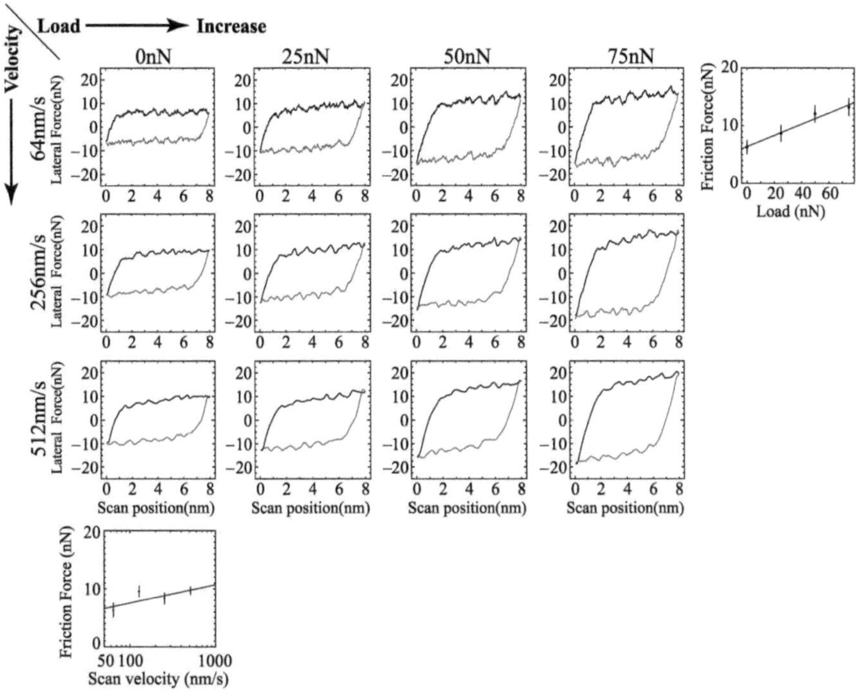

Fig. 4 Lateral forces needed to drive a mica flake on an OMCTS liquid as functions of loading force and scanning velocity, where the *black* and red *lines* indicate one direction and the opposite direction, respectively. The mean friction forces at a loading force of 0 nN are shown on the scale of the logarithm of the scan velocity at the *bottom*. The mean friction forces in relation to loading force at a scan velocity of 64 nm/s are shown on the *right-hand side*

of introducing the OMCTS droplet onto the mica substrate because the OMCTS droplet rapidly volatilizes and disappears. The mean friction forces in relation to loading force for each scan velocity increase linearly; for example, friction coefficient is 0.03, as shown on the right-hand side of Fig. 4. The lateral force loops versus scan velocity for each loading force are also shown in Fig. 4. The mean friction force at a loading of 0 nN increases linearly with the logarithm of the scan velocity, as seen at the bottom of Fig. 4. This behaviour is clearly different from the graphite flake case. The linear increase of the frictional forces in relation to loading force in the mica case reveals that the contact area between the mica surface and the OMCTS liquid surface increases linearly with increasing loading force. This indicates that the surface deformation of mica, which behaves like a hard plate compared to the graphite flake, is significantly smaller than that of graphite, which maintains a contact radius of 1 nm irrespective of loading force.

Keeping in mind the fact that this system consists of friction between the tip and the flake and between the flake and the OMCTS, it was found that the friction in this system includes two different mechanisms. However, it can be concluded that the features of the friction of a flake on OMCTS are only those of friction between the flake and the OMCTS, because they do not include a frictional force map of the tip on the flake. This indicates that the friction between the tip and the flake is greater than that between the flake and the OMCTS.

4 Energy Dissipation and Friction

As shown at the bottom of Figs. 2 and 4, the mean friction forces only slightly increase linearly with the logarithm of the scan velocity V:

$$F = \alpha + \beta \ln V \tag{1}$$

This result can be interpreted within the framework of a modified Tomlinson model, taking into account the thermally activated process [4, 1, 7]. Now, we consider the motion under an effective potential $U_{eff}(x, y) = U_0(x) + V(x, t)$, where $U_0(x)$ and $V(x, t)$ describe the interaction potential between the flake and the OMCTS liquid surface, and the elastic energy stored in a cantilever, respectively. Thus, $V(x, t)$ is given as $1/2k_{eff}(x - x_s(t))^2$, where x, x_s and k_{eff} are the position of the tip, the support position of the microscope and the effective spring constant, respectively. According to the formula derived by Heslot et al. [7], as soon as the work performed by the driving force for moving from the minimum of U_{eff} to the next maximum becomes much greater than the thermal energy ΔE, the friction force F can be given as

$$F(x, t) = 2/a \left\{ \Delta U_0(F_n) + \Delta E \ln \left(\frac{V}{V_0} \right) \right\}, \tag{2}$$

where a, ΔU_0 and V_0 are the space between the minimums, the amplitude modulation of U_0 and the velocity at the minimum U_{eff}, respectively. Here, it should be noted that ΔU_0 is defined as a function of loading force F_n although Heslot et al. [7] defined it as a function of ageing time, because ΔU_0 is related strongly to contact areas and then loading force becomes equivalent to ageing time in a physical meaning. In the case where scan velocity is kept constant, the difference ΔF between the friction forces at different loading forces F_{n1} and F_{n2} is given as follows using Equation (2):

$$\Delta F = \frac{2}{a} \left\{ \Delta U_0(F_{n1}) - \Delta U_0(F_{n2}) \right\}. \tag{3}$$

In the case where the loading force is kept constant, the difference between the friction forces ΔF at different scan velocities V_1 and V_2 is given as follows, using Equations (1) and (2):

$$\Delta F = \beta \ln \frac{V_1}{V_2} = \frac{2\Delta E}{a} \ln \frac{V_1}{V_2}. \tag{4}$$

Using Equation (3), the increase of ΔU_0 per nanonewton loading force, $\mu a/2$, is estimated to be approximately 0.002 eV/nN and 0.3 eV/nN for graphite flake and mica flake cases, respectively. Here, μ is a friction coefficient as $\mu = \Delta F/(F_{n1} - F_{n2})$. Using Equation (4), the thermally activated energy contributing to this slip process, ΔE, was estimated to be approximately 36 $k_B T$and 200 $k_B T$ for graphite flake and mica flake cases using the relation $\beta = 2\Delta E/a$, respectively. The ΔE for graphite flake case is comparable to the chemical energy (approximately 20 $k_B T$) released by a single adenosine triphosphate (ATP) molecule in the molecular motor kinesin, where k_B and $T$are the Boltzmann factor and the absolute temperature, respectively. Hence, it is shown that the dynamics of a graphite flake pulled by the nanotip on an OMCTS liquid can be treated as that of a nanobody on a liquid.

5 Conclusion

Here, the case of one-directional motion, under which a flake is driven on a liquid surface, was presented. The dynamical forces needed to move these bodies increased linearly with the logarithm of scanning velocity. The energy dissipation of graphite and mica were estimated to be approximately 36 $k_B T$ and 200 $k_B T$, respectively. A transition from quasi-periodic to chaotic motions appears in the dynamics of graphite flakes propulsive on an OMCTS liquid. This makes it possible to maintain a small resistance by chaotic motion when the velocity of a graphite flake is increased. This result reveals the dynamical motion of a nanoscale body on the OMCTS liquid irrespective of the graphite flake size because a contact size between the graphite flake and the OMCTS droplet is of the order of nanometer. This study reveals that "viscosity and friction" are phenomena of the complex systems closely related to the nonequilibrium stationary state and "lubrication" arises through "chaotic motions" in the nonequilibrium stationary state.

Generally, the stick-slip oscillations occur in surface force apparatus (SFA) experiments when the contact between the two surfaces is adhesive (or solid-like). Up to now, it has been reported by Drummond et al. [3] that quasi-periodic oscillation appears in stick-slip oscillation on a micrometer scale in the SFA experiment, because the contact area in SFA experiments is of few tens of micrometers. They have measured a complex sequence of periodic regimes separated by aperiodic regimes. They have also pointed out that the number density of adhesive junctions could possibly explain such a complex dynamics. At first sight, the transition between periodic and aperiodic regimes obtained by Drummond et al. [3] can be easily compared to that between quasi-periodic and chaotic regimes obtained by us. Furthermore, unlike measurements by Drummond et al. [3], our measurements show no periodic oscillations. However, we must be careful in making a direct comparison between them. The first difference between these two cases is the system: the adhesive solid–solid interface (hemi-fused state) by Drummond et al. [3] and the solid–liquid interface (a solid nanobody on a liquid) in our experiment. The second difference is the scale of the time and space: the contact area is of micrometer order by Drummond et al. [3] and of nanometer order in our experiment. Thus, we meet difficulties in making a clear comparison with previous works. However, if the bonding between an OMCTS molecule and a graphite flake is assumed to be a single adhesive junction, it may be possible to find a correlation between two different systems and the key to universal understanding of complex dynamics with scaled time and space irrespective of the system.

References

1. Briscoe BJ, Evans DCB (1982) Proc. R. Soc. Lond. A**380**: 389.
2. Dienwiebel M, Verhoeven G, Pradeep N, Frenken J, Heimberg J, Zandbergen H (2004) Phys. Rev. Lett. **92**: 12601.
3. Drummond C, Elezgaray J, Richetti P (2002) Europhy s. Lett. **58**: 503.
4. Eyring H (1935) J. Chem. Phys. **3**: 107.
5. Gransdorff P, Prigogine I (1964) Physica **30**: 351.
6. Guo W, Zhu CZ, Yu TX, Woo CH, Zhang B, Dai YT(2004) Phys. Rev. Lett. **93**:245502.
7. Heslot F, Baumberger T, Perrin B, Caroli B, Caroli C (1994) Phys. Rev. E**49**: 4973.
8. Miura K, Kamiya S (2002) Europhys. Lett. **65**: 610.
9. Miura K, Kamiya S, Sasaki N (2003) Phys. Rev. Lett. **90**: 055509.
10. Miura K, Sasaki N, Kamiya S (2004) Phys. Rev. B**69**: 075420.
11. Rand D, Ostlund S, Sethna J, Siggia ED (1982) Phys. Rev. Lett. **49**: 132.

Hydrogen Concentration, Bonding Configuration and Electron Emission Properties of Polycrystalline Diamond Films: From Micro- to Nanometric Grain Size

Sh. Michaelson, O. Ternyak, R. Akhvlediani, and A. Hoffman

Contents

1 Introduction

Carbon science and technology has developed extensively in the last few decades due to discoveries of various polycrystalline and nanocrystalline diamond film deposition methods, synthesis of nanotubes, fullerenes and carbon fibers, etc. Nanocrystalline diamond films represent a new remarkable material that attracts a lot of

A. Hoffman (✉)
Schulich Faculty of Chemistry, Technion – Israel Institute of Technology, Haifa 32000, Israel
e-mail: choffman@tx.technion.ac.il

Z.M. Wang (ed.), *Toward Functional Nanomaterials*, Lecture Notes in Nanoscale Science and Technology 5, DOI 10.1007/978-0-387-77717-7_7,

attention of the scientific world due to its promising potential in many possible applications, such as tribology [1, 2], field emission [3, 4], electrochemistry [5, 6], protective optical windows [7, 8] as well as its unique ability to incorporate n-type dopants [9] compared to polycrystalline and single crystal diamond.

The microstructure and properties of carbon-based thin films depend on the deposition process and the conditions used, including pressure, gas phase composition, and substrate temperature, as well as the energy of the reactive species (atoms or ions). Nanodiamond films may be deposited by a number of deposition processes, which differ in the growth species and deposition parameters. In fact, each method results in different types of nanodiamond films. These may differ in terms of the diamond particle size, grain boundary nature, hydrogen content, defect density, amorphous or graphitic component of the films, morphological properties of the films, and different chemical and physical properties. Also, the formation mechanisms of the different films are dissimilar. For example, the growth of nanodiamond films deposited by microwave (MW) chemical vapour deposition (CVD) from Ar-reach plasma was suggested to occur from C_2 dimmers on the film surface [10], whereas deposition from hydrogen-reach direct-current (DC) energetic plasma is a sub-surface or sub-plantation process [11].

The ability to deposit diamond films with well-defined crystalline size and nanoscale-ordered surfaces is crucial for possible practical applications. The well-known negative electron affinity (NEA) and high conductivity of diamond surfaces are the properties of fully hydrogenated diamond surfaces [12–14]. Similarly, diamond grain size may influence the electronic and optical properties of the films [15–18]. The properties and associated practical applications of diamond films are significantly modified by their surface structure and chemical composition. However, the chemical and physical characterization of the uppermost surface atomic layers of diamond films presents a great challenge. Common surface-sensitive techniques, such as X-ray photoelectron spectroscopy (XPS), Auger electron spectroscopy (AES), and electron energy loss spectroscopy (EELS), usually deal with a few nanometer surface region rather than with atomic monolayer coverage. Similarly, the characterization of diamond grain size (especially in nanometric scale) requires the use of complicated transmission electron microscopy (TEM) usually combined with X-ray diffraction (XRD) analysis. Therefore, the quest of additional complementary techniques providing information about the grain size and surface properties is of high importance.

In this chapter, we summarize our recent studies related to hydrogen concentration and bonding configuration on the surface and within diamond films as a function of grain size in the tens to hundreds nanometer range [19, 20]. Hydrogen is the major component of the gas mixture usually used for diamond nucleation and growth by CVD methods. It is thus involved in many different processes related to diamond nucleation and growth including (1) formation of clusters necessary for growth (CH_x or C_2H_x), (2) stabilization of diamond clusters and surfaces, (3) preferential etching and removal of non-diamond constituents, and (4) abstraction of hydrogen from the diamond surface to allow incorporation of additional C-containing growth species. It is no wonder that some quantity of hydrogen must be incorporated in diamond CVD films and the questions to be asked are (1) in what quantities, (2) in what locations (in the grains as interstitials, forming defect-H clusters in the grains, on the diamond surfaces, in grain boundaries and in a non-diamond constituent between the grains (e.g., when nanodiamond crystallites are embedded in an a-C/graphitic matrix)), and (3) in what bonding configuration (e.g., bonded to C or as non-bonded H or H_2).

The presence of H in diamond lattice has serious repercussion on its characteristics [21]. While hydrogen incorporation and effects in hydrogenated amorphous carbon films, which may contain up to 45 at% hydrogen [22] are relatively well studied [22–27], only little information is available on hydrogen distribution, concentration and location in diamond films [21, 28–31]. TEM and a combination of Fourier transform infrared (FTIR) absorption and elastic recoil detection analysis (ERDA) were used to evaluate the location of hydrogen in diamond. It has been found that hydrogen in CVD diamond is incorporated on diamond surfaces and in non-diamond regions (graphitic and amorphous carbon). Hydrogen was claimed to be found in grain boundaries [21, 30] and also be trapped in grain defects [31].

In this chapter, we present the studies of diamond films of varying grain size and thickness carried out in our laboratory over the last few years [11, 15, 16, 32, 33]. The films consist of diamond crystallites of sizes ranging from ~5 nm to ~300 nm depending on the deposition methods and conditions [11, 33]. The thickness of continuous films varies from ~70 nm to ~1000 nm depending on the deposition time. The investigated diamond films were grown by three different kinds of CVD methods: (1) hot filament (HF) CVD, (2) direct current glow discharge (dc GD) CVD, and (3) MW CVD. The first two methods utilize hydrogen-rich gas mixtures (CH_4/H_2 ratios were 1/99 and 9/91, respectively); the third one uses a plasma mostly comprised of Ar gas (>95%), while hydrogen atoms originate mainly from methane species. Although the threeinvestigated kinds of diamond films were grown by different techniques resulting in different growth mechanisms, we believe that the final properties of these diamond films should be associated with diamond microstructure and phase composition rather than with a particular deposition method.

The general structure of this chapter is as follows. First, the hydrogen concentration is studied by secondary ion mass spectroscopy (SIMS) as a function of diamond grain size. Then, we correlate the hydrogen retention with the appearance of Raman peaks and clarify the hydrogen bonding configuration within the film bulk. Then, the hydrogen bonding configuration on well-defined film surfaces is studied by high resolution electron energy loss spectroscopy (HR-EELS). The assignment of the peaks is done by means of isotopic exchange. The change of the shape of HR-EEL spectrum with diamond grain size transition from micro- to nanoscale size is reported.

The transition of the grain size in polycrystalline diamond films from micro- to nanoscale may also lead to significant modification of mechanical, tribological, optical, and electronic properties of diamond films. In particular, one of the most intriguing properties of diamond films is their high electron emission under the impact of particle beam and the ability to emit relatively high currents under low electric fields. In principle, diamond films are considered as excellent electron emitters in various applications for the following reasons: (1) NEA of hydrogenated or carbon-covered surface, (2) conductivity can be tuned by doping, surface treatment or phase composition, (3) large mean free path of slow electrons due to wide band gap (5.5 eV), (4) high thermal conductivity, (5) high toughness, (6) large intrinsic breakdown field, and (7) high charge carrier mobility.

The electron emission from diamond films may strongly depend on surface quality, presence of non-diamond carbon phase, defects, doping, and microstructure. In spite of an NEA property of hydrogenated diamond surfaces, electron emission from such surfaces may be hindered by limitations in the electron supply to the film surface. The film thickness and the thickness-dependent microstructure are very important factors, which may control the electron transport, charge accumulation at the diamond surface and potential drop across the film. In different types of electron emission (particle-induced or field-induced), different structural properties of diamond films may be preferable.

HF CVD diamond films, in contrast to the two other types of diamond films, have the well-known columnar microstructure due to preferential growth of some grains and suppression of others. Thus, the grain size of HF CVD diamond films can be controlled only by film thickness, which in turn is being controlled by deposition time without changing the deposition conditions. In the last section of the present chapter, we present the electron emission properties of HF CVD diamond films as a function of film thickness and a constituting grain size. The electron emission, induced by photons, ions, electrons, and electric field, was investigated and compared for diamond film thicknesses ranging from ~70 nm to ~4.7 μm.

2 The Polycrystalline Diamond Films: Brief Description of Deposition Methods and Microstructure of the Films

Three kinds of CVD diamond films were grown on silicon substrates by different methods. The main differences in the underlying mechanism and film microstructure are briefly discussed below.

(1) MW CVD nanodiamond films were grown from a hydrogen-poor gas mixture. The mechanism of the film formation was investigated to a great extent [10] and is considered as a surface process occurring from C_2 molecule as the main growth species. Nanodiamond films grown from C_2 species require a surface pretreatment aimed to introduce diamond growth centers onto which the film grows. The as-deposited films are characterized by a high sp^3 content of ~95% [10] and nanometric grain size (in the present study, in the range 3–30 nm), homogeneously distributed over the whole thickness.

(2) HF CVD diamond films were grown from a hydrogen-rich gas mixture (CH_4/H_2 ratio was 1/99). The evolution of the film occurs on nucleation centers immersed into the substrate by ultrasonic abrasion with polydispersed mixed diamond slurry [33]. This pretreatment results in an initial diamond particle density (DPD) of $\sim 5\times 10^{10}$ cm^{-2}. This high DPD value allows deposition of continuous

films with a minimal thickness of ~70–100 nm [15, 33]. The nucleation is a surface process in which the main growth species are CH_x and hydrogen radicals. The deposited films are characterized by high diamond quality (>98% sp^3) and well-defined diamond facets. The diamond growth occurs in nucleation centers which gradually merge and coalesce into continuous film. The crystal size evolves from ~50–70 nm in the nearly coalescent film (~100 nm film thickness, ~10 min deposition) (Fig. 1a) and reaches nearly constant value of 300–400 nm in micron thick films (deposition time ~1 h) (Fig. 1b). The increased crystalline size observed in the high resolution scanning electron microscopy (HR-SEM) images (Fig. 1) is a result of the suppression of small crystallites by fast growing crystallites which comprise the faceted surface.

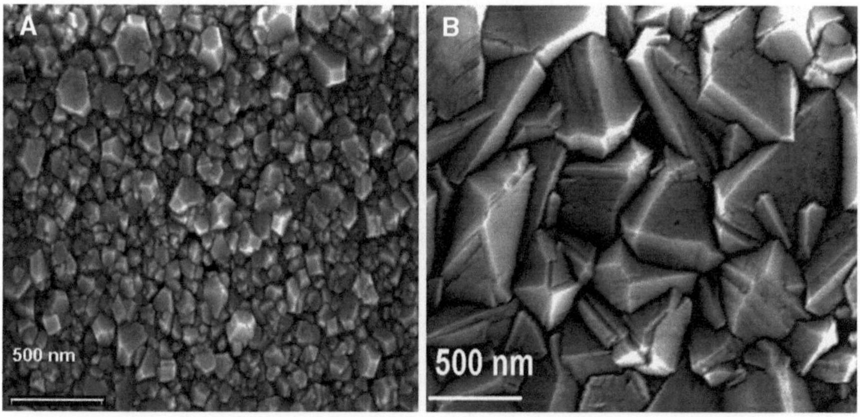

Fig. 1 HR-SEM micrographs of as-deposited HF CVD diamond films. (**a**) Film deposited from CH_4+H_2 gas mixture for 10 min (thickness ~100 nm, grain size ~50–70 nm). (**b**) Film deposited from CH_4+H_2 gas mixture for 60 min (thickness ~700 nm, grain size ~300 nm)

(3) dc GD CVD films were grown from hydrogen-rich plasma (CH_4/H_2 ratio was 9/91) by means of continuous substrate bombardment by positive hydrocarbon and hydrogen ions of ~100–200 eV kinetic energy. No pretreatment procedure is needed to induce diamond nucleation. The nanodiamond nucleation and growth occur in subsurface region of the film resulting in the upper surface region consisting of hydrogenated amorphous carbon. The film evolves on top of a hydrogenated carbonaceous precursor containing a mixture of an amorphous and graphitic carbon formed during the first ~20 min of deposition (~300 nm thickness). When the precursor density reaches ~3 g/cm^3, nanodiamond particles precipitate and grow to a final size of ~5 nm by means of preferential displacement of loosely bonded carbon atoms by energetic ions [11, 32]. The sp^3 content of the film grown for 1 h reaches ~80%. The nanodiamond particles are embedded in hydrogenated amorphous carbon matrix. Table 1 summarizes the deposition methods and the resulting grain size of all the investigated diamond films.

Table 1 Summary of grain size and deposition methods of the investigated diamond films

Film name	Grain size [nm]	Deposition method	Gas composition ($CH_4/H_2/Ar$)
dc-GD-CVD-5 nm [32]	~5	dc glow discharge CVD	9/91/0
HF-CVD-300 nm [33]	300	Hot filament CVD	1/99/0
MW-CVD-3–20 nm [34]	3–20	Microwave CVD	1.4/0/98.6
MW-CVD-10–30 nm [34]	10–30	Microwave CVD	1.4/1/97.6

3 Hydrogen Atom Concentrations in Polycrystalline Diamond Films as a Function of Grain Size Studied by SIMS

The hydrogen concentrations in the three different kinds of polycrystalline diamond films described above were analyzed by SIMS technique [19, 35]. SIMS analysis was carried out in a dynamic mode in a Cameca IMS4f ion microscope using a 14.5 keV Cs^+ ion beam. The sampling area was ~64 μm^2. The basic chamber pressure was 8×10^{-10} Torr, while the ion current was $\sim1\times10^{-8}$ A.

SIMS depth profiles of hydrogen concentration in the different diamond films are shown in Fig. 2. Profile 2A is taken from an ~1.6 μm thick HF CVD sample grown for 2 h from CH_4/H_2 gas mixture. Its initial grain size (close to the film–Si

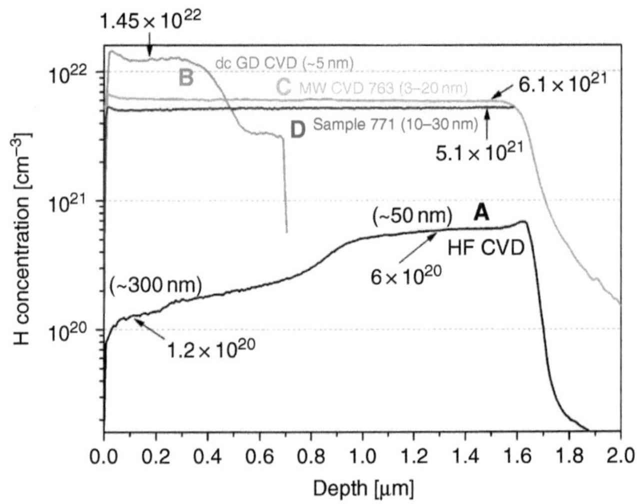

Fig. 2 SIMS profile of hydrogen atoms within three different types of diamond films. (**A**) HF CVD film deposited from CH_4+H_2 gas mixture for 2 h. Note the decrease of H concentration from 6×10^{20} atoms/cm^3 at initial growth region (grain size ~50 nm) to 1.2×10^{20} atoms/cm^3 for thicker film (grain size ~300 nm). (**B**) 700 nm thick nanodiamond film deposited by dc GD CVD. The hydrogen concentration increases from 3.3×10^{21} atoms/cm^3 at the initial stage of film evolution (predominantly graphitic precursor) to 1.45×10^{22} atoms/cm^3 at the region where ~5 nm sized nanodiamond grains are embedded in an amorphous carbon matrix. (**C**) MW CVD diamond film with grain size 3–20 nm. (**D**) MW CVD diamond film with grain size in the range of 10–30 nm

substrate interface) is ~50 nm, which develops during the deposition to a final grain size of ~300 nm, thus allowing a direct measurement of the correlation between the hydrogen trapping and the crystalline size [19, 35]. Note that the sampling depth given in Fig. 2 represents the film surface at a depth of 0 and the Si substrate region at a depth of 1.6 μm. It is very clear that the hydrogen concentration of this film decreases from 6×10^{20} atoms/cm^3 at the region close to the Si substrate (50 nm grain size) to 1.2×10^{20} atoms/cm^3 at the region close to the film's surface formed after a deposition time of ~2 h (~300 nm grain size) [19, 33].

The profile of hydrogen atoms in dc GD CVD film grown for 1 h is shown in Profile 2B. In our previous works [11, 32], we showed that the growth of the dc GD CVD film advances through the following stages: (1) graphitic film with its basal plane perpendicular to the substrate, a low hydrogen concentration and low (~2.2 g/cm^3) density, (2) an increase of the density followed by incorporation of hydrogen, and (3) the formation of a dense (~3 g/cm^3) hydrogen-rich layer in which diamond nucleates and grows. The hydrogen concentration (Profile 2B) in the nanodiamond film deposited by dc GD CVD increases from 3.3×10^{21} atoms/cm^3 at the nano-graphitic precursor region to 1.45×10^{22} atoms/cm^3 at the ~5 nm sized nanodiamond crystallite region [28]. The hydrogen concentration depth profile of the dc GD film reflects the evolution of its structure from an initial low-density, low hydrogen concentration graphitic film through a medium-density hydrogen containing film until a dense hydrogen-containing matrix evolves, in which ~5 nm-sized nanodiamond crystallites precipitate. Our previous measurements of absolute hydrogen concentration in these films by means of ERDA revealed hydrogen concentration of 15–20 at% [28] which confirms the present SIMS measurements.

Profiles 2C and 2D show hydrogen concentration in two nanodiamond films deposited by MW CVD with grain size of 3–20 and 10–30 nm, respectively. The measured hydrogen concentration for samples MW-CVD-3–20 nm and MW-CVD-10–30 nm were 6.1×10^{21} atoms/cm^3 and 5.1×10^{21} atoms/cm^3, respectively. Note the slightly larger hydrogen concentration for MW-CVD-3–20 nm (CH_4/Ar ratio of 1.4/98.6; grain size 3–20 nm) than for MW-CVD-10–30 nm (CH_4/H_2/Ar ratio of 1.4/1/97.6; grain size 10–30 nm), e.g., despite the higher hydrogen concentration in the gas phase, its retention within the film is lower for the latter. Moreover, the hydrogen concentration of MW CVD films (grown from hydrogen-poor gas mixture) largely exceeds that measured for the HF CVD grown films (carried out in a hydrogen-rich atmosphere). Therefore, one may suggest that the hydrogen incorporation in the diamond film is governed by another factor beyond gas mixture composition. The most evident factor that influences the hydrogen atoms retention within the films is diamond grain size. Our results clearly show that the smaller the diamond crystallites, the higher the hydrogen concentration in the films. Note that all samples were mounted at the same time in the SIMS chamber and measured at the same conditions. Therefore, while the absolute hydrogen concentration values in the films may depend on the correct relative sensitivity factor (RSF), the relative measured concentrations can be compared.

Table 2 summarizes the H concentrations in the different films, the atomic percentage of the H (assuming a C density of 10^{23} atoms/cm^3), the crystalline size

Table 2 Hydrogen concentration versus the crystalline size and the surface-to-volume atomic ratio

Film type	H concentration (atoms/cm^3)	H concentration (at%)[a]	Crystalline size (nm)	Surface-to-volume atomic ratio (%)[b]
HF CVD (1 h)	1.2×10^{20}	0.12	300	0.5
HF CVD (10 min)	6×10^{20}	0.6	50	3
MWCVD-3–20 nm	6.1×10^{21}	6.1	3–20	50–7.5
MW-CVD-10–30 nm	5.1×10^{21}	5.1	10–30	15–5
dc-GD-CVD-5 nm	1.45×10^{22}	14.5	5	28

[a] Assuming a C concentration of 1×10^{23} atoms/cm^3
[b] Surface-to-volume atomic ratio of FCC crystallite with the same crystalline size

deduced from HR-SEM and HR-TEM, and the surface-to-volume atomic ratio of FCC crystallites with the same crystalline size. It is very clear that the hydrogen concentration in the diamond films increases with decreasing crystalline size. This result strongly suggests that hydrogen atoms are bonded to the grain *surfaces* within the diamond film's bulk. The trapping of hydrogen in grain boundaries is a natural consequence of the accepted diamond growth mechanism which involves adsorption of a CH_x growth species first and then abstraction of one hydrogen atom from the surface by atomic hydrogen in the gas phase, allowing the diamond carbon atom to be bonded to another CH_x radical. Coalescence of adjacent diamond crystallites most likely does not result in complete hydrogen desorption from touching surfaces leading to the incorporation of hydrogen on the touching crystallite surfaces (grain boundaries).

4 Hydrogen Bonding Configuration in Diamond Film Bulk Studied by Raman Spectroscopy

Raman spectroscopy has a lot to offer as a non-destructive characterization tool. Backscattering geometry, especially with microfocus instruments, allows small regions in the heterogeneous films to be easily examined. Raman spectroscopy is a very powerful technique for the identification of crystalline phases and for studies of defects, structural disorder and stresses in thin films. However, due to quantum size effects, the use of Raman spectroscopy for the unambiguous phase characterization of films composed of nanosize crystalline particles is complicated [36, 37]. The Raman measurements were carried out using a DILOR XY system. The measurements were performed using the Ar line at 514.5 nm in a backscattering geometry. The laser beam spot size was $\sim$1 μm in diameter and the incident power was 10 mW [20].

The Raman spectrum of diamond consists of a single peak at 1332 cm^{-1}, whereas that for graphite has a single peak at 1580 cm^{-1} (usually labelled the G peak) [37, 38]. The diamond peak may be broadened using several wavenumbers by defects, stresses and/or small crystalline size. Usually, in the case of the CVD diamond films, this peak is accompanied by a broadband centered around 1500–1550 cm^{-1}, believed to be due to the sp^2-bonded carbon, possibly in the grain-boundaries [37]. The Raman spectra of polycrystalline graphite and glassy carbon consist of two peaks. The first, located at 1580–1590 cm^{-1} (the G peak), originates from lattice vibrations in the plane of the graphite-like rings [38]. The second peak is located at about 1350 cm^{-1} (the D peak) and occurs in graphitic materials with small crystalline size. This disordered-induced band corresponds to a peak in the vibrational density of states (VDOS) of graphite. In amorphous materials, the lack of long-range order leads to a relaxation of the selection rules governing the Raman scattering process, and all vibrational modes can contribute to the Raman spectra.

Because of the very much larger Raman cross section for graphite sp^2-bonds (5×10^{-5} $cm^{-1}sr^{-1}$) relative to diamond sp^3-bonds (9×10^{-7} $cm^{-1}sr^{-1}$), it is very difficult to observe any spectral Raman features associated with a small amount of sp^3-bonding in the presence of sp^2-bonded material [37, 39]. By the same token, the high sensitivity of the Raman spectra to sp^2-bonding allows sensitive detection of small concentrations of sp^{2-}bonding in a diamond film.

In this section, we clarify the Raman peaks associated with diamond grain boundary vibrations by means of reduced mass change via isotopic modification of the growth mixture. Then, we correlate the shape of the Raman spectrum with hydrogen retention studied by SIMS for the films possessing different grain sizes.

4.1 Clarification of the Hydrogen-Associated Raman Peaks Through Modifications Induced by Isotopic Exchange

The Raman spectra of diamond films display the diamond optical phonon line at 1332 cm^{-1} and other contributions originated in sp^2-carbon-related features (e.g., the 1350 cm^{-1} D line and the 1580 cm^{-1} G line), which relative intensity depends on diamond crystal size, grain quality, etc. [36–38]. An additional feature around 1140 cm^{-1}, commonly observed in nanodiamond films, was initially assigned to sp^3-bonded carbon [40] and often taken as a simple criterion for a nanocrystalline diamond phase [41, 42]. Later it was revealed that the 1140 cm^{-1} peak usually appears together with the broadband at 1480 cm^{-1} and shows significant dispersion, while the diamond mode does not [36]; therefore, its association with sp^3-bonded carbon was questioned. It was shown by Raman [37, 43] and surface-enhanced Raman spectroscopy [44] that these peaks are associated with ν_1 and ν_3 vibration modes of trans-polyacetylene (t-PA) codeposited with diamond phase. This assignment was supported by (1) the disappearance of the peak upon annealing to 1200°C (due to hydrogen release) keeping the diamond peak at 1332 cm^{-1} intact, (2) the shift of the peak position with increasing excitation energy and its disappearance for UV Raman, (3) its absence in the Raman spectra of nanodiamond powders made by detonation, and (4) the peak shift from 1140 to ~860 cm^{-1} upon full hydrogen exchange by deuterium in the CVD plasma used for growth. We confirmed this suggestion by the deposition of diamond films from different isotopic gas mixtures, where hydrogen was replaced by deuterium fully (CD_4+D_2) and partially (CD_4+H_2 and CH_4+D_2) as well as when ^{12}C was replaced by ^{13}C ($^{13}CH_4$+H_2) [20]. Here we briefly summarize our findings. Figure 3 presents the Raman spectra of HF CVD diamond films deposited from ($^{12}CH_4$+H_2), ($^{13}CH_4$+H_2) and ($^{12}CD_4$+D_2) gas mixtures.

Raman spectrum 3A of ~100 nm (crystalline size <50 nm) film, deposited from $^{12}CH_4$+H_2 gas mixture, exhibits the 1332 cm^{-1} diamond peak, the broad contribution centered at 1480 cm^{-1}, and the peak at 1140 cm^{-1}. The additional the broad peak at 1000 cm^{-1} is attributed to the second overtone of Si optical phonon. Exchange of ^{12}C by ^{13}C (Spectrum 3B) shifts the 1332 cm^{-1} peak to 1280 cm^{-1} but does not significantly shift the 1140 cm^{-1} peak. In addition, the peak at 1480 cm^{-1} (Spectrum 3A) is shifted to the 1420 cm^{-1} (Spectrum 3B) when ^{12}C was replaced by ^{13}C. The decrease of the 1332 cm^{-1} diamond peak in the Raman spectra of the HF CVD diamond films due to isotopic exchange of H by D indicates that the crystalline quality of these films depredated due to the replacement, which modified the deposition

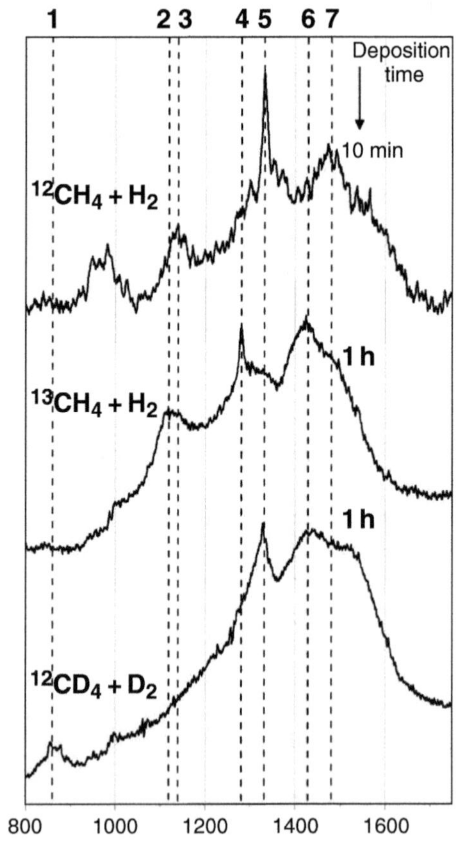

Fig. 3 Raman spectra of submicron diamond films prepared by the HF CVD for different isotopic compositions of methane/hydrogen (1/99) species indicated on the plot. The deposition time was 1 h except for the sample corresponding to the *upper line* which was grown for 10 min

conditions, even though the growth parameters were kept intact. This explains why the 1140 cm^{-1} peak of high-quality diamond film (highest 1332 cm^{-1} peak) is very weak. The same modifications of the Raman spectra were observed when dc GD CVD nanodiamond films are grown from similar isotopic gas mixtures [20].

The presented data clearly show that the 1140 cm^{-1} peak is associated with C–H mode. This peak is shifted to 860 cm^{-1} by isotopic exchange of H in the growth plasma and is very slightly affected by the isotopic exchange of carbon. In contrast, C–C modes are not affected by isotopic substitution of H and are shifted by isotopic substitution of C. The 1480 cm^{-1} peak is shifted to the 1420 cm^{-1} position and is associated with C–C mode. The experimental isotopic shifts of the Raman peaks closely match with the calculated shifts for the specific C–H or C–C modes. Indeed, carbon 1480 cm^{-1} peak shifts to 1420 cm^{-1} in perfect agreement with ^{12}C- and ^{13}C-reduced mass differences:

$$\omega_{^{13}C-^{13}C} = \frac{\omega_{^{12}C-^{12}C}}{\sqrt{\dfrac{\mu_{^{13}C-^{13}C}}{\mu_{^{12}C-^{12}C}}}} = \omega_{^{12}C-^{12}C} \times 0.9608 \tag{1}$$

Whereas the same effect is much less prominent in the case of ^{13}C–H and ^{12}C–H vibration modes:

$$\omega_{^{13}C-H} = \frac{\omega_{^{13}C-H}}{\sqrt{\dfrac{\mu_{^{13}C-H}}{\mu_{^{13}C-H}}}} = \omega_{^{12}C-H} \times 0.997 \tag{2}$$

We determined that the 1140 cm^{-1} Raman peak is associated most likely with the bending mode of hydrogenated sp^2-carbon bonded to diamond grain *surface*. This mode is present on the spectrum of hydrogenated carbon solely if the film is constituted of diamond grains. Our studies of dc GD CVD nanodiamond film evolution on hydrogenated graphitic/amorphous carbon precursor clearly show [45] that this mode is absent from the spectrum of hydrogenated carbon (∼20 min deposition) and appears only when nanodiamond crystallites precipitate in dense hydrogenated amorphous carbon matrix (∼30 min deposition). Since hydrogen atom concentration nearly scales with diamond grain surface ratio, it may be guessed that this mode is related to diamond phase itself, e.g., belongs to sp^2-carbon bonded to diamond grain *surface*. Our work on thermal stability of carbon–hydrogen bonding [45] provides a direct evidence that this mode is stable up to temperatures typically high for hydrogen desorption from diamond grains (up to 1000°C). On the other hand, nearly all kinds of hydrogenated carbon with different sp^2/sp^3 ratio decompose at temperatures up to ∼600°C. In Section 5.2, we correlate this Raman mode with (sp^2)-C–H stretching vibration detected by HR-EELS.

4.2 *The Impact of Diamond Grain Size and Hydrogen Concentration on the Shape of the Raman Spectra*

In Section 3, we showed that substantial amount of hydrogen is trapped within diamond films grown by three different kinds of CVD. Independent of the gas mixture composition, the hydrogen retention within the films scales with the diamond grain size: the smaller grains give rise to the higher hydrogen retention. In this section, we correlate the hydrogen retention with the shape of the Raman peak.

The hydrogen bonding and the crystalline nature of the different films were characterized by Raman (Fig. 4). Spectra 4A and 4B correspond to the MW CVD films grown for 1 h, samples MW-CVD-3–20 nm and MW-CVD-10-30 nm, respectively. Spectra 4C and 4D were measured on the HF CVD diamond films grown for 10 min ($\sim$100 nm film thickness, Spectrum 2C) and 60 min ($\sim$700 nm thickness, Spectrum 2D), respectively; their SEM micrographs are shown in Fig. 1. The Spectrum 4E was measured on dc GD CVD film grown for 60 min. Besides Spectrum 4D, all spectra display most of the characteristic Raman modes of nanodiamond films discussed in Section 4.1: two broad peaks at $\sim$1350 cm^{-1} (D peak) and 1530 cm^{-1} (G peak), associated with sp^2-bonded carbon, and ν_1 and ν_3 modes of t-PA positioned at 1140 and 1480 cm^{-1}, respectively [37, 43]. Spectra 4C and 4D display an additional peak positioned at 1332 cm^{-1} attributed to the diamond optical phonon. Spectrum 2D was measured on a thick HF CVD diamond film and is attributed to high-quality diamond film. Here, the diamond peak at $\sim$1332 cm^{-1} is sharp, well pronounced and large compared to the background contributions of the sp^2-related D and G peaks, reflecting the much larger diamond crystalline size, reduced grain boundary and defects densities. The t-PA vibrational modes at 1140 and 1480 cm^{-1} are practically absent, indicating much smaller amount of C–H bonding at grain boundaries in comparison to the 100 nm thick HF CVD film.

It was shown that the three different classes of diamond films follow the same tendency: the smaller grain size results in the higher hydrogen retention. It is possible, therefore, to associate the shape of the Raman spectrum with diamond grain size and hydrogen content therein. The high hydrogen concentration results in appearance of pronounced t-PA-associated Raman modes at 1140 and 1480 cm^{-1}. In addition, the sharp diamond optical phonon mode at 1332 cm^{-1} is broader for grain size less than 50 nm and for 10–30 nm crystallites the graphitic D peak at 1350 cm^{-1} appears. For the smallest grain size (e.g., $\sim$5 nm in the case of dc GD nanodiamond films, Spectrum E); however, the intensity of 1140 cm^{-1} is diminished relative to carbon-associated D and G peaks in comparison to Spectra B and C. From the SIMS measurements, it is known that the dc GD sample contains the highest (among the investigated films) concentration of the trapped hydrogen atoms. Why is the intensity of carbon–hydrogen bonding less pronounced in more "hydrogen-rich" samples of smaller grain size?

Our results suggest that for diamond films composed of crystallites of $\sim$5 nm, so-called ultra-nanocrystalline (UNC) diamond films, the intensity of hydrogen-associated peaks is diminished relative to the carbon-associated modes. In Section 5.2, additional evidence of this phenomenon comes from HR-EELS

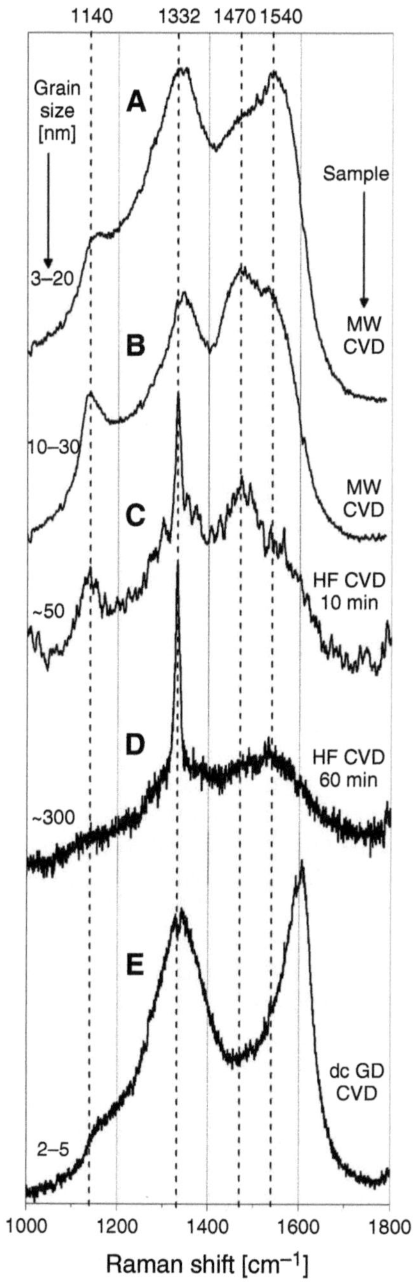

Fig. 4 Raman spectra of the different diamond samples. (**A**) MW CVD sample with grain size of 3–20 nm. (**B**) MW CVD sample with grain size of 10–30 nm. (**C**) HF CVD diamond film (∼50 nm grain size) deposited for 10 min. (**D**) HF CVD diamond film (∼300 nm grain size) deposited for 1 h. (**E**) dc GD CVD diamond film (∼5 nm grain size) deposited for 1 h

measurements of C–H stretching vibrations on UNC diamond surface. A similar tendency was found in IR measurements of diamond C–H stretching mode as a function of diamond grain size [46–48]. The intensity of C–H stretching mode at 2834 cm^{-1} decreases with decreasing of diamond grain size (350 nm $\rightarrow$ 5 nm). In that case, the proposed explanation of this phenomenon (based on density functional calculations) is an absence of well-defined facets on the surface of 5 nm size diamond crystallites [46, 47].

We conclude that the shape of the Raman spectrum is strongly affected by diamond grain size (and, therefore, by hydrogen atoms concentration within the film). While the spectrum of the film consists of diamond grains of the size of a few hundreds of nanometer is dominated by diamond optical phonon loss at 1332 cm^{-1} with minor contribution from mixed sp^2/sp^3 carbon (the band near 1500 cm^{-1}), the smaller diamond crystallites result in enhanced sp^2-carbon–hydrogen peaks near 1140 and 1480 cm^{-1}. However, for UNC diamond films (grain size ~5 nm), the intensity of these peaks decreases and the spectrum is dominated by graphitic D and G peaks at 1350 and 1580 cm^{-1}. The physical explanation of this phenomenon is under current investigation in our lab.

5 Hydrogen Bonding Configuration on Diamond film Surface Studied by HR-EELS

HR-EELS is a most powerful surface characterization probe, which may provide direct information about hydrogen bonding on surfaces [49]. The electron spectroscopy measurements reported in our studies were carried out using an ultra-high vacuum (UHV) system consisting of two chambers connected in tandem. In the first chamber annealing, XPS and ion irradiation experiments were carried out, whereas the second chamber was dedicated to HR-EELS. The samples can be heated in-situ up to ~1000°C and exposed to thermally activated hydrogen. The HR-EELS chamber is equipped with a Delta 0.5 spectrometer (VSI-SPECS) consisting of a double monochromator and a single analyzer housed in an UHV system with a base pressure of $\sim 8\times10^{-10}$ Torr. All spectra were recorded up to loss energies of 700 meV at room temperature in the specular geometry with an incident angle of 55° from the surface normal, incident energy of 5 eV and energy resolution of 5–8 meV (according to full width half maximum (FWHM) of the elastic peak). The XPS measurements were carried out using Al K_α (1486.6 eV) photon irradiation. The surface quality and phase composition of the films were investigated by C(1 s) photoelectron energy loss spectra. A fine-focused Ar^+ ion source IQE 12/38 of SPECS was used for ion irradiation experiments. The beam energy could be varied in the 200 eV to 5 keV range, and typical beam currents were up to 10 μA.

In the following section, we assign the different HR-EELS peaks to specific vibrational modes of well-defined diamond surface by means of isotopic exchange. In Table 3, we compare our assignment with assignments found in the literature [50–70]. In Section 5.2, the impact of diamond grain size on the shape of HR-EEL

Table 3 Assignment of HR-EELS modes relevant to the present discussion

HR-EELS mode position	Assignment based on literature survey	Our present assignment
Hydrogenated diamond (C-12) surface		
~155 meV	C–H bending modes [50–56]; Mixed C–H bend and C–C stretching modes [57–69]	Mixed ^{12}C–H bending and ^{12}C–^{12}C stretching modes
300 meV	First overtone of C–H modes [51, 53–55]; first overtone of C–C modes [52, 57, 58, 62–67].	First overtone of ^{12}C diamond optical phonon
358–362 meV	C–H stretching mode [49–70]	^{12}C–H stretching mode
450 meV	Second overtone of C–H modes [51, 53, 54]; second overtone of C–C modes [62–67].	Second overtone of ^{12}C diamond optical phonon
510 meV	Coupled C–H deformation and C–H stretching modes [50]; coupled C–C stretching and C–H bending modes [67, 70].	Overlapping of both ^{12}C–H bending and ^{12}C–H stretching modes and ^{12}C–^{12}C stretching and ^{12}C–H bending modes
600 meV	Third overtone of C–C modes [67]	Third overtone of diamond optical phonon
Deuterated diamond (C-12) surface		
~110 meV	C–D bending modes [52–54, 57, 58]	^{12}C–D bending modes
150 meV	C–C stretching modes [52, 57, 58, 61]	^{12}C–^{12}C stretching modes
~270 meV	C–D stretching modes [52–54, 57–59, 61, 71, 72]	^{12}C–D stretching modes
300 meV	First overtone of C–C modes [52, 57]	First overtone of ^{12}C diamond optical phonon
~380 meV	–	Coupling of ^{12}C–D bending and ^{12}C–D stretching modes
420 meV	–	Coupling of ^{12}C–^{12}C stretching and ^{12}C–D stretching modes
450 meV	–	Second overtone of ^{12}C diamond optical phonon
Hydrogenated diamond (C-13) surface		
148 meV	–	Mixed ^{13}C–H bending and ^{13}C–^{13}C stretching modes
290 meV	–	First overtone of ^{13}C diamond optical phonon
358 meV	–	^{13}C–H stretching modes
435 meV	–	Second overtone of ^{13}C diamond optical phonon
508 meV	–	Overlapping of both ^{13}C–H bending and stretching modes and ^{13}C–^{13}C stretching and ^{13}C–H bending modes
580 meV	–	Third overtone of ^{13}C diamond optical phonon

spectra is discussed. The influence of hydrogenated sp^2-hybridized carbon most likely originated from diamond grain boundaries is shown.

5.1 The Hydrogen and Carbon Bonding Configuration of Nanoscale-Defined Hydrogenated Polycrystalline Diamond Surface: **The Assignment** *of HR-EELS Peaks*

In the following discussion, we compare the samples grown from isotopic mixtures with the sample grown from the "common" gas mixture ($^{12}CH_4$+H_2) by HF CVD. For simplicity, we shall refer to the last gas mixture as "common" gas mixture and to the sample as the "common" sample.

HR-EEL spectra of the films deposited from isotopic mixtures are shown on Fig. 5. The sample deposited from $^{12}CH_4$+H_2 and $^{13}CH_4$+H_2 was treated *ex situ* in MW-H plasma for 20 min, while the sample deposited from $^{12}CD_4$+D_2 was treated in MW-D plasma for the same time. Each sample was annealed to 450°C–500°C in UHV conditions before measurements. Different vibrational modes are enumerated on top of Fig. 5. Each peak was fitted for contributions from two possible modes: C–H (or C–D) and C–C. Note that in Spectrum 5B ($^{13}CH_4$+H_2) modes #2, #4, #6, and #8 (positioned at 155, 300, 450, and 600 meV, respectively) are shifted towards lower loss energies by Δ, 2Δ, 3Δ, and 4Δ ($\Delta = 5$ meV), respectively, comparing to modes of Spectrum 5A ($^{12}CH_4$+H_2). The mode #7 is nearly unaffected by isotopic carbon substitution (Spectra 5A and 5B, $^{12}CH_4$+H_2 and $^{13}CH_4$+H_2), whereas it completely disappears when hydrogen atom was replaced by deuterium (Spectrum 5C, $^{12}CD_4$+D_2). The most probable isotopic shift of this mode is within complex band around 350–450 meV in CD_4+D_2 (Spectrum 5C). We can thus discuss three families of peaks: (1) pure C–C peaks, (2) pure C–H peaks, and (3) coupling of C–C and C–H peaks. In Table 3, the peak assignment of the HR-EELS of the hydrogenated diamond based on our studies is summarized [74, 75].

Our findings provide direct evidence and are in accordance with the recent work of Lafosse et al. [67], who report on vibrational excitation function of diamond surface. The shape of excitation function is governed (among other things) by the electronic density of states, which gives rise to a peak around 13 eV (incident electron energies) due to the second absolute diamond band gap. In this case, the reflectivity of the electron beam increases and losses are dominated by surface vibrations (e.g., C–H) and not by deeper atoms (C–C modes). Indeed, the intensity of the peaks at 150, 300, 450, and 600 eV substantially decreases for electron energy of 13 eV (the case of high reflectivity) as expected for C–C rather than C–H modes. These facts suggest the assignment of these overtones to the lattice C–C vibrations.

At this point, we should stress the absence of the overtones #4, #6, and #8 from the HR-EEL spectrum of amorphous carbon films reported in the literature [71–73] and in our previous works [68, 69]. Mode #2 is nevertheless present and prominent. On the other hand, these overtones are observed on the well-defined (100), (111) and polycrystalline diamond surfaces [51–54, 57, 58, 62–67].

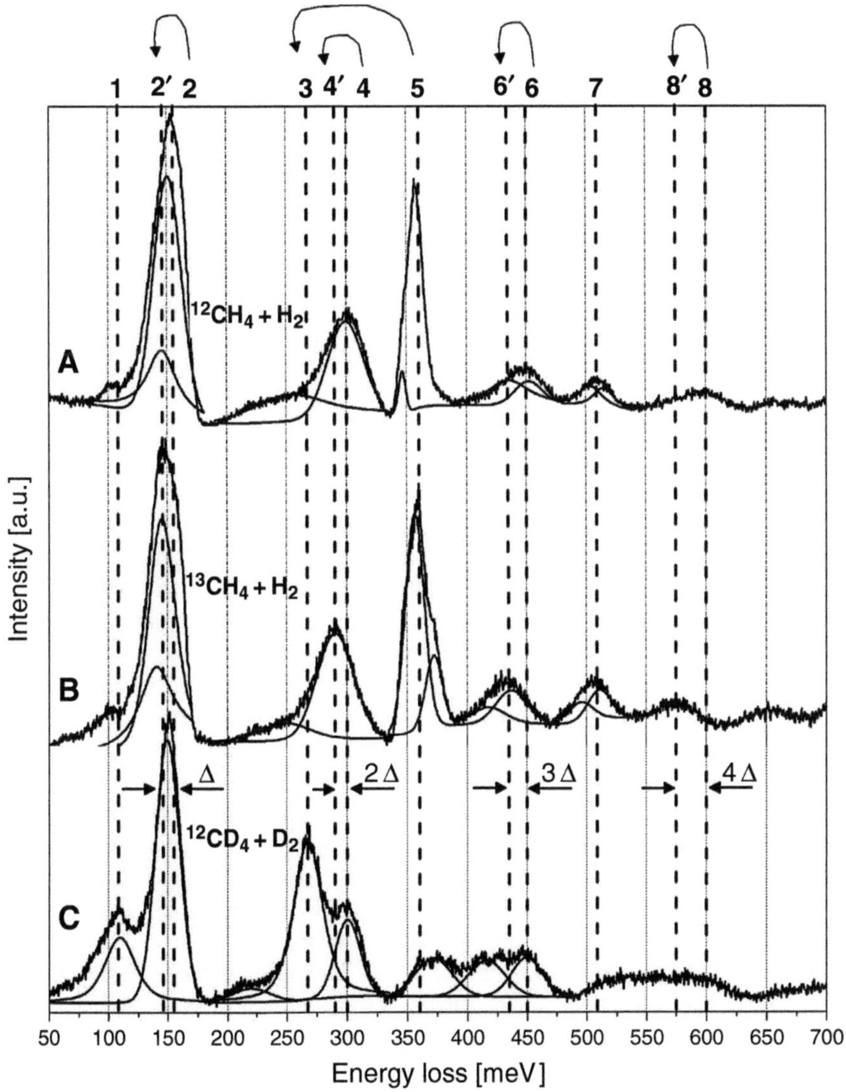

Fig. 5 Three HR-EEL spectra of diamond films deposited from isotopic gas mixtures with peak fitting. Each peak was fitted for two possible contributions of C–C and C–H vibrational modes

In order to clarify the origin of the peaks at 300, 450, and 600 meV, an ion irradiation experiment was done on well-defined diamond surface [74, 75]. Ion irradiation results in partial amorphization of the film surface; therefore, the change in the shape of the spectrum following the ion irradiation may be associated with the change of the crystalline surface structure. Our results show that the intensity of these three overtone HR-EEL peaks (300, 450, and 600 meV) decreases most significantly after the diamond surfaces have been irradiated with low energy Ar ions. In this case, the characteristic diamond loss features (the surface and bulk diamond plasmon) are replaced by amorphous carbon losses [75]. These modes should be, therefore, associated with well-defined diamond surfaces rather than with amorphous carbon surfaces. We thus conclude that the appearance of the three modes #4, #6, and #8 at 300, 450, and 600 meV in the HR-EEL spectrum of diamond should be attributed to the higher losses of the slow electron onto diamond optical phonon and may serve as a fingerprint of nanoscale-defined diamond surfaces. We use this finding in the next section to study the impact of diamond grain size on the shape of HR-EEL spectra.

5.2 The Impact of Diamond Grain Size on the Shape of HR-EEL Spectra

In Section 5.1, the assignment of HR-EELS peaks characteristic to hydrogenated nanoscale-defined diamond surface was summarized. In this section, we use this assignment to study the impact of diamond grain size on the shape of HR-EEL spectrum of diamond surfaces deposited by three different CVD techniques described above (Section 2).

We start with the analysis of C–H stretching mode of *as-deposited* HF CVD diamond film (grain size ~300 nm) as a function of thermal annealing. In Fig. 6, the electronic EEL spectra (electron primary energy, $E_p = 500$ eV) of the as-deposited HF CVD film as a function of annealing temperature is shown. Diamond plasmon losses are nearly unaffected by heating at T<800°C, while heating at T = 1000°C results in appearance of graphitic π electrons (~6 eV energy loss). In Fig. 7, the HR-EEL spectra of the same *as-deposited* HF CVD diamond samples are shown as a function of annealing temperature. The shape of the spectra is nearly unaffected by T<800°C, while T = 1000°C results in elimination of hydrogen-associated features (360 and 510 meV peaks) and addition of ~90 meV peak associated with C = C dimer. Note the correlation between hydrogen desorption detected by HR-EELS (Fig. 7) and graphitic features appearance detected by electronic EELS (Fig. 6). While the hydrogen stretching mode is clearly present in the HR-EEL spectra of diamond annealing up to 800°C, annealing to 1000°C results in its disappearance alongside with the emergence of ~90 meV positioned peak associated with C = C dimer and surface reconstruction. These dimers (which are detected as ~90 meV HR-EELS peaks) contribute to graphitic π plasmon detected by electronic EELS measurements.

The detailed peak-fitting procedure of C–H stretching HR-EELS mode (Fig. 8) reveals at least three different contributions at ~350, ~360, and ~373 meV. The first two are most likely associated with diamond (111) and (100) surfaces, while the last one may be attributed to sp^2-hybridized carbon [76, 77]. Upon annealing to 800°C, the maximum peak position is slightly shifted to smaller loss energies (from 359 to 357 meV), while the sp^2-associated mode at ~373 meV is still prominent.

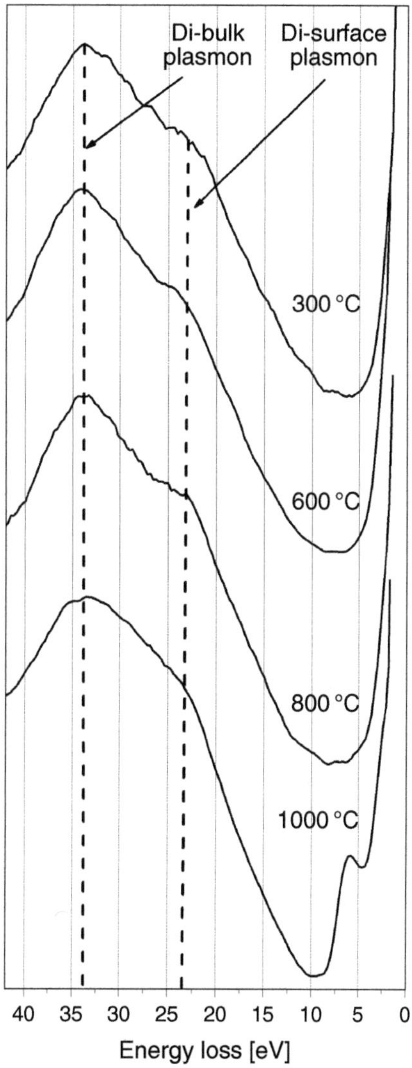

Fig. 6 Electronic EEL spectra of HF CVD diamond films as a function of annealing temperature. Primary electron energy *E*p is 500 eV. The shape of the spectrum is nearly unchanged up to annealing temperature $T_A = 800°C$, while $T_A = 1000°C$ results in appearance of the graphitic $\pi \rightarrow \pi^*$ associated plasmon at ~6 eV. Note the correlation of this graphitic feature with hydrogen desorption detected by HR-EEL spectroscopy

According to thermal stability studies of different kinds of hydrogenated amorphous carbon matrixes [78–84], annealing to a temperature of >700°C results in desorption of hydrogen atoms alongside with hydrocarbon species, independent of the method used for amorphous carbon film preparation. This allows us to attribute the ~375 meV HR-EELS mode to hydrogen bonding to *diamond grain boundary*, which may be stable up to these high temperatures. This suggestion may be confirmed by HR-EELS studies of *nanodiamond* material discussed below.

Now we turn to analyze the vibrational modes on the surface of nanodiamond films deposited by MW CVD. Figure 9 shows the electronic EEL spectra of

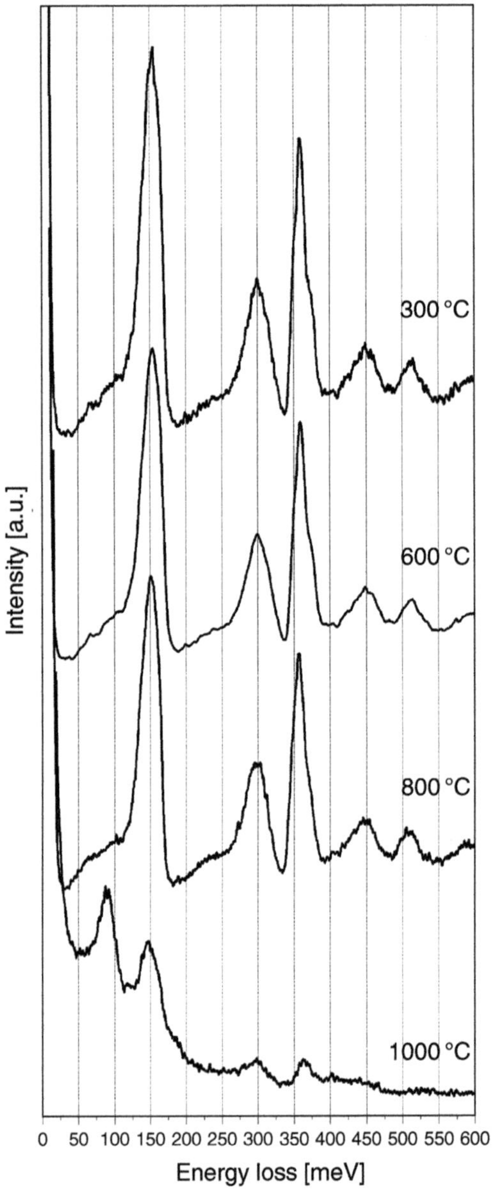

Fig. 7 HR-EEL spectra of as-deposited HF CVD films as a function of annealing temperature. The primary electron energy was 5 eV. The shape of the spectrum is nearly unchanged up to $T_A = 800°C$. The $T_A = 1000°C$ results in complete hydrogen desorption (the absence of ~360 meV peak) alongside with appearance of C = C dimer mode at 92 meV

MW-CVD-3–20 nm sample recorded at primary electron energies of $E_p = 500$ eV before and after *ex-situ* MW hydrogenation. Both spectra are dominated by characteristic diamond losses, namely, bulk and surface diamond plasmons at 33 and 23 eV, respectively. It can be concluded from the comparison of the spectra of *as-deposited* films and the *ex-situ* MW hydrogenated films that the *ex-situ* MW hydrogenation

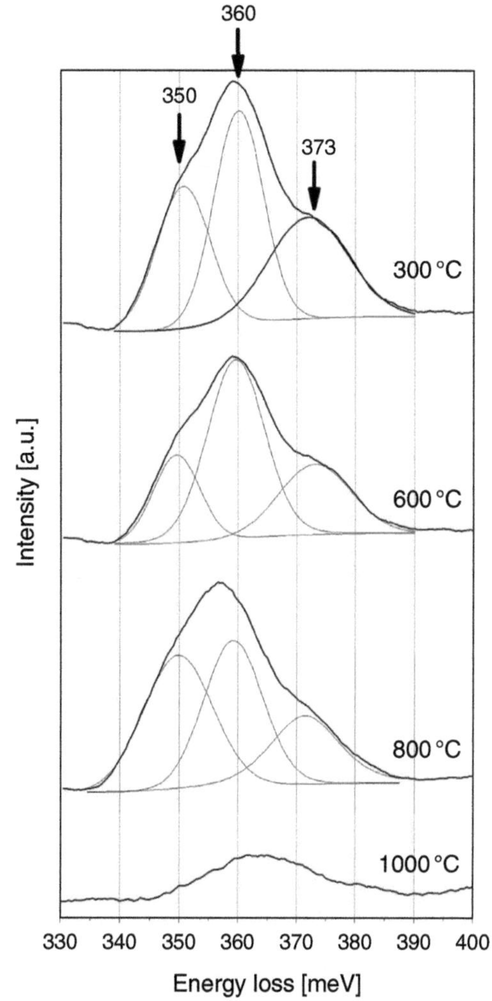

Fig. 8 Detailed analysis of HR-EELS C–H stretching mode of thick as-deposited HF CVD diamond film. The data was derived from Fig. 7 for convenience

results in increasing intensity of the diamond surface plasmon at 23 eV, which may be associated with etching of amorphous carbon component present on the surface of as-deposited film. The direct evidence of this fact comes from HR-EELS analysis of these samples discussed below.

HR-EEL spectra of MW-CVD-3–20 nm nanodiamond film deposited by MW CVD are shown in Fig. 10. The upper line (Spectrum 10A) corresponds to the *as-deposited* sample followed by UHV annealing to 700°C. Detailed analysis of C–H stretching mode is shown in Fig. 10D. The Spectrum 10A is dominated by a mixed C–C stretching and C–H bending mode at 155 meV, the overtones of the diamond optical phonon at 300 and 450 meV, and C–H stretching vibration at 360–375 meV. Note the strong splitting of C–H stretching mode in this spectrum: the two maxima

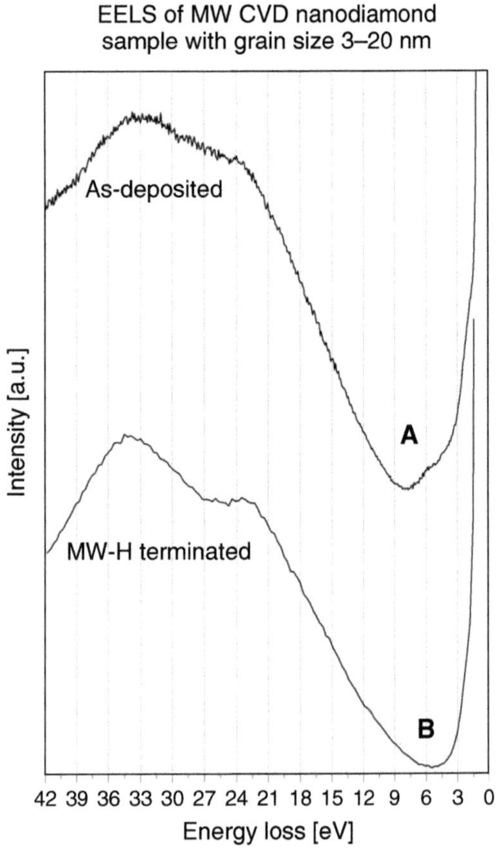

Fig. 9 EEL spectra of MW CVD nanodiamond film with grain size 3–20 nm. The primary electron energy E_p is 500 eV. (**A**) As-deposited sample annealed to 700°C. (**B**) *Ex-situ* MW-H terminated sample annealed to 630°C. Note the enhancement of diamond surface plasmon at energy loss of 23 eV followed *ex-situ* MW hydrogenation

are positioned at ~360 and ~375 meV, reflecting hydrogen bonding to sp^3- and sp^2-carbon, respectively (Spectrum 10D). Following MW hydrogenation and subsequent UHV annealing to 630°C (Spectrum 10B), the diamond overtones underwent enhancement relative to C–H associated modes. The splitting of C–H stretching vibration disappears and now this mode is centered at 360 meV, indicating preferential etching of sp^2-carbon by atomic hydrogen (only sp^3-C–H modes are present, Spectrum 10E). We conclude, therefore, that *ex-situ* exposure to atomic hydrogen results in preferential etching of sp^2-carbon leaving well-defined diamond surface with compelled hydrogenation. Annealing to 1000°C results in hydrogen desorption and appearance of ~90 meV peak (C = C dimer) in a similar way as for HF CVD diamond films with ~300 nm grain size (Fig. 7).

Now we compare the HR-EELS analysis of the as-deposited MW CVD nanodiamond (Spectra 10A and 10D) with the as-deposited HF CVD submicron crystalline samples (Fig. 8). As can be seen in Fig. 8, the HR-EEL spectrum of the as-deposited HF CVD films with ~300 nm grains also display contributions at ~350, ~360, and

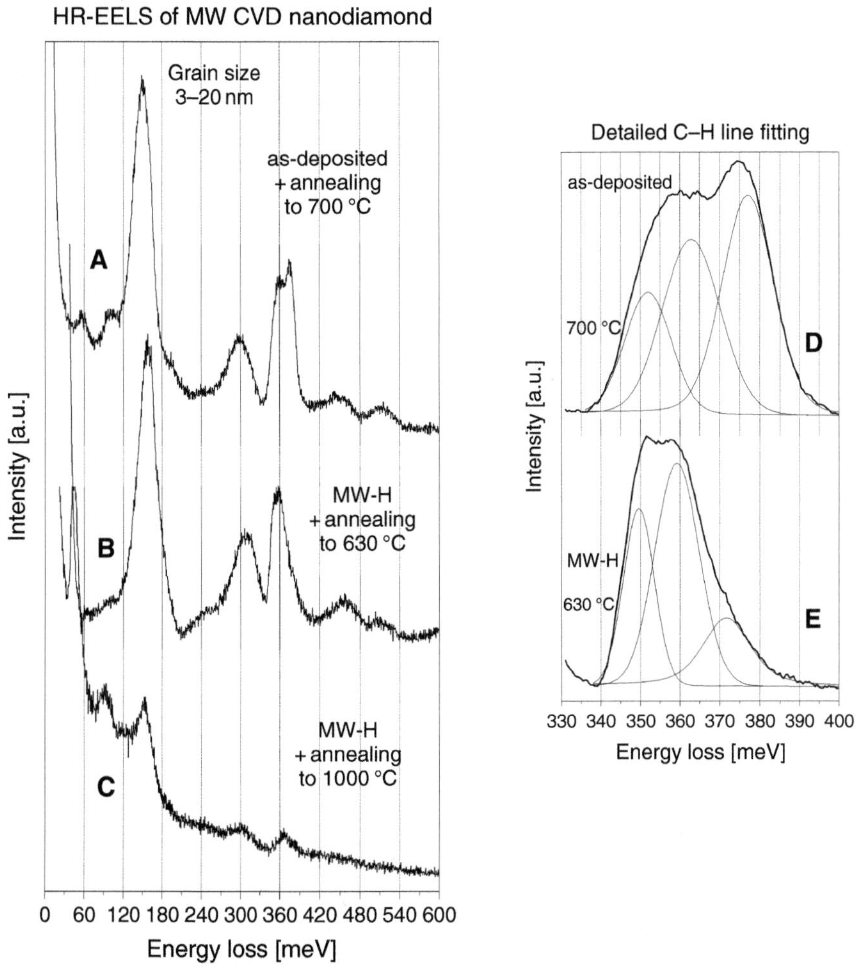

Fig. 10 HR-EEL spectra of MW CVD deposited nanodiamond films. (**A**) As-received sample annealed to 700°C. (**B**) As-deposited sample followed MW-H treatment for 15 min and subsequent annealing to 630°C. (**C**) Followed further annealing to 1000°C. (**D**) and (**E**) Detailed analysis of HR-EELS C–H stretching mode (data originated from Spectra **A** and **B**)

~375 meV within the C–H stretching modes which are stable up to 800°C. Obviously, the density of grain boundaries in the case of nanodiamond films exceeds that in the case of submicron crystallites. Since the contribution of the ~375 meV mode is much more prominent in the case of nanometric grains, this mode may be associated with hydrogen bonded to *diamond grain boundaries*. This is a very exciting feature. In our studies of hydrogen adsorption on ion-irradiated diamond surface [85], we concluded that nearly all kinds of hydrogenated carbon decompose near ~600°C. This is in agreement with other studies of thermal stability of diamond like carbon which undergoes erosion/destabilization at a temperature of about

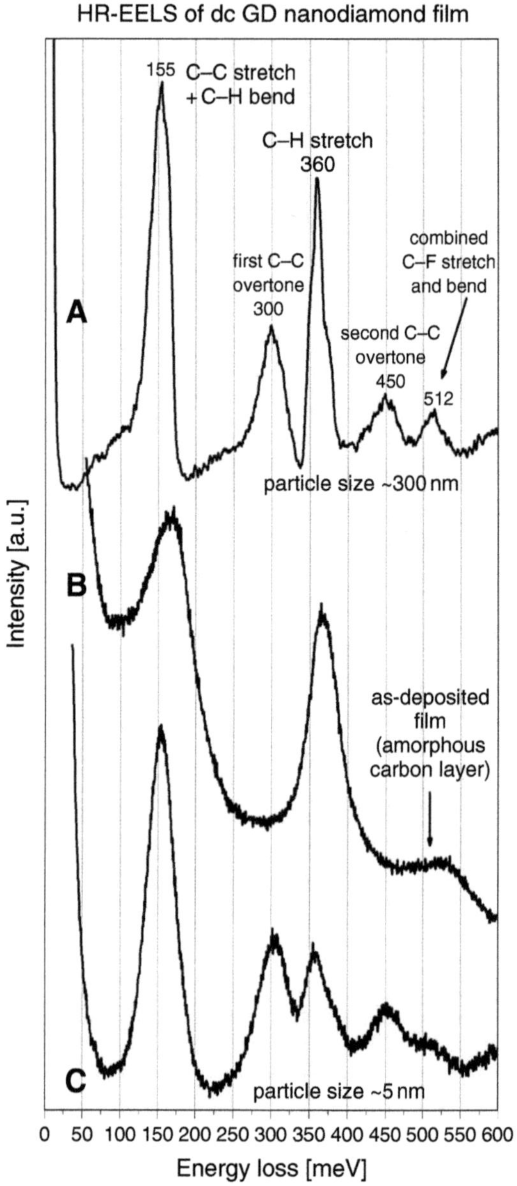

Fig. 11 HR-EEL spectra of carbon films. (**A**) As-deposited HF CVD high-quality diamond film (grain size 300 nm). (**B**) As-deposited dc GD nanodiamond films. (**C**) As-deposited dc GD film followed 2 min MW-H treatment (grain size ~5 nm)

600°C independently on deposition process [78–84]. The present results show that (most likely) the hydrogen bonded to *diamond grain boundaries* (sp^2-hybridized carbon) is stable up to temperatures of at least 800°C. This finding allows one to distinguish the hydrogenated sp^2- hybridized carbon bonded to diamond grains from hydrogenated amorphous carbon by simple annealing procedure.

It is very interesting to note at this stage the possible correlation between the 375 meV HR-EELS mode and the 1140 cm^{-1} Raman peak. It was discussed previously in Section 4.1 that the 1140 cm^{-1} Raman peak is attributed to C–H bending mode of sp^2-carbon and positioned most likely on the diamond grain surface. Both peaks (1140 cm^{-1} Raman and 375 meV HR-EELS mode) are attributed to (sp^2)-carbon–hydrogen vibration. Both peaks are stable up to elevated temperatures where hydrogen bonded to amorphous carbon undergoes desorption. Both peaks are enhanced in the case of nanocrystalline diamond grain size. At this stage, we guess that the 375 meV mode may be attributed to the stretching of the same C–H vibration, which contributes to the 1140 cm^{-1} Raman peak (currently attributed to ν_1 mode of t-PA). Both peaks may serve as indication of nanocrystalline character of diamond films.

Now we turn to analyze HR-EEL spectra of nanodiamond films deposited by dc GD CVD (Spectra 11B and 11C). Typical HR-EEL spectrum of well-defined diamond surface with grains of ~300 nm (Spectrum 11A) is shown for comparison. First, we compare this spectrum of the well-defined diamond surface with the Spectrum 11B of predominantly amorphous carbon layer.

The Spectrum 11B of *as-deposited* dc GD nanodiamond film is dominated by mixed C–C and C–H vibrations (a broadband with maximum near 160 meV), mixed sp^2-/sp^3-carbon–hydrogen stretching modes near 370 meV and broad peak near ~510 meV. It was shortly discussed in Section 2 that the subplantation mechanism leaded to nanodiamond formation leaves the upper surface region consisting of hydrogenated amorphous carbon [11, 32]. Note the absence of the overtones at 300, 450, and 600 meV in this spectrum compared to the Spectrum 11A of well-defined diamond surface. This result further corroborates our previous assumption (Section 5.1) that these overtones originated from diamond optical phonon and are characterized solely to well-defined diamond surfaces.

Spectrum 11C presents the HR-EEL spectrum of nanodiamond film which underwent a short time *ex-situ* exposure to MW hydrogen plasma. This treatment leads to preferential etching of sp^2-carbon leaving hydrogenated nanodiamond grains (~5 nm) exposed to the vacuum. It also results in compelled hydrogenation of nanodiamond crystallites. The Spectrum 11C displays overtones of diamond optical phonon and, therefore, is characteristic of a well-defined hydrogenated diamond surface. Comparing the Spectra 11A and 11C (submicron grains vs. nanocrystallites), one can see the change in relative intensity of pure C–C originated vibrations (300 and 450 meV) to those originated from pure C–H modes (360 and 510 meV). Thus, in the case of nanodiamond surface (Spectrum 11C), the first overtone of diamond optical phonon seems to be enhanced relative to C–H stretching vibration mode (360 meV). The same is true for the second overtone at 450 meV relative to the coupled mode of C–H stretching and C–H bending (or C–C stretching) vibrations at ~510 meV.

The HR-EEL spectra of nanodiamond films are different from those of submicron grain size in the relative intensities of C–C modes compared to C–H vibrations. Figure 12 shows a comparison between the HR-EEL spectra measured for the three kinds of diamond films grown for 1 h by different methods: HF CVD, dc GD CVD

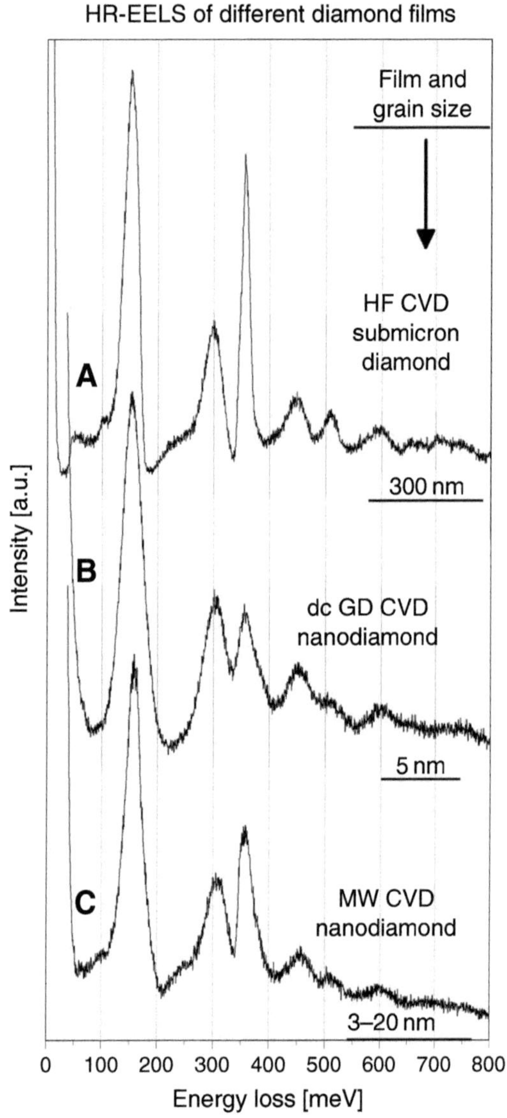

Fig. 12 Comparison of HR-EEL spectra of three different diamond films which underwent MW-H treatment. (**A**) HF CVD film grown for 1 h (grain size ~300 nm). (**B**) Nanodiamond film grown by dc GD method (grain size ~5 nm). (**C**) MW CVD sample (grain size 3–20 nm). Note the different ratios of C–C associated modes relative to C–H vibrations

and MW CVD. In order to compare these spectra, all films were exposed to the same surface treatment in MW hydrogen plasma for ~10 min. Since the first strong peak at ~155 meV is composed of mixed C–C and C–H modes, we should compare only *pure* C–C vibration modes with *pure* C–H modes, namely, overtones of diamond optical phonon at 300 meV with C–H stretching mode at 360 meV. In the case of HF CVD (grain size ~300 nm), the relative ratio of the intensity of 360 meV mode with respect to the intensity of 300 meV phonon overtone is 1.88. In the case of nanodiamond films deposited by MW CVD (grain size lays in 3–20 nm range), the same ratio is 1.2, while for dc GD CVD nanodiamond films (grains ~5 nm) this ratio is 0.9. This ratio is correlated with the diamond grain size, that is, the smaller the diamond grain, the smaller the ratio of the intensity of the pure C–H/C–C associated modes.

We conclude, therefore, that the diamond grain size impacts the shape of HR-EEL spectra. For as-deposited films, the ~375 meV mode of C–H stretching peak is enhanced and attributed to diamond grain boundary contribution. This mode is stable up to elevated temperatures of at least 800°C where all kinds of hydrogenated amorphous carbon are decomposed. In addition, the spectrum of the fully hydrogenated diamond surface (*ex situ* exposed to atomic hydrogen plasma) shows decrease in C–H associated modes relative to pure C–C modes. The last phenomena are correlated with our Raman data (Section 4.1) as well as with IR measurements reported by others [46–48].

6 Enhancement of Electron Emission from Near-Coalescent NanoMeter Thick Continuous HF CVD Diamond Films

In the preceding sections of this chapter, we studied the impact of grain size on compositional and vibrational properties of polycrystalline diamond films deposited by three different CVD methods. In this section, we present the electron emission properties of HF CVD diamond films as a function of film thickness and constituting grain size. The crystal size in such films may vary from ~50 to 70 nm in the nearly coalescent ~100 nm thick films (Fig. 1A) up to ~300–400 nm in micron thick films (Fig. 1B) due to a columnar microstructure of such films. Therefore, the grain size of HF CVD diamond films can be controlled by film thickness, i.e. by deposition time.

The electron emission induced by photons [86], ions [87], electrons [15], and electric fields [88] was consistently investigated and compared for HF CVD diamond film thickness from 70 nm to 4.7 μm [15]. Continuous ~70 nm thick diamond films may be produced when initial nucleation densities are in excess of $\sim 10^{10}$ diamond growth centers per cm^2.

Absolute quantum photoyield (QPY) measurements [86] were performed in a vacuum UV monochromator system using 140–220 nm radiation, which corresponds to a photon energy of 8.9–5.63 eV. The diamond films were hydrogen terminated by MW hydrogen plasma previous to the measurements. Ion-induced electron emission (IIEE) measurements [87] were carried out in a vacuum system using 200 KeV H^+ and 140 KeV Ar^+ incident beams. In addition to hydrogenation, diamond films were heated to 200°C inside the measurement chamber in order to desorb water molecules. Secondary electron emission (SEE) measurements from the as-deposited diamond films were accomplished using SEM, by calibration of the LEO Supra in-lens detector so that the current due to secondary electrons, emitted from the specimen, can be reliably obtained from the measured SEM image. The QPY,

SEE, and IIEE yields are similarly defined as a number of emitted electrons per one incident particle.

For field electron emission (FEE) measurements, the as-deposited negatively biased diamond films were placed in parallel to a flat metal anode inside a vacuum chamber and the current of emitted electrons as a function of bias was measured [88]. The current was limited in these type of measurements (to 5 mA/cm^2) by protective resistance in order to prevent significant chemical and morphological changes in the diamond films. The parameters which define FEE are the turn-on field (the electric field at which the FEE begins) and the threshold field (the electric field required to induce a specific current density, $I_{th,}$ arbitrarily taken as 0.3 μA/cm^2).

Figure 13 summarizes the results of electron emission induced by four different methods [15]. The absolute QPY value at photon energy of 8.85 eV increases with film deposition time from 2.5% (carbonized silicon substrate) to 14% for a continuous 70 nm thick film, followed by a decrease to 12 % and stabilization for continuous films of thicknesses above ~150–200 nm (Fig. 13A). The initial coefficient of IIEE, using 140 keV Ar^+ ions, also reaches its maximum value of 116 for ~100 nm diamond film thickness and decreases to 90–100 for thicker films (Fig. 13B). The SEE shows maximum yield (δ_{max}) at primary electron energy of ~1000 eV. The dependence of δ_{max} on film thickness displays similar behavior as QPY value: it obtains its maximum value of δ_{max} ~10 for film thickness of ~100 nm and drops to a somewhat constant value of δ_{max} ~6.6 for thicker films (Fig. 13C). The threshold field, required for FEE current of 0.3 μA/cm^2, reaches its minimum value of ~8.5 V/μm for ~100 nm thick diamond film and monotonically increases for thicker diamond films (Fig. 13D).

Our experimental results show that the thickness of diamond films and, therefore, the grain size of the crystallites significantly affect the electron emission from HF CVD diamond films induced by photons, ions, electrons and electric fields. In all four cases, the thin 70–100 nm thick diamond films exhibit enhanced electron emission properties. In order to understand this phenomenon, we attempt to correlate between the thickness-dependent properties and the electron emission properties.

The electron emission from diamond films is very sensitive to the surface quality and surface termination, which determine the energy barrier for electron escape from the diamond into the vacuum. EELS spectra show that all diamond films, even 100 nm thick, exhibit surface diamond plasmons, characteristic of good quality crystalline diamond surface [33, 88]. All samples in one set of diamond films of different thicknesses were prepared under the same CVD conditions, exposed to the same atmosphere and treated identically. The XPS spectra of the as-deposited films preheated to 200°C indicate that the surface of all studied diamond films is clean, with small traces of oxygen [88], most probably originated in adsorbated water molecules. Consequently, we believe that the surface state of diamond films is not the reason for the maximum of electron emission yield as a function of diamond film thickness.

An additional very important factor in electron emission processes, besides the surface barrier, is the electron supply to the surface. Accumulation of positive charge at the upper layer of the emitting sample may slow down the secondary electrons or

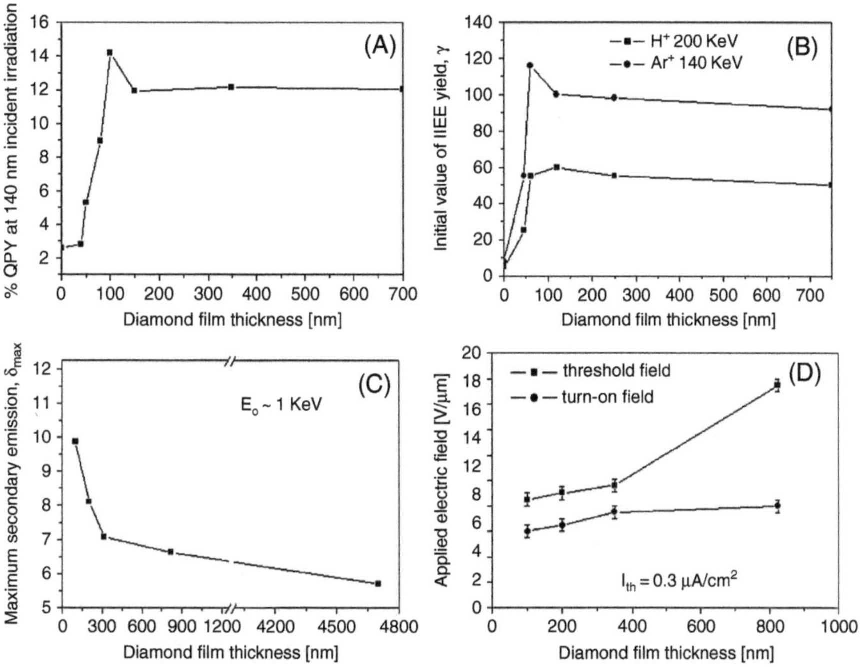

Fig. 13 Electron emission from diamond films of different thicknesses. (**A**) Absolute QPY at 140 nm incident irradiation. (**B**) Initial values of IIEE yield for 140 KeV Ar^+ and 200 KeV H^+ ion irradiations. (**C**) Maximum SEE yield (δ_{max}). (**D**) Threshold field required for FEE current I_{th} of 0.3 $\mu A/cm^2$ as a function of diamond films thickness

even prevent their escape to the vacuum. It has been experimentally shown that the more conductive diamond films also exhibit much better electron emission properties [89–91]. The conductivity may be enhanced by doping or by increase of sp^2 conductive phase percent and grain boundary density. Since the studied diamond films are undoped, the conductive paths extending to the surface are required to reduce the surface positive charge. More conductive paths are expected to present in the thinner diamond films due to the higher content of the grain boundaries and sp^2-bonded carbon phase.

As was mentioned before, the nanometer thick diamond films are distinguished from micron thick films by smaller grains and higher density of grain boundaries. In addition, Raman spectra show a decrease in sp^2 relative content with increasing film thickness (according to ~1550 cm^{-1} peak appearance). However, the quality of all films is high because even 100 nm thick diamond film exhibits the sharp 1332 cm^{-1} Raman peak with FWHM ~9 cm^{-1} [15].

The nature, at the nanometer scale, of the electrically conducting pathways in the diamond films as a function of thickness was experimentally investigated by our group using conducting probe atomic force microscopy (CP-AFM) [15]. The 100 nm thick continuous diamond film, in which the crystallites just touch one another, has been found to be the most conductive due to the existence of short conductive pathways between and along the diamond grains. Thicker films are less conductive and the conductive pathways appear to be mostly on the surface. According to these results, the repletion of the diamond surface with electrons is more efficient in the thin (~100 nm) diamond films than in the thicker films (<1 μm). Therefore, the high conductivity of the undoped, 100 nm thick, continuous, but only near-coalescent, diamond films is likely to be responsible for the enhanced electron emission properties of such films in comparison to the thicker undoped films with columnar structure. QPY, IIEE, and SEE are surface processes and, therefore, the electron supply to the surface is very important in these types of emissions, when the current of the outgoing electrons is even as low as several pA. Field emission is a more complex process. The sp^2-carbon at the grain boundaries not only contributes to the conductivity of the nanometric thick films, but most probably also serves as the emitting "tips" [92, 93], thus facilitating the FEE from these films at low threshold field.

7 Summary

In this chapter, we summarized our studies related to the impact of diamond grain size on the composition and properties of polycrystalline diamond films deposited by three different CVD methods. The films were grown from hydrogen-rich gas mixtures (hot filament and dc GD CVD) and Ar-rich/hydrogen-poor gas mixture (MW CVD). SIMS studies provide direct evidence of the correlation between the concentration of hydrogen atom trapped within the films and the diamond grain size, independently of deposition conditions: the *smaller* diamond grains result in *higher* hydrogen atom retention within the films. The hydrogen atoms are most likely bonded to the internal grain surfaces and within the grain boundary region. High hydrogen concentration in nanodiamond films has impact on the shape of the Raman and HR-EEL spectra of nanodiamond films. Thus, the carbon–hydrogen associated 1140 cm^{-1} Raman peak (C–H bending mode) undergoes enhancement in the case of nanocrystalline films alongside with HR-EELS 375 meV C–H stretching mode. Based on thermal stability studies, we associate these vibrational modes with (sp^2)-C–H vibration originated from diamond surface. Both the Raman and HR-EELS C–H associated peaks undergo intensity decrease relative to pure carbon vibrational modes in the case of ultra-nanocrystalline diamond (grain size ~5 nm). These findings are corroborated by IR measurements of other group [46–48] and are under current investigation in our lab. The electron emission induced by photons, ions, electrons and electric fields was consistently investigated and compared for HF CVD diamond film thicknesses ranging from 70 nm (grain size ~50–70 nm) to 4.7 μm (grain size ~300–400 nm). It was found that regardless of primary excitation, the electron emission exhibits maximum values for near-coalescent nanometer thick continuous diamond films consisting of 50–70 nm grain size.

References

1. S.Y. Luo, J.K. Kuo, B. Yeh, J.C. Sung, C.W. Dai, T.J. Tsai, Mate. Chem. Phys. 2001;72(2): 133–135.
2. L.Y. Huang, K.W. Xu, J. Lu, B. Guelorget, Surf. Coat. Technol. 2002;154(2–3):232–236.
3. N. Jiang, K. Sugimoto, K. Nishimura, Y. Shintani, A. Hiraki, J. Cryst. Growth 2002;242 (3–4):362–366.
4. K.H. Wu, X.R. Wang, S. Liu, E.G. Wang, J. Appl. Phys. 2001;90(9):4810–4814.
5. N.G. Ferreira, L.L.G. Silva, E.J. Corat, Diamond Relat. Mater. 2002;11(3–6):657–661.
6. Q.Y. Chen, D.M. Gruen, A.R. Krauss, T.D. Corrigan, M. Witek, G.M. Swain, J. Electrochem. Soc. 2001;148(1):E44–E51.
7. J.Q. Li, D.Y. He, W.T. Guo, J.H. Zhang, Y.N. Sun, Q.S. Lei, X. Gao, Int. J. Modern Phys. B 2002. 16(6–7):1013–1017.
8. W.B. Yang, F.X. Lu, Z.X. Cao, J. Appl. Phys. 2002;91(12):10068–10073.
9. S. Bhattacharyya, O. Auciello, J. Birrell, J.A. Carlisle, L.A. Curtiss, A.N. Goyette, D.M. Gruen, A.R. Krauss, J. Schlueter, A. Sumant, P. Zapol, Appl. Phys. Lett. 2001;79(10): 1441–1443.
10. D. M. Gruen, Ann. Rev. Mater. Sci. 1999;29:211
11. Y. Lifshitz, Th. Kohler, Th. Frauenheim, I. Guzmann, A. Hoffman, R. Q. Zhang, X. T. Zhou, S. T. Lee, Science 2002;297:1531
12. J. B. Cui, J. Ristein, L. Ley, Phys. Rev. Lett. 1998;81:429.
13. I. L. Krainsky, V. M. Asnin, G. T. Mearini, J. A. Dayton, Phys. Rev. B. 1996;53:R7650.
14. J. Ristein, F. Maier, M. Riedel, M. Stammer, L. Ley, Diamond Relat. Mater 2001;10:416.
15. O. Ternyak, A. Cimmino, S. Prawer, A. Hoffman, Diamond Relat. Mater. 2005;14:272.
16. O. Ternyak, Sh. Michaelson, R. Akhvlediani, A. Hoffman, Diamond Relat. Mater. 2006;15:850.
17. W. Zhu, G.P. Kochanski, S. Jin, Science 1998;282:1471.
18. P. Ascarelli, E. Cappelli, F. Pinzari, M.C. Rossi, S. Salvatori, P.G. Merli, A. Migliori, J. Appl. Phys. 2001;89:689.
19. Sh. Michaelson, O. Ternyak, Y. Lifshitz, A. Hoffman, Appl. Phys. Lett. 2007;90:061918
20. Sh. Michaelson, Y. Lifshitz, O. Ternyak, A. Hoffman, Appl. Phys. Lett. 2006;89:131918
21. P. Reichart, G. Datzmann, A. Hauptner, R. Hertenberger, C. Wild, G. Dollingeer, Science 2004;306:1537.
22. M. Benlahsen, B. Racine, K. Zellama, G. Turban, J. Non-Crystalline Sol. 2001;47:283, 47.
23. B. Racine, M. Benlahsen, K. Zellama, P. Goudeau, M. Zarrabian, G. Turban, Appl. Phys. Lett. 1998;73:3226.
24. J. Robertson, Philos. Mag. Lett. 1988;57:143.
25. J. Robertson, Phys. Rev. Lett. 1992;68:220.
26. W. Jacob, Thin Solid Films 1998;326:1.
27. P. Reinke, W. Jacob, W. Moller, J. Appl. Phys. 1993;74:1354.
28. A. Hoffman, A. Heiman, R. Akhvlediani, E. Lakin, E. Zolotoyabko, C. Cyterman, J. Appl. Phys. 2003. 94, 4589.
29. E. Titus, D.S. Misra, A.K. Sikder, P.K. Tyagi, M.K. Singh, Abha Misra, N. Ali, G. Cabral, V.F. Neto, J. Gracio, Diam. Relat. Mat. 2005. 14, 476.
30. D.F. Talbot-Ponsonby, M.E. Newton, J.M. Baker, G.A. Scarsbrook, R.S. Sussmann, A.J. Whitehead, S. Pfenningen, Phys. Rev. B 1998. 57, 2264.
31. R.D. Maclear, J.E. Butler, S.H. Connell, B.P. Doyle, I.Z. Machi, D.B. Rebuli, J.P.F. Sellschop, E. Sideras-Haddad, Diamond Relat. Mater. 1999. 8, 1615.
32. Y. Lifshitz, X. M. Meng, S.T. Lee, R. Akhvlediani, A. Hoffman, Phys. Rev. Lett. 2004. (93) 056101
33. R. Akhvlediani, I. Lior, Sh. Michaelson, A. Hoffman, Diamond Relat. Matt. 2002;11:545
34. D. Zhou, D. M. Gruen, L. C. Qin, T. G. McCauley, A. R. Krauss, J. Appl. Phys. 1998;84:1981.

35. Sh. Michaelson, Y. Lifshitz, O. Ternyak, R. Akhvlediani, A. Hoffman, Diamond Relat. Mat. 2007. In press
36. A. C. Ferrari, J. Robertson, Phys. Rev. B. 2001;63:R121405.
37. A. C. Ferrari, J. Robertson, Phil. Trans. Royal Soc. A 2004;362:2477.
38. A. C. Ferrari, S. E. Rodil, J. Robertson, Diamond Relat. Mat. 2003;12:905
39. Wada, N., Gaczi, P.J., Solin, S.A., J. Non-Cryst. Solids 1980. 35–6 543
40. R.J. Nemanich, J.T. Glass, G. Lukovsky, R.E. Shreder, J. Vac. Sci. Technol. A 1988;6:1783
41. R. E. Shroder, R. J. Nemanich, J. T. Glass, Phys. Rev. B 1990;41:3738
42. W.A Yarbrough, R. Messier, Science. Washington: Feb 9, 1990;247(4943):688 (9 pages)
43. H. Kuzmany, R. Pfeiffer, N. Salk, B. Günther, Carbon 2004;42:911.
44. Lopez-Rios, T., Mendoza, D., Garcia-Vidal, F.J., Sanchez-Dehesa, J., Pannetier, B. Phys. Rev. Lett., 1998;81:665
45. Sh. Michaelson, A. Hoffman, Diamond Relat. Mater. 2006;15:486
46. C.-F. Chen, C.-C. Wu, C.-L, Cheng, S.-Y. Sheu, H.-C. Chang, J. Chem. Phys. 2002;116:1211
47. Y.-R. Chen, H.-C. Chang, C.-L, Cheng, C.-C. Wang, J.C. Jiang, J. Chem. Phys. 2003;119:10626
48. C.-L, Cheng, C.-F. Chen, W.-C. Shaio, D.-S. Tsai, K.-H. Chen, Diamond Relat. Mater. 2005;14:1455
49. H. Ibach, D.L Mills, Electron Energy Loss Spectroscopy and Surface Vibrations Academic, New York, 1982, p. 199
50. B.J. Waclawski, D.T. Pierce, N. Swanson, R.J. Celotta, J. Vac. Sci. Technol. 1982;21:368
51. B.D. Thoms, J.E. Butler, Phys. Rev. B 1994;50:17450
52. B.D. Thoms, J.E. Butler, Surface Science 1995;328:291
53. B. Sun, X. Zhang, Q. Zhang, Z. Lin, Appl. Phys. Lett. 1993;62:31
54. B. Sun, X. Zhang, Z. Lin, Phys. Rev. B 1993;47:9816
55. S.T. Lee, G. Apai, Phys. Rev. B 1993;48(4):2684
56. T. Aizawa, T. Ando, M. Kamo, Y. Sato, Phys. Rev. B, 1993;48(24):18348
57. P.E. Pehrsson, T.W. Mercer, Surf. Sci. 2000;460:49
58. P.E. Pehrsson, T.W. Mercer, Surf. Sci. 2000;460:74
59. J. Kinsky, R. Graupner, M. Stammler, L. Ley, Diam. Relat. Mat. 2002;11:365
60. K. Okada, T.Aizawa, R. Souda, S. Komatsu, S. Matsumoto, Diam. Relat. Mat. 2001;10:1991
61. T. Aizawa, T. Ando, K. Yamamoto, M. Kamo, Y. Sato, Diam. Relat. Mat. 1995;4:600
62. A. Lafosse, D.T. Billy, J.P. Guillotin, Y. Le Coat, R. Azria, A. Laikhtman, A. Hoffman, Phys. Rev. B 2003;68(23):235421
63. A. Laikhtman, A. Lafosse, Y. Le Coat, R. Azria, A. Hoffman, J. Chem Phys. 2003;119:1794
64. A. Laikhtman, A. Lafosse, Y. Le Coat, R. Azria, A. Hoffman, Surf. Sci. 2004;551:99
65. A. Lafosse, D. Caceres, M. Bertin, A. Hoffman, R. Azria, Surf. Sci. 2005;587:134
66. M. Bertin, A. Domaracka, D. Pliszka, A. Laffose, R. Azria, Surf. Sci. 2006;600:847
67. A. Lafosse, A. Hoffman, M. Bertin, D. Teillet-Billy, R. Azria, Phys. Rev. B 2006;73:195308
68. Sh. Michaelson, A. Hoffman, Diamond Relat. Mater. 2005;14:470
69. Sh. Michaelson, A. Hoffman, Diamond Relat. Mater. 2006;15:486
70. H. Okuyama, S. Thachepan, T. Argua, T. Ando, M. Nishijima Chem. Phys. Lett. 2003. 381 535
71. J. Biener, A. Schenk, B. Winter, C. Lutterloh, U.A. Schubert, J. Küppers, Surf. Sci. Lett. 1993;291:L725
72. J. Biener, A. Schenk, B. Winter, C. Lutterloh, U.A. Schubert, J. Küppers, Surface Science 1994. 228 307–309
73. A. Horn, J. Biener, A. Schenk, C. Lutterloh, J. Küppers, Surface Science 1995. 178 331–333
74. Sh. Michaelson, Y. Lifshitz, A. Hoffman, Appl. Phys. Lett. 2006. 89 223112
75. Sh. Michaelson, Y. Lifshitz, A. Hoffman, Diamond Relat. Mat. (2007, available on-line) In press
76. J. Ristein, R.T. Stief, L. Ley, W. Beyer J. Appl. Phys. 1998;84(7):3836

77. Th. Frauenheim, Th. Kohler, M. Sternberg, D. Porezag, M.R. Pederson, Thin Solid Films 1996. 272 314
78. R. Kalish, Y. Lifshitz, K. Nugent, S. Prawer Appl. Phys. Lett. 1999;74(20):2936
79. R. Kalish, in The Physics of Diamond Int. School of Physics Enrico Fermi, Vol. 135, edited by A. Paoletti and A. Tucciarone, IOS, Amsterdam, 1997, p. 373
80. M. Benlahsen, B. Racine, K. Zellama, G. Turban, J. Non-Cryst. Solids 2001;283:47
81. Ch. Wild, P. Koidl Appl. Phys. Lett. 1987;51:1506
82. J. K. Walters, D.M. Fox, T.M. Burke, O.D. Weedon, R.J. Newport, W.S. Howells, J. Chem. Phys. 1994;101:4288
83. J. K. Walters, J.S. Rigden, R.J. Newport, S.F. Parker, W.S. Howells, Phys, Scr. T 1995;57:142
84. M. Lejeune, M. Benlahsen, R. Bouzerar Appl. Phys. Lett. 2004;84:344
85. M. Bertin, A. Lafosse, R. Azria, Sh. Michaelson, O. Ternyak, A. Hoffman, Appl. Phys. Lett. 2007;90:031914
86. Sh. Michaelson, R. Akhvlediani, O. Ternyak, A. Hoffman, A. Breskin, R. Chechik, Diamond Relat. Mater. 2005;14:546.
87. Sh. Michaelson, V. Richter, R. Kalish, A. Hoffman, E. Cheifetz, R. Akhvlediani, Thin Solid Films 2002;185:420–421
88. O. Ternyak, R. Akhvlediani, W.K. Wong, S.T. Lee, Y. Lifshitz, S. Daren, E. Cheifetz, A. Hoffman, J. Appl. Phys. 2005;98:123522
89. A. Shih, J. Yater, P. Pehrsson, J. Butler, C. Hor, R. Abrams, J. Appl. Phys. 1997;82(4):860
90. Y.D. Kim, W. Choi, H. Wakimoto, S. Usami, H. Tomokage, T. Ando, Diamond Relat. Mater. 2000;9:1096
91. Y. Iseri, M. Honda, Y.D. Kim, T. Ando, W. Choi, H. Tomokage, J. Phys.: Condens. Matter 2004;16:S171
92. F. Lacher, C. Wild, D. Behr, P. Koidl, Diamond Relat. Mater. 1997;6:1111
93. A.V. Karabutov, V.D. Frolov, S.M. Pimenov, V.I. Konov, Diamond Relat. Mater. 1999;8:763.

Super-Resolution Optical Effects of Nanoscale Nonlinear Thin Film Structure and Ultrahigh-Density Information Storage

Jingsong Wei

Abstract Ultrahigh-density optical information storage is very attractive for high-definition digital television, data distribution, memory, etc. The density is required to become higher and higher with the development of information technology, which requires the information bit size to be smaller than the optical diffraction limit. In this chapter, we introduce the super-resolution optical effects of nanoscale nonlinear thin film structure, and the applications in ultrahigh-density optical information storage.

Contents

1 Introduction

Optical storage is very useful for high-definition digital television, data distribution, and memory. With the development of information technology, the capacity of optical storage devices is required to become larger and larger. Correspondingly,

J. Wei (✉)
Shanghai Institute of Optics and Fine Mechanics, Chinese Academy of Sciences, Shanghai 201800, China
e-mail: weijingsong@siom.ac.cn

Z.M. Wang (ed.), *Toward Functional Nanomaterials*, Lecture Notes in Nanoscale Science and Technology 5, DOI 10.1007/978-0-387-77717-7_8,

the density is also required to become higher and higher. In order to improve the density, one has to reduce the information bit size. The size is determined by the optical diffraction limit. That is, $D=1.22\lambda/NA$, where D is the size of the focused spot, λ laser wavelength in vacuum, and NA the numerical aperture of pickup. It is this cutoff frequency that gives an absolute limit on the recording density of an optical system [1–3]. In order to improve the capacity and density of optical storage devices, one has to shorten the laser wavelength and increase the numerical aperture of optical pickup. Table 1 gives the development story of optical storage devices.

Table 1 The development of optical storage devices (optical discs)

	CD	DVD	HD-DVD	BD
Laser wavelength	780 nm	650 nm	405 nm	405 nm
Numerical aperture	0.45	0.60	0.60	0.85~0.90
Size of information bit	830 nm	400 nm	250 nm	150 nm
Capacity of a 5 inch disc	650 MB	4.7 GB	15 GB	27 GB

CD-compact disc, DVD-digital versatile disc, HD-high density, BD-blue ray disc

However, a larger NA will reduce the focal depth of pickup and increase comatic aberration when the disk tilts against the optical axis [4]. So far, for far-field optical storage, it can be seen from Table 1 that the NA of pickup has been very close to the theoretical limit ($NA = 1$). So the improvement of density by increasing NA is quite limited. It is apparent that the most direct way to increase the density is to use a short-wavelength light source; however, even if the light source is changed from near-infrared (780 nm) to blue-purple wavelength (405 nm), the increase of density is only several times. Therefore, one has to propose some new techniques and methods; here we present the ultrahigh-density information storage with optical super-resolution effects.

2 Principle for Breaking Through Optical Diffraction Limit [5]

In quantum mechanics, the coordinate of conjugate dynamic variable $\boldsymbol{r}$ and momentum $\boldsymbol{p}(p_x,p_y,p_z)$ cannot be decided simultaneously. They need to meet the Heisenberg's uncertainty relation

$$\Delta \boldsymbol{r}.\Delta \boldsymbol{p} \geq \hbar \tag{1}$$

where $\hbar$ is the Planck constant. The corresponding components $\Delta \boldsymbol{x},\Delta \boldsymbol{p}_x$; $\Delta \boldsymbol{y},\Delta \boldsymbol{p}_y$; $\Delta z,\Delta \boldsymbol{p}_z$ also meet the Heisenberg's uncertainty relation

$$\Delta \boldsymbol{x}(\Delta \boldsymbol{y}, \Delta z).\Delta \boldsymbol{p}_x(\Delta \boldsymbol{p}_y, \Delta \boldsymbol{p}_z) \geq \hbar \tag{2}$$

According to the Planck–Einstein's relation $\boldsymbol{p} = \hbar \boldsymbol{k}$, where $\boldsymbol{k}$ is the wave vector, the inequality (2) can be rewritten as

$$\Delta \boldsymbol{x}.\Delta \boldsymbol{k}_x \geq 1 \tag{3}$$

Inequality (3) indicates that it is impossible to accurately determine the coordinate and the momentum simultaneously. However,inequality (3) also tells us that the measurement accuracy of one conjugate variable is not limited if the other is not required. Thus, it is possible to break through the optical resolution power (diffraction limit) λ/2.

In what follows, we present the required condition for breaking through the diffraction limit. For wave vector $\boldsymbol{k}(\boldsymbol{k}_x, \boldsymbol{k}_y, \boldsymbol{k}_z)$, if the following inequality is correct

$$\boldsymbol{k}_x > |\boldsymbol{k}| = \boldsymbol{k} \tag{4}$$

where $|\boldsymbol{k}|$ is the modulus of $\boldsymbol{k}$, and two beam of lights, which are from A point and B point, respectively, intersect at O point, then their wave vectors are equal, i.e., $\boldsymbol{k}_1=\boldsymbol{k}_2=\boldsymbol{k}$, and the angle between A and B is 2θ (as shown in Fig. 1). Thus, the difference along x direction between $\boldsymbol{k}_1$ and $\boldsymbol{k}_2$ can be expressed as:

$$\Delta \boldsymbol{k}_x = 2\boldsymbol{k}_x = 2\boldsymbol{k} \sin \theta \tag{5}$$

At $\theta=\pi/2$,we get

$$\Delta \boldsymbol{k}_x = \Delta \boldsymbol{k}_{x.\mathbf{max}} = 2\boldsymbol{k} = 2/\lambda \tag{6}$$

The following inequality can be derived from formulas (4)–(6)

$$\Delta \boldsymbol{k}_x = 2\boldsymbol{k}_x > 2\boldsymbol{k} \tag{7}$$

Inequalities (6) and (7) tell us that $\Delta \boldsymbol{k}_x > 2/\lambda$ at certain conditions. Based on the uncertainty relation (inequality (3)), one can get a very important expression

$$\Delta \boldsymbol{x} < \lambda/2 \tag{8}$$

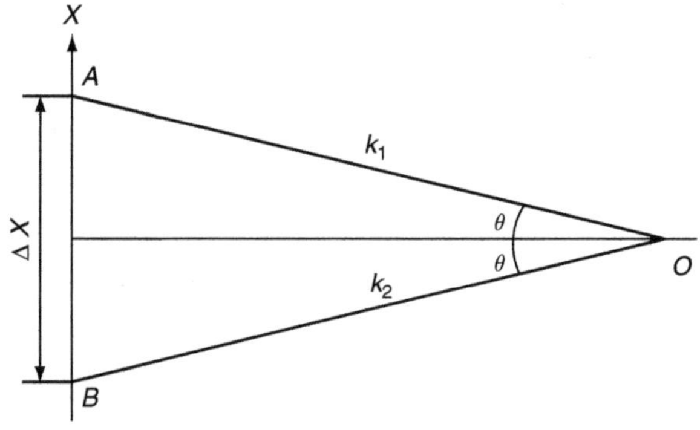

Fig. 1 The resolution power between A point and B point

Inequality (8) indicates that the optical resolution limit $\lambda/2$ can be broken through if the wave vector component $\boldsymbol{k}_x$ is larger than $|k|$. Based on the dispersion relation $\boldsymbol{k}^2=\boldsymbol{k}_x{}^2+\boldsymbol{k}_y{}^2+\boldsymbol{k}_z{}^2$, one can get

$$\boldsymbol{k}_x = \sqrt{\boldsymbol{k}^2 - \boldsymbol{k}_y{}^2 - \boldsymbol{k}_z{}^2} \tag{9}$$

It can be found that $\boldsymbol{k}_x>|\boldsymbol{k}|=\boldsymbol{k}$ only if $\boldsymbol{k}_y$ or (and) $\boldsymbol{k}_z$ is an imaginary number. In other words, the requirement that the transversal resolution power Δx breaking through the diffraction limit is that the longitudinal wave vector $\boldsymbol{k}_z$ ($\boldsymbol{k}_y$) is an imaginary number. This means that the electromagnetic field is strongly localized along the y or (and) z direction.

One can assume that the wave vector $\boldsymbol{k}$ can be written as ($\boldsymbol{k}_x,\boldsymbol{k}_y,i\boldsymbol{k}_z$); thus, the optical field $\boldsymbol{u}(\boldsymbol{x},\boldsymbol{y},z;t)$ can be expressed as

$$\boldsymbol{u}(\boldsymbol{x},\boldsymbol{y},z;t) = \boldsymbol{u}(\boldsymbol{x},\boldsymbol{y},z)\exp\left[i\left(\boldsymbol{k}_x.x + \boldsymbol{k}_y.y - \omega\cdot t\right) - \boldsymbol{k}_z z\right] \tag{10}$$

Equation (10) indicates that $\boldsymbol{u}(\boldsymbol{x},\boldsymbol{y},z;t)$ is an evanescent field, it propagates along the (x, y) plane, and attenuates exponentially along the z direction. This electromagnetic field is also called as the near-field light since it only exists in the close to the (x, y) plane. It is very significant to apply the near-field light to the ultrahigh-density optical information storage due to the below-diffraction-limited resolution power.

3 Super-Resolution Optical Effects of Nanoscale Nonlinear Thin Film Structure and Ultrahigh-Density Information Storage

In the last few years, all kinds of near-field optical information storage techniques were extensively researched and developed since they can reduce the information bit size, and increase the storage density and capacity. For example, Betzig et al. [6] realized the magnetic domain with a size of 60 nm in the magnetooptical thin film with the near-field optical probe technique; Terris et al. [7] designed and fabricated the solid immersion lens flight pickup, and realized the information density of 3.8×10^8 bits·inch^{-2} and the data transfer rate of 3.3×10^6 bits·s^{-1}. In this section, we mainly present and discuss the super-resolution effects of nanoscale nonlinear thin film structure and the corresponding ultrahigh information storage techniques.

Recently, the giant optical nonlinearity of thin films is always a hot subject [8]. For example, Fukaya et al. [9] found the optical switching properties of Sb thin film under pulse laser irradiation. Choi et al. [10] measured the nonlinearity of polycrystalline Si films using near-field scanning optical microscopy, and found that the real part of the coefficient reaches up to 0.02esu. Liu et al. [11] prepared the Bi thin films with the laser ablation method, the nonlinearity ceofficient is about 0.124 cm^2·kW^{-1}. Making use of the giant optical nonlinearity of multilayer thin film structure, the below-diffraction-limited information bits can be realized.

3.1 Super-Resolution Optical Storage Stemming from Internal Multi-Interference of Nonlinear Thin Film Structure

3.1.1 The Super-Resolution Principle [12]

In this section, we deal with a special geometry of the nonlinear thin film structure, which is made up of *dielectric layer/nonlinear thin film/dielectric layer*. Laser beam is focused by a normal optical lens system onto the surface of the structure (as shown in Fig. 2), where the nonlinear thin film is sandwiched between two dielectrics of constant optical indices n_1 and n_2, respectively. The Gaussian laser beam is incident upon the structure. We calculate the transmitted light beam at the other side of the structure. Let us start with the reflectance of the nonlinear thin film structure under internal multiple interference schemes:

$$R(r) = \frac{(n_1 - n_2)^2 \cos^2 \frac{\delta(r)}{2} + \left(\frac{n_1 n_2}{n} - n(r, I)\right)^2 \sin^2 \frac{\delta(r)}{2}}{(n_1 + n_2)^2 \cos^2 \frac{\delta(r)}{2} + \left(\frac{n_1 n_2}{n(r,I)} + n(r, I)\right)^2 \sin^2 \frac{\delta(r)}{2}} \tag{11}$$

with the phase factor

$$\delta(r) = \frac{4\pi}{\lambda} n(r, I) L \cos\theta \tag{12}$$

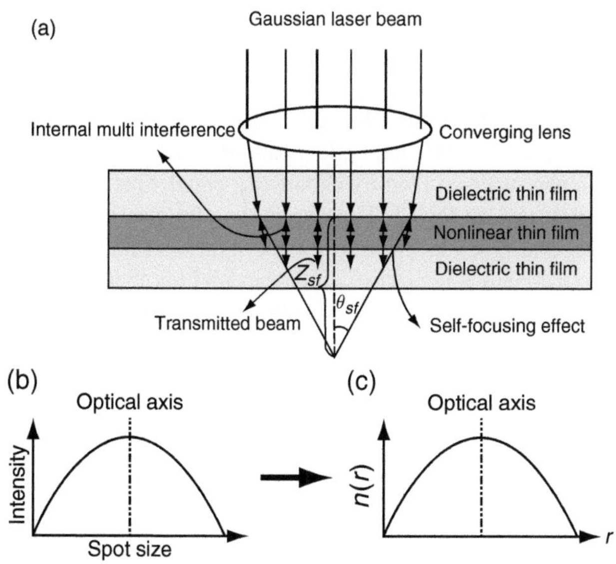

Fig. 2 Schematic of a super-resolution in nonlinear thin film structure.
(a) The path of the internal multi-interference and the self-focusing effects;
(b) Gaussian light intensity distribution;
(c) induced refractive index profile within the spot

where r coordinator is chosen along the film from the center of the light spot, L the thickness of the nonlinear thin film, and θ the incident angle (in what follows we shall choose normal incidence as $\theta=0$). In Equation (11) the nonlinear refractive index $n(r,I)$ reflects the optical Kerr effect and is a function of the radial position r and the local light intensity $I=E(r)E*(r)$ as

$$n(r, I) = n_0 + n_I[E(r)E*(r)] = n_0 + n_I I \tag{13}$$

where the coefficients n_0 and n_I vary with the materials and are to be determined by experiments. If the absorption of structure can be neglected, the transmittance is finally calculated from the reflectance as:

$$T(r) = 1 - R(r) = \frac{4n_1n_2}{(n_1+n_2)^2\cos^2\frac{\delta(r)}{2} + \left(\frac{n_1n_2}{n(r,I)} + n(r,I)\right)^2\sin^2\frac{\delta(r)}{2}} \tag{14}$$

In the second part, let us consider the self-focusing effect. Self-focusing is a property of positive nonlinear optical Kerr medium. When a laser beam is incident on the material, the light penetrates into the medium and modifies the refractive index along the optical path. The nonlinear modification is described as in Eq. (13). When the light beam travels through the medium, it is focused after a distance. So the nonlinear medium acts like an optical lens. The self-focusing effect is significant only when the medium has a considerable thickness. For ultrathin film, the self-focusing effect becomes insignificant. In what follows, we analyze the self-focusing effect of thin films.

The schematic of the self-focusing effect is also shown in Fig. 2, where the light is focused at a distance from the film z_{sf} with an angle θ_{sf}. We apply directly the Fermat's principle, namely, that the optical path of all rays traveling from the wavefront at the entrance surface to the focus must be equal. One takes into account the marginal and chief rays in Fig. 2. The nonlinear refractive index is $n(r,t)=n_0+n_I I$. For marginal ray that goes at the edge of the light spot and has the lowest light intensity, the optical path is $n_0 z_{sf}/\cos\theta_{sf}$. For the chief ray that goes through the center point of the light spot and has the maximum of the light-induced refractive index $n_0+n_I I_0$, the optical path is $(n_0+n_I I_0)z_{sf}$. According to the Fermat's principle, the two paths are equal

$$(n_0 + n_I I_0)\cdot z_{sf} = \frac{n_0 z_{sf}}{\cos\theta_{sf}} \tag{15}$$

which resolves the angle

$$\cos\theta_{sf} = \frac{n_0}{n_0 + n_I I_0} \tag{16}$$

If the light spot radius is w_0, one has eventually the focus distance

$$z_{\mathrm{sf}} = w_0 \cot\theta_{\mathrm{sf}} = \frac{n_0 w_0}{\sqrt{(n_0 + n_I I_0)^2 - n_0^2}} \tag{17}$$

Let us now have an analysis on the self-focusing angle θ_{sf} in comparison with the diffraction angle θ_{diff} known as

$$\theta_{\mathrm{diff}} = \frac{1.22\lambda}{n_0 w_0} \tag{18}$$

The diffraction effect becomes insignificant when the self-focusing effect overwhelms the light diffraction, i.e., $\theta_{\mathrm{sf}} > \theta_{\mathrm{diff}}$, which corresponds (see Eq. 16) to

$$\theta_{\mathrm{sf}} > \theta_{\mathrm{diff}} \Rightarrow \cos\theta_{\mathrm{sf}} = \frac{n_0}{n_0 + n_I I_0} < \cos\left(\frac{1.22\lambda}{n_0 w_0}\right) \tag{19}$$

or equivalently

$$I_0 > \frac{n_0\left(1 \cos\frac{1.22\lambda}{n_0 w_0}\right)}{n_I \cos\frac{1.22\lambda}{n_0 w_0}} = I_{\mathrm{diff}} \tag{20}$$

Equation (20) indicates that if the incident intensity is large enough $I_0 > I_{\mathrm{diff}}$, the ray optics becomes valid, and the diffraction can be neglected.

Here, we only consider the strongly nonlinear materials with high refractive index. We choose the relevant constants from some experimental reports [10, 13], where the polycrystalline silicon thin film shows a large third order nonlinear susceptibility $\chi^{(3)}=0.02$esu, and a high optical index $n_0=4.21$. The other constant n_I can be calculated by

$$n_I = \frac{12\pi^2}{n_0}\chi^{(3)} \tag{21}$$

For polycrystalline silicon thin film, one obtains $n_I=4.4572\times10^{-9}$ m.W^{-1}. Assuming the laser wavelength $\lambda=633$ nm, the spot radius $w_0=1$ μm, and the laser power $P=5$ mW, one has the intensity of the primary optical spot $I_0=P/\pi w_0^2=1.592\times10^9$ W·m^{-2} and the diffraction intensity $I_{\mathrm{diff}}\approx1.6117\times10^7$ W·m^{-2}. Comparing the two intensities I_0 and I_{diff}, one realizes immediately that $I_0>>I_{\mathrm{diff}}$, which means that for strongly nonlinear media with high refractive index and under high laser power, the diffraction effect is negligible. Furthermore, substituting the chosen constants into Equation (17), one obtains $z_{\mathrm{sf}}=402$ nm. If the thickness of the nonlinear thin film we consider is restricted within nanometric scale, say <100 nm, we have the condition $z_{\mathrm{sf}}>>L$. In other words, the self-focusing effect can also be neglected for ultrathin films. Overall, since both self-focusing and diffraction are negligible for strong radiative and ultrathin films, one has to consider only the internal multiple interference effect outlined in Eq. (14).

For a symmetric nonlinear thin film structure $n_1=n_2$, according to the expression in Eq. (14), the maximum transmittance occurs when

$$L(r) = \frac{k\lambda}{2n(r, I)}, \quad k = 0, \pm 1, \pm 2 \cdots \tag{22}$$

On the other hand, for Gaussian beam, the intensity distribution is

$$I(r) = I_0 \exp\left(-\frac{2r^2}{w_0^2}\right) \tag{23}$$

Substituting Eq. (23) and (13) into (22), one has

$$L(r) = \frac{k\lambda}{2} \frac{1}{n_0 + n_I I_0 \exp\left(-\frac{2r^2}{w_0^2}\right)} \quad k = 0, \pm 1, \pm 2, \cdots \tag{24}$$

In order to get the maximum transmittance in the center of spot and obtain an effective minimum spot at the exit surface of nonlinear thin film structure, one has to realize the maximum transmittance ($T = 1$) at the center of spot, that is, for $r = 0$ at the center point, one has

$$L(0) = \frac{k\lambda}{2} \frac{1}{n_0 + n_I I_0} \quad k = 0, \pm 1, \pm 2, \cdots \tag{25}$$

For the first order, $k = 1$ and $L(0) = 27.9$ nm, which is much shorter than the self-focusing distance. Finally, with the various parameters given in the previous paragraphs, we have calculated the light spot with the following formula for the intensity mapping of the transmitted light spot

$$I_{\text{trant}}(r) = T(r) \cdot I(r) = \frac{4n_1 n_2 I_0 \exp\left(-\frac{2r^2}{w_0^2}\right)}{(n_1 + n_2)^2 \cos^2 \frac{\delta(r)}{2} + \left[\frac{n_1 n_2}{n(r,I)} + n(r, I)\right]^2 \sin^2 \frac{\delta(r)}{2}} \tag{26}$$

and the calculated results are presented in Fig. 3. In Fig. 3, the three-dimensional images are given together with the cross section in one dimension. The first image is given in Fig. 3(a) for the primary light spot according to the distribution in Equation (23). The second is shown in Fig. 3(b) for the light-modulated nonlinear transmittance distribution as in Equation (14). The last is presented in Fig. 3(c) for the transmitted spot as described in Equation (26). One compares the primary and the transmitted spot in Fig. 3(a) and (c), respectively, and realizes that the spot is reshaped and the size of the spot is reduced to about one-third. Accordingly, the full width at half maximum is reduced from 1.6 μm in Fig. 3(a) to 0.6 μm in Fig. 3(c). Thus the super-resolution can be obtained by the internal multiple interference effect with the nonlinear thin film structure.

Fig. 3 Optical field distributions as a function of radial distance. (a) Primary spot morphology and intensity distribution; (b) transmittance distribution; (c) reshaped spot morphology and intensity distribution

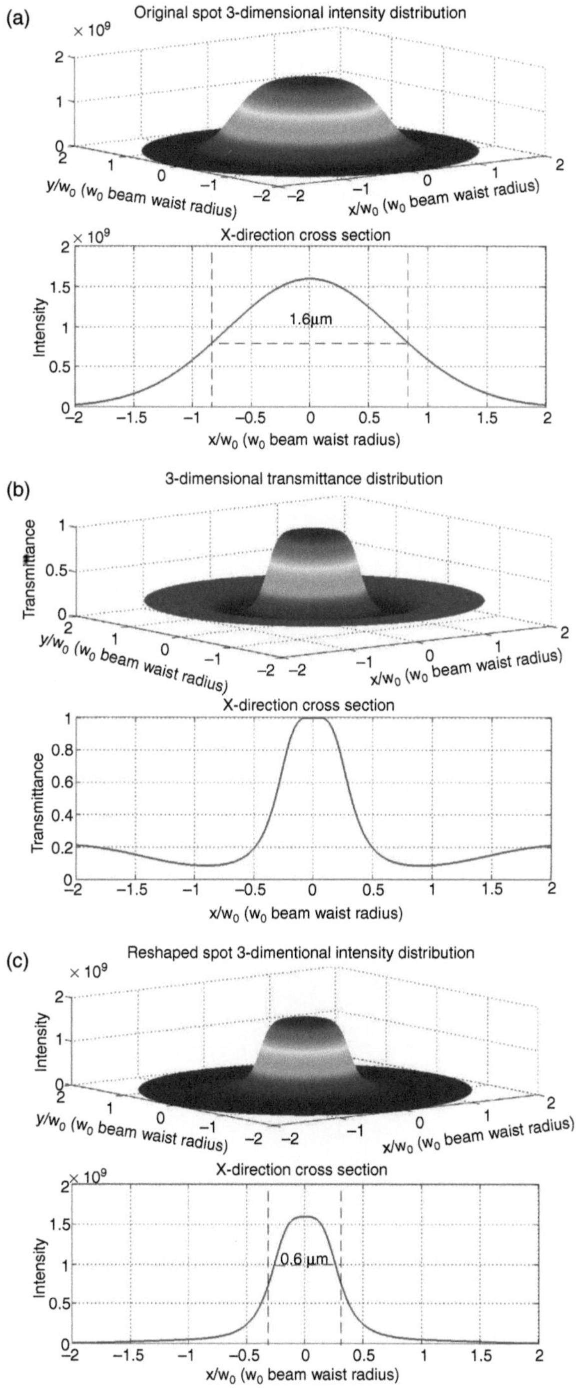

3.1.2 The Super-Resolution Optical Recording [12]

We first prepared a single $Ge_2Sb_2Te_5$ thin film with a thickness of about 30 nm by magnetron sputtering method. The $Ge_2Sb_2Te_5$ thin film is a typical medium used in phase change optical data recording [14]. When a laser pulse with high power density exceeding 10^9 W·m^{-2} irradiates onto the medium, the phase change from crystalline to amorphous state occurs, and the area of phase change is restricted by the optical spot size. In other words, the information bit size is determined by the spot size. Then, we prepared a $SiN/Si/SiN/Ge_2Sb_2Te_5$ sample, where the SiN thin films were prepared by radio frequency reactive sputtering method, and Si thin film was deposited by radio frequency sputtering method. Here the SiN (20 nm)/Si(28 nm)/SiN(20 nm) is a nonliear thin film structure. The numerical aperture of the optical pickup system is NA=0.6, and the laser wavelength is $\lambda = 650$ nm, as is shown in Fig. 4. With these parameters, one expects the diffraction limit to be $D = 1.321$ µm. We have carried out the optical recording experiments with both the

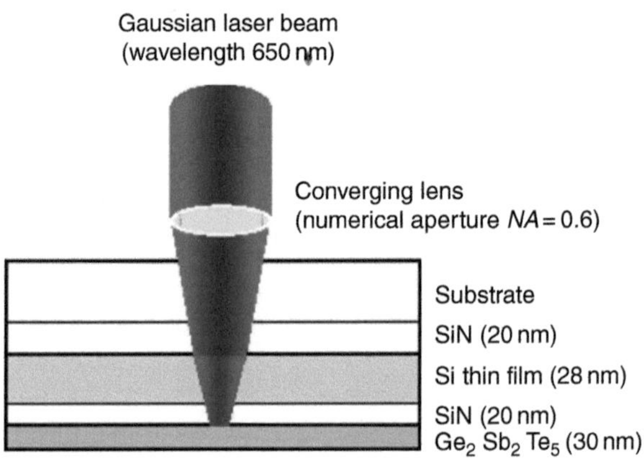

Fig. 4 Schematic for the optical recording experiments

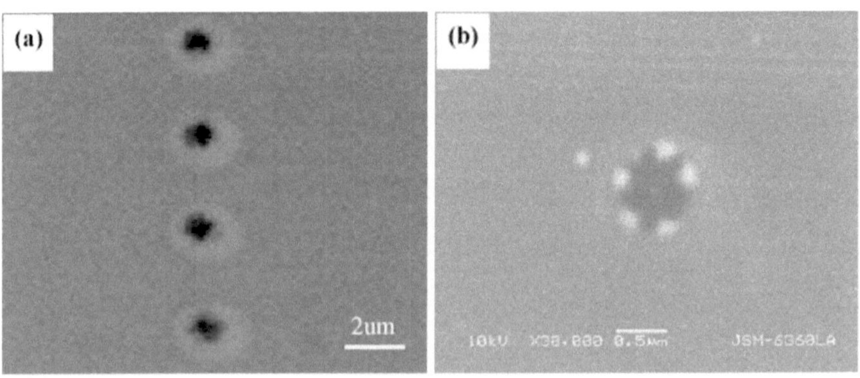

Fig. 5 Experimental results of super-resolution optical information recording.
(a) The single $Ge_2Sb_2Te_5$ thin film
(b) SiN/Si/SiN/$Ge_2Sb_2Te_5$ structure

samples, and the results are presented in Fig. 5. In Fig. 5(a), the recording mark is obtained with a single $Ge_2Sb_2Te_5$ thin film. The size is about 1.5 μm, which is only determined by the spot size. In Fig. 5(b), the optical recording marks are formed on the $SiN/Si/SiN/Ge_2Sb_2Te_5$ multilayer thin film structure. The mark size is about 0.5 μm. Comparing the two images in Fig. 5 (a) and Fig. 5(b), one finds that the mark size with the nonlinear thin film structre is one-third of that with a single phase change recording thin film. Since the parameters involved in the experiments and the calculations are about the same, one concludes that the super-resolution optical information storage can be realized via the internal multiple interference within the nonlinear thin film structure.

3.2 Super-Resolution Optical Storage Stemming from Self-Focusing Thermal Lens Effect with a Nonlinear Thin Film Structure

3.2.1 Thermal Lens Principle for Measuring the Temperature Coefficient of Refractive Index

In general, the optical nonlinearity of materials is from several physical mechanisms, such as magnetostriction and electronic polarization; however, the main contribution is from the thermal effect for some phase change materials. The thermally induced temperature coefficient of refractive index can be measured by an *in situ* ellipsometer equipped with a heating cell (such as Ellipso Tech, Elli-633-FH) in the temperature range varying from room temperature to random temperature. Here, we introduce a time-resolved dual beam mode-mismatched thermal lens method for measuring the variations of refractive index with the temperature [15]. The measurement setup is shown in Fig. 6. A wavelength of 532 nm laser beam, which is operated in the TEM_{00} mode giving a Gaussian intensity distribution, is used as a pumping light whose power can be changed by an attenuator. The pumping beam passes through one converging lens, and is focused to the sample surface. A cw

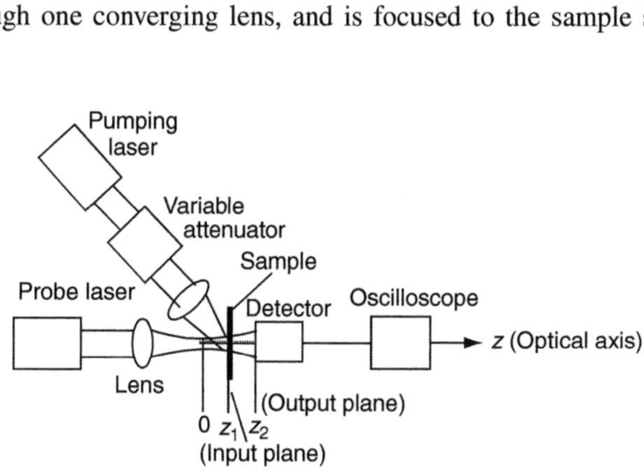

Fig. 6 The setup of time-resolved dual beam mode-mismatched thermal lens

He–Ne laser, whose power is less than 0.5 mW, is used as a probe beam that passes through the other converging lens, and goes through the center of the focused pumping spot. The location of waist of the probe spot is taken as the origin along the z-axis. The sample is located at z_1 and a detector is centered in the probe beam at a position z_2. The aperture of the detector is small compared with the probe beam diameter at this location. The transient variation of the probe light transmitted from the sample can be recorded by an oscilloscope. The diameter of the sample is large compared with that of the pumping spot so that the sample can be considered as infinite in radial direction; heat conduction through the ends of the sample can be neglected, and the temperature variation can be taken as purely radial.

The temperature variation $\Delta T(r,t)$ of sample heated by the pumping spot as a function of radius r and time t can be calculated as [16, 17]

$$\Delta T(r,t) = \frac{0.48 P_e b_e}{\pi \cdot c \cdot \rho \cdot w_e^2} \int_0^t \left(\frac{1}{1 + 2t'/t_c} \right) \exp\left(-\frac{2r^2/w_e^2}{1 + 2t'/t_c} \right) dt' \tag{27}$$

where

$$\begin{aligned} t_c &= \frac{w_e^2 C\rho}{4K} = \frac{w_e^2}{4D} \\ D &= K/C\rho \end{aligned} \tag{28}$$

where t_c is the characteristic buildup time of the thermal lens. D, K, C, and ρ are the thermal diffusivity, thermal conductivity, specific heat, and density of the sample. w_e is the pumping spot, p_e the pumping laser power, and b_e the optical absorption coefficient at the pumping laser wavelength. Equation (27) is a desired expression for the temperature calculation in the sample.

For a homogeneous sample, the optical path length $s(r,t)$ is a function of temperature

$$s(T) = n(T).L(T) \tag{29}$$

The relative optical path length variation between two sides of sample induced by the pumping laser with respect to the axis can be written as

$$\Delta s(r,t) = [n(r,t).L(r,t) - n(0,t).L(0,t)] + [\Delta L(0,t) - \Delta L(r,t)] \tag{30}$$

where $[\Delta L(0,t) - \Delta L(r,t)]$ is the optical path length, and

$$\Delta L(r,t) = \left(\frac{\partial L}{\partial T} \right)_{T_0} . \Delta T(r,t) \tag{31}$$

using the Taylor series expansion; Eq. (30) can be rewritten as

$$\Delta s(r,t) = L_0 \left[\frac{(n_0 - 1)}{L_0} (\frac{\partial L}{\partial T})_{T_0} + (\frac{\partial n}{\partial T})_{T_0} \right] . [\Delta T(r,t) - \Delta T(0,t)] \qquad (32)$$

where L_0 and n_0 are the thickness and refractive index, respectively, of the sample at the room temperature T_0, and

$$\frac{\mathrm{d}s}{\mathrm{d}T} = \left(\frac{n_0 - 1}{L_0} \right) \cdot \left(\frac{\partial L}{\partial T} \right)_{T_0} + \left(\frac{\partial n}{\partial T} \right)_{T_0} \qquad (33)$$

is the temperature coefficient of the optical path length of the sample. The first term in Eq. (33) represents the sample thickness change and the second term, the refractive index change. If the sample thickness change with temperature is negligible then

$$\frac{\mathrm{d}s}{dT} \approx \frac{\mathrm{d}n}{dT} \qquad (34)$$

The refractive index as a function of r and t can be obtained by substituting Eq. (27) into the following expression

$$n(r,t) = n_0 - \frac{\mathrm{d}n}{\mathrm{d}T} \Delta T(r,t) \qquad (35)$$

Equation (35) assumes a decrease of refractive index with temperature. According to the diffraction theory of aberration and Huygens' principle, the complex phase amplitude of a wave at a point on the output plane is the result of a superposition of Huygens' wavelets emanating from all points on the input plane. It is written as

$$U_0(t) = \frac{i}{\lambda_p} \int_0^{\infty} \cdot \int_0^{2\pi} U_i(r,t) \cdot \left(\frac{1 + \cos\alpha}{2} \right) \cdot \frac{\exp(-i \frac{2\pi}{\lambda} \cdot |z_2 - r|)}{|z_2 - r|} r \cdot \mathrm{d}r \cdot \mathrm{d}\theta \qquad (36)$$

where $U_i(r,t)$ is the complex phase and amplitude of the wave at the input plane. $U_0(t)$ is the complex phase and amplitude of wave on the axis or the beam center at the output plane where the detector is located, λ_p and α are the wavelength of the probe laser beam and the angle between the sample and the probe beam, respectively, in our designed setup, $\alpha = 0$.

Assuming the beam to be composed of spherical wave with curvature radius R and the thermal lens effect to be absent in the sample (shown in Fig. 7 (a)), the amplitude factor in the input plane can be written as

$$|U_i| = B \exp\left(-\frac{r^2}{w_p^2} \right) \qquad (37)$$

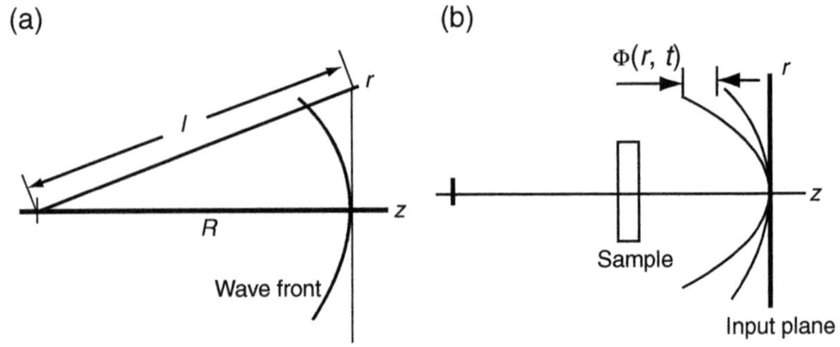

Fig. 7 Phase distribution at the input plane.
(a) Without the lensing medium and
(b) with the lensing medium

where B is a constant, and w_p is the radius of the probe beam at the location of sample; the phase at points on the input plane is

$$\frac{2\pi}{\lambda_p}l = \frac{2\pi}{\lambda_p}(R^2 + r^2)^{\frac{1}{2}} \approx \frac{2\pi}{\lambda_p}(R + r^2/2R) \tag{38}$$

The approximation is valid since the beam is confined to a narrow region about the axis, namely, $R>>r$. Based on Eq. (38), the relative phase lag is $\pi \cdot r^2/(\lambda_p R)$. Following the textbook written by Born and Wolf, there is an aberration or additional phase lag in the spherical wave due to the existence of the thermal lens effect (shown in Fig. 7 (b)). An expression for this additional phase lag can be obtained from the optical path length in the sample. According to Eqs. (30) and (34), the optical path length variation about the axis can be written as

$$\Delta s(r, t) = \Phi(r, t) = L[n(r, t) - n(0, t)] \tag{39}$$

Substituting Eq. (35) into (39) and multiplying both sides by $2\pi/\lambda_p$ give

$$\frac{2\pi}{\lambda_p}\Phi(r, t) = \frac{2\pi}{\lambda_p}\frac{\mathrm{d}n}{\mathrm{d}T}L[\Delta T(0, t) - \Delta T(r, t)] \tag{40}$$

The complex phase and amplitude at the input plane can be written as

$$U_i(r, t) = B\exp\left(-\frac{r^2}{w_p^2}\right) \cdot \exp\left[-i\frac{2\pi}{\lambda_p}\left(\frac{r^2}{2R} + \Phi(r, t)\right)\right] \tag{41}$$

By substituting Equation (41) into (36), Sheldon et al. [17] obtained the following diffraction integral

$$U_o(t) = C \int_0^\infty \{1 - i\frac{\vartheta}{t_c} \cdot \int_0^t \tau \cdot [1 - \exp(-2\tau \cdot u)] \cdot dt'\} \cdot \exp[-(1 + i \cdot \xi) \cdot u] \cdot du \quad (42)$$

where $u = \frac{r^2}{w_p^2}$, $\xi = \frac{z_1}{z_c}$, $\vartheta = \frac{0.24 P_e L}{\lambda_p} \frac{b_e}{K} \frac{dn}{dT}$, $\tau(t') = \frac{1}{1+2t'/t_c}$, $z_c = \frac{\pi w_p^2}{\lambda_p}$, z_c is called as the confocal distance of the probe beam.

In finding the intensity variation $I_0(t)=|U_0(t)|^2$ all terms of order ϑ^2 are negligible, and a more convenient form for the equations is the fractional intensity change, $[I(t) - I(\infty)]/I(\infty)$. The result is

$$\frac{I(t) - I(\infty)}{I(\infty)} = \frac{1 - \vartheta \tan^{-1} \frac{2\xi}{3+\xi^2+(9+\xi^2)(t_c/2t)}}{1 - \vartheta \tan^{-1} \frac{2\xi}{3+\xi^2}} - 1 \quad (43)$$

The total fractional intensity change found by setting t=0 is

$$\frac{I(0) - I(\infty)}{I(\infty)} = \frac{1}{1 - \vartheta \tan^{-1} \frac{2\xi}{3+\xi^2}} - 1 \quad (44)$$

According to Eqs. (43) and (44), the thermal lens effect is sensitive to the location z_1 of the sample on the axis. Setting the derivative of Eq. (44) with respect to ξ being equal to zero leads to the predication that the effect is optimized when $\xi = \pm\sqrt{3}$ or the sample is located at $\sqrt{3}\, z_c$ in front of or behind the waist of the probe beam. Note that there is no thermal lens effect if the sample is at the waist. When it is behind the waist at negative value of z_1 the effect is inverted. Let $\xi = \sqrt{3}$, then the probe beam intensity change with time at the detector can be written as

$$I(t) = I(0)\left[1 - \frac{\vartheta}{2} \tan^{-1} \frac{2\sqrt{3}m}{[(1 + 2m)^2 + 3]\,(t_c/2t) + 1 + 2m + 3}\right]^2 \quad (45)$$

where $m=(w_p/w_e)^2$, m is the degree of mode-mismatch of the probe beam and pumping beam at the sample. $I(0)$ is the value of $I(t)$ when t or ϑ is zero. The detected probe beam intensity variation on the axis can defined as

$$\frac{I(t) - I(0)}{I(0)} = \left[1 - \frac{\vartheta}{2} \tan^{-1} \frac{2\sqrt{3}m}{[(1 + 2m)^2 + 3](t_c/2t) + 2m + 4}\right]^2 - 1 \quad (46)$$

Here w_e and w_p can be determined by a spot size measurement method; ϑ can be determined by curve fitting to Eq. (46). Accordingly, dn/dT can be further obtained.

3.2.2 Measurement of Temperature Coefficient of Sb Thin Film [15]

In order to verify the thermal-induced super-resolution effect of nonlinear thin film structure, based on the optical switching properties of Sb thin film under focused pulse laser irradiation, we prepared the Sb thin film with a thickness of about 15 nm. The temperature coefficient was measured using the setup shown in Fig. 6. The pumping laser power is fixed at 5 mW, and its radius at the sample w_e is 42.35 μm. The probe beam is focused to a spot with a radius w_p of 25.2 μm; accordingly, the confocal distance of z_c (according to the Equation (42)) is 778 μm, the optimum location of the sample $z_I(=\sqrt{3}z_c)$ is 1.365 mm, and m is 0.3656.

Figure 8 shows the dependence of $[I(t)–I(0)]/I(0)$ on time. It can be seen from Fig. 8 that $[I(t)-I(0)]/I(0)$ is zero when the pumping time is less than 23.5 μs, and then increases from 0 to 0.6 up to $t = 28$ μs, finally keeps at 0.6 at t exceeding to 28 μs. According to Eq. (28) and the thermal-physical parameters of Sb thin films shown in Table 2, the characteristic buildup time of the thermal lens t_c is equal to 22.35 μs. According to Eqs. (27), (42), and (46) and Fig. 8, the refractive index change with temperature dn/dT can be obtained by fitting, and the results are given in Fig. 9. It can be found from Fig. 9 that the refractive index is always at 3.12 when temperature is below 868 K, and linearly increases to 4.52 at about 928 K, and then stays unchanged above 928 K. Fig. 9 is also approximately expressed as

$$\begin{aligned} n(T) &= 3.12 T < 868\,K \\ n(T) &= 3.12 + 0.0235(T - 868\,K) \quad 928\,K > T > 868\,K \\ n(T) &= 4.52\; T > 928\,K \end{aligned} \tag{47}$$

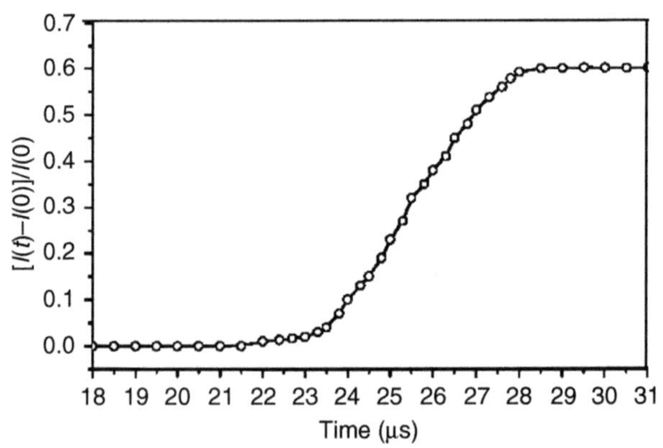

Fig. 8 The dependence of $[I(t)–I(0)]/I(0)$ on time

Table 2 Thermal-physical parameters of Sb thin films [18, 19]

$K(W{\cdot}m^{-1}{\cdot}K^{-1})$	$D\,(m^2{\cdot}s^{-1})$	$C\,(J{\cdot}kg^{-1}{\cdot}K^{-1})$	$\rho(Kg{\cdot}m^{-3})$	$b_e\,(m^{-1})$
24.3	2.01×10^{-5}	190	6380	1.337×10^8

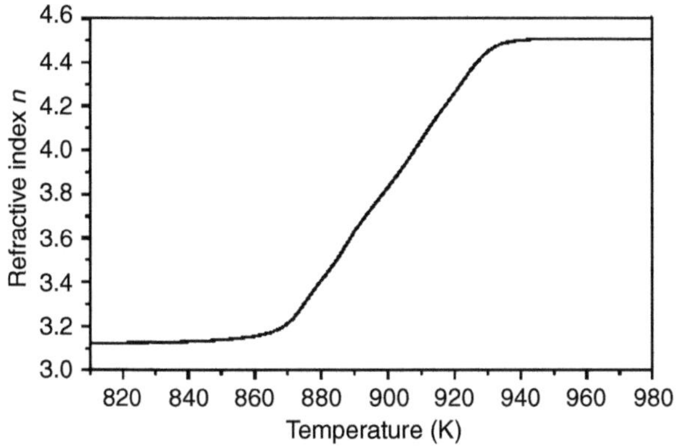

Fig. 9 The dependence of refractive index on temperature

Equation (45) and Fig. 9 indicate that the change of refractive index with temperature of Sb thin film occurs only at 928 K>T>868 K. Due to the melting temperature of Sb bulk being about 930 K, the refractive index change with temperature, dn/dT, is a thermal lens effect, and is induced by the structural transformation between crystalline and melted states. The transformation temperature ranges from 868 K to 928 K, i.e., the initial melting temperature is 868 K, and the thoroughly melting temperature is about 928 K. It should be noticed that for Sb thin film the refractive index change from structural transformation near the melting temperature is much larger than other physical mechanisms. The phenomena can also be found in other phase change materials, such as $Ge_2Sb_2Te_5$ thin films and *AgInSbTe* thin films [20]. Generally speaking, the physical origin of the *dn/dt* is from the difference between the volume expansion and the electronic polarization. A general thermal change of refractive index can be explained by [21]

$$\frac{\mathrm{d}n}{\mathrm{d}T} = \frac{(n^2 - 1)(n^2 + 2)}{6n}(\varphi - \beta) \tag{48}$$

where ϕ and β are the polarizability coefficient and the expansion coefficient, respectively. In the process of melting, the electronic polarization will increase when Sb thin film changes into the more dissociated molten state, where the decrease in the size of atomic groupings introduces the larger electronic polarizability. It should be stressed that the Sb thin film is not only a semimetal with a large refractive index, but also an aquosity substance with a negative expansion coefficient β [22]. These result in the larger temperature coefficient of refractive index in the process of melting.

3.2.3 Super-Resolution Self-Focusing Thermal Lens Model of Nonlinear Thin Film Structure [23, 24]

In Section 3.1, we give a super-resolution model via beam reshaping induced by internal multiple interference effect within nonlinear thin film structue. For the model, the self-focusing effect is negligible due to the self-focusing distance being much larger than the thin film thickness, i.e., $z_{sf}>>L$. However, when the self-focusing effect is dominant within the nonlinear structure, i.e., $z_{sf} \sim L$, the spot reshaping induced by internal multiple interference effect becomes negligible. Here we take Sb thin film for example, and propose a super-resolution model with self-focusing thermal lens effect.

For a nonlinear structure SiN/Sb/SiN, the diameter of the sample is larger than that of the focused laser beam, so the sample can be considered as infinite in radial direction, and heat conduction through the ends of the sample is negligible. The temperature variation induced by a focused Gaussian laser beam can be calculated by Eq. (27). The laser energy is mainly absorbed by the Sb thin film. We assume that a laser beam with a wavelength of 632.8 nm goes through a converging lens whose numerical lens is 0.60, and is focused on the Sb thin film in the structure. The calculated temperature distribution is presented in Fig. 10(a).The central temperature in the spot can reach up to 928 K under the laser power P of 2 mW and the irradiation time t of 1.45 ns; so the dynamic aperture is formed in the region whose temperature is between 868 K and 928 K, and the radius R_0 of aperture is approximately equal to 80 nm (as shown in Fig. 10(a)). According to [25] and Eq. (27), the dynamic aperture is actually a structural transition region from crystalline to melted states of Sb thin film. The fraction of the melted is presented by a greyscale gradient in Fig. 10(b), and the radial distribution of the refractive index $n(r,t)$ is also presented. Thus, the self-focusing thermal lens is formed within the dynamic aperture (as shown in Fig. 10(c)), and its focal length can be calculated by the following expression

$$F(t) = \frac{\pi \cdot n_0 \cdot K \cdot R_0^2 \cdot (R_0^2 + 8D \cdot t)}{0.24PL(\mathrm{d}n/\mathrm{d}T)(8D \cdot t)} \tag{49}$$

where R_0 is the radius of the self-focusing thermal lens, $n_0(=3.12)$ is the initial refractive index of Sb thin film, and $F(t)$ is the focus distance. For the thermal lens formed in Fig. 10(c), the corresponding experimental parameters are shown in Table 3. The focus distance F is 22.5 nm. The size d of the spot focused by the thermal lens can be obtained by the following equation

$$d = 0.61\lambda \frac{F}{R_0} \tag{50}$$

Table 3 The parameters of Sb thin film self-focusing thermal lens

n_0	w_0	P	t_c	L	dn/dT
3.11	80 nm	2 mW	1.45 ns	30 nm	0.0235

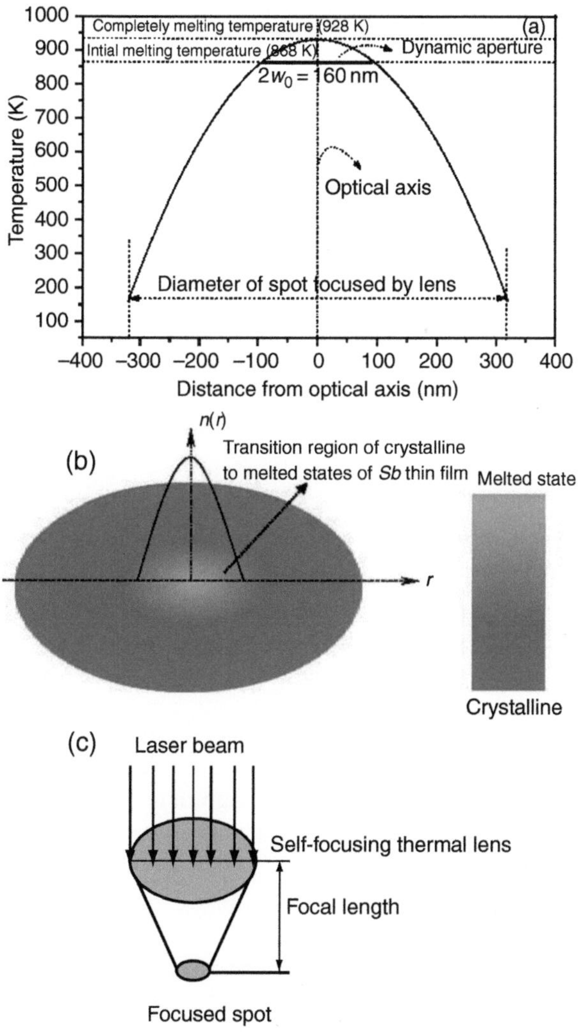

Fig. 10 Self-focusing thermal lens model.
(a) Dynamic aperture with a radius of 80 nm;
(b) morphology of dynamic aperture and refractive index profile;
(c) formation of the self-focusing thermal lens

One can find from Eq. (50) that the diameter d of the focused spot by the thermal lens is about 100 nm. The axial change of the radius of the thermal lens spot can be obtained by

$$w(z)^2 = w_1^2(1 + \frac{z^2}{z_c^2}) \tag{51}$$

where $w_1 = d/2$, z is the distance from the focus of thermal lens, and $z_c (= \pi w_1{}^2/\lambda^2)$ is the confocal distance of the thermal lens. Equation (51) is plotted in Fig. 11; the

Fig. 11 Change of spot size through self-focusing thermal lens along z-axis

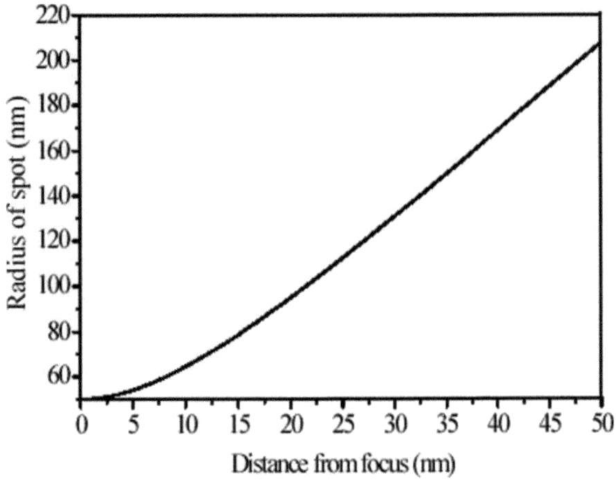

spot size is sensitive to the distance along z-axis. The intensity profile of the spot through the thermal lens can be calculated by

$$I(r) = \frac{0.48P_t}{\pi \cdot w_1^2} \exp\left(\frac{-2r^2}{w(z)^2}\right) \tag{52}$$

where $P_t(=P\times 15.6\%)$ is the transmitted laser power through the thermal lens because the transmittance of the dynamic aperture (thermal lens) is about 15.6%. The intensity change of the spot along z-axis is given in Fig. 12; it can be seen from Fig. 12 that the self-focusing thermal lens can not only reduce the radius of the spot, but also increase the intensity.

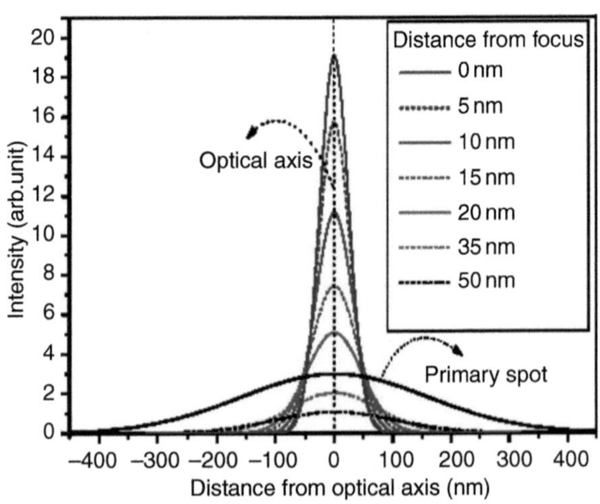

Fig. 12 Intensity profile of spot through self-focusing thermal lens along z-axis

3.2.4 Near-Field Optical Simulation of Self-Focusing Thermal Lens [26]

In order to get a better understanding of the super-resolution effect of self-focusing thermal lens, near-field optical distribution was calculated and simulated with the aid of the finite-difference-time-domain (FDTD) method. In the simulated model, we assume the extinction coefficient to be infinite outside the thermal lens and to be zero within the thermal lens. The refractive index n is set to be 3.12 outside the thermal lens and 4.52 at the center of the thermal lens. The refractive index within the thermal lens is assumed to obey the Gaussian profile (as shown in Fig. 13). With regard to the simulation of the near-field optical distribution, the center of the thermal lens is defined as the origin of the Cartesian coordinate, and the whole simulation space is divided into 1.4×10^6 cells with a volume of $\Delta x\times\Delta y\times\Delta z$, where Δx, Δy, and Δz are 4 nm. The incident light from the input plane (at z=–32 nm) is a homogeneous plane wave with a wavelength of 632.8 nm and its polarization direction is along the x-axis. The out plane is at $z = 0$ nm.

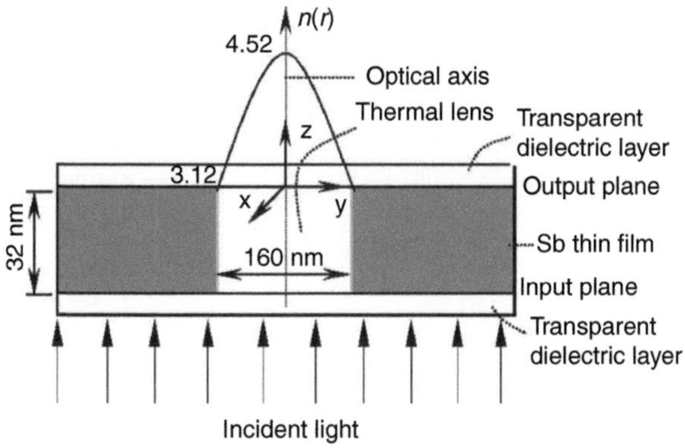

Fig. 13 The physical model for simulation

Figure 14 shows the near-field optical intensity distribution along the zy cross-section direction. One can find that a focused optical spot with a size of 80–120 nm is formed within the Sb thin film thermal lens. The initial incident intensity distribution is homogeneous. However, the optical energy is gradually concentrated to the center of the thermal lens, and the homogeneous intensity profile is converted to an approximate cosine distribution along the z direction, which causes the spot size to be smaller than the wavelength of incident light. The intensity increases with the z value, reaches a maximum at z=–8 nm, and then decreases with a further increase of z value. The results above indicate that the strongest spot in the zy cross section is formed at z=–8 nm. The full width at half maximum (FWHM) of the spot can be determined by normalizing the data from Fig. 14(b). The normalized results are shown in Fig. 15. It can be seen that the FWHM always increases from 80 nm to 120 nm with the z value varying from –32 nm to 40 nm. It should be noted that the FWHM of approximately 100 nm is actually unchanged within several nanometers from the output plane of the thermal lens.

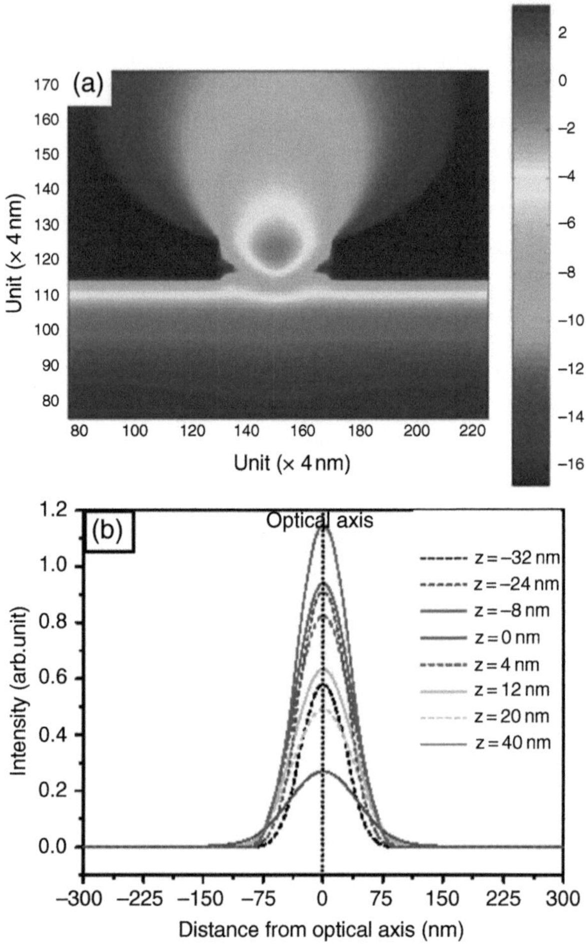

Fig. 14 Near-field optical distribution at *zy* cross section. (a) Optical intensity profile represented by color gradient and (b) optical intensity distribution curves for different *z*-value

The intensity distribution in the *xz* cross section is shown in Fig. 16. Seeing Fig. 16(a), an enhanced spot occurs near the output plane of thermal lens. The spot size is comparable with that of the thermal lens, and independent of the *z* value. The curves of intensity change along *z*-axis shown in Fig. 16 are generally consistent with those in Fig. 14. However, the difference between Fig. 16 and Fig. 14 is that a border enhancement phenomenon characterized by the relatively enhanced amplitude with *z* value occurs. It reaches a maximum at z=–8 nm, and then quickly decreases, and completely disappears when *z* value exceeds 12 nm.

Figure 17 shows the shape and the intensity of the spot in *xy* cross section, where the intensity is highlighted using the color gradient. Figure 17 (a) and (b) provides

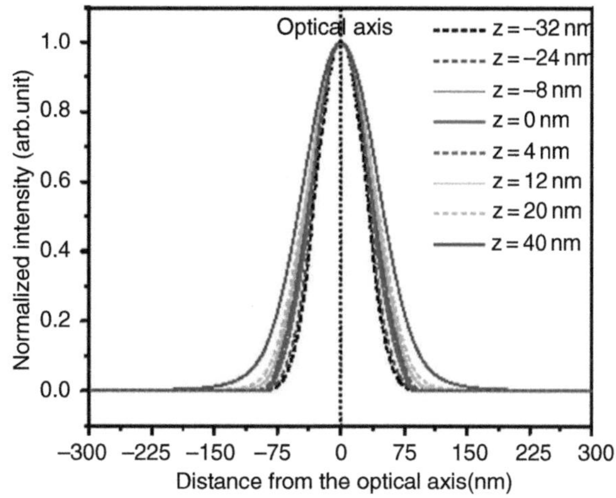

Fig. 15 Normalized intensity distribution curves at zy cross section for different z value

the spot shape at input plane ($z = -32$ nm) and output plane ($z = 0$ nm), respectively. Comparing Fig. 17 (a) with Fig. 17 (b), it is found that a focused elliptical spot is formed at the center of the output plane of the thermal lens, and the border enhancement is obvious at the output plane. With the increase of distance from the output plane, the spot gradually changes to a round shape, and the corresponding intensity also decreases. The border enhancement amplitude quickly decreases with the increase of the distance from the output plane. From Fig. 17 (d)–(f), the border enhancement completely disappears, and the focused and enhanced spot occurs at $z > 12$ nm.

3.2.5 Super-Resolution Optical Information Storage with Self-Focusing Thermal Lens [26]

In order to realize the super-resolution optical information storage with self-focusing thermal lens, we prepared a sample with a structure of "SiN(30 nm)/$Ge_2Sb_2Te_5$(15 nm)/SiN(15 nm)/Sb(30 nm)/SiN (100 nm)" by magnetron sputtering method. In this structure, the Sb thin film acts as a nonlinear self-focusing thermal lens. The space between the Sb thin film and the $Ge_2Sb_2Te_5$ recording layer is controlled by the thickness of the middle SiN thin film; the thickness is about 15 nm in the sample. This can guarantee that the super-resolution optical spot obtained by Sb nonlinear self-focusing thermal lens can be effectively coupled with the $Ge_2Sb_2Te_5$ recording thin film in the near-field range. The static recording experiements are conducted in a setup with a laser wavelength of 650 nm and a pickup with a NA at 0.85. The focused spot diameter of the setup can be calculated by $D = 1.22\lambda/NA$ and is approximately 933 nm. In our experiment, the recording laser power and pulse width are 8 mW and 100 ns, respectively. The recording marks are shown in Fig. 18. Figure18 (a) presents the recording marks obtained in a single $Ge_2Sb_2Te_5$ thin film. It can be seen that the recording mark size is around 1 μm. On the contrary, the

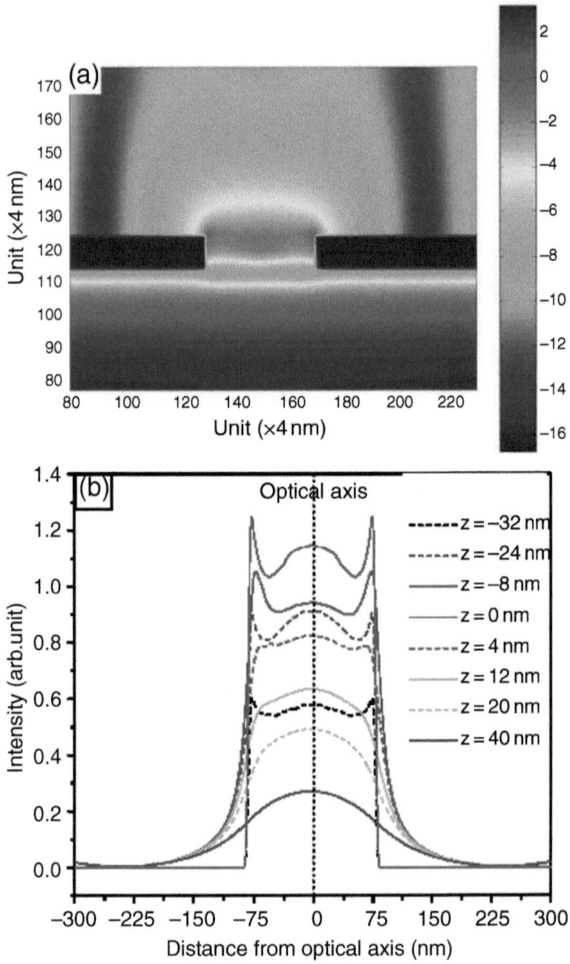

Fig. 16 Near-field optical distribution at *zx* cross section. (a) optical intensity profile represented by color gradient and (b) optical intensity distribution curves for different *z* values

recording mark size using the nonlinear self-focusing thermal lens under the same recording conditions is only 200–250 nm (as shown in Fig. 18 (b)), which is about one-fourth the size obtained in the single $Ge_2Sb_2Te_5$ thin film. The results above demonstrate that, in the course of optical recording, the dynamic self-focusing thermal lens is produced within the Sb thin film. The recording laser passes through the thermal lens, and is further focused and coupled with the $Ge_2Sb_2Te_5$ thin film. This indicates that the self-focusing thermal lens can enhance the optical spot intensity and reduce the spot size; therefore, it is very useful for ultrahigh-density optical information storage.

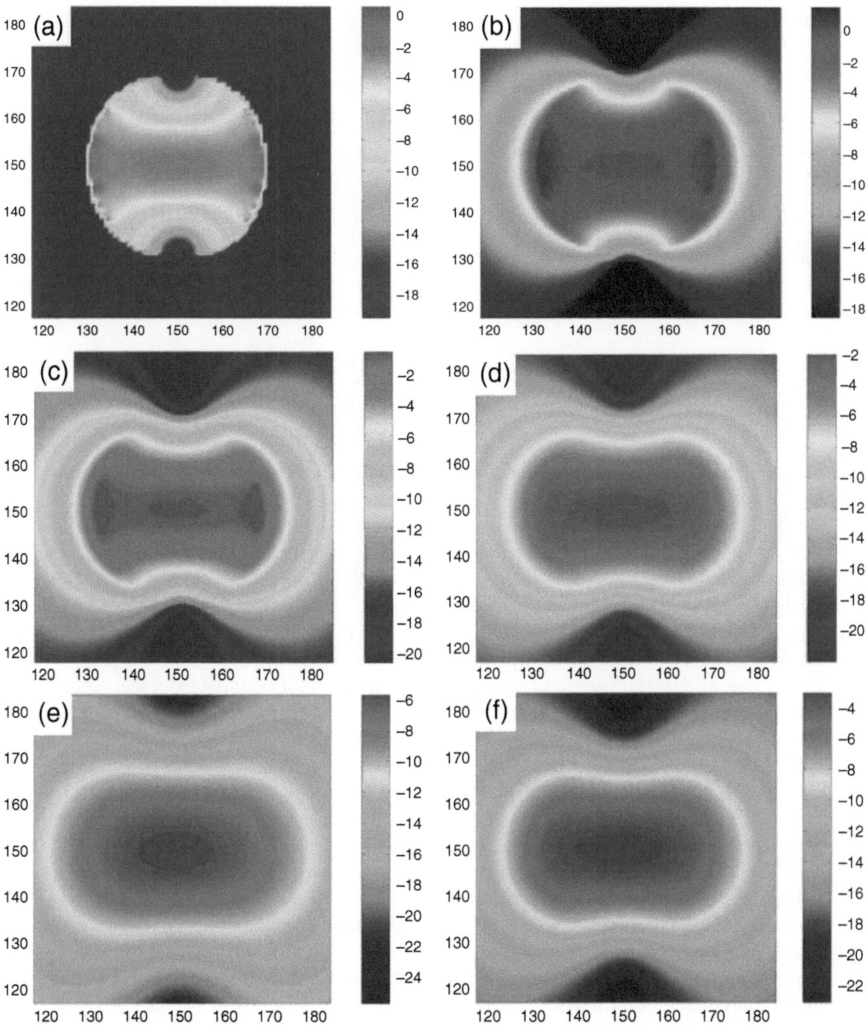

Fig. 17 Optical spot shape and intensity distribution. (a) z=–32 nm; (b) z=0 nm; (c) z=4 nm; (d) z=12 nm; (e) z=20 nm; and (f) z=40 nm

Fig. 18 Static optical recording marks. (a) In the single $Ge_2Sb_2Te_5$ thin film; (b) in the SiN/Sb/SiN/$Ge_2Sb_2Te_5$ sample

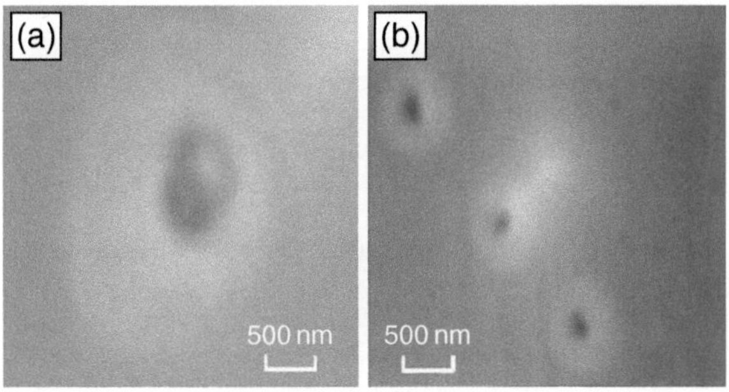

4 Conclusion

The basic principles of super-resolution optical effects of nanoscale nonlinear thin film structure are described, and their applications in ultrahigh-density information storage are presented theoretically and experimentally. The experimental results indicate that the super-resolution optical information storage with nanoscale nonlinear thin film structure is a very potential and practical technqiue.

Acknowledgments The work is partially supported by Shanghai Key Basic Research Project (No. 06DJ14007), National Natural Science Foundation of China (Nos. 50772120, 60507009, and 60490290), and Chinese Academy of Sciences (No. KJCXZ. YW. NANO. 06).

References

1. Wei J, Gan F (2005) Dynamic readout of subdiffraction-limited pit arrays with a silver superlens. Appl. Phys. Lett 87: 211101.
2. Wei J, Gan F (2002) Novel approach to super-resolution pit readout. Opt. Eng. 41: 1073–1074.
3. Wei J et al. (2002) Study on read-only optical disk with *Sb* mask super-resolution. Chin. Sci. Bull. 47: 1064–1066.
4. Wang H, Gan F (2001) A novel approach to super-resolution. Opt. Eng. 40: 851–855.
5. Wei J et al. (2002) Near-field optical recording and its recent progress. Progress in Physics 22: 188–197 (in Chinese) (also reading the references wherein).
6. Betzig E et al. (1992) Near-field magneto-optics and high-density data storage. Appl. Phys. Lett. 61: 142–144.
7. Terris BD et al. (1996) Near-field optical data storage. Appl. Phys. Lett. 68: 141–143.
8. Wei J, Xiao M (2006) Optical transmission larger than one (T>1) through ZnS-SiO_2/$AgOx$/ZnS-SiO_2 sandwiched thin films. Appl. Phys. Lett. 89: 101917.
9. Fukaya T et al. (1999) Optical switching property of a light-induced pinhole in antimony thin film. Appl. Phys. Lett. 75: 3114–3116.
10. Choi Y et al. (2001) Direct observation of self-focusing near the diffraction limit in polycrystalline silicon film. Appl. Phys. Lett. 78: 856–858.
11. Liu DR et al. (2002) Giant nonlinear optical properties of bismuth thin films grown by pulsed laser deposition. Opt. Lett. 27: 1549–1551.
12. Wei J et al. (2006) Super-resolution with a nonlinear thin film: Beam reshaping via internal multi-interference. Appl. Phys. Lett. 89: 223126.
13. Wei J et al. (2002) Readout of read-only super-resolution optical disc with *Si* mask. Chin. Phys. 11: 1073–1075.
14. Wei J, Gan F (2003) Theoretical explanation of different crystallization processes between as-deposited and melted-quenched amorphous $Ge_2Sb_2Te_5$ thin films. Thin Solid Films 441: 292–297.
15. Wei J, Gan F (2003) Time-resolved thermal lens effect of *Sb* thin films induced by structural transformation near melting temperature. Opt. Commun. 219: 261–269 (also reading the references wherein).
16. Baesso ML et al. (1994) Mode-mismatched thermal lens determination of temperature coefficient of optical length in Soda lime glass at different wavelengths. J. Appl. Phys. 75: 3732–3737.
17. Sheldon SJ et al. (1982) Laser-induced thermal lens effect: a new theoretical model. Appl. Opt. 21: 1663–1669.
18. Beaton CF, Hewitt GF (1998) Physical property data for the design engineer. Hemisphere Publishing Corporation, pp 365.
19. Gray DE (1963) American Institute of physics handbook. McGraw HillBook Company Inc. pp. 4–207.

20. Kim SY et al. (2006) Ehanced readout signal of elliptic-bubble super-resolution near-field structure by temperature-dependent complex refractive index of phase-change medium. Jap. J. Appl. Phys. 45: 1390–1393.
21. Prod'home L (1960) A new approach to the thermal change in the refractive index of glasses. Phys. Chem. Glasses 1: 119–122.
22. Gilvarry J J (1956) The Lindemann and Grüneisen Laws. Phys. Rev. 102: 308–316.
23. Wei J, Gan F (2003) Thermal lens model of *Sb* thin films in super-resolution near-field structure. Appl. Phys. Lett. 82: 2607–2609.
24. Wei J et al. (2003) Working mechanism of *Sb* thin films in super-resolution near-field structure. Opt. Commun. 224: 269–273.
25. Serna R et al. (1993) Melting kinetics of *S*b under nanosecond UV laser irradiation. J. Appl. Phys. 73: 3099–3101.
26. Wei J et al. (2005) Optical near-field simulation of *Sb* thin film thermal lens and its application in optical recording. J. Appl. Phys. 97: 073102.

Spin-Transfer and Current-Induced Spin Dynamics in Spin Valves: Diffusive Transport Regime

Martin Gmitra and Józef Barnaś

Abstract Current-induced dynamics observed in point contacts or in spin valve structures is one of the most challenging concepts that can bring these nanomagnetic systems toward direct applications. The most promising applications are the nonvolatile memories and nano-oscillators which can be used in telecommunications. Despite the clear application proposals, the full understanding of the microscopic mechanism of current-induced spin dynamics is still an open question. The transfer of angular momentum from conduction electrons to local magnetization is the key microscopic effect. In this chapter, we describe the spin transfer in layered metallic systems (spin valves) and current-induced dynamics via the spin transfer. We present the relation between the normal and inverse current-induced switching, and normal and inverse current-perpendicular-to-plane giant magnetoresistance CPP-GMR. It turns out that these effects are related to the ferromagnet–normal-metal interface asymmetries that in some cases may lead to a nonstandard angular dependence of the spin transfer torque. We also present analysis of the current-induced dynamics within the macrospin model for standard spin valves as well as for asymmetric valves that have nonstandard angular dependence of spin-transfer torque. The asymmetric spin valves are promising candidates for current-induced microwave nano-oscillators in zero magnetic field.

Contents

M. Gmitra (✉)
Institute of Physics, P.J. Šafárik University in Košice, Slovak Republic
e-mail: martin.gmitra@upjs.sk

Z.M. Wang (ed.), *Toward Functional Nanomaterials*, Lecture Notes in Nanoscale Science and Technology 5, DOI 10.1007/978-0-387-77717-7_9,

1 Introduction

Competition between spin and charge degrees of freedom in magnetic layered structures including 3*d* ferromagnetic films leads to two related effects in electronic transport. First, magnetic structure has a significant impact on electronic transport, leading to the so-called giant magnetoresistance (GMR) effect [1, 2, 3]. Second, charge current flowing in such structures is associated with a spin current and the latter may change the magnetic state of the system [4, 5, 6]. Physical origins of both the effects are very similar – both follow from the presence of two spin-dependent channels for electronic transport in 3d ferromagnetic metals. The concept of two-channel conduction in transition metals was introduced long time ago by Mott [7] and developed later by Campbell and Fert [8] and also by others [3, 9, 10]. In this chapter, we will deal only with the second phenomenon, i.e., with the influence of charge and spin currents on the magnetic state of a system.

The concept of spin transfer and associated spin torque has been introduced in pioneer works by Slonczewski [4] and Berger [5]. Since then, current-induced magnetic switching (CIMS) of thin ferromagnetic films due to spin transfer from conduction electrons to localized moments has been studied in a number of experiments, most of them performed on current-perpendicular-to-plane (CPP) spin valves [6, 11–13]. Starting from parallel configuration of the magnetizations in a system consisting of two magnetic films of different thicknesses separated by a nonmagnetic spaces, charge current exceeding a certain critical value was shown to reverse the magnetic moment of the thinner magnetic layer to set up an antiparallel configuration [6, 14, 15]. In turn, a current in the opposite direction can switch the structure back to the parallel configuration. Moreover, with an externally applied field, the current can generate a steady precession of the magnetization, detected by oscillations of the current in the microwave frequency range [16]. Thus, spin current can cause not only switching to another static magnetic configuration, but also transition to steady precessional modes, where the energy is pumped from conduction electrons to localized magnetic moments [16–22]. In typical Co/Cu/Co spin valves, steady precessions have been observed for external magnetic fields applied in the layer plane and larger than the anisotropy field, and for currents exceeding a certain critical value [16, 22]. For lower fields, the current drives the switching either to antiparallel or to parallel states, depending on the initial state of the system.

Recently, asymmetric spin valves, with the two magnetic films having different bulk and interface spin asymmetries and also different spin diffusion lengths, have been studied theoretically in the limit of diffusive transport [23, 24]. It has been shown that the torque due to spin transfer can vanish for a certain noncollinear magnetic configuration, which results from an inverse spin accumulation built up in the nonmagnetic spacer layer. As a result, an appropriate spin current can destabilize both magnetic configurations for one orientation of current and stabilize both configurations for the opposite current. The former case is of particular interest for applications, as the spin current can excite precessional modes in the absence of external magnetic field [24].

In the concept introduced by Slonczewski [4], as well as in most theoretical models [25, 26, 27, 28, 29, 30], the current-induced torque exerted on the magnetization of a magnetic layer is related to the absorption of the transverse component of spin current by the layer (transverse means perpendicular to the magnetization axis of the layer). From the CPP-GMR experiments, one knows that spin polarization of the current is due to spin-dependent reflections at interfaces and spin-dependent scattering within the magnetic layers. In turn, recent experiments [12, 31] have shown that the switching currents can be modified and even reversed by doping the magnetic layers with impurities of selected spin-dependent scattering cross sections. Thus, both CPP-GMR and CIMS depend on spin asymmetries of the two transport channels in the system, and in this way also on spin accumulation.

Since most of the available experimental data on CPP-GMR have been successfully accounted for within the model of Valet and Fert [32], the natural attempt toward a unified description of CPP-GMR and CIMS relies on an extension of this model to include also the spin torque. This has been done quite recently [23], and our main objective in this chapter is to present such a unified description based on the diffusive transport model. Most of the parameters in this description, like the interface and bulk spin asymmetry coefficients, interface resistances, layer resistivities, and spin diffusion lengths, can be evaluated directly from the analysis of CPP-GMR experimental data [33]. Two additional parameters, namely the real and imaginary parts of the mixing conductance [28, 29, 30, 34], are also needed. These can be derived from quantum-mechanical calculations [27, 35] of the spin current transmission across a particular interface.

2 Spin Current and Spin Accumulation in Layered Systems

In the spirit of the two-channel model [7, 8], we assume that magnetic properties of ferromagnetic films are determined by narrow and spin split bands of quasi-localized $3d$ electrons, whereas transport properties are determined by free-like s-electrons. For simplicity, we assume the same concentration of the conduction s-electrons in all the layers and neglect any of their spin polarization at equilibrium. The presence of narrow d-bands has a significant impact on the transport properties as they provide an additional and strongly spin-dependent channel for scattering of conduction electrons. In collinear configuration, one may assume that the spin channels are independent. However, when, magnetic moments of the layers are not collinear, the distribution function $\check{f}$ inside the films is generally a nondiagonal 2×2 matrix in the spin space. Spatial variation of the distribution function can be described by the diffusion equation. We assume the distribution function to be uniform in the plane of the films, and varies only along the axis x normal to the films. Let us first consider the ferromagnetic layers.

2.1 Magnetic Films

The diffusion equation for arbitrary spin quantization axis takes the form [34]

$$\check{D}\frac{\partial^2 \check{f}}{\partial x^2} = \frac{1}{\tau_{\rm sf}}\left[\check{f} - \check{1}\frac{{\rm Tr}\{\check{f}\}}{2}\right], \tag{1}$$

where $\check{D}$ is the 2×2 diffusion matrix in the spin space, $\check{1}$ is the 2×2 unit matrix, and $\tau_{\rm sf}$ is the spin-flip relaxation time. As already mentioned in the introduction, the internal exchange field inside ferromagnetic metals is strong enough to suppress the component of the distribution function corresponding to the spin orientation normal to the local magnetization. Thus, the distribution function is diagonal when the spin quantization axis is parallel to the local spin polarization of the ferromagnetic system. As shown in [23], Eq. (1) then leads to the following two equations for the electrochemical potentials $\bar{\mu}_\uparrow$ and $\bar{\mu}_\downarrow$ of spin-majority and spin-minority electrons, respectively:

$$\frac{\partial^2(\bar{\mu}_\uparrow - \bar{\mu}_\downarrow)}{\partial x^2} = \frac{1}{l_{\rm sf}^2}(\bar{\mu}_\uparrow - \bar{\mu}_\downarrow), \tag{2}$$

$$\frac{\partial^2(\bar{\mu}_\uparrow + \bar{\mu}_\downarrow)}{\partial x^2} = \eta\frac{\partial^2(\bar{\mu}_\uparrow - \bar{\mu}_\downarrow)}{\partial x^2}, \tag{3}$$

where $l_{\rm sf}$ is the spin diffusion length,

$$\frac{1}{l_{\rm sf}^2} = \frac{1}{2}\left(\frac{1}{l_\uparrow^2} + \frac{1}{l_\downarrow^2}\right), \tag{4}$$

with $l_\uparrow^2 = D_\uparrow \tau_{\rm sf}$ and $l_\downarrow^2 = D_\downarrow \tau_{\rm sf}$, η is defined as

$$\eta = -\frac{D_\uparrow - D_\downarrow}{D_\uparrow + D_\downarrow}. \tag{5}$$

The above equations are equivalent to those derived by Valet and Fert [32] from the Boltzmann kinetic equations.

The electrochemical potentials can be written in the form

$$\check{\bar{\mu}} = \bar{\mu}_0 \check{1} + g \check{\sigma}_z, \tag{6}$$

with

$$\bar{\mu}_0 = \frac{1}{2}(\bar{\mu}_\uparrow + \bar{\mu}_\downarrow) \tag{7}$$

and

$$g = \frac{1}{2}(\bar{\mu}_\uparrow - \bar{\mu}_\downarrow). \tag{8}$$

The latter quantity, g, describes spin splitting of the electrochemical potential, i.e., the spin accumulation. Equations (2) and (3) lead to the following explicit expressions for $\bar{\mu}_0$ and g:

$$\bar{\mu}_0 = \eta[A\exp(x/l_{\mathrm{sf}}) + B\exp(-x/l_{\mathrm{sf}})] + Cx + G, \tag{9}$$

and

$$g = A\exp(x/l_{\mathrm{sf}}) + B\exp(-x/l_{\mathrm{sf}}), \tag{10}$$

where A, B, C, and G are constants to be determined later from the appropriate boundary conditions.

For an arbitrary quantization axis, the particle and spin currents are determined by the 2×2 matrix $\check{j}$ in the spin space

$$\check{j} = -\check{D}\frac{\partial \check{f}}{\partial x} = -\rho(E_F)\check{D}\frac{\partial \check{\bar{\mu}}}{\partial x}, \tag{11}$$

where $\rho(E_F)$ is the density of states at the Fermi level per spin (per unit volume and unit energy). When the quantization axis is parallel to the local spin polarization, it is convenient to write the matrix $\check{j}$ in the form

$$\check{j} = \frac{1}{2}\left(j_0\check{1} + j_z\check{\sigma}_z\right), \tag{12}$$

where $j_0 = (j_\uparrow + j_\downarrow)$ is the total particle current density and $j_z = (j_\uparrow - j_\downarrow)$ is the z-component of the spin current. Thus, one finds

$$\frac{1}{\rho(E_F)}j_0 = -C(D_\uparrow + D_\downarrow), \tag{13}$$

and

$$\frac{1}{\rho(E_F)}j_z = -C(D_\uparrow - D_\downarrow) - \frac{2\widetilde{D}}{l_{\mathrm{sf}}}[A\exp(x/l_{\mathrm{sf}}) - B\exp(-x/l_{\mathrm{sf}})], \tag{14}$$

where

$$\widetilde{D} = 2\frac{D_\uparrow D_\downarrow}{D_\uparrow + D_\downarrow}. \tag{15}$$

The particle current density j_0 is related to the charge current I *via* $I = ej_0$, where e is the electron charge ($e < 0$). Thus, positive charge current corresponds to negative particle current.

2.2 Nonmagnetic Films

Solution of the diffusion equation for the distribution functions inside nonmagnetic films leads to the following expression:

$$\check{\mu} = \bar{\mu}_0 \, \check{1} + \mathbf{g} \cdot \check{\sigma}, \tag{16}$$

where $\check{\boldsymbol{\sigma}} = (\check{\sigma}_x, \check{\sigma}_y, \check{\sigma}_z)$ are the Pauli matrices. All the three components of $\mathbf{g}$ are nonzero in a general case. The general solutions for $\bar{\mu}_0$ and $\mathbf{g}$ have the forms

$$\bar{\mu}_0 = Cx + G, \tag{17}$$

$$\mathbf{g} = \mathbf{A} \exp(x/l_{\mathrm{sf}}) + \mathbf{B} \exp(-x/l_{\mathrm{sf}}). \tag{18}$$

The spin and charge currents are then given by

$$\check{j} = \frac{1}{2}(j_0 \check{1} + \mathbf{j} \cdot \check{\boldsymbol{\sigma}}), \tag{19}$$

with

$$\frac{1}{\rho(E_F)} j_0 = -2CD \tag{20}$$

and

$$\frac{1}{\rho(E_F)} \mathbf{j} = -\frac{2D}{l_{\mathrm{sf}}} [\mathbf{A} \exp(x/l_{\mathrm{sf}}) - \mathbf{B} \exp(-x/l_{\mathrm{sf}})], \tag{21}$$

where now $D_\uparrow = D_\downarrow \equiv D$. All the constants, $\mathbf{A}$, $\mathbf{B}$, C, and G, may be different in different layers.

3 Boundary Conditions and Torque

To determine the unknown constants that enter the general expressions for electric current and distribution functions inside all the magnetic and nonmagnetic parts of any layered structure, we need to specify the boundary conditions, which have to be fulfilled by the distribution function and currents at each interface. Such boundary conditions were derived by Brataas et al. [34] within the phenomenological description. According to these boundary conditions, the charge and spin currents across the normal-metal–ferromagnet interface, calculated on the normal-metal side in the coordinate system with the axis z along the local quantization axis in the ferromagnet, can be written as [34]:

$$e^2 j_0 = (G_\uparrow + G_\downarrow)(\bar{\mu}_0^F - \bar{\mu}_0^N) + (G_\uparrow - G_\downarrow)(g_z^F - g_z^N), \tag{22}$$

$$e^2 j_z = (G_\uparrow - G_\downarrow)(\bar{\mu}_0^F - \bar{\mu}_0^N) + (G_\uparrow + G_\downarrow)(g_z^F - g_z^N), \tag{23}$$

$$e^2 j_x = -2\mathrm{Re}\{G_{\uparrow\downarrow}\} g_x^N + 2\mathrm{Im}\{G_{\uparrow\downarrow}\} g_y^N, \tag{24}$$

$$e^2 j_y = -2\mathrm{Re}\{G_{\uparrow\downarrow}\} g_y^N - 2\mathrm{Im}\{G_{\uparrow\downarrow}\} g_x^N, \tag{25}$$

where $\mathbf{g}^N$ ($\mathbf{g}^F$) is the spin accumulation on the normal-metal (ferromagnetic) side of the interface, $G_\uparrow$ and $G_\downarrow$ are the interfacial conductances in the spin-majority and spin-minority channels, respectively, and $G_{\uparrow\downarrow}$ is the spin-mixing conductance of the interface, which comes into play only in noncollinear configurations. The mixing conductance $G_{\uparrow\downarrow}$ is generally a complex parameter with the imaginary part being usually one to two orders of magnitude smaller than the real part [36]. We emphasize that the boundary conditions are valid in the absence of spin-flip scattering at the interface.

The above mentioned boundary conditions can be specified as follows: (i) particle current is continuous across all interfaces (in all layers and across all interfaces it is constant and equal to j_0), (ii) the spin current component parallel to the magnetization of a ferromagnetic layer is continuous across the interface between magnetic and nonmagnetic layers, and (iii) the normal component (perpendicular to the magnetization of a ferromagnetic film) of the spin current vanishes in the magnetic layer and there is a jump of this component at the interface between magnetic and nonmagnetic films, described by Eqs. (24) and (25). These boundary conditions have to be fulfilled at all interfaces. The number of the corresponding equations is then equal to the number of unknown constants, which allows one to determine the spin accumulation and the charge and spin currents in the whole structure.

Distribution functions and spin currents in a magnetic film are written in a coordinate system whose axis z is along the local spin polarization. In turn, the distribution functions and spin currents inside a nonmagnetic film are written in the system whose axis z coincides with the local quantization axis in one of the two adjacent ferromagnetic films. Since magnetic moments of the two ferromagnetic films are noncollinear, it is necessary to transform the distribution function and spin current from one system to the other. Thus, if the solution for electrochemical potentials in a given coordinate system has the form (16), then the solution in the coordinate system rotated by polar θ and azimuthal φ angles, see Fig. 1, is still given by Eq. (16), but with $\mathbf{g}$ replaced with $\mathbf{g}'$ given by

$$g'_x = g_x \sin\varphi - g_y \cos\varphi, \tag{26}$$

$$g'_y = g_x \cos\varphi \cos\theta + g_y \sin\varphi \cos\theta - g_z \sin\theta, \tag{27}$$

$$g'_z = g_x \cos\varphi \sin\theta + g_y \sin\varphi \sin\theta + g_z \cos\theta. \tag{28}$$

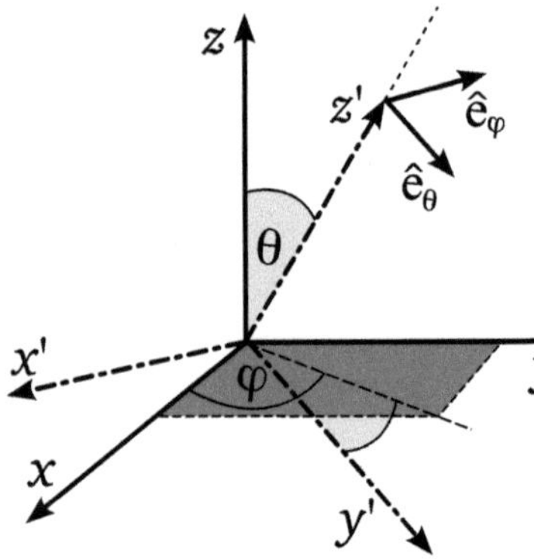

Fig. 1 Two coordinate systems associated with the two magnetic films adjacent to a nonmagnetic layer. Transformation from the $O(xyz)$ to $O(x'y'z')$ systems is given by $\mathbf{g}' = \check{R}_{x'}(-\theta)\check{R}_z(\varphi - \pi/2)\mathbf{g}$, where $\check{R}_q(\alpha)$ is the matrix of rotation by the angle α around the axis q in the counterclockwise direction when looking toward origin of the coordinate system. Unit vectors along the positive torques τ_θ and τ_φ acting on the spin aligned along the z' axis are also indicated

Similar relations also hold when transforming spin current $\mathbf{j}$ from one coordinate system to the other one.

Since the perpendicular component of the spin current is absorbed by the magnetic layers, the corresponding torque $\boldsymbol{\tau}$ per unit square, exerted on a ferromagnetic film, can be calculated as

$$\boldsymbol{\tau} = \frac{\hbar}{2}\left(\mathbf{j}_{\perp L} - \mathbf{j}_{\perp R}\right), \tag{29}$$

where $\mathbf{j}_{\perp L}$ and $\mathbf{j}_{\perp R}$ are the normal (to the magnetization) components of the spin current at the left and right interfaces of the magnetic film, respectively, calculated on the normal-metal side of these interfaces. In the simple case (as in Fig. 2), where

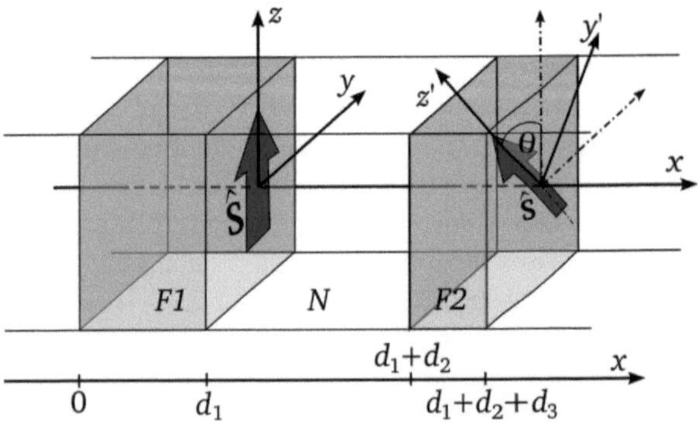

Fig. 2 Schematic plot of the three-layer system consisting of thick (F1) and thin (F2) ferromagnetic films of thicknesses d_1 and d_3, respectively, separated by a nonmagnetic (N) spacer layer of thickness d_2. The system is sandwiched between nonmagnetic leads which are assumed to be semi-infinite. The arrows indicate orientation of the net spin of the magnetic films, with θ being the angle between the spins $\hat{\mathbf{s}}$ and $\hat{\mathbf{S}}$

there is no additional magnetic layer outside the F1/N/F2 trilayer and no transverse spin current at the outer edges of the trilayer, $\boldsymbol{\tau}$ is simply given by $\tau = -\hbar \mathbf{j}_{\perp R}/2$ for F1 and $\boldsymbol{\tau} = \hbar \mathbf{j}_{\perp L}/2$ for F2, where $\mathbf{j}_{\perp R}$ and $\mathbf{j}_{\perp L}$ have to be calculated in the nonmagnetic spacer layer (N) at the right interface of F1 and left interface of F2 (left and right interfaces of N), respectively.

4 Torque in a Spin valve Structure

The spin valve structure F1/N/F2 under consideration consists of two magnetic layers – left (thicker) and right (thinner) – separated by a nonmagnetic layer. Magnetic moment of the thicker layer is assumed to be fixed, whereas that of the thin layer (also called the free or sensing layer) is assumed to freely rotate upon applied magnetic field or electric current. The three-layer system is sandwiched between semi-infinite nonmagnetic leads. Thicknesses of the magnetic layers are d_1 and d_3, respectively, whereas of the spacer layer is d_2. The structure is shown schematically in Fig. 2, where for simplicity it has been assumed that both ferromagnetic layers are magnetized in their planes ($\varphi = \pi/2$). The magnetization of the thin layer is rotated by an angle θ around the axis x (normal to the films) as shown in Fig. 2. The local quantization axis in both ferromagnetic layers is along the local net spin, while as the global quantization axis we choose the local one in the thick ferromagnetic layer F1. According to our definition, charge current I is positive when it flows along the axis x from left to right, i.e., from the thick toward the thin magnetic layer (electrons flow then from right to left).

The in-plane component $\boldsymbol{\tau}_\theta$ of the torque acting on the thin magnetic layer can be written as (in-plane means in the plane determined by the vectors $\hat{\mathbf{s}}$ and $\hat{\mathbf{S}}$)

$$\boldsymbol{\tau}_\theta = a\, I\, \hat{\mathbf{s}} \times (\hat{\mathbf{s}} \times \hat{\mathbf{S}})\,, \tag{30}$$

where $\hat{\mathbf{s}}$ and $\hat{\mathbf{S}}$ are the unit vectors along the spin polarization of the thin and thick magnetic layers, respectively, and a is a parameter. The torque τ_θ is defined in such a way that the positive torque tends to increase the angle θ ($\theta \in \langle 0, \pi \rangle$), whereas the negative torque tends to decrease the angle θ between spin moments of the layers (see Fig. 1). Equation (30) can be also written as

$$\tau_\theta = a\, I\, \sin\theta\,. \tag{31}$$

The torque τ_θ can be calculated from Eq. (29) as

$$\tau_\theta = \frac{\hbar}{2}\, j'_y|_{x\to x_0} = \frac{\hbar}{2}\left(j_x \cos\varphi \cos\theta + j_y \sin\varphi \cos\theta - j_z \sin\theta\right)|_{x\to x_0}\,, \tag{32}$$

where $x_0 = d_1 + d_2$ is the position of the active interface in the spin valve, whereas j'_y and j_x, j_y, j_z are the components of spin current in the nonmagnetic thin layer written in the local system of the thin and thick magnetic layers, respectively, and calculated at the very interface between the nonmagnetic and thin magnetic layers. By comparison of Eqs. (31) and (32), one finds a general expression for the parameter a as a function of the spin currents. Moreover, by taking into account Eqs. (24) and (25), one can relate the torque directly to the spin accumulation in the nonmagnetic layer, taken at the interface with the thin magnetic layer,

$$\tau_\theta = \frac{\hbar}{e^2}\left[\mathrm{Re}\{G_{\uparrow\downarrow}\}\, g'_y + \mathrm{Im}\{G_{\uparrow\downarrow}\}\, g'_x\right]|_{x\to x_0}\,. \tag{33}$$

Taking into account the rotations for spin accumulation (Eqs. (26, 27, 28)) and Eq. (31) for the torque, one can write the general expression for the parameter a in the form

$$a = \frac{\hbar}{e^2}\Bigg[\mathrm{Re}\{G_{\uparrow\downarrow}\}\left[\cot\theta(\tilde{g}_x\cos\varphi + \tilde{g}_y\sin\varphi) - \tilde{g}_z\right] + \frac{\mathrm{Im}\{G_{\uparrow\downarrow}\}}{\sin\theta}(\tilde{g}_x\sin\varphi - \tilde{g}_y\cos\varphi)\Bigg]\Bigg|_{x\to x_0}, \tag{34}$$

where $\tilde{\mathbf{g}} = \mathbf{g}/I$.

The out-of-plane component $\boldsymbol{\tau}_\varphi$ of the torque may be generally written as

$$\boldsymbol{\tau}_\varphi = b\, I\, \hat{\mathbf{s}} \times \hat{\mathbf{S}}, \tag{35}$$

or in the scalar form, $\tau_\varphi = -b\, I\, \sin\theta$. The torque τ_φ is defined as positive (negative) when it tends to increase (decrease) the angle φ, and can be calculated from the formula

$$\tau_\varphi = \frac{\hbar}{2} j'_x|_{x\to x_0} = \frac{\hbar}{2}(j_x\sin\varphi - j_y\cos\varphi)|_{x\to x_0}\,. \tag{36}$$

From Eqs. (35) and (36), one can find directly the expression for the parameter b as a function of the spin currents. Using now Eq. (24) for the spin current j_x across the active interface and the rotation transformation for spin accumulation, one can write the general expression for the parameter b in the form

$$b = -\frac{\hbar}{e^2}\Bigg[\frac{\mathrm{Re}\{G_{\uparrow\downarrow}\}}{\sin\theta}(\tilde{g}_x\sin\varphi - \tilde{g}_y\cos\varphi) - \mathrm{Im}\{G_{\uparrow\downarrow}\}\left[\cot\theta(\tilde{g}_x\cos\varphi + \tilde{g}_y\sin\varphi) - \tilde{g}_z\right]\Bigg]\Bigg|_{x\to x_0}. \tag{37}$$

Equations (32) and (36) are the final formulae for the torque components expressed in terms of the spin currents. In turn, Eqs. (34) and (37) are the corresponding formulae for the parameters a and b, expressed in terms of the spin accumulation. For numerical calculations, one can use either the former equations or equivalently the latter ones.

5 CIMS and Its Relation to CPP-GMR

The CPP-GMR effect plays a significant role in CIMS experiments, mainly as a tool that allows us to determine changes in magnetic configuration of the system due to spin transfer. However, both the effects have similar physical origin and therefore should be correlated. Generally, the GMR is related to the spin dependence of electronic conduction in ferromagnetic metals and alloys, where the current at low temperatures is carried by two independent spin-majority and spin-minority channels. Furthermore, for the N/F interface one can introduce spin-dependent interface resistance. To describe the above-mentioned spin asymmetry, it is convenient to introduce the bulk and interfacial spin asymmetry factors for ferromagnetic layers according to the standard definitions [32],

$$\rho_{\uparrow(\downarrow)} = 2\rho^*(1 \mp \beta) \tag{38}$$

and

$$R_{\uparrow(\downarrow)} = 2R^*(1 \mp \gamma), \tag{39}$$

where $\rho_\uparrow$ and $\rho_\downarrow$ are the bulk resistivities for spin-majority and spin-minority electrons, respectively, and $R_\uparrow$ and $R_\downarrow$ are the interface resistances per unit square for spin-majority and spin-minority electrons, whereas β and γ are the bulk and interfacial spin asymmetry coefficients. The formula in (38) will also be used for nonmagnetic layers (with $\beta = 0$). The conductances $G_\uparrow$ and $G_\downarrow$, see Eq. (23) and (22), are then equal $G_\uparrow = 1/R_\uparrow$ and $G_\downarrow = 1/R_\downarrow$.

The key bulk parameters that enter the description, i.e., mean free paths and diffusion constants can be expressed in terms of a free electron model by the parameters defined in Eq. (38) and the relevant Fermi energy E_F (in the following numerical calculations we assume the same Fermi energy for both magnetic and nonmagnetic layers). The diffusion parameters $D_{\uparrow(\downarrow)}$ are then calculated from the formulae,

$$D_{\uparrow(\downarrow)} = \frac{1}{3} v_F \lambda_{\uparrow(\downarrow)}, \tag{40}$$

where $v_F = \sqrt{2E_F/m_e}$ is the Fermi velocity of electrons, and the mean free paths $\lambda_{\uparrow(\downarrow)}$ are

$$\lambda_{\uparrow(\downarrow)} = \frac{m_e v_F}{ne^2 \rho_{\uparrow(\downarrow)}}, \tag{41}$$

with m_e denoting the electron mass and $n = (1/6\pi^2)(2m_e E_F/\hbar^2)^{3/2}$ being the density of electrons per spin. Apart from this, $\rho(E_F)$ (see Eq. (11)) is given by $\rho(E_F) = (1/4\pi^2)(2m_e/\hbar^2)^{3/2} E_F^{1/2}$. For such a description (based on free electron-like model), one finds $\lambda_\downarrow/\lambda_\uparrow = (1-\beta)/(1+\beta)$, and the parameter η defined by Eq. (5) is determined by β *via* the simple relation

$$\eta = -\beta\,. \tag{42}$$

For nonmagnetic layers, we use the same definitions, but now $\lambda_{\uparrow(\downarrow)}$ and $D_{\uparrow(\downarrow)}$ are independent of the spin orientation (the corresponding β is equal to zero). Finally, the spin diffusion lengths will be assumed as independent parameters and will be taken from GMR experiments.

5.1 Spin-Transfer Torque in Co/Cu/Co Spin Valve

Consider now a concrete spin valve, Co(20)/Cu(10)/Co(8), where the numbers in brackets denote the layer thicknesses, which is sandwiched between semi-infinite Cu electrodes. The corresponding parameters used for calculations, such as bulk resistivities ρ^*, bulk spin asymmetry factors β, spin-flip lengths l_{sf}, interface resistances R^*, spin asymmetry factors γ, and mixing conductances $G_{\uparrow\downarrow}$, have been taken from various GMR measurements and are gathered in Tables 1 and 2. The mixing conductance $G_{\uparrow\downarrow}$ could be derived from the angular dependence of the CPP-GMR experiment. However, there is still no reliable experimental information on $G_{\uparrow\downarrow}$ from GMR [37, 38]. Therefore, we assume the value calculated in a free electron model corrected by certain factors taken from *ab-initio* calculations. Thus, in the following numerical calculations we assume $\mathrm{Re}\{G_{\uparrow\downarrow}\} = 0.542 \times 10^{-15}\,(\Omega\,\mathrm{m}^2)^{-1}$. As for the imaginary part, $\mathrm{Im}\{G_{\uparrow\downarrow}\}$, we determine it assuming the

Table 1 Bulk parameters used in calculations, obtained from CPP data (see the relevant references and/or references therein)

Magnetic layer				Nonmagnetic layer		
Material	$\rho^*(\mu\,\Omega\,\mathrm{cm})$	β	l_{sf} (nm)	Material	$\rho^*(\mu\,\Omega\,\mathrm{cm})$	l_{sf} (nm)
Co [40]	5.1	0.51	60	Cu	0.5	1000
Py [41]	16.0	0.77	5.5	Au	2.0	60
$Ni_{97}Cr_3$ [42]	41.5	−0.35	3.3	Cr	10.0	50
IrMn	150	0	1.0	Ru [43]	9.5	14

Table 2 Interfacial parameters used in calculations, deduced from the available literature. The resistance is given in $\mathrm{f}\,\Omega\,\mathrm{m}^2$ and mixing conductances in $(\mathrm{f}\,\Omega\,\mathrm{m}^2)^{-1}$

Interface	R^*	γ	$\mathrm{Re}\{G_{\uparrow\downarrow}\}$	$\mathrm{Im}\{G_{\uparrow\downarrow}\}$
Co/Cu	0.5	0.77	0.542	0.016
Py/Cu	0.5	0.7	0.39	0.012
Co/Au	0.5	0.7	0.39	0.012
Co/Ru	0.5	−0.2	0.26	0.008
Ni_{Cr}/Cu	0.27	0.15	0.26	0.008
Ni_{Cr}/Cr	0.5	−0.2	0.3	0.009

same ratio $\mathrm{Im}\{G_{\uparrow\downarrow}\}/\mathrm{Re}\{G_{\uparrow\downarrow}\}$ as that following from *ab initio* calculations. Thus, we assume $\mathrm{Im}\{G_{\uparrow\downarrow}\} = 0.016 \times 10^{-15}\,(\Omega\,\mathrm{m}^2)^{-1}$.

Numerical results on the in-plane and out-of-plane components of the torque as well as on the corresponding parameters a and b are shown in Fig. 3(a) and (b), respectively, where the torque is normalized to current. Within the linear model assumed here, the spin accumulation and spin currents are linear functions of the charge current, therefore, the curves in Fig. 3 are the same for arbitrary magnitude of the charge current I. Note that the sign of torque changes when I is reversed.

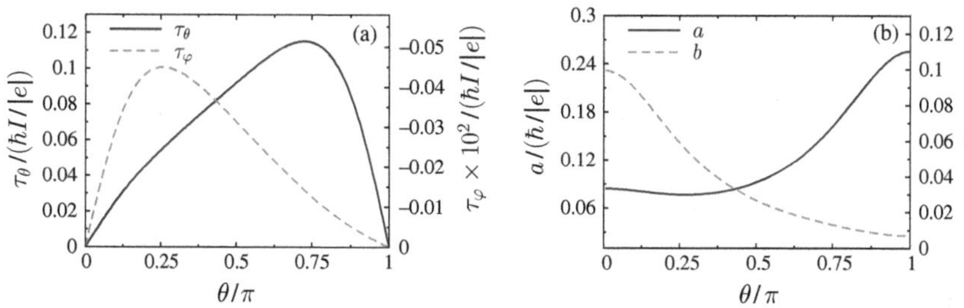

Fig. 3 Spin transfer characteristics for the Co(20)/Cu(10)/Co(8) spin valve: (a) normalized torque τ_θ and τ_φ acting on the sensing layer, Co(8), calculated as a function of the angle θ for the parameters typical of the Co/Cu system, see Tables 1 and 2; (b) the corresponding angular dependence of the parameters a and b

Figure 3 implies that a positive current ($I > 0$) tends to destabilize the parallel configuration and can switch it to an antiparallel one when current exceeds a certain threshold value. On the other hand, a negative current tends to destabilize the antiparallel configuration. This behavior can be defined as a normal CIMS [6, 12]. Calculation of the critical currents will be described later. What we want to emphasize here is the physical picture emerging from the plots of Fig. 3(b). The main point is that the angular dependence of the spin accumulation amplitude $g = |\mathbf{g}|$ in the nonmagnetic spacer is similar to that of the coefficient a in the expression (31) for the torque. This similarity results from the relation between the transverse spin current and spin accumulation in the boundary conditions involving the mixing conductance, Eqs. (24) and (25).

The τ_φ component of the torque, i.e., the component coming from the imaginary part of $G_{\uparrow\downarrow}$, is shown in Fig. 3(a), and is rather small, much smaller than the in-plane component (it would vanish for $\mathrm{Im}\{G_{\uparrow\downarrow}\} = 0$). Therefore, it plays a minor role in the switching phenomenon.

5.2 Limiting Case of Real Mixing Conductance

As it has been already mentioned before, the imaginary part of the mixing conductance is usually small. When $\mathrm{Im}\{G_{\uparrow\downarrow}\} = 0$, the formulae (34) and (37) acquire a simpler form. First of all, the out-of-plane components of the spin accumulation and spin current vanish, $g'_x = 0$ and $j'_x = 0$, so $g_x \sin\varphi = g_y \cos\varphi$. The in-plane torque can be then written in the form

$$\tau_\theta = \frac{\hbar}{e^2} G_{\uparrow\downarrow} g'_y|_{x\to x_0} = \frac{\hbar}{e^2} G_{\uparrow\downarrow} \left(g_y \frac{\cos\theta}{\sin\varphi} - g_z \sin\theta \right)\Bigg|_{x\to x_0}, \tag{43}$$

where $G_{\uparrow\downarrow}$ is real. In turn, the out-of-plane torque vanishes then exactly,

$$\tau_\varphi = 0\,. \tag{44}$$

When the spin accumulation in the spacer layer N at its interface with the layer F2 forms an angle θ_g with the axis z, then one finds $g'_y = g \sin(\theta_g - \theta)$ and the torque may be written in the form

$$\tau_\theta = \frac{\hbar}{e^2} G_{\uparrow\downarrow} g \sin(\theta_g - \theta)\,, \tag{45}$$

where g is the absolute value (amplitude) of the spin accumulation at the N/F2 interface. We note that positive current induces positive torque τ_θ. According to Eq. (34), the parameter a may be then expressed in the form

$$a = \frac{\hbar}{e^2} G_{\uparrow\downarrow} \frac{g}{I} \frac{\sin(\theta_g - \theta)}{\sin\theta}\,. \tag{46}$$

The formulae (45) and (46) relate the torque and the parameter a to the amplitude g of spin accumulation in the spacer layer N at its interface with F2, and to $\sin(\theta_g - \theta)$. It is interesting to look at the angular variation of these parameters, shown in Fig. 4 for the Co(20)/Cu(10)/Co(8) spin valve. The angular variation of the spin accumulation amplitude g, the parameter a, angle θ_g, and the factor $\sin(\theta_g - \theta)/\sin\theta$ of Eq. (46) are shown there as a function of θ. The amplitude g of the spin accu-

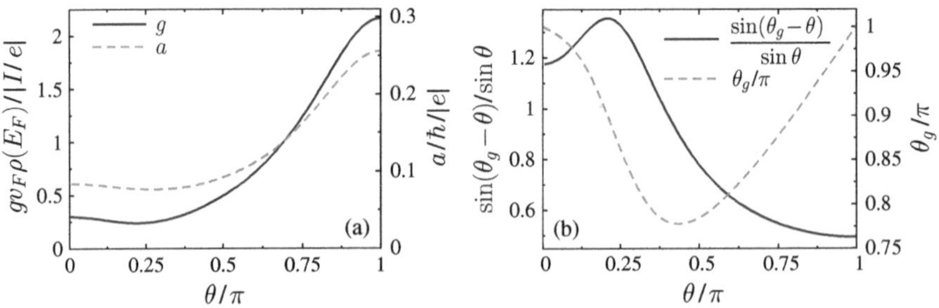

Fig. 4 Spin transfer in the limiting case of real mixing conductance in the Co(20)/Cu(10)/Co(8) spin valve: (**a**) spin accumulation amplitude g and parameter a, calculated as a function of the angle θ; (**b**) the corresponding angular dependence of $\sin(\theta_g - \theta)/\sin\theta$ and θ_g/π

mulation varies from its small value in the parallel state to its higher value in the antiparallel state, as expected when F1 and F2 have the same spin asymmetries. Similar angular variation takes place also in the case of the parameter a. In turn, θ_g is mainly determined by spin polarization of the thick magnetic layer and weakly departs from $\theta_g = \pi$ for $0 < \theta < \pi$. Consequently, $\sin(\theta_g - \theta)/\sin\theta$ is also weakly dependent on the angle θ.

Analogically, one can find that similar angular variation of g and a also occurs in the general case with nonzero but small imaginary part of $G_{\uparrow\downarrow}$. The explanation is the same as above, but only a little more complex because **g** has also a normal component. The general conclusion is that the torque is closely related to the spin accumulation in the nonmagnetic spacer layer N. From the above one can conclude that the ratio between the switching currents for the parallel → antiparallel and antiparallel → parallel transitions, which reflects approximately the ratio between the derivatives $\partial\tau_\theta/\partial\theta$ at $\theta = 0$ and $\theta = \pi$, simply reflects the ratio between the spin accumulation amplitudes in both magnetic configurations. In other words, the different magnitude of g for parallel and antiparallel states results in different critical currents necessary to destabilize both collinear states. In particular this ratio is inverted when the sign of the spin asymmetries is inverted in one of the magnetic layers, as will be discussed in the next section. Moreover, the torque amplitude can be enhanced or reduced by increasing or decreasing spin accumulation in the nonmagnetic spacer. This has been confirmed by the experiments of [12], where the torque was either enhanced or reduced by introducing spin-flip scatterers.

5.3 Controlled Normal and Inverse CIMS

The discussion presented in Sections 5.1 and 5.2 was devoted to systems where minority electrons are scattered more strongly than majority ones, i.e., when both the ferromagnetic layers in a spin valve structure have positive bulk and interfacial spin asymmetries. In a CPP-GMR experiment, the resistance of the spin valve is the lowest in high magnetic fields, when both the layer magnetizations are parallel. The resistance increases when the configuration deviates from the parallel one and is the largest in the antiparallel configuration. Such behavior is described as normal GMR. Moreover, such structures also show a normal CIMS, i.e., positive current (flowing from the thicker film towards the thinner one) switches magnetic configuration from parallel to antiparallel one while negative current switches the system back to the parallel configuration. If one considers now a spin valve with reversed spin asymmetries for both magnetic layers and the corresponding interfaces, i.e., when the majority electrons are scattered strongly by both magnetic layers, the GMR effect is still normal [33, 39]. However, the CIMS changes now from normal to inverse, i.e., positive current switches the sensing layer to parallel configuration while negative current switches it back to the antiparallel configuration. The problem of how manipulation of the spin asymmetries affects the magnetoresistance and CIMS was studied experimentally by AlHajDarwish et al. [12]. In this section we present the angular dependence of the torque acting on the sensing layer for representative combinations of the bulk and interface spin asymmetries and demonstrate controlled inversion of the CIMS. Moreover, the correlations between the type of CPP-GMR and types of CIMS will also be discussed in the context of spin accumulation at the corresponding active interfaces N/F2.

The parameters describing the bulk and interface spin asymmetries, β and γ, respectively, can be generally different for F1 and F2 magnetic layers. Similarly, parameters corresponding to the nonmagnetic electrodes can also be different. In the following, however, we assume that the nonmagnetic leads are made of the same material (Cu), whereas spin valves may generally be different, as recently studied experimentally [12]. For numerical calculations we have used parameters obtained from the GMR measurements, see Tables 1 and 2. The following situations are discussed:

1. $\beta_1 = \beta_2 > 0$, $\gamma_1 = \gamma_2 > 0$, which corresponds to F1 and F2 of the same material with positive spin asymmetries for both bulk resistivities and interfacial resistances. In this case, spin-majority electrons are less scattered both inside the layers and at the interfaces.
2. $\beta_1 = \beta_2 < 0$, $\gamma_1 = \gamma_2 < 0$, which corresponds to the same material for F1 and F2, with negative spin asymmetries for both bulk resistivities and interfacial resistances. Spin-majority electrons are then scattered in bulk and by interfaces more strongly than spin-minority electrons.
3. $\beta_1 > 0$, $\gamma_1 > 0$, $\beta_2 < 0$, $\gamma_2 < 0$, which corresponds to different materials for F1 and F2, with positive spin asymmetries for F1 and negative spin asymmetries for F2.
4. $\beta_1 < 0$, $\gamma_1 > 0$, $\beta_2 > 0$, $\gamma_2 > 0$, which corresponds to different materials for F1 and F2, with negative bulk and positive interface spin asymmetries for F1 and positive both spin asymmetries for F2.

One of the systems in category (i) is the Co/Cu/Co structure, which has been extensively studied experimentally. We have devoted previous sections to present the numerical results for the Co(20)/Cu(10)/Co(8) spin valve. The torque τ_θ is then

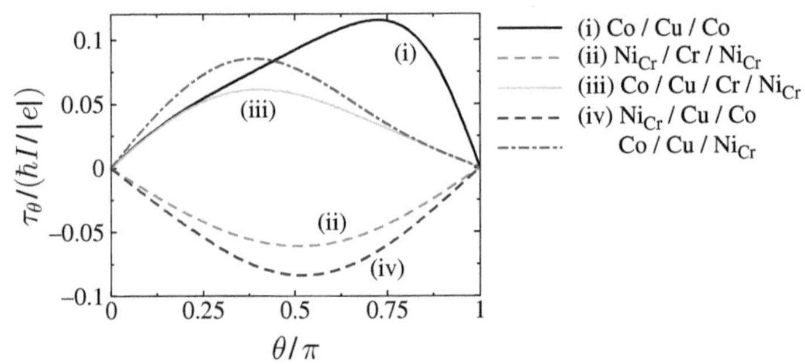

Fig. 5 Angular dependence of the torque τ_θ acting on the sensing layer F2 in the spin valves as indicated. The spin valves are sandwiched between semi-infinite Cu leads

positive for $0 < \theta < \pi$, see Fig. 3 and 5. Since positive torque tends to increase θ, it destabilizes the parallel alignment and stabilizes the antiparallel state when current exceeds a certain critical current density. The current drives then the parallel $\rightarrow$ antiparallel switching. We will discuss the critical current densities in Section 6.1.

In contrast to the Co/Cu/Co spin valve, the ferromagnetic layer of $Ni_{Cr} = Ni_{97}Cr_3$ and the Ni_{Cr}/Cr interfaces both have negative scattering anisotropies. Thus, the spin valve $Ni_{Cr}(20)/Cr(10)/Ni_{Cr}(8)$ belongs to category (ii). Such a system exhibits the normal GMR effect. However, the CIMS becomes the inverse, see Fig. 5, i.e., positive current stabilizes the parallel configuration (low resistive) and destabilizes the antiparallel one (high resistive). Replacing Ni_{Cr}, for instance, by $Fe_{95}Cr_5$, studied in [12], leaves the system in the same category.

Combining positive spin asymmetries for F1 and negative spin asymmetries for F2 leads to inverse GMR. The calculations of spin torque for the spin valve $Co(20)/Cu(10)/Cr(10)/Ni_{Cr}(8)$ confirm the experimentally observed normal CIMS [12], see Fig. 5. Thus, positive current destabilizes the parallel (high resistive) state and drives the system to antiparallel (low resistive) state.

Consider now a spin valve where all the interfacial spin asymmetries are positive and bulk spin asymmetry of the fixed layer F1 is negative while that of F2 is positive. Such a system allows to study crossover from normal to inverse GMR when thickness of F1 increases [39]. At some thickness of the F1 layer, the bulk spin asymmetry compensates the one due to interface scattering and the overall spin asymmetry changes sign. For the $Ni_{Cr}(20)/Cu(10)/Co(8)$ spin valve, one finds inverse GMR effect and also inverse CIMS. When Ni_{Cr} is used for the sensing layer F2 the GMR effect becomes normal while CIMS is still inverse, see Fig. 5.

Table 3 lists the signs of spin asymmetries for both F1 and F2 layers, sign of τ_θ as well as type of the corresponding CPP-GMR [12] and CIMS. The four cases cover all combinations of normal and inverse magnetoresistance and CIMS. The magnetoresistance is normal if the overall spin asymmetries of F1 and F2 have the same sign, and inverse if the asymmetries are opposite. For the spin valves described by the parameters listed in Tables 1 and 2, we found that type of CIMS is controlled by the spin asymmetry of the thick layer F1. This is supported also by calculations for the $Co(20)/Cu(10)/Ni_{Cr}(8)$ valve, where normal CIMS has been found, see Fig. 5, and magnetoresistance remains inverse. The results presented in Fig. 5 are consistent with the experimental observations in [12], but are in contradiction to ballistic models where only interface scattering was assumed [44].

As we discussed in Section 5.1, τ_θ is much larger than τ_φ and plays a crucial role in CIMS. However, to conclude on the sign of τ_θ acting on the sensing layer and to determine the type of CIMS by analyzing the general expressions for τ_θ, Eqs: (31, 32, 33 and 34), even assuming zero imaginary part of mixing conductance at the active N/F2 interface, Eq. (45), is not as straightforward as one could expect. Taking into account the fact that g_x and g_y tend to zero for both collinear configurations, one could conclude from Eq. (34) that the sign of τ_θ is controlled mainly by the term $-I\,\tilde{g}_z$ near the parallel state and $I\,\tilde{g}_z$ near the antiparallel state. This, however, is true only when both CPP-GMR and CIMS have the same sign, i.e., for systems in the categories (i) and (iv).

The analysis should be extended to investigate the sign of the derivative $\partial\tau/\partial\theta$ at the active interface N/F2 for $\theta = 0$ and $\theta = \pi$. If the derivative is positive for the parallel state ($\theta = 0$) and negative for the antiparallel one ($\theta = \pi$), the CIMS is normal. The CIMS is inverse when the derivatives change their signs. Considering Eqs. (31) and (34), one can find the following formula for $\partial\tau_\theta/\partial\theta$,

$$\left.\frac{\partial\tau_\theta}{\partial\theta}\right|_{\mathrm{P(AP)}} = \frac{\hbar I}{e^2}\left.\left[\pm\mathrm{Re}\{G_{\uparrow\downarrow}\}\left(\frac{\partial\,\tilde{g}_y}{\partial\theta} - \tilde{g}_z\right) + \mathrm{Im}\{G_{\uparrow\downarrow}\}\frac{\partial\,\tilde{g}_x}{\partial\theta}\right]\right|_{\mathrm{P(AP)}}, \qquad (47)$$

where the spin accumulation and mixing conductance correspond to the N/F2 interface. The first term on the right-hand side is the dominant one. Thus, the sign of the torque derivative depends on $\tilde{g}_z$ and the derivative of $\tilde{g}_y$. However, the transverse components of the spin accumulation for both collinear configurations vanish, and the derivative can reach the amplitude of $\tilde{g}_z$. The sign of g_z can be simply deduced within the two-channel model. To conclude on the sign of $\partial\,\tilde{g}_y\,/\partial\theta$, one has to consider transport characteristics of noncollinear configurations, which are implicitly included in the solution of the set of boundary equations discussed in Section 3. For this a numerical analysis is necessary.

From the results presented above it follows that it is the spin asymmetry of the thick magnetic film that determines whether the switching effect is normal or inverse. When this spin asymmetry is positive (negative), one finds a normal (an inverse) switching phenomenon. It is also interesting to note that when the spin asymmetries of both magnetic films have the same sign, the structure shows normal GMR effect, whereas when they are opposite, the corresponding GMR effect is inverse as shown in many CPP-GMR measurements [39].

5.4 Nonstandard Angular Dependence of the Spin Torque

For the parameters used in numerical calculations described above, the current-induced torque vanishes in collinear configurations and one of them (either parallel or antiparallel) is unstable. This leads to either normal or inverse switching phe-

Table 3 Characteristics of the CIMS in the four studied cases. Correlation between the signs of the torque and GMR is also given

	Spin asymmetries		Sign of torque ($0 < \theta < \pi$)		
Situation	Thick layer	Thin layer	Torque	GMR	CIMS
(i)	$(\beta, \gamma) > 0$	$(\beta, \gamma) > 0$	$\tau_\theta > 0$	Normal	Normal
(ii)	$(\beta, \gamma) < 0$	$(\beta, \gamma) < 0$	$\tau_\theta > 0$	Normal	Inverse
(iii)	$(\beta, \gamma) > 0$	$(\beta, \gamma) < 0$	$\tau_\theta < 0$	Inverse	Normal
(iv)	$(-\beta, \gamma) > 0$	$(\beta, \gamma) > 0$	$\tau_\theta < 0$	Inverse	Inverse

nomena. An interesting situation can occur when the magnitudes of spin asymmetry in the thick and thin layers are different. In Fig. 6(a), we show an example of the torque calculated for $\beta_1 = 0.2$ and $\gamma_1 = -0.2$, $\beta_2 = 0.51$ and $\gamma_2 = 0.77$, and the rest of parameters are taken as for case (i) studied in Section 5.3. Let us consider first the case of $I > 0$. The torque is negative when θ increases from zero, then comes back to zero at $\theta = \theta_c$ and becomes positive between θ_c and π. This means that the torque stabilizes both the parallel and antiparallel states. In other words, it increases the damping of the system in both configurations. This is of interest for some devices, for example, for stabilizing the configuration of read heads based on CPP-GMR against fluctuations or, at least, avoiding spin transfer-induced fluctuations. The angle θ_c significantly depends on the spin asymmetries of the fixed layer as shown in Fig. 6(b). The larger the θ_c, the greater the torque derivative $\partial\tau_\theta/\partial\theta$ at $\theta = 0$. Thus, a practical demand for the most effective device for stabilizing both P and AP configuration is to have θ_c close to $\pi/2$.

The behavior of spin torque in Fig. 6(a) for $I < 0$ has a different interest. Now, above some threshold value of the current, the spin transfer torque can destabilize both collinear states of the spin valve. The only solution is then a steady state precession or a stable static state with noncollinear orientation of the magnetic moments.

In trilayer spin valves which have been studied up to now, steady microwave oscillations have been generally observed when a magnetic field is applied and in a given range of current density [16]. The possibility of having microwave oscillations at zero field would be of great interest for several devices. Xiao et al. [45] have predicted that microwave oscillations can be observed at zero field and in a certain range of current density when (for asymmetric structures) the spin transfer and damping torques have remarkably different angular dependence, so that their sum

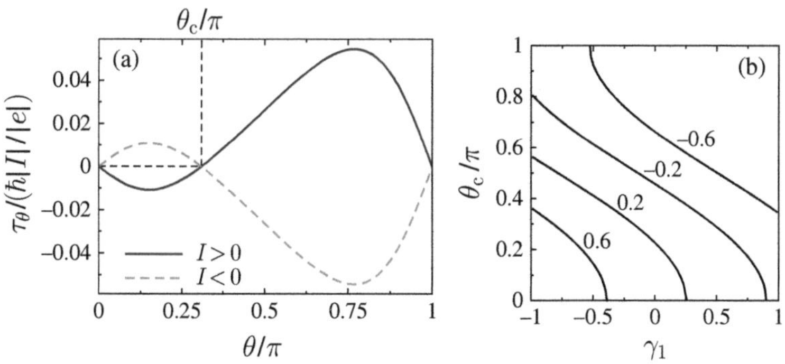

Fig. 6 Nonstandard variation of the spin transfer torque with θ: (**a**) torque for the case when the spin asymmetry factors of the thick ferromagnetic layer, $\beta_1 = 0.2$ and $\gamma_1 = -0.2$, are significantly different from those of the thin one, $\beta_2 = 0.51$ and $\gamma_2 = 0.77$. τ_θ is normalized to $|I|$ and the two curves correspond to $I > 0$ and $I < 0$, as indicated. The angle θ_c corresponds to the point where the torque vanishes; (**b**) variation of the angle θ_c with the spin asymmetry factor γ_1 for several values of β_1 indicated on the curves. The other parameters as for standard Co(20)/Cu(10)/Co(8) spin valves

can exhibit a wavy angular dependence. In our structure of Fig. 6(a), the existence of steady precessions at zero field has a different origin and comes from the wavy angular dependence of the spin torque itself. Such a structure (for $I < 0$) could be of interest for the generation of microwave oscillations at zero magnetic field. In Section 6.3 we will study in detail the dynamics of induced oscillatory modes and conditions for their observation.

5.5 Correlation Between Spin Transfer Torque and CPP-GMR

We have learned that proper manipulation of the spin asymmetry parameters can qualitatively change the angular dependence of spin transfer torque. From the technological point of view, however, manipulation of spin asymmetries is very difficult, and therefore another approach to prepare spin valves with nonstandard torque is desired. It has been shown [24] that for this purpose it is sufficient to take advantage of two different ferromagnetic layers. If one combines permalloy and cobalt in a Co/Cu/Py valve, the spin transfer torque acting on the Py layer has nonstandard dependence. This has a significant impact on the stability of magnetic configuration. As a result, precessional states in zero magnetic field were predicted in the Co/Cu/Py nanopillars [24] and later experimentally confirmed [46]. One may naturally expect that the nonstandard behavior of the spin transfer torque may be associated with some anomalous angular behavior of the CPP-GMR.

Within the diffusive approach [23, 32, 34], spatial dependence of the average electrochemical potential in a ferromagnetic layer has the general form given by equations (9) and (10). Similar formulas also hold for normal-metal layers, but with $\beta = 0$, see equations (17) and (18). Having the electrochemical potential, one can calculate the driving field as $E(x) = (1/e)(\partial\bar{\mu}/\partial x)$ [32]. The presence of N/F interfaces gives rise to additional voltage drops due to spin accumulation in their vicinity. The total voltage drop can then be written as $\Delta V = \sum_i \Delta V_i$, where ΔV_i is the voltage drop in the ith layer of the spin valve (voltage drops at interface resistances will be included to the ferromagnetic layers). When the index i corresponds to a ferromagnetic layer, $\Delta V_i = \Delta V_i^{\rm SI} + \Delta V_i^{\rm spl}$. If, however, i corresponds to a nonmagnetic layer, $\Delta V_i = \Delta V_i^{\rm SI} + I\rho_i d_i$, where ρ_i is the bulk resistivity of the normal metal, d_i is the corresponding layer thickness, and I is the current density. The voltage drops due to spin accumulation (in magnetic and nonmagnetic layers) read

$$\Delta V_i^{\rm SI} = \int_{x\in d_i} [E(x) - E_0]\,{\rm d}x, \tag{48}$$

where the corresponding electric field E_0 is taken far from the interface. Apart from this, $\Delta V_i^{\rm spl} = I\,R_i^{\rm spl}$ for ferromagnetic films, where $R_i^{\rm spl} = [(1/R_{i\uparrow}) + (1/R_{i\downarrow})]^{-1}$, $R_{i\sigma} = R_{i\sigma}^{\rm L} + d\rho_{i\sigma} + R_{i\sigma}^{\rm R}$ (for $\sigma = \uparrow, \downarrow$), with $\rho_{i\sigma}$ being the corresponding spin dependent bulk resistivity and $R_{i\sigma}^{\rm L}$ ($R_{i\sigma}^{\rm R}$) denoting the interfacial resistances (per unit square) associated with the left (right) interface of the ith (ferromagnetic) film.

The total resistance of the system (per unit square) is $R = \Delta V/I$, while the magnetoresistance, $\Delta R(\theta) = R(\theta) - R(0)$, describes a change in the total system resistance when magnetic configuration varies from a noncollinear to parallel one. What one needs to calculate are the ΔV_i^{SI} contributions only. It is convenient to define reduced magnetoresistance as

$$r(\theta) = \frac{R(\theta) - R_{\mathrm{P}}}{R_{\mathrm{AP}} - R_{\mathrm{P}}}. \tag{49}$$

Several theoretical approaches have been proposed to describe the angular variation of GMR [47]. The explicit form for the reduced magnetoresistance, $r(\theta) = \sin^2(\theta/2)/[1 + \chi \cos^2(\theta/2)]$ has been obtained with the magneto-circuit theory [30] and diffusive approach [48]. In measurements on Py/Cu/Py valves [49], the parameter χ has been considered as a fitting parameter, and the $r(\theta)$ dependence describes the experiment relatively well. However, this formula breaks down for asymmetric spin valves, where a global minimum of the system resistance may appear for a noncollinear configuration [49].

The minimum in resistance (and also in GMR) in a noncollinear configuration appears only in asymmetric spin valves. Angular dependence of the reduced magnetoresistance in Cu/Co(d_1)/Cu(10)/Co(8)/Cu valve is shown in Fig. 7(a) for thicknesses $d_1 = 16$ nm and 60 nm. For $d_1 = 60$ nm, the reduced magnetoresistance exhibits a minimum for $\theta \simeq \pi/3$. To explain the presence of this minimum, one has to analyze the angular dependence of the voltage drops ΔV_{SI} within the F layers. This angular dependence in the thin F layer is a monotonic function, whereas in the thick F layer a minimum appears in a noncollinear configuration. The thicker the fixed layer the more pronounced minimum in the angular dependence of the corresponding ΔV_{SI} develops. Since the resulting voltage drop is a sum of individual contributions, the minimum in a noncollinear configuration appears when the decrease in the thick F layer overcomes the increase in the thin F layer. The global minimum arises as a result of the spatial depletion of electrical field in the thick F layer, which is a consequence of spin accumulation discontinuity at the N/F interface, controlled by the mixing conductances. The non-monotonic behavior of the reduced GMR is more pronounced in spin valves that are more asymmetric, see Fig. 7(b).

We see that by proper manipulation of the layer thicknesses, one can change qualitative behavior of GMR. In Fig. 7(c) and (e) we show the diagrams that present the regions of layer thicknesses, where the non-monotonic behavior of the reduced GMR can be observed (gray regions). For the Co/Cu/Co spin valves [Fig. 7(c)] as well for the Py/Cu/Py ones, the diagrams are symmetric with respect to $d_1 = d_2$, and the non-monotonic angular variation of the GMR (global minimum at a noncollinear configuration) can be noticed for spin valves with significantly different layer thicknesses. Here d_1 and d_2 refer to the F layers' thicknesses. Contrarily, in Co/Cu/Py spin valves, where an additional asymmetry appears due to different mag-

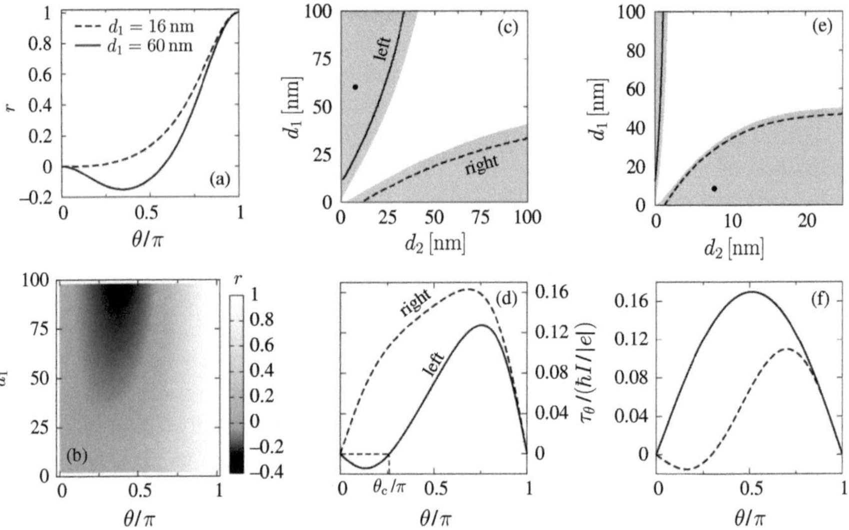

Fig. 7 Transport characteristics of Co/Cu/Co and Co/Cu/Py spin valves sandwiched between semi-infinite Cu leads for positive current density $I > 0$ (current flows from the layer of thickness d_1 to the layer of thickness d_2): **(a)** angular dependence of the reduced magnetoresistance for Co(d_1)/Cu(10)/Co(8) spin valve and **(b)** reduced magnetoresistance as a function of θ and d_1. Diagrams illustrating presence of a global magnetoresistance minimum at noncollinear configurations – *gray regions* in **(c)** and **(e)** – and angular spin transfer torque dependence **(d)** and **(f)**. Diagrams for the **(c)** Co(d_1)/Cu(10)/Co(d_2) and **(e)** Co(d_1)/Cu(10)/Py(d_2) spin valves. The solid and dashed lines denote critical thicknesses where $\partial\tau_\theta/\partial\theta|_{\theta\to 0} = 0$ for the torque exerted on the left F layer of thickness d_1 and right layer of thickness d_2, respectively. **(d)** and **(f)** angular dependence of the spin transfer torques acting on the left F layer of thickness d_1 (*solid line*) and right F layer of thickness d_2(*dashed line*) for systems corresponding to the dots in the upper panel **(c,e)**

netic materials, a non-monotonic angular variation of GMR can be observed even for comparable layer thicknesses, see Fig. 7(e). This is mainly due to strong asymmetry in the spin diffusion lengths of Co and Py, but difference in the bulk as well as interface spin asymmetries of the Co and Py also contributes to the non-monotonic behavior.

Experimental observations on Py/Cu/Py spin valves revealed a weak nonmonotonic angular variation of the GMR [49]. This has been attributed to the absorption of transverse spin accumulation in a noncollinear configuration, which reduces the resistance. Such absorption also gives rise to the spin transfer torque acting on the F layer, which in asymmetric spin valves can exhibit an anomalous (nonstandard) angular dependence. In systems with a nonstandard spin transfer torque, the transverse component of spin current (accumulation) at the active N/F interface vanishes at a certain noncollinear configuration. The presence of a GMR minimum in a noncollinear configuration can be thus related to the nonstandard spin transfer torque. We know that spin transfer torque generally consists of two in-plane and out-of-plane components. Since the latter component is much smaller (due to small imaginary part of the mixing conductances [35]) than the former one, in the following discussion we will consider only the in-plane component, τ_θ Let us discuss the angular dependence of the spin transfer torque and GMR in asymmetric spin valves. In Fig. 7(d) we show the angular dependence of spin transfer torque in the Co(60)/Cu(10)/Co(8) spin valve acting on the Co(60) (solid line) and Co(8) (dashed line). The torque exerted on the Co(8) layer destabilizes P and stabilizes AP configuration, whereas the torque acting on the Co(60) vanishes at a noncollinear configuration and stabilizes both P and AP configurations. The torques in Fig. 7(d) correspond to the system indicated by the dot in Fig. 7(c). This point is below the critical line, given by $\partial\tau_\theta/\partial\theta|_{\theta\to 0} = 0$, which identifies the region where a nonstandard spin transfer torque acting on the Co(d_1) layer appears. When the layer thicknesses are chosen above the critical line, but still in the gray region, the torque acting on the particular F layer vanishes only for the collinear configurations, but reduced GMR still exhibits a global minimum at a noncollinear configuration. Since the critical lines are close to the boundary of the non-monotonic angular GMR behavior (gray regions), one may say that the nonstandard spin transfer torque is correlated with the non-monotonic angular variation of GMR.

At the critical angle θ_c, where the torque τ_θ vanishes, the transverse component of spin accumulation at the active interface disappears. In a general case, however, the angle θ_g between the spin moment of the F layer and spin accumulation vector at the normal-metal side of the N/F interface is nonzero. Angular dependence of the spin transfer torque can then be expressed as a function of θ_g, see Section 5.2. Noncollinear configuration of the F layer magnetizations leads to the discontinuities of the spin accumulation at the F/N interfaces. From this we deduce that if the thickness of one of the F layers is smaller than the corresponding spin diffusion length and thickness of the second F layer is larger than the appropriate spin diffusion length, then the spin accumulation is predominately determined by the latter F layer. One finds then nonstandard torque and non-monotonic GMR angular behavior. This behavior is mostly controlled by the mixing conductance of the interface between the spacer layer and the F layer whose thickness is smaller than the corresponding spin diffusion length.

What stems from the above results? The non-monotonic behavior of the resistance (and consequently also GMR) is generally accompanied by the nonstandard angular dependence of spin transfer torque. These two features seem to be characteristic of asymmetric systems in the diffusive transport regime. Moreover, the Co/Cu/Py system is a good candidate to test whether the diffuse approach used to analyze CPP-GMR in collinear configurations is well justified. On the other hand, the asymmetric spin valves can be considered for current-driven devices, where novel properties originate from the nonstandard spin transfer torque angular dependence.

6 Spin Transfer-Induced Dynamics – Macrospin Model

The general expression for spin transfer torque given in Section 4 is a consequence of the total angular momentum conservation and absorption of the transverse spin current component at an active interface. The arisen spin transfer torque exerted on the magnetic moment of the sensing layer may drive it out of equilibrium. The model we use (the so-called macrospin model) is based on the assumption that magnetization (or net spin) of the sensing layer is spatially uniform and behaves as a single magnetic moment, whose dynamics can be described by the Landau–Lifshitz–Gilbert equation including the spin transfer torque $\boldsymbol{\tau}$. Additionally, we assume that the thick magnetic layer does not undergo the dynamics in the current densities and external fields of interest. Thus, dynamics of the sensing layer can be described by the Landau–Lifshitz–Gilbert equation of the form

$$\frac{d\hat{\mathbf{s}}}{dt} = -|\gamma_g|\mu_0\,\hat{\mathbf{s}} \times \boldsymbol{H}_{\rm eff} - \alpha\,\hat{\mathbf{s}} \times \frac{d\hat{\mathbf{s}}}{dt} + \frac{|\gamma_g|}{M_s d_3}\,\tau\,, \tag{50}$$

where $\hat{\mathbf{s}}$ is the unit vector along the spin moment of the sensing layer, γ_g the gyromagnetic ratio, μ_0 the magnetic vacuum permeability, d_3 the thickness of the sensing layer, see Fig. 2, $\mathbf{H}_{\rm eff}$ an effective magnetic field for the sensing layer, α the damping parameter, and M_s the saturation magnetization. The effective field $\mathbf{H}_{\rm eff}$ includes an external magnetic field $H_{\rm ext}$, the uniaxial magnetic anisotropy field H_a, and the demagnetization field $\mathbf{H}_d$:

$$\mathbf{H}_{\rm eff} = -\boldsymbol{H}_{\rm ext}\,\hat{\mathbf{e}}_z - H_a(\hat{\mathbf{s}}\cdot\hat{\mathbf{e}}_z)\,\hat{\mathbf{e}}_z + \mathbf{H}_d\,, \tag{51}$$

where $\hat{\mathbf{e}}_z$ is the unit vector along the axis z (in-plane). The last term in Eq. (50) stands for the torque due to spin transfer, $\boldsymbol{\tau} = \boldsymbol{\tau}_\theta + \boldsymbol{\tau}_\varphi$, with $\boldsymbol{\tau}_\theta = \tau_\theta\,\hat{\boldsymbol{e}}_\theta$, and $\boldsymbol{\tau}_\varphi = \tau_\varphi\,\hat{\boldsymbol{e}}_\varphi$ given in Section 4. We recall that $\hat{\mathbf{S}}$ is here the unit vector along the spin moment of the thick magnetic layer ($\hat{\mathbf{S}} = \hat{\mathbf{e}}_z$), see Fig. 2, while $\hat{\mathbf{e}}_\theta$ and $\hat{\mathbf{e}}_\varphi$ are the unit vectors of a coordinate system associated with the polar θ and azimuthal φ angles which describe orientation of the vector $\hat{\mathbf{s}}$, see Fig. 1.

6.1 Critical Currents

Depending on the current direction (the sign of τ), the spin transfer torque can act against the Gilbert damping or effectively can increase the damping torque. In other words, the spin transfer can either destabilize or stabilize a given collinear configuration. Let us assume a positive spin transfer torque acting on the sensing layer. The dynamics of $\hat{\mathbf{s}}$ is damped by sustained dissipation of energy (Gilbert damping), so the spin transfer torque can overcome the energy dissipation only when current exceeds a certain critical value, leading finally to CIMS. Thus, by analyzing the stability of the stationary (fixed) points of the system, one can find the conditions for critical currents.

A qualitative picture of the nonlinear dynamics can be obtained from the local phase portraits of the corresponding linearized system in the vicinity of fixed points. Accordingly, we linearize Eq. (50) and calculate eigenvalues of the corresponding Jacobian. For a two-dimensional problem (as in our case) the eigenvalues depend only on the trace and determinant of the Jacobian [50]. Apart from this, the eigenvalues are current dependent and determine stability of the fixed points. More specifically, a given point is stable when real parts of all the eigenvalues are negative, and becomes unstable when at least one of them is positive. In what follows the dynamics described by the Eq. (50) can have several fixed points given by the stationary solution $\boldsymbol{\tau} = \mu_0 M_s d_3 \hat{\mathbf{s}} \times \boldsymbol{H}_{\text{eff}}$. However, only two of them, P and AP, are trivial with position independent of current.

Consider first the P state and assume positive determinant of the corresponding Jacobian. Since the eigenvalues of the linearized problem can be complex numbers, the nonzero imaginary parts give rise to periodic components of the fundamental solutions. The condition of vanishing trace determines the critical current I_c^{P} which destabilizes the P state and switches system away from the P state,

$$I_c^{\text{P}} = \frac{\alpha \mu_0 M_s d_3}{a - b\alpha} \left(H_a + H_{\text{ext}} - H_{\text{d}z} + \frac{H_{\text{d}x} + H_{\text{d}y}}{2} \right), \tag{52}$$

where the parameters a and b have to be calculated in the limit of $\theta \to 0$. Similarly, one can find the critical current that destabilizes the AP state in the form

$$I_c^{\text{AP}} = -\frac{\alpha \mu_0 M_s d_3}{a - b\alpha} \left(H_a - H_{\text{ext}} - H_{\text{d}z} + \frac{H_{\text{d}x} + H_{\text{d}y}}{2} \right), \tag{53}$$

for the parameters a and b calculated in the limit of $\theta \to \pi$. The signs of critical currents depend on the parameters a and b taken at $\theta \to 0$ and $\theta \to \pi$, respectively. In the case of standard Co/Cu/Co spin valves, the parameter b is at least two orders smaller than a, and is positive for arbitrary θ, see Fig. 3(b).

6.2 Dynamics in Symmetric Spin Valve

Nonlinear nature of the dynamics described by Eq. (50) is complex and requires numerical treatment. In this section we analyze dynamical behavior of the Cu/Co(30)/Cu(10)/Co(4)/Cu spin valve. For the sensing layer, Co(4), we assume the Gilbert damping parameter $\alpha = 0.003$, saturation magnetization $M_s = 17.8$ kOe, anisotropy field $H_a = 560$ Oe, and the demagnetization field $H_{\text{d}x} = 0.65 M_s$, $H_{\text{d}y} = H_{\text{d}z} = 0$ to model easy plane geometry.

Figures 8(a) and (b) show dynamical behavior of the net spin component $\hat{s}_z$ for external magnetic fields 200 Oe and 800 Oe, driven by the spin transfer for the current density 10^8 A/cm^2, which is over six times larger than the critical values given by Eqs. (52) and (53). The initial bias of $\theta = 1^\circ$ and $\varphi = \pi/2$ has been assumed there. The results clearly demonstrate the current-driven parallel to antiparallel switching for the applied field $H_{\text{ext}} = 200$ Oe. For magnetic field of $H_{\text{ext}} = 800$ Oe, the spin transfer destabilizes the P state, and induces precessional modes, see Fig. 8(b). The insets to Fig. 8(a) and (b) show the time dependence of the corresponding reduced magnetoresistance signal r,

$$r = [1 - \cos^2(\theta/2)]/[1 + \cos^2(\theta/2)]\,. \tag{54}$$

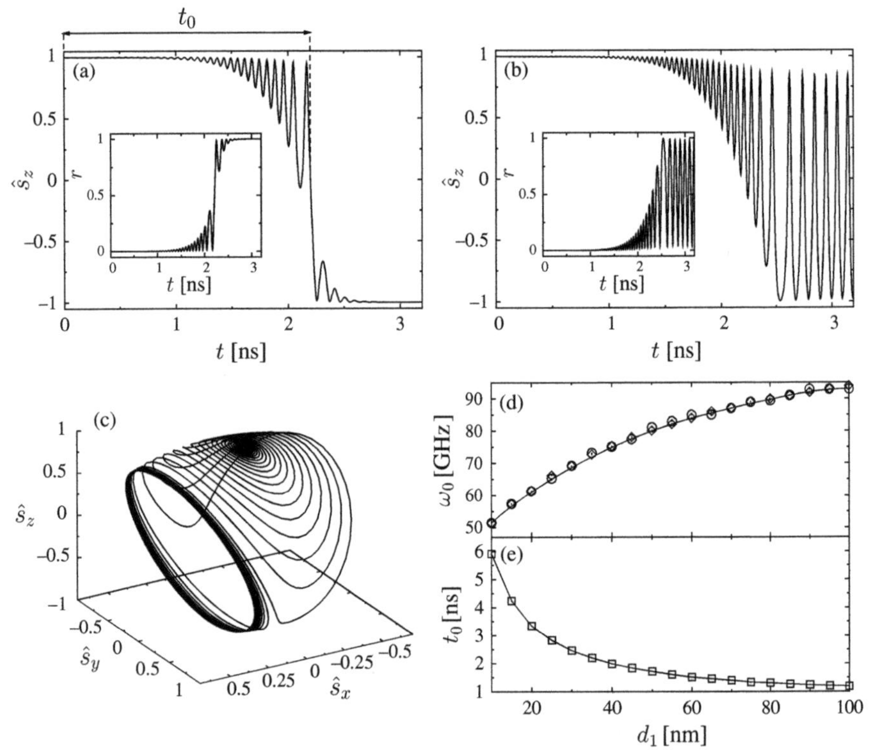

Fig. 8 Current-driven dynamics of the sensing layer in the Cu/Co (d_1)/Cu(10)/Co(4)/Cu spin valve, calculated for the current density $10^8\ \mathrm{A/cm^2}$: (**a**) time dependence of the $\hat{s}_z$ component for $H_{\mathrm{ext}} = 200\,\mathrm{Oe}$ and $d_1 = 30\,\mathrm{nm}$. The inset shows the time dependence of the corresponding magnetoresistance signal; (**b**) time dependence of $\hat{s}_z$ for $H_{\mathrm{ext}} = 800\,\mathrm{Oe}$; (**c**) trajectory of $\hat{\mathbf{s}}$ for $d_1 = 30\,\mathrm{nm}$ and $H_{\mathrm{ext}} = 800\,\mathrm{Oe}$; (**d**) fundamental frequency ω_0 of magnetoresistance signal calculated as a function of the thickness d_1 of the fixed layer (*circles*). The *diamond* points represent the first harmonics in the Fourier spectra, scaled by a factor of 2. Each point has been obtained by averaging 100 events. The events are determined from the Fourier spectra as a result of the current-driven switching from randomized initial nearly parallel configurations corresponding to the temperature of 50 K; (**e**) switching time t_0 versus d_1. Each point is a mean value of 1000 events obtained for the current-driven switching, similarly as for the ω_0 dependence

One can note that the above formula is valid only for structures with normal GMR. Figure 8(c) illustrates the dynamic behavior of the vector $\hat{\mathbf{s}}$ for the case presented in Fig. 8(b). The spin rotates off the z direction in a spiral motion, before it approaches a steady precessional state, which is a closed loop nearly parallel to the film plane. The precessional state is sustained due to the balance of dissipation of the energy *via* Gilbert damping and the energy pumping in *via* the spin transfer torque [19, 51]. The fundamental frequency ω_0 of the magnetoresistance oscillations depends on the thickness d_1 and increases with increasing d_1, as shown in Fig. 8(d) for $H_{ext} = 800\,Oe$. The dependence of the frequency ω_0 on the thickness d_1 reflects the increasing trend in the spin transfer torque amplitude. One can also note that the switching time t_0 decreases with increasing thickness d_1. This is due to the increasing spin accumulation in the nonmagnetic spacer near the interface with the sensing layer, which results in an enhanced spin transfer torque. The switching time t_0 is defined here as the point where $\hat{s}_z$ crosses zero for the first time. The dependence of the switching time t_0 on the thickness d_1 is shown explicitly in Fig. 8(e). The manipulation of sensing layer thickness affects the critical currents directly, see Eqs. (52) and (53), and implicitly through the parameters a and b. An increase in the sensing layer thickness d_3 leads to a decrease of a near the P state and to an increase of a near the AP state. In turn, the parameter b is diminished independently of the magnetic configuration of the system.

The inset to Fig. 9(a) shows the current dependence of the switching time t_0 from the P configuration to steady precessional regime for the magnetic field as indicated in the figure caption. The current dependence of t_0 reveals simple rational behavior. However, the slope of the inverse plot does not depend on the applied magnetic field, but we found its significant dependence on the initial bias (not shown). The initial bias affects the switching time which logarithmically diverges when bias tends to zero ($\theta \to 0$). From the later follows that duration of the transient regime can be significantly affected by the initial bias, influenced for instance by thermal noise.

The fundamental frequency ω_0 (first harmonics) of the magnetoresistance oscillations is shown in Fig. 9(a) as a function of reduced current density I/I_0, where

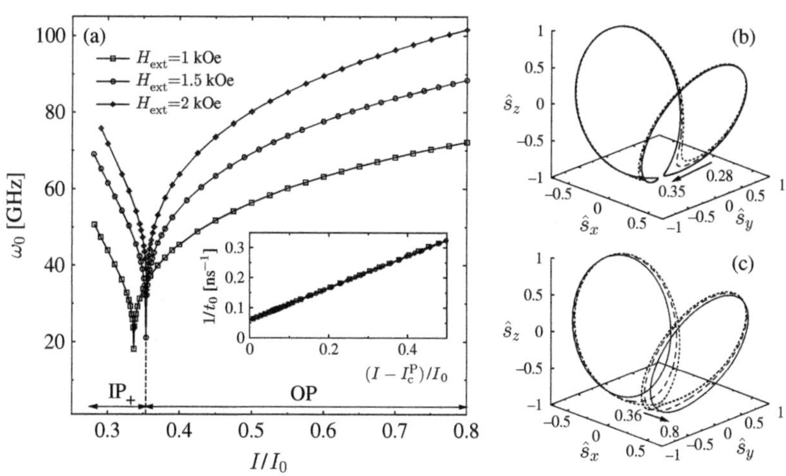

Fig. 9 Current-induced precessional regimes in the Cu/Co(30)/Cu(10)/Co(4)/Cu spin valve: (**a**) fundamental frequency of the magnetoresistance signal calculated as a function of the reduced current density, and for magnetic fields as indicated. The inset shows the corresponding current dependences of the switching times, where critical current density, I_c^P, is given by Eq. (52); (**b**) steady large angle in-plane orbits for $I/I_0 = 0.28, 0.3, 0.35$ and for $H_{ext} = 1.5\,kOe$; (c) out-of-plane orbits for $I/I_0 = 0.36, 0.5, 0.8$. The arrows indicate change in orbits as current increases

$I_0 = 10^8$ A/cm^2, and for magnetic fields as indicated. The frequency first decreases with increasing current, and then increases revealing a profound minimum. To understand the current dependence of the frequency, it is instructive to analyze steady spin orbits of the sensing layer. Figure 9(b) shows saddle-shaped large angle orbits for $H_{\text{ext}} = 1.5\,\text{kOe}$. Since the vector $\hat{\mathbf{s}}$ passes then the layer plane, such modes will be referred to as large angle in-plane (LIP) precessions also known as "clamshell" mode [16, 52]. The arrow indicates change in the orbits as the current increases. Along the segments where $\hat{\mathbf{s}}$ moves almost in the layer plane, the spin precesses mainly around H_{a} and H_{ext}, with the angular velocity proportional to $|\gamma_{\text{g}}|(H_{\text{a}} + H_{\text{ext}})$. Along the remaining part (vector $\hat{\mathbf{s}}$ moves almost perpendicularly to the layer plane), the angular velocity is larger and proportional mainly to $|\gamma_{\text{g}}| H_{\text{d}x}$. With increasing I, the average orbital speed decreases while the arc length of the orbit increases. Consequently, ω_0 decreases with increasing I.

At a certain value of current, at which the segments touch each other near the $-z$ axis, the LIP orbit bifurcates into two so-called out-of-plane (OP) orbits, see Fig. 9(c). The orbits are interchangeable with respect to the $x \to -x$ and $y \to -y$ reflections. The mirror symmetry with respect to the axis x has been reported in [21]. In our case, the additional mirror symmetry with respect to the y axis is due to a nonzero value of b. Further increase in current enhances the spin torque and tends to push the orbit away from the layer plane. The arc length decreases then and orbital speed increases due to larger demagnetization field. Consequently, ω_0 increases with increasing I.

The frequency ω_0 increases with the increasing external magnetic field in both LIP and OP cases. We have compared the field dependence of the recorded ω_0 for selected current densities to the Kittel's formula for small-angle elliptical precession of a thin film ferromagnet. For the OP regime, we have found an increase in effective demagnetization term with increased current. The formula fits very well to the $\omega_0(H_{\text{ext}})$ dependence, however, the effective demagnetization term differs from that considered in simulation. Similar discrepancies have been discussed in experiments [16, 52, 22]. The discrepancy can be associated with an additional effective field induced due to spin transfer, which pumps energy to the system and facilitates such precessions. In turn, for LIP precessions the $\omega_0(H_{\text{ext}})$ dependence reveals features that cannot be fitted to the Kittel's formula.

To get a more complete picture of the spin valve dynamical behavior, it is instructive to construct a dynamical stability diagram. Such a diagram – magnetoresistance as a function of the current and applied magnetic field – see Fig. 10(a), captures different modes driven by the spin transfer. The diagram was obtained by sweeping current density I in the steps of size ΔI in a constant magnetic field. Before changing to the next value of I, we integrate Eq. (50) for a "waiting time" Δt. Over that time the instantaneous magnetoresistance is averaged and reported for each I. The sweeping rate is then defined as the ratio $\Delta I / \Delta t$. Simulations were started for biased P state about one degree in the layer plane. For low current densities, the dark region corresponds to the low-resistive P state. By increasing current up to values exceeding the critical value I_{c}^{P}, the spin valve is switched to the AP state (light region) when $H_{\text{ext}} < H_{\text{a}}$ For fields exceeding the anisotropy field, the increasing

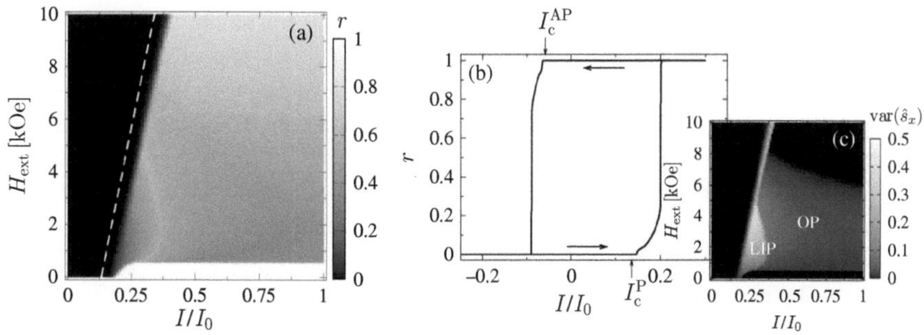

Fig. 10 Dynamical behavior of the Cu/Co(30)/Cu(10)/Co(4)/Cu spin valve scanned by increased current with the sweeping rate 4.5×10^5 A/cm^2 s at a constant applied field: (**a**) dynamical phase diagram of magnetoresistance and critical current given by Eq. (52) (*dashed line*); (**b**) current-driven hysteretic magnetoresistance behavior at $H_{ext} = 200$ Oe. The arrows indicate the direction of current changes. The critical currents given by Eqs. (52) and (53) are also indicated; (**c**) dynamical scan of the variance of $\hat{\mathbf{s}}_x$ as a function of the current and applied field. The regions of the steady precessional modes are indicated

current drives the spin valve *via* the LIP regime to the OP one. The white dashed line corresponds to I_c^P given by Eq. (52). The critical line is in satisfactory agreement with the numerical results. However, one can note some discrepancy, which appears because of finite sweeping rate ($4.5 \times 10^5\,\mathrm{A/cm^2\,s}$ that is much faster than usually in experiments). In the case of very fast sweeping rate, the spin during the "waiting time" Δt is unable to approach the steady in-plane regime near $+z$ axis (IP$_+$), and the averaging gives magnetoresistance lower than that one would get when "waiting" over relaxation time. Moreover, the relaxation time to the steady IP$_+$ regime ($I > I_c^P$) diverges when $I \to I_c^P$.

For very high external fields, comparable to the saturation magnetization of the sensing layer, the spin valve reveals current-driven switching between P and AP states – very similar to that at low fields. The main difference is in the spin trajectory for the sensing layer. The spin precesses now in a tight spiral from the P state to the final AP state, mainly around the external magnetic field. This can be observed only if current exceeds the critical value, $I > I_c^P$, when spin transfer pumps a sufficient amount of energy to the system, which allows to overcome the external field and damping.

The relevant quantity in the analysis of steady-state dynamics of the sensing layer is the variance of the corresponding spin component, var$(\hat{s}_x) = \langle(\hat{s}_x - \langle\hat{s}_x\rangle)^2\rangle$, where averaging is taken over the "waiting time." In Fig. 10(c), the spin variance is scanned by sweeping current in a constant field. Due to the geometry of steady-state spin trajectories, see Fig. 9(b) and (c), the variance of $\hat{s}_x$ for the in-plane regimes significantly differs from that for OP regime. Thus, the scan of var$(\hat{s}_x)$ can precisely distinguish different dynamical regimes. The dark regions in Fig. 10(c) correspond to the static states (the variance of $\hat{s}_y$ and $\hat{s}_z$ is also zero, not shown). The LIP regime separates the P state and OP regime, as well as the P state and high-resistance static states (HSS) for high applied fields. Further increase in external field leads to current-driven transition to AP state through a series of steady-state precessional regimes (IP_+ and OP) and HSS states. It is known that hysteresis is present when a system is driven between two steady states, and at least one of them is a fixed point. The HSS states are given by stationary solution of Eq. (50), so the hysteresis between the P state and the HSS state as well as between the HSS and AP states can be observed.

Typical hysteretic reversal from the low-resistive P state to the high-resistive AP state, driven by the spin transfer torque in the external field $H_{ext} = 200\,\mathrm{Oe}$, is shown in Fig. 10(b). We have found that critical currents $I_c^P = 0.136\,I_0$ and $I_c^{AP} = -0.059\,I_0$, calculated within stability analysis, Eqs. (52) and (53), correspond well to the numerical results. The right corner in the hysteresis loop indicates that the transition from P to AP states goes through the IP_+ states. The transition back to the P state is followed by the in-plane precessional regime, where the spin precesses mainly around the $-z$ axis. The shape of the hysteresis loop is significantly affected by the sweeping rate and temperature. It has been shown [21] that decreased sweeping rate leads to smaller hysteresis. The sweeping rate in numerical analysis is usually much larger than in experiments. Therefore, one can expect that LIP regime becomes narrower and completely disappears for low fields at finite temperatures [21] as it was observed in experiments [16, 52].

From the above follows that steady precessional regimes in symmetric spin valves are stabilized (we assume initial P state) by the interplay of: (i) spin transfer, which for $I > I_c^P$ destabilizes the P state and stabilizes the AP one; (ii) Gilbert damping that sustainedly dissipates energy pumped *via* spin transfer; (iii) positive external field which prefers the P state. From the application point of view, manipulation with the magnetic field is inconvenient, so it is highly desired to have devices that can operate in the absence of external magnetic field.

6.3 Dynamics in Asymmetric Spin Valves

In Section 6.2, we presented dynamical behavior in spin valves that contain fixed and sensing layers made of the same material. In Section 5, we showed that a proper choice of the material for magnetic layers may reverse the CIMS. Moreover, in Section 5.4, we discussed a model situation, where appropriate manipulation of the spin asymmetries produces a nonstandard angular dependence of the torque. The purpose of this section is to show that this nonstandard angular dependence may lead to a stable precessional regime in the absence of external magnetic field, which can be of great interest from the application point of view.

Asymmetric spin valves contain magnetic films having quite different bulk and/or interface spin asymmetry factors and different spin diffusion lengths. The spin valve Co/Cu/Py is an example of such structures. Two characteristics of an asymmetric system are important from the application point of view. First, the angle θ between magnetic moments of the layers, at which the spin torque vanishes, should be well resolved and sufficiently far from $\theta = 0$ (P configuration) and $\theta = \pi$ (AP configuration). Second, the critical current that destabilizes both P and AP configurations should be sufficiently low. Some enhancement of the spin transfer torque (decrease in the critical current) can be obtained for instance by reducing the spin-flip length in a nonmagnetic layer adjacent to the sensing layer [53], or constructing an antisymmetric spin valve [54, 55]. In turn, the asymmetry needed for the inverse spin accumulation enhancement in the spacer layer requires significant back-scattering in the spin-down channel at the normal-metal ferromagnet interface layer and, especially for the Co/Cu/Py spin valve, a relatively thin Co layer to decrease its overall spin asymmetry. Therefore, we assume an asymmetric spin valve including a synthetic antiferromagnetic structure pinned to an antiferromagnetic lead (substrate) to fix the Co layer. Along with the torque enhancement, such a structure significantly reduces the dipolar field acting on the sensing layer, provides lower switching fields, and supports single-domain stabilization – even for low aspect ratio of sensing layer [56].

In the following numerical analysis, we assume parameters typical for the spin valve IrMn/Co(d_1)/Ru(d_2)/Co(d_3)/Cu(d_4)/Py(d_5)/Cu, see Fig. 11(a). The system consists of pinned and reference cobalt layers (thicknesses d_1 and d_3, respectively) separated by a ruthenium layer (thickness d_2), copper spacer layer of thickness d_4, and sensing permalloy layer of thickness d_5. The reference layer is considered as a spin current polarizer. In turn, the P and AP configurations refer to relative alignment of the magnetic moments of reference and sensing layers. Strong antiferromagnetic coupling is assumed between the Co layers across the Ru spacer layer – the reason why such a structure is referred to as a synthetic antiferromagnet (SAF) [57]. The SAF is coupled to IrMn *via* exchange biasing, and therefore does not undergo dynamics in current densities and external magnetic fields of interest. Fixing the SAF to IrMn also allows one to use a thinner reference layer, which is required for enhancement of the inverse spin accumulation in the spacer layer.

In Fig. 11(b) we show the angular dependence of the torques τ_θ and τ_φ, respectively, calculated for the IrMn/Co/Ru/Co/Cu/Py/Cu nanopillar. To preserve the definition of the positive torque acting on the sensing Py layer in the P configuration, we change in the following the definition of positive current. Thus, since now

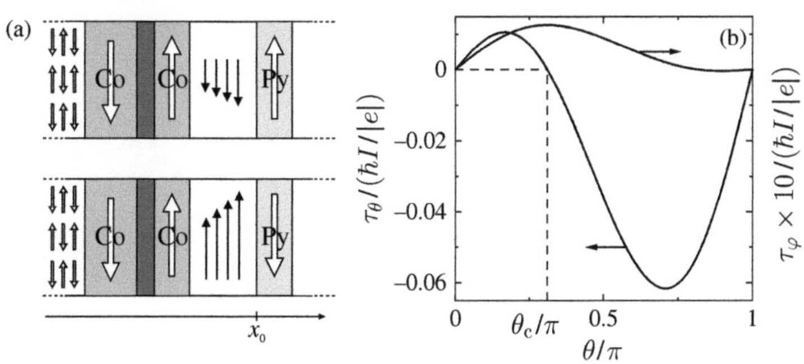

Fig. 11 Spin transfer torque in IrMn/Co(6)/Ru(2)/Co(4)/Cu(8)/Py(4)/Cu spin valve: (**a**) schematic structure of the spin valve and spin accumulation in thin spacer layer for the parallel and antiparallel configurations of the reference and sensing layers, and for electrons flowing along the axis x; (**b**) angular dependence of the torque τ_θ (*left scale*) and τ_φ (*right scale*) acting on the sensing Py layer

positive current corresponds to electrons flowing from the Co to Py layer. From Fig. 11(b) follows that positive current destabilizes both P and AP magnetic configurations and negative current stabilizes both the configurations. The destabilization of the collinear configurations can be intuitively understood within the two-channel model. The high bulk resistivity and quite short spin-diffusion length of Py results in high spin down-channel resistivity in both collinear configurations, which causes significant reflection of the spin-down electrons back to the spacer layer. This gives rise to negative spin accumulation, as shown schematically in Fig. 11(a). This scenario is very similar to that of structures with inverse CIMS discussed in Section 5. The main difference is that transverse spin current component incident in non-collinear configurations on the active interface ($x = x_0$) vanishes due to mixing both spin channels. Therefore, the mixing conductances of both Co/Cu and Cu/Py interfaces have a significant influence on θ_c.

The angular dependence of τ_θ shows that τ_θ vanishes at $\theta = \theta_c \neq 0, \pi$. The parameter θ_c/π is plotted in Fig. 12(a) and (b) as a function of the magnetic layer thicknesses. It follows from the plots that θ_c becomes particularly large for thick sensing and thin reference layers. However, relatively thick sensing layers require higher critical currents. The dependence of θ_c on the spacer thickness is rather weak, so thicker spacers can be used to eliminate possible residual interlayer exchange interaction.

Figure 13(a) shows dynamical phase diagram of the reduced magnetoresistance calculated as a function of the external magnetic field and current density. Simulations have been performed for the initial P state and current was swept in a constant magnetic field. The corresponding phase diagram for the initial AP state is shown in Fig. 13(b). In both cases the white and black dashed lines correspond to the critical currents given by Eqs. (52) and (53), which destabilize the P and AP configurations, respectively. The demagnetization field for sensing Py layer of a flat ellipsoid shape with the radii 140 nm and 70 nm, and with a reduction factor of 0.5,

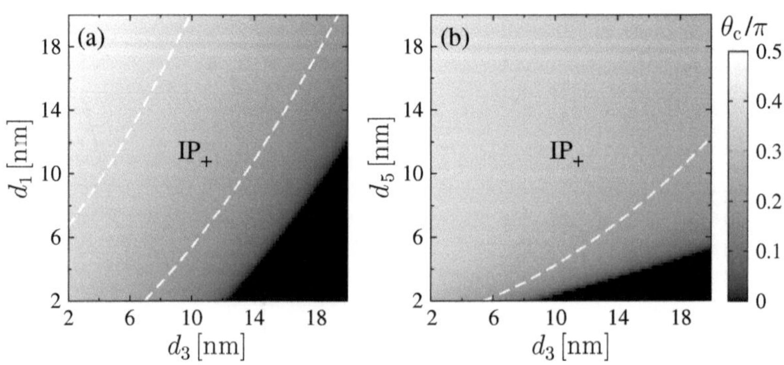

Fig. 12 The θ_c/π parameter in IrMn/Co(d_1)/Ru(2)/Co(d_3)/Cu(8)/Py(d_5)/Cu spin valve: (**a**) θ_c/π as a function of d_1 and d_3; (**b**) θ_c/π is presented as a function of d_5 and d_3. The other thicknesses are $d_1 = 6$ nm, $d_3 = 4$ nm, and $d_5 = 4$ nm. The critical thicknesses for the IP_+ steady precessional modes (*white dashed lines*) for $H_{ext} = 0$ and $I = 10^8$ A/cm^2 are also indicated

was assumed [58]. Apart from this, for the saturation magnetization we assumed $M_s = 10053.1$ Oe, for the anisotropy field $H_a = 2.51$ Oe, and for Gilbert damping $\alpha = 0.01$.

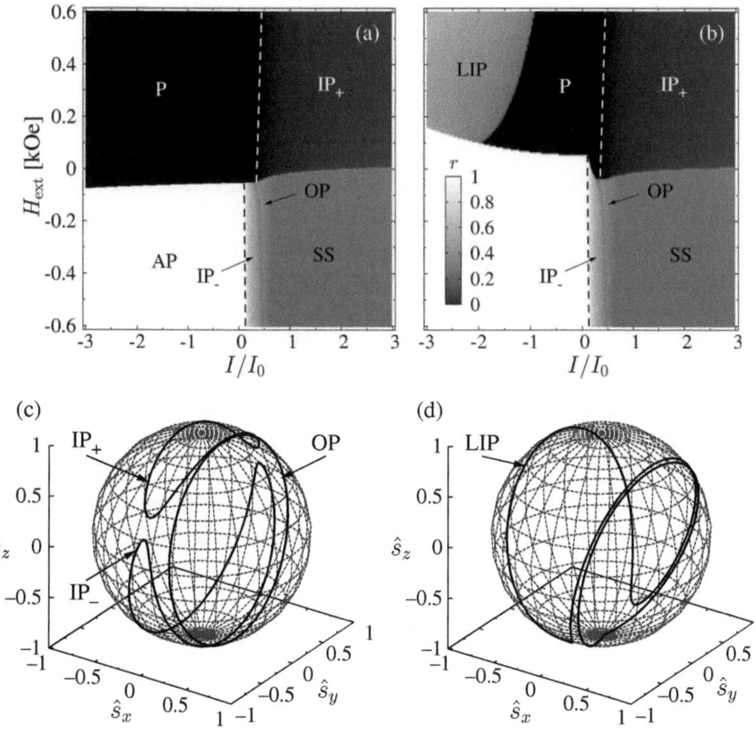

Fig. 13 Dynamical phase diagram of magnetoresistance for the IrMn/Co(6)/Ru(2)/Co(4)/Cu(8)/Py(4)/Cu spin valve: (**a**) scanned by increasing and (**b**) decreasing current with the sweeping rate 3.3×10^4 A/cm^2 s in a constant applied field. The corresponding critical currents given by Eqs. (52) and (53) (*dashed lines*) are indicated; (**c**) representative in-plane IP_+ and IP_- as well as OP stable orbits; (**d**) steady LIP orbit and *one-wing* orbit after bifurcation observed at large fields

For the initial P state, Fig. 13(a), we find stable AP state for $I < I_c^{AP}$ and negative magnetic fields (below a certain threshold value). When current increases in a constant magnetic field and becomes positive, the AP state is driven due to spin transfer to the in-plane precessional regime (IP_-), with the steady orbits near the $-z$ axis, and then to the OP regime. When the current increases further, the system is driven to the steady static (SS) states. For low fields the SS states involve HSS states with $\hat{\mathbf{s}}$ near the $\pm x$ axis, which are close to the magnetic energy maxima. The presence of steady static states gives rise to current-driven hysteretic behavior (discussed below).

For $I < I_c^{P}$ and positive or small negative fields, the P state is stable. With increasing current there is a transition from the P state to steady in-plane precessional mode, IP_+. The precessional modes can be observed in a broad current and field regions – also in the absence of magnetic field. The region where the precessional modes exist without field is of particular interest from the application point of view. This region depends on other parameters of the system. For instance, in Fig. 12(a) and (b) the regions bounded by the white dashed lines specify the layer thicknesses where the zero-field IP_+ modes occur for $I/I_0 = 1$ and initial P state. In the areas below the IP_+ regions in Fig. 12(a) and (b), the initial P state is stable for thick reference layers (system behaves as symmetric spin valve – dark regions). In turn, for the region above the IP_+ regime in Fig. 12(a), the system is driven to SS states.

For the AP initial state, Fig. 13(b), the diagram for positive currents is similar to the diagram for the initial P state. The difference for $I < I_c^{AP} < I_c^P$ is due to different stability conditions of the P and AP states, i.e., the result of different $\partial\tau_\theta/\partial\theta$ near the P and AP states, see Fig. 11(b). For the considered flat ellipsoid in zero magnetic field, one can assume $H_{dx} > H_{dy} > H_{dz}$, and taking into account $H_a < H_{dz}$, the coercive field calculated near the P (AP) state for $I = 0$ is given by $H_c^{P(AP)} = \pm(H_{dz} - H_{dy} - H_a)$. The P state is stable for $H_{ext} > H_c^P$ and the AP one is stable for $H_{ext} < H_c^{AP}$. This symmetrical hysteresis in field is broken when $I \neq 0$, and simple stability analysis provides quite complicated formula for H_c. Moreover, the initial AP state for positive fields and quite negative currents is driven to steady LIP precessional regime, see Fig. 13(d). In large magnetic fields, the LIP orbit bifurcates to *one-wing* orbits in a sufficiently large negative current. This can be observed in a current-driven frequency domain experiment as a profound frequency minimum, also observed in symmetric spin valves, see Fig. 9.

When considering the SAF structure we assumed that dipolar coupling between the reference and sensing layers is completely eliminated. However, one can speculate that some residual coupling is still present and affects the steady precessional regime. A coupling due to dipolar field acting on the sensing layer can be assumed as an additional negative field. In Fig. 13 we see that precessional regimes also exist even for negative fields.

Let us study in detail the spin valve in the absence of external magnetic field. Figure 14(a) shows the reduced magnetoresistance as a function of the current density. The arrows indicate direction of the current changes. The initial AP configuration is stable for $I/I_0 < 0.1$. Further increase in current drives the system *via* the IP_- modes to the P state. The steady orbits of the IP_- oscillations are shown in Fig. 14(e) for several current values. The corresponding oscillation frequency ω_0 decreases with increasing I, as shown in Fig. 14(b).

The P state is stable up to $I = 0.36I_0$, and above this point the system is driven to the IP_+ modes. Such precessions of decreasing frequency with increasing current in symmetrical spin valves are called "clamshell" [16, 52] or "in-plane" [21] modes. But the main difference between symmetrical pillars and those studied in this section concerns the way the spin transfer acts on $\hat{\mathbf{s}}$. In this case, increasing current leads to an increase in τ , and the orbits become narrower and elongated in the z direction due to τ_θ, see Fig. 14(c). At a certain current density, when $s_z = \cos\theta_c$, τ_θ vanishes and orbital velocity reaches minimum. The damping torque is small and the orbit bifurcates at $I = 1.95I_0$ due to a nonzero τ_φ,

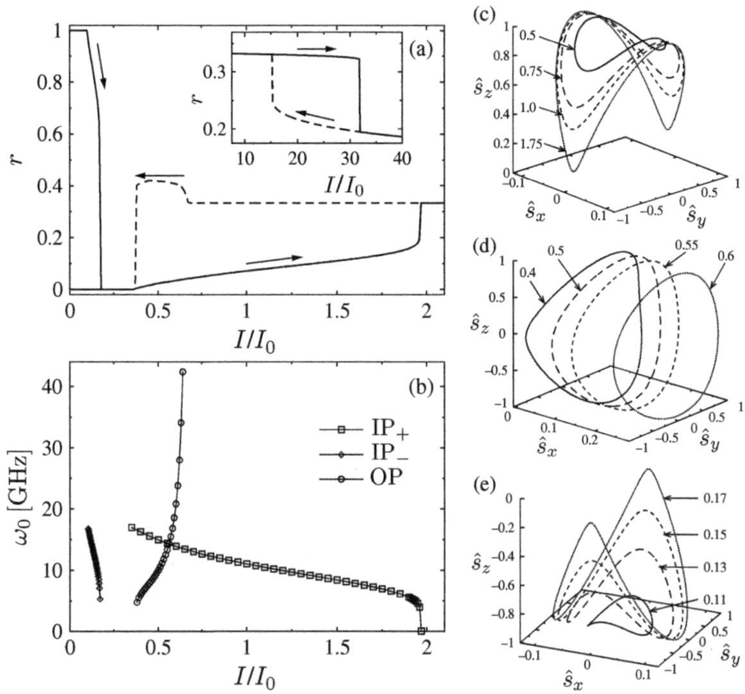

Fig. 14 Dynamical behavior of IrMn/Co(6)/Ru(2)/Co(4)/Cu(8)/Py(4)/Cu spin valve in the absence of external magnetic field: (**a**) current-driven hysteretic behavior of the magnetoresistance in zero field, scanned with increasing (*solid line*) and decreasing (*dashed line*) current with the sweeping rate 1.2×10^4 A/cm^2 s; (**b**) Fundamental frequency ω_0 of the magnetoresistance oscillations calculated as a function of current; (**c**) in-plane steady state orbits IP$_+$ near the $+z$ axis; (**d**) out-of-plane (OP) orbits; (**e**) in-plane orbits (IP$_-$) near the $-z$ axis presented for indicated current values

$$\tau_\varphi = -\frac{\hbar}{e^2}\left[\operatorname{Re} G_{\uparrow\downarrow} + \frac{(\operatorname{Im} G_{\uparrow\downarrow})^2}{\operatorname{Re} G_{\uparrow\downarrow}}\right](g_x \sin\varphi - g_y \cos\varphi)\Bigg|_{\substack{\theta\to\theta_c \\ x\to x_0}}, \qquad (55)$$

and the spin is driven *via* the transient regime to the HSS regime. The nonzero torque τ_φ at $\theta \to \theta_c$ assists in the transition and thus reduces the critical switching current. Here, $G_{\uparrow\downarrow}$ is the mixing conductance for the active Cu/Py interface. When $\operatorname{Im} G_{\uparrow\downarrow}$ vanishes for the Co/Cu interface, one finds $\tau_\varphi = \tau_\theta = 0$ at $\theta = \theta_c$, and the region of IP$_+$ modes extends to high current densities.

The HSS states are stable up to a large current density and give rise to the current-driven hysteresis between HSS and SS states of lower resistance, see the inset of Fig. 14(a). Switching between the steady states is driven *via* transient regimes, where spin transfer pumps energy to the system (T_+) or dissipates energy (T_-). The hysteresis can be formally written as a sequence of transitions AP $\to$ IP$_-$ $\to$ P $\to$ IP$_+$ $\to$ T$_+$ $\to$ HSS $\to$ T$_-$ $\to$ SS for increasing current and SS $\to$ T$_+$ $\to$ HSS $\to$

$T_- \rightarrow OP \rightarrow P$ for decreasing current. The OP modes exist roughly in the region $0.4 < I/I_0 < 0.6$ only when system is driven from the SS state. The fundamental frequency of the OP oscillations decreases with decreasing current, see Fig. 14(b). For the current densities close to $I = 0.5I_0$, both the IP_+ and OP modes have similar frequencies. However, both the regimes are well separated by irreversible paths, as follows from the presence of hysteresis. We note that finite temperature can induce "telegraph" jumps between the regimes, which leads to vanishing of the hysteresis. A periodical modulation of the current, around the value where the bistability occurs, induces controlled jumps between both the oscillatory regimes increasing the singal to noise ratio, which is know in bistable systems as the stochastic resonance.

7 Conclusions

The unified description of GMR and CIMS, based on diffusive transport equations and macroscopic boundary conditions, allows one to study transport characteristics of spin valves as a function of layer magnetization configuration. Most of the parameters of this description can be evaluated directly from the analysis of CPP-GMR experimental data or derived within first-principles calculations of the transmission of spin currents across a particular interface.

The spin transfer torque exerted on the sensing layer has a significant influence on its dynamics. Results obtained for symmetric spin valves correspond well to previous macrospin calculations, and are in good qualitative agreement with experimental measurements. Moreover, we have shown that spin polarized current in certain asymmetrical structures can destabilize both P and AP configurations of a spin valve, and can drive microwave oscillations in the absence of magnetic field. This takes place only for one orientation of the bias current and for a given initial state.

The macrospin model describes well many dynamical aspects of spin valves. However, several experiments on Co/Cu/Co spin valves [16] pointed out the presence of a W-state with the resistance between P and AP state, and with only small microwave signals for large positive current and applied field above that destabilizing the AP state. Since the macrospin model predicts the presence of IP_+ or OP states, the authors speculated that macrospin model does not explain the W-state due to possible incoherent magnetization movement within the sensing layer. Contrarily, the measurements on Py/Cu/Py stacks [52] have not reported the W-phase. To explain the discrepancy, some authors performed detail micromagnetic simulations, but the question of critical nanoelement size that has to be considered needs to be answered. Such micromagnetic simulations [19, 59, 60] reported transition from quasi-monodomain to multidomain and further to chaotic state with increasing current densities. Moreover, study of the lateral size of the sensing layer with respect to considered nanoelement, which is treated as monodomain, revealed the transition from steady precessional regime to chaotic regime induced by spin transfer [61].

Recently, the pump-probe x-ray microscopy technique using x-ray magnetic circular dichroism effect was used for direct observation of the current-induced magnetization dynamics in spin valves [62]. The direct spatiotemporal imaging confirmed that magnetization switching follows an inhomogeneous evolution. Thus, to study the dynamics in spin valves, one needs to work out more elaborated models with detailed micromagnetic description and considering the Oersted field, which shifts micromagnetic elements toward smaller sizes. This can be used when simple macrospin model fails. However, the model presented in this chapter explains, at least qualitatively, many features of the observed dynamics in spin valve structures.

Acknowledgments This work, as part of the European Science Foundation EUROCORES Programme SPINTRA, was supported by funds from the Ministry of Science and Higher Education as a research project in years 2006–2009 and the EC Sixth Framework Programme, under Contract N. ERAS-CT-2003-980409. The work was also supported by Slovak Ministry of Education within the project MVTS POL/SR/UPJS/07 and Slovak Grant Agency VEGA under Grant No. 1/2009/05.

References

1. M.N. Baibich, J.M. Broto, A. Fert et al., Phys. Rev. Lett. **61** (1988) 2472
2. G. Binasch, P. Grünberg, F. Saurenbach et al., Phys. Rev B **39**, 4828 (1989)
3. J. Barnaś, A. Fuss, R.E. Camley et al., Phys. Rev. B **42**, 8110 (1990)
4. J. C. Slonczewski, J. Magn. Magn. Mater. **159**, L1 (1996); **195**, L261 (1999)
5. L. Berger, Phys. Rev. B **54**, 9353 (1996)
6. J.A. Katine, F.J. Albert, R.A. Buhrman et al., Phys. Rev. Lett. **84**, 3149 (2000)
7. N.F. Mott, Adv. Phys. **13**, 325 (1964)
8. I.A. Campbell, A. Fert: Transport Properties of Ferromagnets. In: *Ferromagnetic Materials*,Vol. 3, ed by E.P. Wohlfarth, North-Holland, Amsterdam (1982), pp 747–804
9. P.M. Levy, S. Zhang, A. Fert., Phys. Rev. Lett. **65**, 1643 (1990)
10. S. Zhang, P.M. Levy, A. Fert., Phys. Rev. B **45**, 8689 (1992)
11.. J. Grollier, V. Cros, A. Hamzic et al., Appl. Phys. Lett. **78**, 3663 (2001)
12. M. AlHajDarwish, H. Kurt, S. Urazhdin et al., Phys. Rev. Lett. **93**, 157203 (2004)
13. M. Tsoi, J.Z. Sun, M.J. Rooks et al., Phys. Rev. B **69**, 100406(R) (2004)
14. M. Tsoi, V. Tsoi, J. Bass et al., Phys. Rev. Lett **89**, 246803 (2002)
15. M. Tsoi, J.Z. Sun, S.S.P. Parkin., Phys. Rev. Lett. **93**, 036602 (2004)
16. S.I. Kiselev, J.C. Sankey, I.N. Krivorotov et al., Nature (London) **425**, 380 (2003)
17. A.N. Slavin, V.S. Tiberkevich., Phys. Rev. B **72**, 094428 (2005)
18. J.Z. Sun., Phys. Rev. B **62**, 570 (2000)
19. Z. Li, S. Zhang., Phys. Rev. B **68**, 024404 (2003)
20. H. Xi, Z. Lin., Phys. Rev. B **70**, 092403 (2004)
21. J. Xiao, A. Zangwill, M.D. Stiles., Phys. Rev B **72**, 014446 (2005)
22. W.H. Rippard, M.R. Pufall, S. Kaka et al., Phys. Rev. Lett. **92**, 027201 (2004); Phys. Rev. B **70**, 100406(R) (2004)
23. J. Barnaś, A. Fert, M. Gmitra et al., Phys. Rev. B **72**, 024426 (2005)
24. M. Gmitra, J. Barnaś., Phys. Rev. Lett. **96** 207205 (2006); Appl. Phys. Lett. **89**, 223121 (2006); Phys. Rev. Lett. **99** 097205 (2007)
25. X. Waintal E.B. Myers, P.W. Brouwer., Phys. Rev. B **62**, 12317 (2000)
26. J. Slonczewski., J. Magn. Magn. Mat. **247**, 324 (2002)
27. M.D. Stiles, A. Zangwill., Phys. Rev. B **66**, 014407 (2002); J. Appl. Phys. **91**, 6812 (2002)
28. A. Brataas, Yu.V. Nazarov, G.E.W. Bauer., Phys. Rev. Lett. **84**, 2481 (2000)

29. D.H. Hernando, Yu.V. Nazarov, A. Brataas et al: Phys. Rev. B **62**, 5700 (2000)
30. A.A. Kovalev, A. Brataas, G.E.W. Bauer., Phys. Rev. B **66**, 224424 (2002)
31. S. Urazhdin, N.O. Birge, W.P. Pratt et al., Appl. Phys. Lett. **84**, 1516 (2004)
32. T. Valet, A. Fert., Phys. Rev. B **48**, 7099 (1993)
33. J. Bass, W.P. Pratt Jr., J. Magn. Magn. Mater. **200**, 274 (1999)
34. A. Brataas, Yu.V. Nazarov, G.E.W. Bauer., Eur. Phys. J. B **22**, 99 (2001)
35. K. Xia, P.J. Kelly, G.E.W. Bauer et al., Phys. Rev. B, **65**, 220401 (2002)
36. M. Zwierzycki, Y. Tserkovnyak, P.J. Kelly et al., Phys. Rev. B **71**, 064420 (2005)
37. W.P. Pratt, private communication
38. P. Dauguet, P. Gandit, J. Chaussy et al., Phys. Rev. B **54**, 1083 (1996)
39. C. Vouille, A. Barthélémy, F. Elokan Mpondo et al., Phys. Rev. B **60**, 6710 (1999)
40. A.C. Reilly, W. Park, R. Slater et al., J. Magn. Magn. Mater. **195**, L269 (1999)
41. S.D. Steenwyk, S.Y. Hsu, R. Loloee et al., J. Magn. Magn. Mater. **170**, L1 (1997)
42. W. Park, R. Loloee, J.A. Caballero et al., J. Appl. Phys. **85**, 4542 (1999)
43. K. Eid, R. Fonck, M. AlHaj Darwish et al., J. Appl. Phys. **91**, 8102 (2002)
44. X. Waintal, E.B. Myers, P.W. Brouwer et al., Phys. Rev. B **62**, 12317 (2000)
45. J. Xiao, A. Zangwill, M.D. Stiles., Phys. Rev. B **70**, 172405 (2004)
46. O. Boulle, V. Cros, J. Grollier et al., Nature Phys. **3**, 492–497 (2007)
47. A. Vedyayev97, N. Ryzhanova, B. Dieny et al., Phys. Rev. B **55**, 3728–3733 (1997)
48. A. Shpiro. P.M. Levy, S. Zhang, Phys. Rev. B **67**, 104430 (2003)
49. S. Urazhdin, R. Loloee, W.P. Pratt, Jr., Phys. Rev. B **71**, 100401 (2005)
50. S. Wiggins: *Introduction to Applied Nonlinear Dynamical Systems and Chaos*, Springer, New York, (1990) pp 35–36
51. G. Bertotti, C. Serpico, I.D. Mayergoyz et al., Phys. Rev. Lett. **94**, 127206 (200)
52. S.I. Kiselev, J.C. Sankey, I.N. Krivorotov et al., Phys. Rev. B **72**, 064430 (2005)
53. J. Manschot, A. Brataas, G.E.W. Bauer., Appl. Phys. Lett. **85**, 3250 (2004)
54. L. Berger., J. Appl. Phys. **93**, 7693 (2003)
55. Y. Jiang, G.H. Yu, Y.B. Wang et al., Appl. Phys. Lett. **86**, 192515 (2005)
56. K. Inomata, N. Koike, T. Nozaki et al., Appl. Phys. Lett. **82**, 2667 (2003)
57. S.S.P. Parkin, D. Mauri., Phys. Rev. B **44**, 7131 (1991)
58. J.A. Osborn., Phys. Rev. **67**, 351 (1945)
59. X. Zhu, J.G. Zhu, R.M. White., J. Appl. Phys. **95**, 6630 (2004)
60. K.J. Lee, A. Deac, O. Redon et al., Nat. Mater. **3**, 877 (2004)
61. D. Berkov, N.L. Gorn., Phys. Rev. B **71**, 052403 (2005); Phys. Rev. B **72**, 094401 (2005)
62. Y. Acremann, J.P. Strachan, V. Chembrolu et al., Phys. Rev. Lett. **96**, 217202 (2006)

Self-Organized Surface Nanopatterning by Ion Beam Sputtering

Javier Muñoz-García, Luis Vázquez, Rodolfo Cuerno, José A. Sánchez-García, Mario Castro, and Raúl Gago

Abstract The production of self-organized surface nanopatterns by ion beam sputtering (IBS) at low (<10 keV) and intermediate (10–100 keV) energies has emerged in the last decade as a promising bottom-up nanostructuring tool. The technique is remarkably universal, being applicable to metals, semiconductors or insulators, and it enables large degree of control over the main pattern features with high throughput (it requires low process time and can be used over extended areas). However, there is a wide scatter in the experimental results obtained as a function of system type and process parameters. In parallel, diverse theoretical models have been developed that differ in their capabilities to reproduce such a wide range of experimental features. We provide an overview of the most recent studies on the production of nanoripple, nanohole and nanodot periodic nanostructures by IBS, with special attention to the comparison between experiments and (continuum) models, and with a focus on those issues that remain open or, at least, ambiguous. The pattern properties to be considered are those of potential increased technological importance, such as the variation of size, shape, distance and ordering of the nanostructures as a function of parameters such as ion energy, target temperature and sputtering time (i.e., fluence). Finally, reported and proposed applications of IBS nanopatterns are briefly presented showing, in this way, the high-potential functionality of IBS nanostructured surfaces.

Contents

R. Gago (✉)
Instituto de Cìencia de Materiales de Madrid, Consejo Superior de Investigaciones Científicas, E-28049 Madrid, Spain
e-mail: rgago@icmm.csic.es

Z.M. Wang (ed.), *Toward Functional Nanomaterials*, Lecture Notes in Nanoscale Science and Technology 5, DOI 10.1007/978-0-387-77717-7_10,

1 Introduction

Nanotechnology and Nanoscience are inducing a turning point in fields from Condensed Matter Physics and Materials Science to Chemistry and Biology. This is due to the new types of behaviors and properties displayed by nanostructures, ranging across traditional disciplines [83]. Interest in these systems is also triggered by the possibility of characterizing them morphologically using scanning probe microscopy (SPM) techniques [11] and by the advance in surface structure determination methods. Within this more specific context, there is an increasing need for the development of techniques and methods for the patterning of materials at the nanoscale. Moreover, this goal must be achieved in an efficient, fast and low-cost manner for eventual technological applications to be compatible with mass production. In addition, the processes involved should allow control over the size, shape and composition of the produced nanostructures.

Within this trend, there is also a great interest in developing methods for producing confined nanostructures, which would display quantum confinement effects on surfaces. Of special interest would be those systems in which these surface nanostructures form an ordered pattern.

Among the technological issues behind surface nanostructuring, we can highlight:

- New advances in magnetic storage technology [152].
- The production of surfaces and thin films with well-controlled shape at the nanoscale for applications in catalysis [189].

- The design of nanoelectronic devices based on the use of quantum devices [3], for example for developing multispectral detector arrays by exploring novel detection techniques, nanopatterning and control of the production of self-assembled quantum structures.
- The possibility of functionalization of these surface nanostructures that allows the selective attachment of specific molecules [92].
- The design of nanosensors based on surface-enhanced Raman scattering (SERS) effects or localized surface plasmon resonance [186].
- The use of these nanopatterned surfaces as templates to transfer these patterns to highly functional surfaces [9] that cannot be patterned directly. Also, the use of these patterns as templates can be applied to reduce fabrication steps or to increase productivity.

There are different approaches for surface nanostructuring. Among the so-called *top-down* methods we can mention lithographic techniques (nanoimprint lithography [76], nanosphere lithographic techniques [186], soft lithographic methods [184] and focused ion beam (FIB) techniques [117]). Other approaches are based on the use of SPM to induce nanostructures on a surface through different processes such as tip-induced oxidation [18], tip-induced e-beam lithography [155], dip-pen nanolithography [116], the application of strong electric fields between tip and sample [106], and the mere use of the probe as a stylus or pen to write at the nanoscale on the sample surface [156]. However, these methods present different limitations, such as proximity effects, low resolution or the need of parallel processing because of the (relatively) small-processed area. In the case of SPM-based techniques, mainly Scanning Tunneling Microscopy (STM) and Atomic Force Microscopy (AFM), the use of an array of tips aims to compensate for this last limitation.

Potential alternatives to overcome the limitations of the *top-down* approach are provided by the *bottom-up* methods, which are mainly based on self-organized processes. In this field, most efforts have been traditionally focused on the production of self-organized nanostructures occurring in semiconductor heterostructure growth [159,170] in which strain relief mechanisms take place. This simpler method seems to be a highly cost-efficient route toward large-scale arrays of nanostructures, although it presents some disadvantages such as enabling less control on structure size and shape than other methods and the need to work under ultra-high vacuum conditions.

In recent years, special interest is being paid to the study of self-organized nanopattern formation on surfaces by *ion beam sputtering* (IBS) techniques [55]. In general, two types of surface nanostructures can be induced by IBS: (i) nanoripples and (ii) nanodots. In both cases, the pattern formed by these nanostructures can have dimensions ranging from a few up to hundreds of nanometers. These patterns can be produced on different materials, amorphous or crystalline, in just a few minutes and over areas of several square millimeters. The diversity of materials processed and the similar morphologies obtained indicate the universality of the process. In addition, it allows the eventual control of the induced nanostructures by changing the sputtering parameters, such as ion energy, dose, substrate temperature and ion incidence geometry. Moreover, it can be used to produce functional surfaces and isolated structures [177]. Thus, IBS becomes a versatile, fast and cheap technique for surface nanopatterning.

The ability of IBS to induce submicrostructures on surfaces was reported more than forty years ago by Navez et al. [130]. In that work, they reported the production of periodic nanoripples on glass surfaces. In Fig. 1b we show ripples obtained on a silicon surface immersed in an argon plasma. These ripples indeed remind us of those formed on sand dunes by the wind or on the sand bed close to the water edge by the water flow (Fig. 1a). As we will show below, this similarity is more than merely visual as the theories dealing with both types of (macro and nano) structures share many conceptual aspects. In the pioneering work of Navez and coworkers the authors did report important morphological observations, such as the change of the ripple orientation when the ion incidence polar angle is larger than a threshold value. Later, similar findings were reported for other materials that have been reviewed elsewhere [21,112,127,177]. Usually, the length scales involved in these patterns lie in the submicrometer scale, rather than in the nanometer scale. Moreover, the traditional field for ripple applications had been that of optical gratings [86]. From the theoretical point of view, the first advances for understanding IBS nanostructuring was made by Sigmund [154] as he showed that local surface minima should be eroded at a faster rate than local maxima (i.e., the sputtering rate depends on the local surface curvature), leading to a surface instability, which is the origin of the nanostructuring process. Based on this work, Bradley and Harper (BH) proposed later the first continuum model describing ripple formation [13]. In recent years,

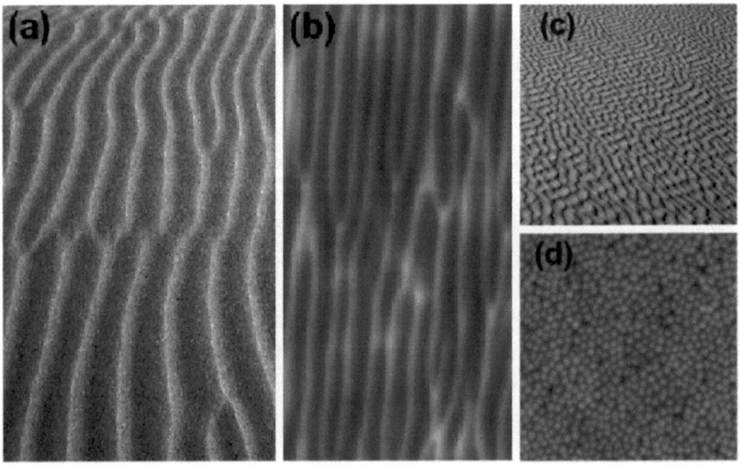

Fig. 1 (**a**) Ripples on a sand dune in Morocco. Photograph courtesy of J. Rodríguez and E. Blesa. (**b**) $3.7 \times 6.7\,\mu m^2$ top view AFM image of a Si surface immersed in argon plasma. (**c**) "Dots" on a sand dune in New Mexico, USA. Copyright Bruce Molnia, Terra Photographics. Image Courtesy Earth Science World Image Bank http://www.earthscienceworld.org/images. (d) $1 \times 1\,\mu m^2$ top view AFM image of a GaSb surface irradiated by 0.7 keV Ar^+ ions under normal incidence

other continuum models have been proposed [44,111,137]. These models account for different experimental behaviors, such as the presence or absence of saturation for the ripple amplitude, ripple orientation, ripple dynamics as well as the existence or not of kinetic roughening.

However, a clear turning point occurred when in 1999 Fackso and coworkers [55] reported on the IBS production of GaSb nanodot patterns, which also display short-range order. This work shifted the focus of the research to the nanometer scale. The evident technological implications [34,107] of the possibility to induce nanostructures on semiconductor surfaces on a relatively large area (several square millimeters) and in just a couple of minutes motivated further interest in these processes. Thus, nanodot production has been reported in different materials such as InP [64], Si [70] and Ge [197]. With respect to the theoretical understanding of the nanodot IBS production, the theories must share most of the concepts with those developed for IBS-induced ripples. An example is shown in Fig. 1c and d, in which "dotted" structures formed on both sand dunes and IBS nandot structures are displayed. Once more, this similarity appears that, rather than being just visual, is, in fact, deeper as will be showed in Section 4.3.

The production of dot patterns has changed the focus of the theoretical investigations. Now, for dot patterns the interest is more directed toward dot size and pattern order because of their technological implications. In the first case, it is interesting to know how the dot size changes with the different IBS parameters, especially ion energy, target temperature and sputtering time (i.e., fluence). In the second case, it is necessary to know under which conditions the pattern order is enhanced. Thus, issues such as pattern wavelength coarsening and order enhancement have become relevant. Accordingly, new theoretical descriptions have been proposed [25,35,59,87,112,123,178].

The aim of this chapter is to give an overview of ripple and dot IBS nanopatterning that focus mainly on those issues that remain open or, at least, ambiguous. This will be more evident in the case of nanodot patterns on amorphizable targets due, probably, to their relative novelty. In fact, we will treat mainly the case of this type of patterns since previous reviews were devoted mainly either to ripple patterns [21,112] or to the case of metals [177]. Stress will be paid to review the observed behaviors and the predictions of the main existing continuum models. When possible, we will contrast the experimental data with the theoretical predictions.

The chapter will be divided as follows. After this introduction and for the sake of completeness, we will present the basics of the physics behind the ion sputtering process. Then, we will give an overview of experimental findings on surface nanopatterning by IBS on amorphous materials and single-crystal semiconductors. We will review briefly the results for the special case of single-crystal metal surfaces as they were already extensively reviewed by Valbusa and coworkers [177]. In the next section, a review of the theoretical studies of these processes will be presented, with special emphasis on the various continuum models proposed so far. In the last section of the chapter we will propose possible applications of these nanopatterns and, finally, we will state the issues that still remain open, from our point of view, both theoretically and experimentally.

2 Fundamentals of Ion Sputtering

2.1 Introduction to Ion Sputtering

Physical sputtering is defined as the removal of atoms from a solid surface due to energetic particle bombardment. This phenomenon was first reported by Grove [75] in 1852, although others had probably noted the effect while studying glow discharges. Grove studied the discharge generated by the tip of a wire positioned close to a polished silver surface and noted a ring-like deposit onto the silver surface when it was acting as an anode in the circuit. However, it was not until the early 1900's when the effect was attributed to the impingement of positive ions accelerated toward the cathode by the electrical field [160]. The sputtering process resembles the macroscopic sandblasting process but with ionized atoms instead of sand grains as bombarding species. In fact, the erosion efficiency with the incident angle of the abrasive particles [62] is analogous to that observed in ion sputtering experiments.

The sputtering process is important from both a fundamental and a practical point of view. On the one hand, the understanding of the process is relevant to describe the basic interactions of ions with matter. On the other hand, as shown below, the process has found a broad range of applications. The improvement in experimental methods, as well as in the theoretical description of the process, has promoted the rapid maturation of the field during the last few decades.

2.2 Applications of Ion Sputtering

The applications of ion sputtering to Surface Science and Technology are very diverse [127]. First, the removal of atoms from the surface can be used as an effective method for *surface cleaning* [167]. This may be aimed, for example, at removing the undesirable contamination layer present on a surface (mostly containing oxygen and carbon impurities), which may affect the intrinsic electrical or optical properties. This cleaning process may be crucial for the optimal performance of a semiconductor device [149].

The sputtered material from the target can also be transferred to a substrate for *thin film deposition* (sputter deposition methods) [143]. This application of sputtering was first addressed by Plücker in 1958 [138] after he observed the formation of a platinum film inside a discharge tube with a metallic mirror-like appearance. Sputter deposition is a non-equilibrium process where material ejection occurs at hyperthermal energies (in the few eV range) [165]. This energy range is much higher than in thermal evaporation methods, leading to coatings with better properties (higher density, etc.). The sputtering phenomenon is the base of many industrial coating activities, being scalable from small pieces up to processing of large areas (it is commonly used to coat structural window panels). It is relevant for the development of magnetron sputtering techniques [143], where high erosion/growth rates are obtained by confining the plasma discharge close to the cathode through a magnetic field.

The ejected particles during sputtering can also be identified for compositional analysis of the target material. This technique is known as *secondary ion mass spectrometry* (SIMS) [191] and is based on the principle that a small fraction of the sputtered particles are ionized. These particles are guided and detected by means of a mass spectrometer and provide very high depth resolution (in the monolayer range) and elemental sensitivity down to parts per billion (ppb).

The progressive etching of the target material through sputtering finds usage for *sample thinning*. Such a process is often incorporated into specimen preparation for transmission electron microscopy (TEM) where, after mechanical milling, grazing incidence ions impinge onto the remaining part of the sample until the desirable electron transparency is achieved (a few hundreds of nanometers for the energies commonly used for imaging). The progressive erosion of the surface is also an extended route for *depth profiling* with surface analytical techniques by a sequence of sputtering and analysis steps [133]. In this case, high depth resolution can be achieved by erosion depths of a few monolayers but, as may occur in SIMS, by-products of ion bombardment such as atomic intermixing or preferential sputtering deteriorate the resolution of the analysis.

The application of material etching through sputtering is more extended in the fabrication of *integrated circuit devices*. Here, pattern or feature formation taking part in implementing the design of the device is imprinted onto the substrate in several steps of ion beam etching or "drilling." This submicron architecture is defined by means of lithography, masks or FIB methods.

Finally, as mentioned in the Introduction, sputtering can be used to alter the *surface morphology* due to its dependence on the local surface curvature [177]. In this case, depending on the conditions (angle of incidence energy, etc.), it can induce either surface *smoothening* or *roughening*. These competing processes may be used to design the surface geometry required for certain applications. For example, smooth surfaces are needed in optical components such as mirrors or lenses [67] or for tribological surfaces. For this purpose, sputtering should preferentially erode prominent topographies by, for example, grazing incidence bombardment, or promote ion-induced surface diffusion processes. On the contrary, roughening of a surface may enhance, for example, the surface activity (more effective surface is available), catalytic or other chemical processes, as well as control the biological response of the surface. Also, the competition between roughening and diffusion can lead to the evolution of correlated features and induce the spontaneous formation of a regular pattern [177]. This *self-organized pattern* formation is the main subject of this chapter and will be described in detail in subsequent sections.

2.3 Quantification of the Sputtering Process

The sputtering process depends on several parameters such as the nature of the target, type of ion, ion energy, incidence angle, ion flux (rate of incoming ions toward the target per unit area) and total fluence (or dose). For quantification of the ion–target interaction as a function of the different parameters, early investigations focused on the average number of target atoms ejected per incident ion. This magnitude is defined as the total sputtering yield (S) or, if measured in terms of a specific energy or angular interval, it is described as differential yield. However, there are other observables such as the effect on surface morphology or ion-induced lattice damage that can be used for quantification of the sputtering effect. The understanding of the latter effects has only been possible with the recent development of proper characterization tools for surface imaging or structure determination.

For multicomponent targets, the yield for each particular element should be taken into account. When the difference in S among species in the substrate is large, a *preferential sputtering* of the target elements with greater S occurs. This implies that the composition of the sputtered particles differs from that of the multicomponent target and, as a consequence, induces an enrichment of the low S component at the outermost layers of the target.

2.4 Experimental Measurements of the Sputtering Yield

For a given ion and energy, there is no direct relation between the atomic number of the target element and the resulting S [4]. However, there are general experimental observations on the behavior of S as a function of other sputtering parameters, mainly ion flux, energy and incidence angle. An important experimental observation is that ion erosion occurs even in the limit of low ion flux and that S is independent of the incoming flux. In addition, for a given projectile, S varies rather smoothly with incident ion energy until reaching a broad maximum (in the 10–100 keV range) and, then, gradually dropping to zero for very high energies (in the MeV range) [190]. This behavior is found in almost all projectile–target combinations. The variation of S with the ion incidence angle, θ, which is measured with respect to the surface normal is considerable. In particular, S increases from normal to oblique incidence until reaching a maximum around 60–70° and, then, drops sharply at glancing incidence angles [132]. The increase of S with θ is explained by the higher energy deposited in the surface region, a simple geometrical projection giving a $\cos^{-1}\theta$ relation. The reduction of S at glancing incidence is due to the increase in the reflection of the incident ions [181]. The variation of the sputtering yield with the angle (relation between the maximum sputtering yield and the value at normal incidence) depends strongly on the ion–target combination, being larger for light ions. The general trends of S as a function of ion energy and incidence angle are illustrated in Fig. 2.

It has been reported that S can considerably increase with ion fluence [4]. The steady state is reached only after the target has been eroded to a depth of the order of the projected ion range (maximum penetration depth of the ions in the direction normal to the surface). Obviously, this observation could be explained by the ion-induced modification of the subsurface due to ion implantation. However, the analysis of the ejected particles also suggests that the yield increase may come from a mechanism of gas trap release [182].

Regarding other parameters, it is generally thought that the temperature, apart from small changes in the sublimation energy, does not influence sputtering behavior

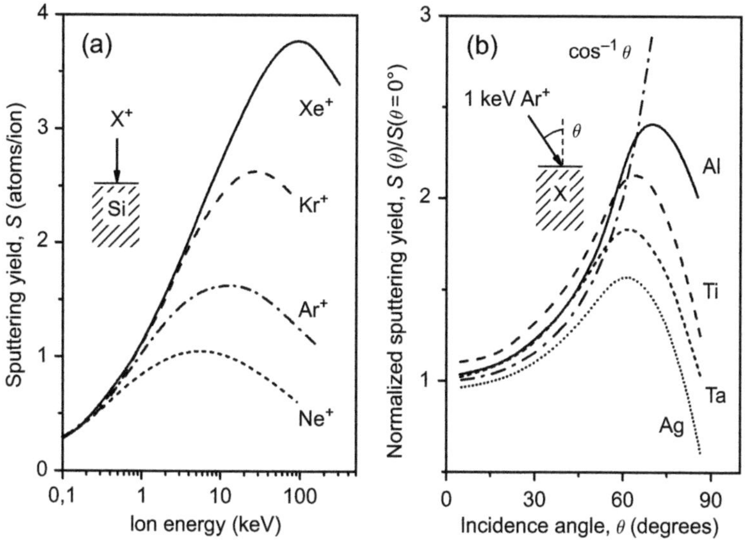

Fig. 2 (**a**) The observed energy dependence of the sputtering yield, S, of Si for normal incidence bombardment with different ions (adapted from [190]). (**b**) Normalized sputtering yield as a function of the polar angle of incidence for 1 keV Ar^+ on different materials (adapted from [132]).

significantly. Regarding crystallinity, the value of S may differ significantly as a function of energy among single crystal, polycrystalline or amorphous targets of the same element. In single crystals, the yield may also depend on the surface orientation. A satisfactory explanation for these behaviors relies on the concept of channeling (the collision probability is reduced if the ion is guided through a crystal channel), the contribution of this effect being less important in the low-energy range (below a few kiloelectron Volts).

In the previous description, energetic monoatomic particles impinging on the target surface have been considered. This is typically the case of sputtering experiments with noble gas ions. However, the sputtering process can be further complicated when the incoming ions react chemically with the target species or by bombardment with polyatomic or molecular ions. In the first case, *chemical sputtering* may also contribute to the increase of S by formation of volatile compound products [145]. An example of this process has been observed by the increase of S during bombardment of Si with F or Cl ions as compared with the values obtained for noble gas ions bombardment [166]. Conversely, when a non-volatile compound forms by reaction of the incoming ion and the target element (for example, SiO formation by O bombardment of Si), the value of S is generally reduced. In the case of polyatomic or molecular ions, it should be considered that the molecules fragment upon impact and S should account for the relative contribution of the individual constituents. In this case, a fair approximation considers that the individual ions have the same velocity as the incoming molecule and transfer their energy independently to the surface.

Regarding the emitted particles, the spatial distribution of sputtered atoms is generally described as a cosine distribution. However, as a function of energy, this distribution can change into an under-cosine or over-cosine for low or high ion energy, respectively, or can be somehow more directional if the incident particle impinges at a high angle with respect to the surface normal [134]. Usually, the direction of the sputtered cone is opposite to the incident direction and is thus called a *forward-peaked distribution*.

Because of the energetic nature of the sputtering process, emitted atoms tend to have energies E well exceeding the thermal energy that would be consistent with the target temperature. The energy of the sputtered particles is typically peaked at a few eV and has a long tail that falls as E^{-2} and goes up to the order of the incident ion energy [165]. The peak or average kinetic energy depends strongly on the mass of the target atom, as well as on the energy and mass of the incident particle.

2.5 Theory of Sputtering

The many-body considerations of the sputtering process pose a formidable challenge to theoreticians. The first models to explain the sputtering process at atomic scale were based in terms of evaporation from a hot spot, i.e., "thermal spike" [160]. Although the angular distribution of sputtered atoms follows the distribution from a liquid phase and the presence of thermal spikes has been described by Molecular Dynamics (MD) simulations [52], this mechanism gives very low S values due to the short-lived nature of the spikes [52] and it can not account for the flux independence in the sputtering yield.

The theoretical understanding of ion sputtering processes was considerably improved by the introduction of collision cascades. This advance was achieved through a better description of the ion–solid interaction, which was triggered by the building of particle accelerators in the 1950–60s. When an energetic ion penetrates the surface of a solid target, it travels through the solid losing its energy through collisions with the nuclei and electrons of the target atoms. The range (maximum penetration depth) of the ions can be determined to a good approximation considering the nuclear ($s_n = [dE/dx]_n$) and electronic ($s_e = [dE/dx]_e$) stopping independently [129]. Except for light ions (H, He), the nuclear stopping dominates at low energies (below a few kiloelectron Volts) since the ions get closer to the atoms. If the energy transfer to the target atoms is larger than the threshold given by the displacement energy, lattice displacements take place, which precisely occur mostly when nuclear stopping dominates the energy losses.

The concept of nuclear stopping was thus used by Sigmund [153] to develop a linear cascade model for sputtering. Here, the projectile shares its kinetic energy with the target atoms initially at rest in a series of random binary collisions, a process in which fast recoils are created. The recoils knock on other target atoms creating a collision cascade in which more atoms are in motion, but progressively with slower speed. If the energy is deposited in the near-surface region, atoms located close to the surface may gain sufficient energy and the appropriate momentum to leave the surface and be sputtered away. The collision cascade occurs in a time interval of picoseconds after the impact, when the recoil energies at the edges of the cascade have come below the threshold of displacement energy (of the order of a few tens of electron Volts). The cascade is damped out by energy dissipation through several mechanisms, such as creation of phonons.

By assuming a Gaussian energy distribution centered at the ion range, an analytical expression for the sputtering yield, S, can be deduced as follows:

$$S = \alpha N S_n(E)/E_B \tag{1}$$

Here, *N is the target atom density*, E_B is the binding energy of a surface atom and α contains material and geometric parameters. The product $N S_n(E)$ represents the energy loss per unit length as the ion travels in the target. Sigmund's theory accounts satisfactorily for many of the observations in amorphous or polycrystalline elemental target materials, such as the angular and energy dependences of S, as well as the angular and energy distributions of the sputtered particles. However, it is a continuum theory and, therefore, it does not address inherent inhomogeneity of materials. It cannot be applied to crystalline structures since it neglects channeling effects and, finally, it does not consider other relevant issues such as the starting surface morphology. In addition, purely repulsive forces between the ion and the target atoms are only valid for ion energies above a few kiloelectron Volts.

The correlation of sputtering with the surface morphology is a crucial issue for the content of this chapter. The effect of microrough surfaces was already posed by Sigmund in 1973 [154]. Differences in energy deposition, which are induced by topographical features, may cause a significant reduction of the local sputtering yield at local maxima and, inversely, an increase at local morphological minima. As a consequence, small irregularities on a relatively smooth surface may result enhanced by ion bombardment. This implies that microscopically flat surfaces are unstable under high-dose ion bombardment unless atom migration acts as a dominating smoothing effect. Conversely, sharp cones appear to be surprisingly stable under bombardment leading to surface nanostructuration by IBS. This prediction is the keystone for the theoretical understanding of roughening or pattern formation induced by IBS, as will be shown in the following sections.

Regarding the fundamental description of the sputtering process, the shortcomings of the continuum theory have been addressed by the development of computer simulations, providing a direct view of the collision events and incorporating the role of intrinsic material properties. In this case, discrete systems are considered where the ions and target atoms interact in a deterministic way, like in MD simulations, or in a random nature, like in the case of binary collision approximation (BCA). In these models, the interaction is driven by repulsion forces that determine the ion trajectory in the solid. In the BCA method, the interaction is treated sequentially until the energy of the projectile and recoils are thermalized. This method was implemented in the TRIM code [200,201], where a screened Coulomb potential is considered for faster computation. The BCA is valid when the individual collisions are separated in space (fulfilled only for energy above a few kiloelectron Volts) and fails when there is cascade overlap, as occurs for low ion energies. In this regime, attractive forces among the particles become more relevant. Even at high energy, the BCA does not provide good description of the collision cascade as the particles slow down.

In the low-energy range, a proper description should consider the simultaneous movement of the projectile and recoils. Since typical processes during sputtering occur in the picoseconds regime, they can be properly described by MD simulations. In this way, experimental observations of crystalline targets have been successfully reproduced, such as the yield as a function of energy or the angular distribution of sputtered atoms [80]. However, MD calculations are time-consuming in terms of computation and the interaction potential may not be accurate. In addition, the rather small length scales as well the short temporal window analyzed in MD simulations preclude a dynamical description of the sputtering process by MD and make necessary the implementation of Monte Carlo simulations.

2.6 Experimental Considerations for Ion Sputtering

There are many considerations that should be taken into account for precise and reproducible sputtering experiments. The large number of parameters involved and the variety of experimental configurations may imply a large dispersion of results that makes it difficult to present an unified picture. As an illustration of this complexity, most experiments for the determination of S before the 1970s are not reliable since they were not done under sufficient vacuum conditions [127]. Most experiments of surface nanostructuration by IBS are done by means of low-energy broad ion beam guns (Kaufman, cold cathode, hollow cathode, etc.), medium-energy ions from ion implanters, FIBs or directly by plasma immersion of the target. In most cases, ion beams are generated by a glow discharge and then extracted by means of proper beam optics. The nature of the discharge as well as the beam optics will define the range of ion flux and the ion energy that can be achieved. In all cases, several considerations should be taken into account such as energy spread, ion flux, vacuum environment or the eventual presence of multiple charge or mass ions. Moreover, a precise knowledge of the irradiation dose is required (direct ion current measurement with a Faraday Cup, collecting the sputtered material from the target or by measuring the etched volume by masking the target). Other relevant issues may be the unintentional temperature increase induced during the ion bombardment (target cooling is desirable) or the control over the starting material conditions (surface morphology, electrical properties, size, etc.) since small differences among experiments may lead to inconsistent results.

Low-energy ion beams are difficult to handle in many cases due to space-charge blow-up. This is normally solved by neutralizing the beam with thermal electrons from a hot filament, by placing the ion source very close to the target or decelerating the energetic ions in front of the target. The presence of a neutralizer may also be necessary for the processing of insulating materials. All these aspects may affect the beam divergence or beam focusing, which has been shown to play a relevant role in surface nanostructuring [43,194].

Finally, broad ion beams are normally used to incorporate large-area processing. However, beam scanning in the case of small-diameter beams, as in the case of FIB or ion implanter, can also result in the formation of similar patterns. In this case, the beam scanning may incorporate new effects in the nanostructuring process [38,43], mainly driven by heating effects on the target.

3 Experimental Observations of Surface Patterning by IBS

As mentioned above, in the past few years a relatively large amount of experimental works have appeared on the production of nanostructures and nanopatterns by IBS. Roughly, the experimental reports can be divided into two groups: (i) those that merely report on the production of some IBS nanopattern and (ii) those works in which some systematic pattern analysis is done, related with the dependence on some of the experimental variables. In this section, we will try to summarize the most relevant results of the second type of works, with special emphasis on these open issues described in the Introduction. We will distinguish in this analysis two main categories depending on the nature of target material. Thus, we will describe separately the experimental findings for amorphous or amorphizable materials, such as single-crystalline semiconductors, and for crystalline materials that are not amorphized by the ion beam (i.e., metals). This division is grounded on the fact that continuum models are mainly based on the random distribution of the target atoms, which is the case for the first type of substrates. Although we will see that both types of materials share many characteristics when they are subjected to IBS, the physical processes that actually take place can naturally lead to important differences in the pattern formation. An excellent recent review of IBS patterning on metal surfaces can be found in the work by Valbusa and coworkers [177]; the experimental section mainly focuses on IBS nanostructuring on amorphous and amorphizable materials.

3.1 IBS Patterning Formation on Amorphous or Amorphizable Surfaces

Under this section, we include all the target materials that are amorphous (i.e., glass). In addition, we consider crystalline targets whose atom layers close to the surface become amorphous upon the action of the impinging ions (i.e., Si, GaSb, InP, etc.). It is for these materials that Sigmund's theory [153] of IBS is strictly applicable.

In order to present the different results on the IBS nanostructuring of these materials, we will divide them into two main groups: (i) nanoripple patterns and (ii) nanodot patterns. This is so because usually they are obtained under different experimental conditions and because the work on ripple patterning dates back to the 1960s whereas that performed on dot patterns is more recent.

3.1.1 Ripple Formation by Off-Normal Ion Incidence

Ripple structures are induced by IBS under off-normal conditions. However, this usually happens within a limited window of angles (for the Si case, see [19,183]). The ripple pattern has been reported in a huge amount of materials. To mention just a few examples, we can quote observations on glass [130], SiO_2 [63,114], Si [19,49,53,71,79,176], Ge [28,197], GaAs [48,89,110], InP [110,51], highly oriented pyrolytic graphite (HOPG) [78] and diamond [47].

Ripple Morphology

The ripple morphology is usually characterized by AFM, STM and TEM techniques because they provide enough resolution to reveal the ripple topography at the nanometer scale. When the ripple dimensions are relatively large, it can also be assessed by Scanning Electron Microscopy (SEM). Whereas from SPM techniques we can obtain morphological data such as ripple amplitud, surface roughness and even pattern ordering (see below), with TEM we can access structural properties such as the thickness of the ion-induced amorphized layer as well any possible structure (e.g., bubbles) induced by the incorporation of the bombarding ions into the target [39]. One example of the information obtained by TEM is shown in Fig. 3. Here, we can observe that the amorphized layer thickness is quite different if the ripple slope faces the incoming beam or not, being larger for the face that is in front of the ion beam. Besides, some cavities corresponding to bubble-like features, due to the Ar^+ incorporation to the target are clearly visible on the front slope.

Fig. 3 Cross-sectional TEM image of a Si surface sputtered with 120 keVAr^+ beam. The *arrow* indicates the ion beam direction. The *horizontal bar* corresponds to 500 nm. Micrograph courtesy of T. K. Chini

In contrast, the data typically obtained by SPM techniques are displayed in Fig. 4. In this figure, we show a typical ripple pattern obtained on a silicon surface by Ar^+ ion bombardment at $\theta = 10°$. In the cross section, we define two magnitudes that are usually studied, namely, the typical wavelength, λ, of the pattern and the amplitude of the ripple structures, A. In particular, the analysis of the former is very interesting as different models predict different behaviors of λ, for instance, with the ion fluence. Thus, some models imply that λ increases with sputtering time, that is, that the pattern coarsens with time. Therefore, the eventual coarsening of λ is an object of study both experimentally and theoretically.

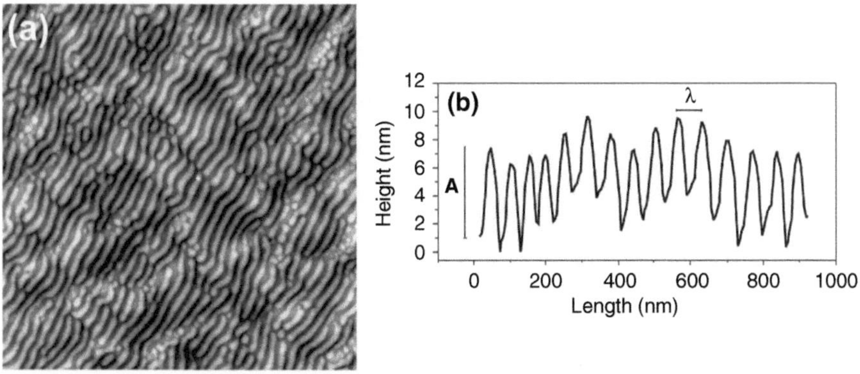

Fig. 4 (**a**) $2 \times 2\ \mu m^2$ AFM image of a Si surface bombarded by an Ar^+ 1 keV ion beam at $\theta = 10°$ for a fluence of 9×10^{17} ions cm^{-2}. (**b**) Cross section taken perpendicularly to the ripples in which λ and A are indicated.

Ripple Orientation

The dependence of the ripple orientation with θ is one of the most relevant morphological properties of ripple formation by IBS. There is a critical angle, θ_c, such that for $\theta < \theta_c$ the ripples are perpendicular to the beam direction while for $\theta > \theta_c$ the ripples run parallel to it. This fact, which was already observed on glass in the seminal work by Navez et al. [130], is a classic behavior that was explained already by the theory proposed by BH [13]. This behavior has been observed for many target materials such as glass, SiO_2 [63] and HOPG [77]. One of these examples is shown in Fig. 5 for the case of fused silica.

Fig. 5 $1 \times 1\ \mu m^2$AFM image of a fused surface bombarded by an Ar^+ beam with 0.8 keV with $\Phi = 400\ \mu A\ cm^{-2}$ (**a**) for 20 min at $\theta = 60°$ and (**b**) for 60 min at $\theta = 80°$. The *dark arrow* indicates the projection of the ion beam onto the target surface. Figure reprinted with permission from [63]

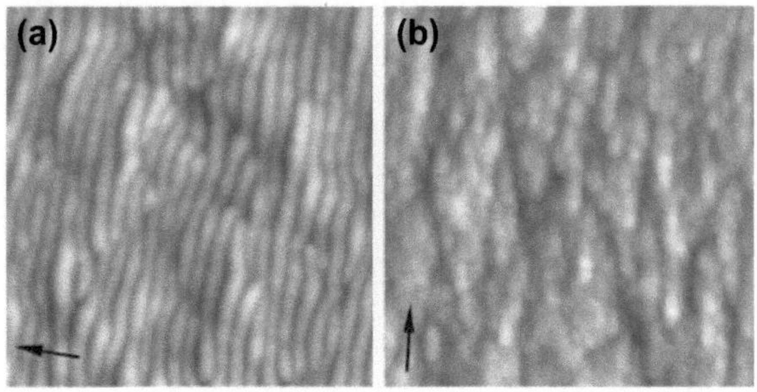

Ripple Pattern: Dependence on Ion Energy and Type

- *Ion type*—Ripple patterns have been produced by bombarding the target surface by different ions. The most frequently used species is Ar^+ due to its low cost, inertness and relatively high mass. Among the noble gases, Kr^+ and Xe^+ have also been employed [195]. Also, ripple patterns have been induced using beams

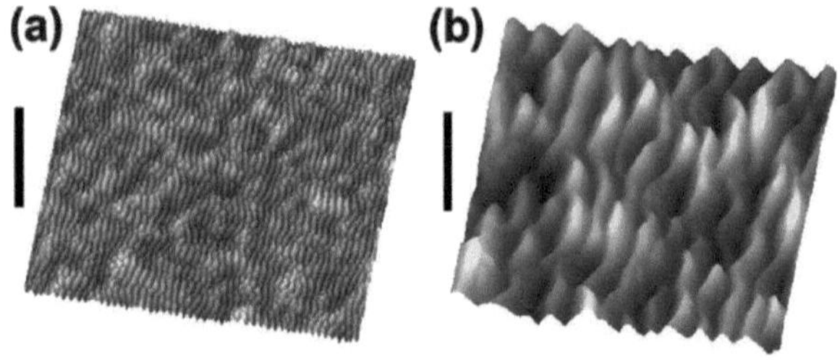

Fig. 6 (**a**) $3 \times 3\ \mu m^2$ AFM image of a Si surface bombarded by a 1 keV Ar^+ ion beam at $\theta = 10°$ for ion dose of 2×10^{15} ions cm^{-2}. (b) $4 \times 4\ \mu m^2$ AFM image of a Si surface bombarded by a 40 keV Ar^+ ion beam at $\theta = 50°$ for ion dose of 2×10^{18} ions cm^{-2}. The *vertical bars* indicate 150 nm and 900 nm, respectively

of Cs^+ [110], Ga^+ [79], O_2^+ [108,157] and N_2^+ ions [157]. The last two cases imply, as explained before, that reactive sputtering effects can take place. Therefore, ripple formation by IBS is quite generally a process virtually independent of the target materials and the bombarding ions.

- *Ion energy*—The study of the dependence of the pattern wavelength with the ion energy, E, is quite interesting because it can be used to further check the consistency of the experimental erosion system with the assumptions of Sigmund's theory [153] on which the different continuum models proposed so far are based.

Regarding the energy of the ions, the following distinction is usually made: (a) low-energy range, which is normally applied to ion energies smaller than 2–3 keV and (b) medium-energy range, which is applied to quite a wide range of energies ranging from 10 keV up to 100 keV or even higher values. Despite the wide range of energies, most of the works addressing the influence of the energy on the ripple pattern report a qualitatively similar behavior, namely, that the typical wavelength increases with energy following a power dependence of the type $\lambda \sim E^m$, where usually $0 < m \leq 1$. Figure 6 shows two examples of ripple production on silicon targets with low-energy (a) and medium-energy (b) Ar^+ ion beams. Clearly, the wavelength is quite different but the surface morphology is nonetheless similar.

This behavior has been found for Si targets irradiated at low energy for different ion species: O_2^+ [1,176], Ar^+ [195], Kr^+ [195] and Xe^+ [198], and also for sputtering experiments at medium energy employing O_2^+ ions [91] and Ar^+ ions under ion beam scanning conditions [38]. Also, this dependence has been observed for SiO_2 targets bombarded by low-energy Ar^+ ions [175], for HOPG (bombarded by medium-energy Ar^+ ions) [78] and diamond (bombarded by medium-energy Ar^+ ions) [47].

However, discrepancies arise when considering the values reported for the exponent m. It should be noted that this study becomes especially difficult for low-energy IBS experiments in which, usually, the sampled energy range is quite narrow as recently pointed out in [195]. Furthermore, sometimes it is difficult to assess whether the experimental data follow a linear or, rather, a different power-law behavior. Thus, for different targets irradiated by low-energy ions, values of m in the 0.2—0.8 range have been reported [1,175,176,195].

For the medium-energy ion experiments, the energy range sampled is usually wider due to the use of ion implanters. Therefore, the assessment of the quantitative dependence between λ and E becomes more reliable. However, discrepancies still exist among the different values reported since m values in the 0.45–1 range have been obtained [38,78,91]. Particularly, interesting is the case of HOPG targets: when they were bombarded by Ar^+ ions a linear relationship was found, but when the ion was Ke^+ a power-law behavior was observed with an exponent value $m \approx 0.7$ [78].

Finally, quite a different behavior was found for Si targets bombarded by low-energy Ar^+ ions [16]. In this work, λ was found to decrease with energy when the target surface was held at 717°C, whereas no clear trend was observed for ripples produced at 657°C. Also, an inverse relationship between λ and ion energy was reported by Chini et al. for IBS of Si surfaces without beam scanning [38]. Suppression or application of the ion beam scanning could influence the actual temperature at the target surface. This fact would be in agreement with a similar inverse behavior observed for the experiments performed by Brown and Erlebacher at higher temperature [16].

Ripple Pattern Evolution with Time or Ion Fluence

The study of the ripple morphological evolution with irradiation time is usually done through the dynamics of two magnitudes: the ripple wavelength and the surface roughness. The former is related with the ripple lateral dimension and the latter with the vertical one (i.e., ripple amplitude). This study is quite important because eventual control of the pattern morphological properties would enable applications for technological purposes.

- *Ripple coarsening*—The existence or not of ripple coarsening can be, as mentioned above, a touchstone for the continuum models. In fact, ripple coarsening is not predicted by the seminal BH theory [13] or some of its non-linear extensions [44,59,111,112,178]. In contrast, it is predicted by more recent theories [25,123]. Thus, the analysis of ripple coarsening becomes a relevant issue. Moreover, since there are systems that display ripple coarsening and others that do not, it is necessary to assess experimentally the differences between them in order to elucidate which physical phenomena are behind the coarsening process. This coarsening process typically reflects in a power-law dependence such as $\lambda \sim t^n$ where t is irradiation time (fluence) and n is a coarsening exponent.

  An example of coarsening behavior (see Fig. 7) shows two AFM images of a Si(001) surface irradiated by 40 keV Ar^+ ions at 50° for different times together with typical surface cross sections of the surface morphology for both cases. In all images, the ripples run along the perpendicular direction with respect to the projected ion beam direction (which runs along the horizontal axis of the

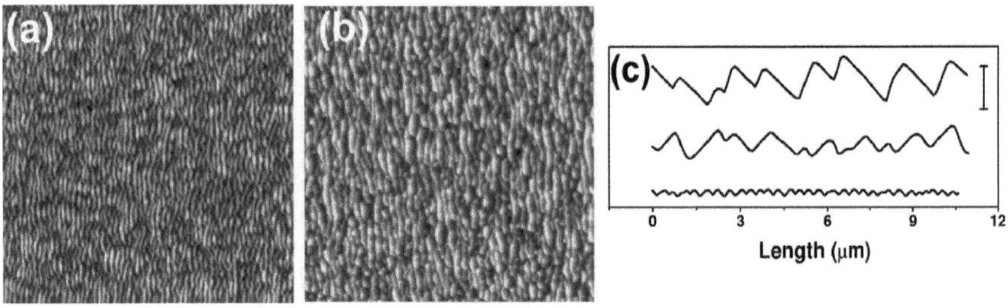

Fig. 7 *Top* view AFM images of a Si(001) surface irradiated by 40 keV Ar^+ ions at 50° for different times (ion fluences): **(a)** 20 × 20 μm^2, irradiation time 2 h (1.6 × 10^{18} ions cm^{-2}). **(b)** 50 × 50 μm^2, irradiation time 16 h (1.3 × 10^{19} ions cm^{-2}). **(c)** Typical surface profiles taken along the projection of the ion beam of the samples irradiated for 2 h *(bottom profile)*, 8 h (ion fluence of 6.5 × 10^{18} ions cm^{-2}; *middle profile*) and 16 h *(top profile)*

top view AFM images). From the AFM images, coarsening of the typical ripple wavelength is already evident. This is better appreciated in panel (c), together with the clear surface roughening that also implies a clear increase in the amplitude of the ripple morphology.

For the case of Si targets, ripple coarsening for low-energy ion irradiation experiments with 0.5 keV Ar^+ ions impinging at 60° is found [17]. Moreover, this coarsening process has been observed for a relatively wide substrate temperature range, from 600°C up to 748°C [16]. Also, a coarsening process was observed for 1.5 keV Ar^+ ions impinging at 45° and room temperature [103] and 1 keV O_2^+ ions impinging at 52° [108] for both high and low ion fluxes.

In contrast, coarsening was not observed for irradiation experiments performed at 582°C and at 67.5° with 0.75 keV Ar ions [53]. This was also the case for experiments done by bombarding the Si surface at $\theta = 15°$ with a 1.2 keV Ar^+ ion beam. The same behavior was found when the ion species were Xe^+ or Kr^+ [195].

For the medium-energy ion range, ripple coarsening has been reported for Si surfaces bombarded at $\theta = 45°$ by 20 keV and 20 keV Xe^+ ions where target temperatures were maintained between 100 and 300 K [19]. Similarly, it has been observed for Si targets irradiated either by 40 keV Ar^+ ions at $\theta = 60°$ [49] or by 30 keV Ga^+ ions at $\theta = 30°$ [79].

In other materials, ripple coarsening has been observed for HOPG and diamond surfaces. In the first case, the irradiation process was performed by 5 keV Xe^+ ions at $\theta = 60°$ and 70° [77] whereas in the last case Ga^+ ions impinged at $\theta = 60°$ with energies of 50 keV and 10 keV [47]. Also, ripple coarsening has been reported for fused silica bombarded by an 0.8 keV Ar^+ ion beam at $\theta = 60°$ [63] and for glass targets irradiated by 0.8 keV Ar^+ ions, which were generated in a defocused electron cyclotron resonance plasma with an angle of incidence of 35° [174]. Finally, ripple coarsening was also observed for InP targets irradiated at $\theta = 41°$ by 0.5 keV Ar^+ ions [51] but not for GaAs surfaces bombarded at $\theta = 41°$ by 10.5 keV O_2^+ ions [89].

With regard to the value of the exponent n, different values have been reported. Thus, on silicon targets a value of 0.5 for irradiation at 30 keV has been reported [79]. Also for high energies, 60 keV, two regimes with values $n_1 = 0.64$ and $n_1 = 0.22$ were observed. In contrast, for low-energy irradiation experiments an exponential dependence, rather than a power-law, was found for a relatively wide range of substrate temperatures [16]. When 0.8 keV Ar^+ ions were employed to irradiate fused silica [63] and glass surfaces [174], coarsening exponents of 0.15 and 0.95 have been found, respectively.

Finally, only for 60 keV Ar^+ ion irradiation at $\theta = 60°$ of GaAs surfaces, a disordered dot morphology has been reported previous to, and later coexisting with, the ripple morphology [48]. For long sputtering times (i.e., ion fluences of 3×10^{18} ions cm^{-2}), only the ripple morphology remained, without any nanodot structure superimposed.

- *Surface roughening*—The study of the surface roughness, W—defined as the mean square deviation of the local height with respect to its mean value—of the patterns can be very useful for both technological (e.g., developing metal surfaces for SERS applications) and fundamental purposes. In principle, W should be proportional to the ripple amplitude (i.e., the peak to valley height difference), A, in case the patterns were perfectly periodic. However, in real patterns there are height fluctuations among ripples that imply that A and W are not completely equivalent. Although most part of the studies deal with W, some of them analyze A rather than W.

  The most frequently observed behavior is that W initially increases steeply (usually increasing exponentially with ion fluence or sputtering time) to either saturate or grow at a slower pace, usually following a power-law dependence such as $W \sim t^{\beta}$. In the latter case, it is interesting to measure the value of β since it can be contrasted with predictions from theoretical models.

  For the first case, i.e., exponential increase followed by saturation, we can mention experiments on Si surfaces irradiated by low-energy Ar^+ ions [53], [195].

  The second case, i.e., sharp increase followed by power-law behavior, has also been observed in many systems: Si targets bombarded by medium-energy Ar^+ ions [49,91] and also for HOPG surfaces bombarded under similar conditions [78]. For Si targets irradiated by 60 keV Ar^+ ions an initial value of $\beta_1 = 0.76$ was reported, although this relatively high value could be compatible with an exponential increase, whereas for longer times $\beta_2 = 0.27$ was found [49]. In addition, for 16.7 keV O_2^+ ions $\beta = 0.38$ was observed after the initial sharp increase of W [91]. Also, for HOPG surfaces irradiated by 5 keV Xe^+ ions a β value compatible with the Kardar–Parisi–Zhang [88] universality class was reported [77,78].

  Finally, there are works where only power-law behaviors were reported. Most of these systems studied present a β value such that $0.45 \leq \beta \leq 1$ [19,49,63,174] (in the second case, for low ion flux conditions). These behaviors could be due to partial analysis of the initial exponential increase that, when analyzed in a limited temporal range, can be interpreted in terms of a power-law dependence with an exponent $\beta \gtrsim 0.5$ as remarked.

- *Shadowing effects*—An important issue that should be taken into account for studying ripple evolution is geometrical shadowing. These effects appear for long sputtering times and when relatively large θ values are employed. The important role of shadowing effects was highlighted in [19]; subsequently, Carter gave a simple estimation of the conditions under which these effects begin to operate [20]. In particular, he proposed that shadowing operates for ripples with an amplitude (here taken as proportional to W) to wavelength ratio such that

$$2\pi W/\lambda \gtrsim \tan\left(\frac{\pi}{2} - \theta\right) \tag{2}$$

In Eq. (2), as θ increases (i.e., approaching grazing incidence) the right-hand side making shadowing effects more likely to appear. These effects have been considered recently for Si irradiation by 60 keV Ar^+ ions at $\theta = 60°$ [49,50]. According to (2), shadowing effects should appear when $W/\lambda \gtrsim 0.09$, which occurs after 800 s of irradiation under the sputtering conditions described in [49]. Interestingly, this value is very close to the threshold time for which ripple coarsening begins to be observed. Thus, shadowing effects can influence largely the ripple dynamics. This fact can be important because usually shadowing is not incorporated into continuum models. Moreover, these models are usually derived under a small slope approximation and the abrupt morphologies generated when shadowing processes appear cannot be described under such approximation. A remarkable exception is the continuum equation proposed in [35] that seems to correctly describe steep surface features.

Ripple Pattern Dependence on Target Temperature

The study of the pattern evolution with target temperature can contribute to further increase our knowledge of the main physical mechanisms governing the IBS pattern formation. In particular, temperature can affect the surface diffusivity, which can lead to changes in the morphology of the pattern. However, studies are scarce due, probably, to their experimental complexity. For low-energy ions, there are two reports on silicon surfaces irradiated by Ar^+. In the first one, in which 0.75 keV Ar^+ ions impinged at $\theta = 67.5°$ onto the surface [30,53], λ coarsened with the target temperature in the 460–600°C range following an Arrhenius law [112]:

$$\lambda \sim \frac{1}{T^{1/2}} \exp\left(\frac{-\Delta E}{2k_B T}\right). \tag{3}$$

These data are shown in Fig. 8. The value of λ obtained at the lowest temperature was measured by AFM because it was out of range of the light spectroscopy measurements. This analysis led to a value for the activation energy, ($\Delta E = 1.2 \pm 0.1$ eV), for surface mass transport on ion-bombarded Si(001).

The same behavior was found for Si(111) irradiated at 60° by 0.5 keV Ar^+ ions in the 500–750°C range [16]. This study led to $\Delta E = 1.7 \pm 0.1$ eV. Similarly, for SiO_2 targets irradiated by 0.5–2 keV Ar^+ ions an Arrhenius behavior was observed for $T > 200$°C [175].

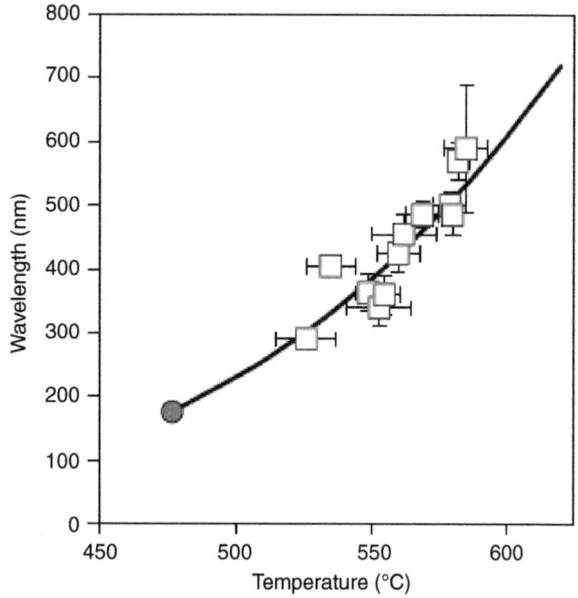

Fig. 8 Ripple wavelength λ versus target temperature for a Si(001) substrate irradiated at $\theta = 67.5°$ by 0.75 keV Ar^+ ions. *Open squares* correspond to light scattering measurements, whereas the *filled circle* was obtained by AFM. The *solid line* corresponds to a fit following the Arrhenius law (3). Figure reprinted from [53] (http://link.aps.org/abstract/PRL/v82/p2330) with permission

Finally, there are two reports on the variation of the pattern with target temperature for medium-energy IBS experiments. In the first one, HOPG surfaces were irradiated by 5 keV Xe^+ ions in the 573–773°C range [78].

For this system, the ripple wavelength also followed the Arrhenius law giving a value of $\Delta E = 0.14$ eV. The second experiment was on bombardment of GaAs targets by 17.5 keV Cs^+ ions at $\theta = 25°$ [110], in which the ripple wavelength was analyzed in the 0-100°C range. The data obtained have been later analyzed in [113]. For $T > 60$°C, the behavior was well described by Eq. (3) with an activation energy for surface self-diffusion of $\Delta E = 0.26$ eV. For lower temperatures, a slight decrease of λ was observed following a $T^{-1/2}$ law. However, within error bars, these experimental data are also compatible with a temperature-independent behavior, in agreement with models for effective surface diffusion effects of erosive origin [111,112]. In any case, the present experiment [110,113] provides a clear example of the existence of different temperature regimes for ripple formation under IBS.

In contrast, Carter and Vishnyakov did not find any change of λ with temperature in the sampled 100–300 K range when Si targets were irradiated by 10–40 keV [19]. These findings are consistent with the existing theories since for relatively low temperatures ion-induced surface diffusion processes, which do not depend on the target temperature, dominate over thermally activated ones [111].

Ripple Pattern Dependence on Ion Flux

The ripple pattern morphology, in particular its wavelength, can also depend on the ion flux, Φ, i.e., the number of incoming ions per area and time units.

Among the scarce studies of this behavior, most of them did not find any change of the ripple wavelength with ion flux. This was the case for Si surfaces bombarded by either 1.5 keV O_2^+ ions at $\theta = 40°$ [176] or low-energy Ar^+ ions at different ion fluences [16] or 2 keV Xe^+ ions at $\theta = 20°$ [198]. In addition, the same behavior was found for fused silica targets irradiated at $\theta = 60°$ at different angles by a low-energy Ar^+ ion beam [63]. Also, no change of λ with the ion flux was obtained for diamond surfaces irradiated by 50 keV Ga^+ FIB at $\theta = 57°$ [47]. All these studies were done at room temperature, except for the one on fused silica that was performed at 12°C.

In contrast, for Si surfaces bombarded by 0.75 keV Ar^+ ions at 67.5° to normal in the 500–600°C range [53], λ was found to decrease with Φ, as $\lambda \sim \Phi^{-1/2}$. Also, a decrease of the ripple wavelength with ion flux was reported for Si surfaces bombarded at room temperature by 1 keV $O^+{}_2$ ions at $\theta = 52°$, although in this case the quantitative dependence was not addressed [108].

Ripple Pattern Order

The size of ordered ripple domains is another essential property of these patterns for potential technological applications. However, it is somehow difficult to assess. In principle, there may be several methods to evaluate it. A first one is based on the data obtained by AFM. From these data, it is straightforward to obtain the Power Spectral Density (PSD) of the surface morphology, $h(\mathbf{r}, t)$. The PSD is defined as $\mathrm{PSD}(k, t) = \langle h(\mathbf{k}, t)h(-\mathbf{k}, t)\rangle$, where $h(\mathbf{k}, t)$ is the Fourier transform of $h(\mathbf{r}, t) - \bar{h}(t)$ with $\bar{h}(t)$ being the space average of the height and $k = |\mathbf{k}|$. This PSD curve usually presents a peak denoting the existence of a characteristic mode whose associated length scale is identified with the ripple wavelength λ. It has been proposed that the pattern lateral correlation length, ζ, which gives an estimation of the average size of the ordered domains, can be obtained from the full width at half maximum of the PSD peak [193]. This method was employed by Ziberi et al. [195] for estimating the range of order of ripple patterns produced by 1.2 keV Ar^+, Kr^+ and Xe^+ ions at $\theta = 15°$ on silicon surfaces. Although in these experiments coarsening was not observed, (thus λ was constant for all the ion fluences), in all cases ζ was observed to increase with the ion fluence (especially for the largest ion fluences $\zeta \geq 11\lambda$ was found). In this work, the authors also studied the change of ζ with the ion energy for the three different ions employed. Whereas for Xe^+ and Kr^+ ions ζ and λ were found to increase in the same way with ion energy, for the case of Ar^+ ions a maximum for the ζ/λ ratio was observed for an ion energy of 1.2 keV. It should be noted that for the largest energy employed in all cases, 2 keV, $\zeta \approx 10\lambda$, irrespectively of the ion species.

The previous method for assessing the pattern order degree has the disadvantage of the local character of SPM techniques. However, there is another method, based on grazing incidence diffraction (GID) or small angle scattering (GISAXS) synchrotron techniques, that provides better sampling statistics. However, to our knowledge, there is no published work using this technique for such purposes on amorphous materials where it has been used in ripple crystallinity assessment [82].

Ripple Propagation

The transversal collective motion of ripples ν has only been studied in two works by simultaneous real-time monitoring of the ion-induced ripple morphology by SEM. In the first one [79] a Si surface was irradiated by 30 keV Ga^+ ions at $\theta = 30°$; it was found that ripples initially propagated along the ion beam projection direction with a velocity of $\nu = 0.33$ nm s^{-1} to slow down later on as they coarsened. In the second case, in which glass surfaces were irradiated by 30 keV Ga^+ ions [2], ripples also propagated along the projection of the ion beam direction. In this case, the ripple evolution was shown in real time. Moreover, ripples did not initiate the propagation until most of them were completely formed, to finally reach an uniform propagation velocity.

3.1.2 Nanodot Patterning in Amorphous/Amorphizable Materials

Dot IBS nanopatterns are produced when the anisotropy caused by the oblique incidence of the ion beam is suppressed. Basically, there are two ways to eliminate this anisotropy: (i) by IBS under normal incidence, which is the most frequently employed technique [55,70] and (ii) by IBS under oblique incidence but with simultaneous rotation of the target [64]. More recently, nanodot patterns have been produced on Ge surfaces by a 2 keV Xe^+ ion beam impinging on the Ge surface at $\theta = 20°$ [197], possibly related with the role of the critical angle $\theta = \theta_c$ mentioned at the beginning of Section 3.1.

These patterns are usually characterized by a highly uniform dot-size distribution and short-range in-plane ordering. These two properties make them very interesting for potential technological applications. Although the first report on the production of such IBS nanopatterns [55] is relatively recent, dating from 1999, many groups are investigating the mechanisms leading to their formation. Thus, up to now, these patterns have been produced in different materials: GaSb [12,55,66,185], InP [64], InAs [68], InSb [57], Si [70,71,196] and Ge [197].

As occurred for the ripple patterns, the formation of the dot patterns under different experimental conditions and for a relatively wide range of materials suggests that this process does not depend on the specific ion–target interactions.

Another important issue regarding the target material is its crystallinity. Thus, it has been proved that, for GaSb targets, IBS nanodot patterns can be produced on both crystalline [55] and amorphous surfaces [58]. In addition, these patterns have been produced by IBS on both Si(001) and Si(111) surfaces [72]. In this work, it was found that, although the surface crystallinity does not affect the pattern formation, it can affect to some extent the pattern dynamics. In particular, it was observed that Si(111) surfaces have a faster dynamics in terms of pattern coarsening (see Fig. 9) and ordering than Si(001) surfaces. Again, the fact that the surface crystallinity does not determine the formation of the pattern is compatible with Sigmund's theory [153] for which the surface is considered as amorphous, either originally or as induced by the ion beam action.

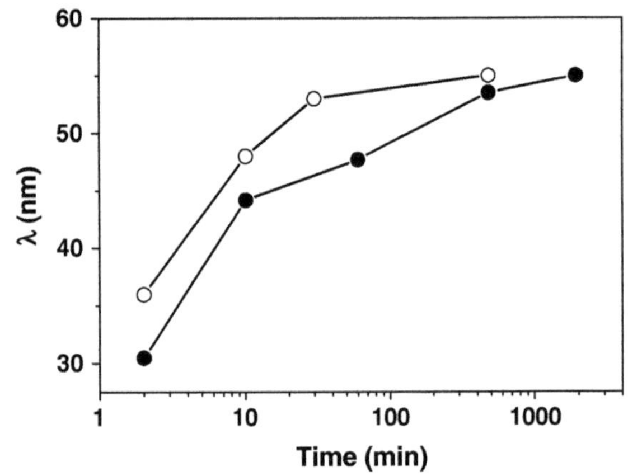

Fig. 9 Dependence of λ with sputtering time obtained by GID for Si(001) (•) and Si(111) (◦) surfaces irradiated under normal incidence by 1.2 keV Ar^+ ions. Note the faster coarsening dynamics for the IBS pattern induced on Si(111) surface

In the following, we will review the main experimental findings on the properties of these patterns depending on the different experimental parameters. One of the most studied issues is the variation with physical parameters of the basic pattern length scale, λ, which now corresponds to the average dot-to-dot distance, usually proportional to the dot size. It is worth noting that the interest in these studies on the dot shape and size lies in the required control of these parameters, particularly the dot size, for developing technological applications.

It should be noted that, whereas the dot size is usuallyaffected by the AFM tip (AFM being the routine technique for characterizing the surface morphology), the dot-to-dot distance is not usually affected by tip convolution effects [65]. The pattern wavelength is usually determined from the radially averaged PSD of AFM images as is shown in Fig. 10. In panel (a) of this figure, we display a typical AFM image of a nanodot pattern induced onto a Si(001) surface. Panel (c) shows the radially averaged PSD function corresponding to this image, in which we can observe a dominant peak corresponding to the basic wavelength λ. At higher k values, we can detect other minor peaks or shoulders indicating the high lateral ordering and size homogeneity of dots [199]. Conversely, panel (c) also illustrates the power-law behavior of the PSD at small k values, which signals height disorder between dots at long distances. This behavior is very frequently found for this type of experiment.

Similarly to the case of ripples, TEM analysis can provide us useful information regarding the morphology and structure of the dot patterns. Thus, different dot morphologies obtained by TEM are presented in Fig. 11. In (a) we observe the conical morphology of crystalline GaSb dots produced under normal irradiation and target rotation [69]. In contrast, in (b) we observe GaSb dots with a sinusoidal shape obtained under irradiation at $\theta = 75°$ and target rotation. In these cases, the amorphous layer thickness was $\simeq 4$ nm [69]. Finally, in (c) we observe Si nanodots produced under normal irradiation without target rotation displaying, rather, a lenticular shape with an amorphous layer thickness of $\simeq 2$ nm [70].

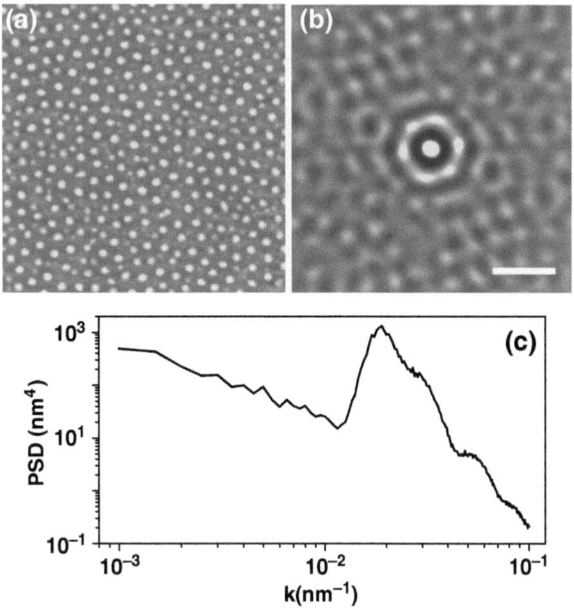

Fig. 10 (**a**) $1 \times 1\ \mu m^2$ AFM image of a Si(001) substrate irradiated under normal incidence by 1.2 keV Ar^+ ions. (**b**) Two-dimensional auto-correlation function in which the short-range hexagonal order is observed. The *horizontal bar* corresponds to 68 nm. (**c**) Radially averaged PSD for image of (**a**). The main peak corresponds to the $k = 2\pi/\lambda$ value corresponding to the dominant pattern wavelength

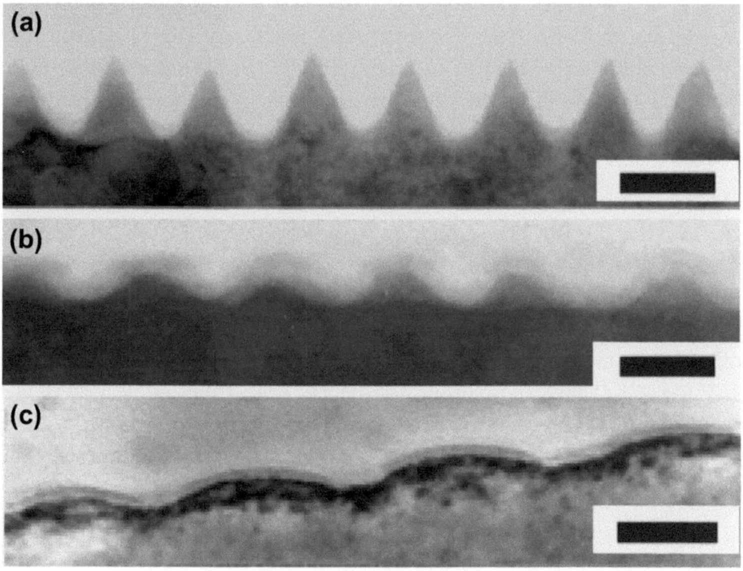

Fig. 11 Cross-sectional TEM images of (**a**) GaSb dot nanopatterns obtained under 0.5 keV Ar^+ irradiation with a fluence of 10^{19} ions cm^{-2} at normal incidence with target rotation. (**b**) GaSb dot nanopatterns obtained under 1.2 keV Ar^+ irradiation with a fluence of 10^{19} ions cm^{-2} at $\theta = 75°$ with target rotation. (**c**) Si dot nanopatterns obtained under 1.2 keV Ar^+ irradiation with a fluence of 9×10^{17} ions cm^{-2} at normal incidence. The *horizontal bars* correspond to 50 (**a**), 25 (**b**) and 30 nm (**c**), respectively. Figures (**a**) and (**b**) taken from [68] with permission

Nanodot Pattern Dependence on Ion Energy and Type

- *Ion type*—To date, all the experiments in which nanodot patterns have been reported are done with low-energy Ar^+ ion beams, except for one work in which Ne^+, Kr^+ and Xe^+ ion beams were also employed [199]. In this work, oblique ion beam incidence and simultaneous target rotation were employed. Although some characteristics of the pattern differed, the authors concluded that there was almost no difference in the morphological evolution of the mean size of dots when using different ions. In contrast to the experimental findings of the same group on ripple formation, the use of Ne^+ ions did lead to the production of nanodot patterns with an experimental behavior similar to that observed when Ar^+ ions were employed. Recently, experiments have been performed using 1 keV O_2^+ ions [168]. However, in this work a sort of nanodot chain was produced that seem more similar to nanoripples with a superimposed nanodot morphology.
- *Ion energy*—Regarding the energy range used in the nanodot experiments, there is a marked difference with respect to the studies realized on nanoripple IBS production (see above). Namely, for nanodot experiments only low-energy ions have been used, with $E \leq 2$ keV. As remarked earlier in Section 3.1, working in this energy range can lead to relatively large errors in the determination of the power-law dependence $\lambda \sim E^m$, especially in the value of m [66]. However, this was not the case for the study by Facsko et al. for IBS nanodot patterns on GaSb since they sampled 15 energy values within this range [57]. They obtained a value of $m = 0.5 \pm 0.02$. In the same study, the authors estimated a similar m value for InSb surfaces from three experimental data points.

In Fig. 12, we present our results on GaSb surfaces irradiated at different ion energies. Clearly, the dot size coarsens with ion energy. In fact, our experimental data are consistent with an exponent of $m = 0.5$ as in the case of Facsko and coworkers [57]. Also we must note the evident surface roughening with ion energy given that the vertical scale is the same in all three AFM images.

Another experimental study was done by Frost and coworkers for Si surfaces with Ne^+, Ar^+, $Kr^+$and Xe^+oblique ion beams onto rotating Si targets [199]. In all cases, they found an increase of λ with ion energy. Unfortunately, they did not estimate the value of m. However, from a visual inspection of Fig. 6 of [199] it is clear that $m_{\mathrm{Ne}} \approx m_{\mathrm{Ar}} > m_{\mathrm{Kr}} > m_{\mathrm{Xe}}$.

Nanodot Pattern Evolution with Sputtering Time or Ion Fluence

The study of the influence of sputtering time (i.e., ion fluence) on the pattern morphology has attracted the interest of the researchers already since the seminal work by Facsko and coworkers [55]. As occurred for the ripple morphologies, two main dynamics are studied: that of the pattern wavelength and the evolution of surface roughness.

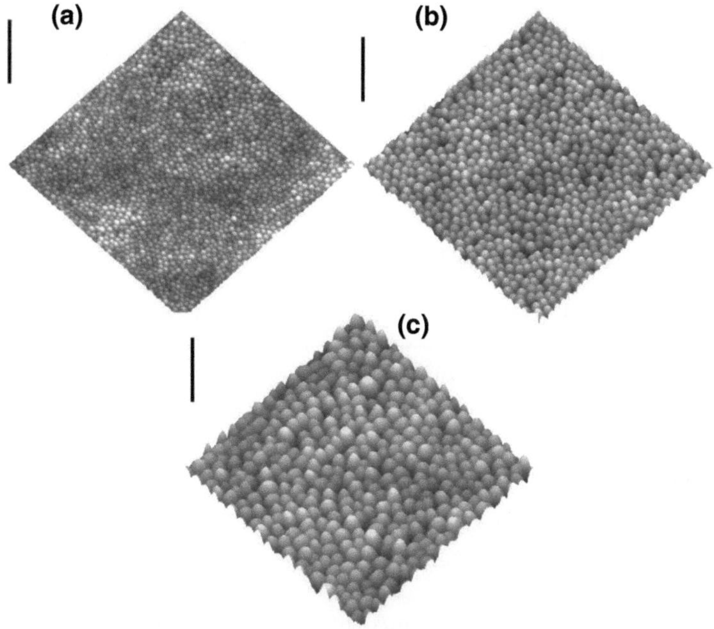

Fig. 12 $1 \times 1 \mu m^2$ AFM images of a GaSb substrate irradiated with Ar^+ ions under normal incidence with an approximate fluence of 10^{17} ions cm^{-2} at (**a**) 0.3 keV, (**b**) 0.7 keV and (**c**) 1.2 keV. The *vertical bars* correspond to 200 nm. Note the surface roughening and dot-size coarsening with increasing ion energy

- *Dot nanopattern coarsening*—Basically, the same behavior is observed for many of the various experimental systems: initially, λ increases to saturate afterwards. However, the dynamics of this process is quite different depending on the target material and the ion current density. Thus, for GaSb [12,185] and InP [64,66] surfaces the saturation regime is attained for an ion dose close to 10^{18} ions cm^{-2}. It should be noted that this saturation was attained for an ion dose of 1.7×10^{17} ions cm^{-2} when InP targets were irradiated under normal incidence by Ar^+ ions without rotation, but with a flux six times smaller [169] than that employed in [64]. For Si surfaces saturation takes place for a considerably larger ion dose, at 4×10^{19} ions cm^{-2} [72]. In another study where the Si surface was intentionally seeded with molybdenum, λ can be estimated to saturate at an ion dose of 7×10^{17} ions cm^{-2} [135]. The different dynamics of GaSb and Si surfaces under normal incidence Ar^+ irradiation and rotating InP targets under oblique Ar^+ bombardment is shown in Fig. 13, where results obtained on these systems are displayed.

  Measurement of the exponent n in the power-law dependence $\lambda \sim t^n$ before saturation has been done for four systems. (i) For Si surfaces irradiated under normal incidence by 1.2 keV Ar^+ ions with $\Phi = 0.24$ mA cm^{-2}, Gago et al. found $n \simeq 0.2$ [70]. (ii) For GaSb targets sputtered by 0.5 keV Ar^+ ions under

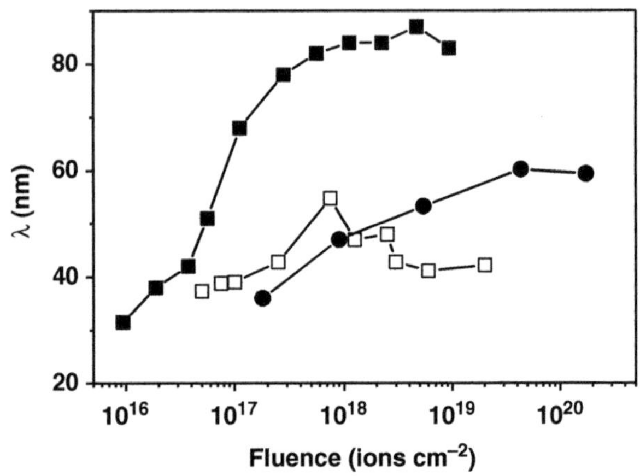

Fig. 13 Dependence of the characteristic pattern wavelength with ion fluence for GaSb at $\Phi = 0.8$ mA cm^{-2} (□), InP at $\Phi = 0.15$ mA cm^{-2} (■) and Si at $\Phi = 0.24$ mA cm^{-2} (•) surfaces. The data for GaSb and InP have been adapted from [12] and [64], respectively

normal incidence with $\Phi = 0.8$ mA cm^{-2}, Xu et al. reported $n = 0.14 \pm 0.03$ [185]. (iii) For rotating InP targets irradiated by oblique 0.5 keV Ar$^+$ ions with $\Phi = 0.15$ mA cm^{-2}, Frost et al. reported a value of $n = 0.26 \pm 0.04$ [64,66]. (iv) For InP targets irradiated under normal incidence by 1 keV Ar$^+$ ions with $\Phi = 0.0233$ mA cm^{-2}, Tan et al. reported a value of $n = 0.23 \pm 0.01$ [169].

In addition, for Si surfaces irradiated by 0.5 keV Ar$^+$ ions Ludwig et al. did observe a coarsening process with sputtering time but they did not estimate the value of the coarsening exponent [109]. However, they did not observe the saturation regime up to ion doses of 4.8×10^{17} ions cm^{-2}, which agrees with other experimental reports on Si surfaces. Besides, the authors did find that the coarsening exponent value increased with the ion energy in the 100–200 eV range. However, the opposite behavior, i.e., absence of coarsening, was observed by Frost group for rotating Si targets irradiated by oblique beams of Ne$^+$, Ar$^+$, Kr$^+$ and Xe$^+$ ions [196,199].

- *Surface roughening*—The surface roughness gives us a measure of the nanodot height as well as the height fluctuations among dots. For GaSb surfaces [12], W initially increases rapidly with sputtering time to reach later a maximum value. In this first region Xu and Teichert obtained $\beta = 0.87 \pm 0.12$ [185]. Again, this high value could be an indication that the surface roughness increases exponentially rather than follow a power law. For longer times, the surface roughness decreases to attain a saturation value.

  For the case of InP targets irradiated under normal incidence conditions a first regime for which $\beta_1 = 0.74 \pm 0.03$ was reported [169]. This regime led to another one in which $\beta_2 = 0.09 \pm 0.03$, also roughly compatible with roughness saturation. In contrast, for rotating InP targets [64], the surface roughness increases for the whole temporal range that was sampled. A first value $\beta_1 = 0.8 \pm 0.1$ was measured, once more possibly compatible with an exponential time dependence, while for longer times $\beta_2 = 0.27 \pm 0.06$ was obtained.

For Si targets, both fixed and rotating, a similar behavior of the roughness has been reported, namely a sharp initial increase followed by saturation [70,135,199]. In the latter study, the behavior observed for Ar^+, Kr^+ and Xe^+ ions was found to be consistent with an initial regime during which the roughness increased exponentially.

The main findings of the above works are displayed in Fig. 14 where we plot the evolution of *W* with sputtering time for the GaSb, InP and Si systems. In this plot, we have represented the *x*-axis in logarithmic scale in order to display the three systems in a single graph, which have different dynamics. It can be appreciated that, although the temporal evolution is different for each system, there is always a sharp initial increase of the roughness before reaching either a stationary value or a regime with a slower growth. In the inset, we display the same plot only for the initial stages of the sputtering process (i.e., $t < 170$ s); now the *y*-axis is the one in logarithmic scale so that for all three systems the initial roughness seems to increase exponentially with time as the straight lines suggest in the plot.

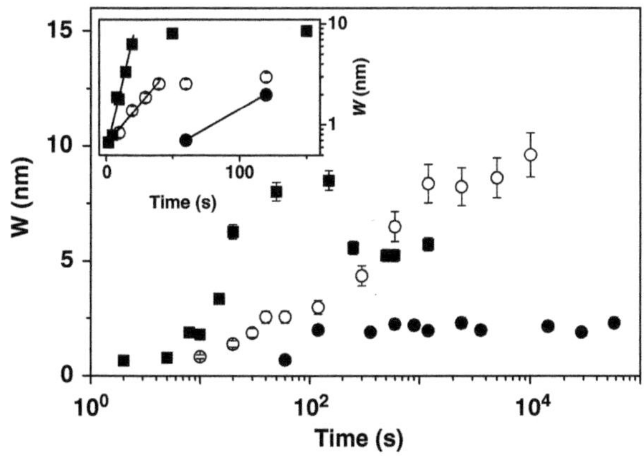

Fig. 14 Surface roughness versus sputtering time for IBS dot patterning of GaSb irradiated under normal incidence (■), InP bombarded under oblique incidence and simultaneous target rotation (○) and Si irradiated under normal incidence (•). Note that the temporal axis is in logarithmic scale. Inset: same graph but restricted to short sputtering times, i.e., $t < 170$ s. Note that now the *y*-axis is in logarithmic scale. The *solid straight lines* are guides to eye to indicate the exponential dependence of the initial roughness increase for the GaSb, InP and Si systems, respectively

Finally, we have also studied [70] how the surface roughness changes with the substrate temperature for Si targets irradiated under normal incidence. We observed that *W*, which was constant for temperatures up to 400 K, decreased to reach a saturation value at 550 K where the pattern vanished [73].

Nanodot Pattern Dependence on Target Temperature

To the best of our knowledge, there are four studies on the dependence of the pattern morphology on the target temperature, each one on a different material. Thus, we will describe separately the main findings of these studies.

1. GaSb surfaces irradiated under normal incidence by 0.5 keV Ar^+ ions [57]: for this system the pattern wavelength did not change with target temperature between −60°C and 60°C. This behavior was interpreted as a confirmation that the main relevant smoothing process is non-thermal under these experimental conditions; the main relaxation process is, rather, due to ion-induced effects [111].
2. Rotating InP surfaces irradiated by 0.5 keV Ar^+ ions impinging at 30° [66]: in this case, quite a striking complex behavior was found since pattern symmetry changed in the temperature range between 268 and 335 K from short-range hexagonal to square patterns symmetry. In addition, the characteristic wavelength increased with temperature [68]. This behavior is shown in Fig. 15.
3. Si targets irradiated by 1.2 keV Ar^+ under normal incidence [73]: it was observed, by both AFM and GISAXS techniques, that the nanopattern wavelength was constant up to 425 K and, then, decreased in the 425–525 K range. For higher temperatures, the pattern vanished and the surface became featureless. This behavior is not explained by any of the existing theories on IBS nanostructuring. In Fig. 16 we display AFM images of IBS-induced dot patterns at 300 K (a) and 425 K (b) showing qualitatively how both λ and the dot size become smaller with increasing substrate temperature. This shrinking process becomes evident in panels (c) and (d) in which we show the PSD and GID curves measured on both patterns, respectively. In both cases, the main peak shifts to higher k and q values, i.e., the characteristic length scale diminishes as substrate temperature increases [73].
4. Fixed InP surfaces irradiated under normal incidence [169] by 1 keV Ar^+ion beam, which was scanned over the target surface, at $\Phi = 0.0233$ mA cm^{-2} with a fluence of 1.05×10^{18} ions cm^{-2} at three temperatures, namely −110°C, 23°C and 36°C. Under these conditions, the pattern was only obtained at $\theta = 23°$. In parallel, surface roughness increased markedly with temperature from 0.5 nm up to 76.8 nm.

In addition, there is one study [60] where it is found that for low ion fluxes the dot size changes with increasing temperature according with the existence of Ehrlich–Schwoebel (ES) energy barriers [54,150] whereas for high ion fluxes the dot size decreases with temperature. However, Si dots seem to arrange differently from other IBS nanodot structures induced on Si surfaces; specifically a clear pattern cannot be visualized.

Nanodot Pattern Dependence on Ion Flux

For GaSb [57,66] and InP [66] surfaces, it was found that the pattern wavelength was independent of the ion current density or ion flux for the different ranges sampled,

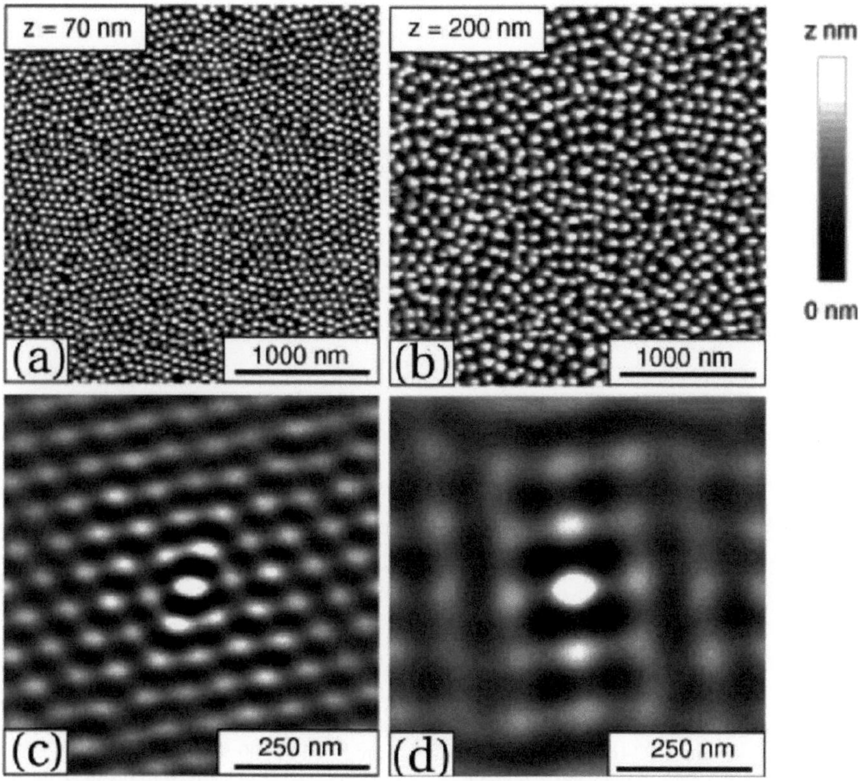

Fig. 15 AFM images of a rotating InP target irradiated at $\theta = 30^\circ$ by 0.5 keV Ar^+ at $\Phi = 300\ \mu A\ cm^{-2}$ at different target temperatures: (**a**) 268 K and (**b**) 335 K. The corresponding two-dimensional auto-correlation functions are displayed in (**c**) and (**d**), respectively. Figure reprinted from [68] with permission

namely $10^{15} - 4 \times 10^{15}\ cm^{-2}s^{-1}$ and $6.2 \times 10^{14} - 5 \times 10^{15}\ cm^{-2}s^{-1}$, respectively. For fixed InP targets irradiated under normal incidence by 1 keV Ar^+ ions with a fixed fluence at 23°C, Tan and Wee [169] found that at low ion fluxes there was not any dot pattern but it appeared at $j_{ion} = 0.0174\ mA\ cm^{-2}$. The pattern in-plane hexagonal order increased when ion flux was increased up to $j_{ion} = 0.0233\ mA\ cm^{-2}$. In another study, where dot structures did not form a clear pattern, it was proposed that for ion fluxes below 220 $\mu A\ cm^{-2}$ ES energy barriers [54,150] dominate while for higher flux values the dot size decreased as $\sim 1/\Phi^{1/2}$ [60].

In-Plane Order of Nanodot Pattern

The nanodot patterns usually present short-range hexagonal in-plane order. The analysis usually employed for assessing their degree of order is made through the two-dimensional auto-correlation function $C(\mathbf{r}, t)$ of the AFM images

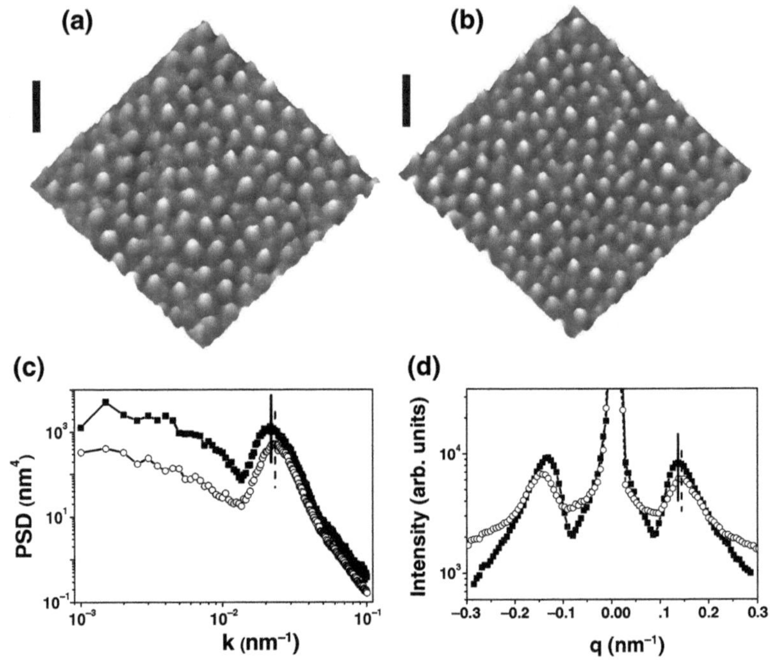

Fig. 16 500 × 500 nm^2 AFM images of a Si(001) substrate irradiated at normal incidence by 1.2 keV Ar^+ at different target temperatures: (**a**) 300 K, (**b**) 425 K. The vertical bars indicate 25 nm. PSD (**c**) and GID (**d**) curves for the patterns produced at 300 K (■) and 425 K (○). The vertical solid and dashed lines indicate the $k = 2\pi/\lambda$ (PSD) and $q = 2\pi/\lambda$ (GID) values for the 300 K and 425 K systems, respectively

$$C(\mathbf{r}, t) = \left\langle \frac{1}{L^d} \int \left[h(\mathbf{x} + \mathbf{r}, t) h(\mathbf{x}, t) - \bar{h}^2(t) \right] d\mathbf{x} \right\rangle, \tag{4}$$

which is a measure of how well a structure matches a space-shifted version of itself [193]. An example is shown in Fig. 10 where an AFM image of the dot pattern is displayed in panel (a) together with its corresponding two-dimensional auto-correlation (panel b). Here, six bright spots are clearly observed indicating the short-range hexagonal ordering of the dot pattern.

When the symmetry of the pattern is large, it can also be assessed through the two-dimensional Fast Fourier Transform (FFT) of the AFM images [197], although this is not the most common case, as IBS patterns usually give ringed FFTs [197].

Besides the symmetry of the pattern, a further issue is to quantify its degree of order in the pattern. As mentioned for the ripple patterns, two main approaches exist: (i) through size of the mean peak of the PSD of AFM images and (ii) using synchrotron techniques such as GISAXS and GID.

Based on PSD data, Bobek and coworkers observed that the range of order of the nanopattern for GaSb surfaces increased appreciably after 100–200 s of Ar^+ ion irradiation [12]. Also with this type of analysis Frost et al. [68] found that the range of in-plane order decreased with increasing temperature, in the 268–335 K range, for rotating InP surfaces sputtered at $\theta = 30^\circ$ by 0.5 keV Ar^+ ions.

Two studies have been performed for Si surfaces. For rotating Si substrates bombarded at $\theta = 75^\circ$ either by 0.5 keV Ar^+ ions or by 1 keV Kr^+ ions or by 1 keV Xe^+ ions [199], Ziberi and coworkers obtained that the lateral correlation length, ζ, increased with sputtering time up to a value close to 170 nm, in the window sampled, when Ar^+ ions were employed. In contrast, for both Kr^+ and Xe^+, ζ saturated at $\simeq$ 145 nm and $\simeq$ 120 nm, respectively. Saturation took place earlier for Xe^+. Moreover, the change of ζ with ion energy depended on the ion species. Thus, for Kr^+ and Xe^+ the order increased with ion energy until the normalized correlation length, ζ/λ, saturated for 1 keV. In contrast, for Ar^+ ions ζ had a maximum value at 0.5 keV.

The second system consists of a fixed Si target bombarded at normal incidence by 1.2 KeV Ar^+ ions. In this work, we used both AFM and synchrotron techniques to study the pattern dynamics in terms of coarsening and ordering as functions of sputtering time. In Fig. 17, we display the data obtained on Si by our group [72] and by Frost et al. [68] and Ziberi et al. [199] in terms of the normalized ordered domain size (i.e., the ratio of the lateral correlation length to the pattern wavelength) versus ion dose. For the rotating Si target, the order seems to increase with ion dose irrespective of the ion species, although saturation is observed for 1 keV Kr^+ ions but not for 0.5 keV Ar^+. In principle, it appears that the order of the pattern is larger for the rotating substrate configuration than for the fixed configuration when Ar^+ ions are employed. However, it should be noted that for 1 keV Ar^+ ions the order for the rotating configuration dropped to less than half that obtained for 0.5 keV Ar^+ ions.

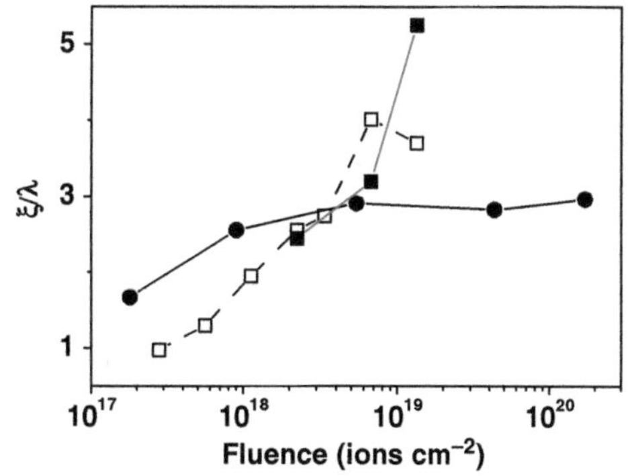

Fig. 17 Normalized ordered domain size, ζ/λ, versus ion dose for fixed Si(001) target sputtered by 1.2 keV Ar^+ ions under normal incidence with $\Phi = 240\ \mu A\ cm^{-2}$ (•); rotating Si target sputtered by 0.5 keV Ar^+ ions (■) and 1 keV Kr^+ ions with $\Phi = 300\ \mu A\ cm^{-2}$ (□). Data for the rotating substrates have been adapted from [199]

Regarding the quantification of the range of order in the pattern, SPM-based techniques pose a problem due to their locality. These surface characterization techniques sample a relative small area of the surface and lack enough statistics to a certain extent. Thus, in a recent work [72], we have used AFM and synchrotron-based techniques (namely GISAXS and GID) to quantify the ordering of the same IBS patterns. The corresponding results for the normalized ordered domain size are presented in Fig. 18. Clearly, AFM data saturate at a value very close to 3 while GID data roughly increase monotonously in the sampled temporal window up to a value close to 10. This difference is due to the improved sampling statistics of GID with respect to AFM [72]. Thus, although AFM is a well-suited technique for routine characterization of IBS nanopatterns, GID and GISAXS can provide us with more reliable quantitative data.

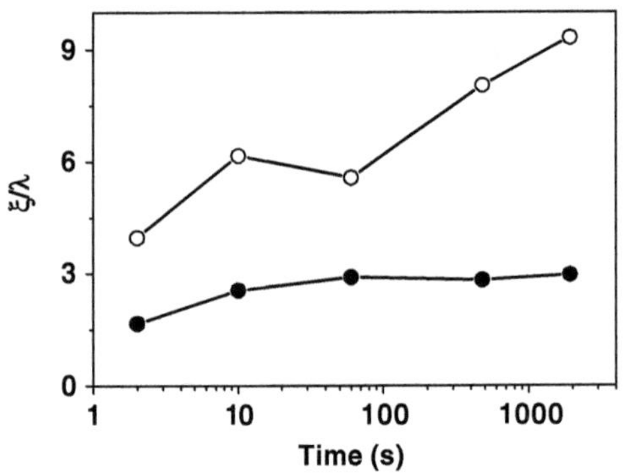

Fig. 18 Normalized ordered domain size, ζ/λ obtained by AFM (•) and GID (○) versus sputtering time for a Si(001) target sputtered by 1.2 keV Ar^+ ions under normal incidence with $\Phi = 240\ \mu A\ cm^{-2}$

3.2 *Nanohole or Nanopit Patterning*

In principle, hole nanopatterns can also be induced by IBS. The fact that nanodots or nanoholes are produced on the target surface depends only on the anisotropies of the collision cascades [87]. After the ion sputtering theory by Sigmund [153] this shape depends only, for a given target material and ion species, on the ion energy. However, despite such a theoretical prediction, there is not clear evidence, up to now, of hole or pit pattern production by IBS. Two main pit structures have been produced by IBS: the so-called cellular structures [33,37,69,84] and hole structures on semiconductor heterostructures [105]. The former look like network of hole-like structures, with a relatively wide distribution of hole sizes, resembling those obtained in plasma etching of silicon [192]. In contrast, those induced on semiconductor heterostructures display a more homogeneous hole size distribution with typical hole diameter of 170 ± 30 nm and hole spacing of 190 ± 40 nm.

Recently, we have achieved the production of nanohole patterns by IBS on silicon surfaces. These patterns display a characteristic wavelength similar to that obtained on the standard IBS experiments on dot pattern production. As an example, we present in Fig. 19a a typical AFM image of a nanohole IBS pattern. The hole structures are 2–3 nm deep and have a lateral size in the 30–40 nm range. It should be noted that both the hole depth and the lateral size can be underestimated because of tip convolution effects. The inset of Fig. 18a displays the auto-correlation of an

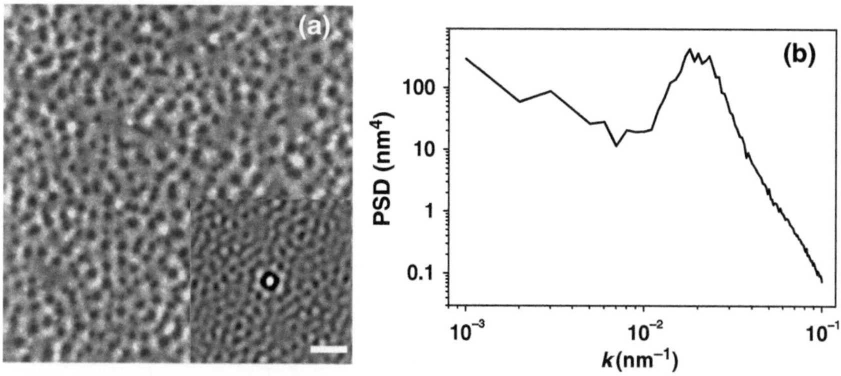

Fig. 19 (**a**) $1 \times 1\,\mu m^2$ AFM image of a nanohole pattern induced on Si surfaces irradiated by 1 keV Ar^+ ions under normal incidence for a fluence of 4×10^{17} ions cm^{-2}. Inset: two-dimensional autocorrelation function corresponding to a surface area of $740 \times 740\,nm^2$ in which the short-range hexagonal order is observed. The *horizontal bar* corresponds to 165 nm. (**b**) Radially averaged PSD function corresponding to the image of panel (**a**). The peak observed corresponds to an average inter-hole distance of $\approx$ 50 nm

AFM image taken over an area of $740 \times 740\,nm^2$. Although there is some distortion, the short-range hexagonal order is clear. Finally, in panel (b) the radially averaged PSD function of image (a) is presented. Similarly to the results obtained above for nanodot patterns, the PSD presents a clear peak that is associated with the characteristic pattern wavelength that in this case is close to 50 nm.

3.3 General Considerations

In general, the behaviors found for ripple and dot nanopattern formation are analogous. Nevertheless, the ripple patterns present specific properties, due to the anisotropy of the process, such as ripple orientation, ripple propagation and shadowing. Analyzing the behaviors reported for the changes of both pattern morphologies with other parameters, the following map of trends can be drawn (we collect the main experimental observations in Tables 1 and 2):

- IBS patterns are produced with different ion species, even with those implying reactive sputtering.
- IBS patterns are produced on a large variety of substrates: amorphous, semiconductors and metals.
- The pattern symmetry reflects that of the experimental set-up (i.e., anisotropic or isotropic).
- In general, the pattern in-plane order increases with fluence (i.e., sputtering time).
- Frequently, the pattern wavelength coarsens with time to finally saturate.

Table 1 Summary of experimental pattern behaviors for oblique incidence IBS onto amorphizable targets.

Oblique incidence	Temperature	Flux	Time	Energy
λ	const. (low T) [19, 114] $T^{-1/2}e^{-\Delta T/2K_BT}$ (high T) [78, 112, 114, 176]	const. (low Φ) [16, 47, 63, 176, 198] decreases (high Φ) [30, 53]	const. (short t) t^n (intermediate t): $n = 0$ (i.e., no coarsening) [53, 195]; $0.15 \leq n \leq 1$ [16, 19, 49, 63, 79, 91, 174] const. (large t)	E^m: $0.2 \leq m \lesssim 1$ [1, 38, 63, 79, 91, 175, 176, 195]; $m < 0$ [16, 38]
W	decreases [78]	const. [63, 198] decreases [108]	$e^{\omega t}$ (short t) t^{β} (intermediate t) const. (large t)	const. (with beam scanning) increases (without beam scanning) [38]
ν	NR	NR	$t^{-0.75}$ (short t) const. (large t) [79]	NR

NR stands for *Not reported*

Table 2 Summary of experimental pattern behaviors for normal incidence and rotating target IBS of amorphizable targets.

Normal incidence				
	Temperature	Flux	Time	Energy
λ	const. [57] decreases [73]	const. [57]	t^n (i.e., coarsening) [12, 70, 71, 135, 168, 185]	$E^{0.5}$ [57]
W	decreases [73]	NR	$e^{\omega t}$ (short t) t^β (intermediate t) const. (large t)	NR
Order	Hexagonal	Hexagonal [57]	Hexagonal (enhancement with time) [12, 72]	Hexagonal [57]

Oblique incidence with rotating substrate				
	Temperature	Flux	Time	Energy
λ	increases with T [64]	NR	$t^{0.26}$ (intermediate t) [64] const. (i.e., no coarsening) [196, 199]	Increases [64, 99]
W	increases with T [64]	increases [64, 197]	t^β with $\begin{cases} \beta = 0.8, \text{short } t \\ \beta = 0.27, \text{large } t \text{ [64]} \end{cases}$	Increases [64, 99]
Order	Hexagonal (low T) Square (high T) [67]	NR	Disordered at short t Hexagonal (enhanced order) (large t) [196, 199]	Increases [199]

NR stands for *Not reported*

- The surface roughness initially increases sharply to later on either saturate or increase at a lower rate.
- At low temperatures, the pattern wavelength does not depend on target temperature and ion flux, but it increases with ion energy.
- Ripple patterns:
  - Ripples run perpendicular or parallel to the ion beam projection direction depending on the ion beam incidence angle.
  - Ripples propagate with a non-uniform velocity.
  - Shadowing effects appear in the ripple temporal evolution for long sputtering times and large θ values.
  - Ripples are produced by ions with energies ranging from just a few hundreds of electron volts up to 10^5 eV.
  - At high temperatures, λ increases with target temperature following the Arrhenius behavior and decreases with ion flux and ion energy.
- Dot patterns:
  - Patterns are produced mainly with in-plane hexagonal order but also with square symmetry (only for oblique incidence and rotating targets).
  - To date, dot patterns are only produced by low-energy ions.
  - Different dependences with target temperature have been found implying either no change or increase or decrease of λ with increasing target temperature. In addition, for InP the pattern symmetry changes from hexagonal to square with increasing temperature.

In general, the behaviors for ripple patterns seem to be better established than for dot patterns. This may be related to the fact that dot pattern production by IBS is relatively recent. In particular, the dependence of the morphological properties of these patterns with temperature, ion energy and fluence has to be systematically addressed in order to obtain a more general picture of the nanopatterning process. On the other hand, both types of IBS patterns share many behaviors, which is consistent with the corresponding theoretical models.

Also, there is a further experimental issue to be addressed: the possible influence of technical parameters on IBS pattern production. This important problem has been only studied by Ziberi and coworkers [194]. They obtained that the settings of the Kaufman ion-gun, the one that they employed, can affect the IBS nanopattern. In particular, they found that the divergence of the ion beam as well as the angular distribution of the ions within the ion beam influenced the pattern formation. If this happens already for a given ion gun, we should likely expect some differences also when using other ion gun types. Thus, it is convenient to perform systematic experiments on different targets and under different experimental conditions with the same equipment. This approach would allow a more direct comparison and contrast of the different experimental findings, which would also contribute to improve contrast between these results and the theoretical predictions.

3.4 Pattern Formation in Single-Crystal Metals by IBS

The general observation of ripple and nanodot formation by off-normal and isotropic ion beam irradiation is only valid in the case of amorphous materials or for those that amorphize upon ion bombardment, such as single-crystal semiconductors. The case of metals should be treated through a different approach since the above properties do not occur. For example, ripples can be produced at normal ion incidence, or isotropic patterns may be formed under off-normal irradiation [146]. A large number of experimental results reported in metals have been compiled by Valbusa et al. [177].

The peculiar behavior of metal surfaces relies on the nature of the metallic bond. Due to its non-directional character, metals do not amorphize upon ion bombardment, at least for low fluence and up to energies of a few kiloelectron Volts [177]. Therefore, the surface retains its properties, the ion bombardment being only responsible for producing vacancies or vacancy aggregates at the surface [41], which increase the already large surface diffusion in these systems. Another relevant constraint for surface diffusion in metals comes from the presence of ES energy barriers for adatoms to descend steps [54,150]. These contributions add more complexity to surface nanostructuring by IBS in the case of metals.

The *diffusive regime* [177] for pattern formation in metals appears when the ES energy barrier or a preferential diffusion path determines the final pattern characteristics. In this regime, the erosion process is limited to incorporating a larger number of diffusive particles (such as atoms or vacancies) that align along the more thermodynamically favorable directions. This means that, in contrast with amorphous or semiconductor materials, the intrinsic properties of the surface reflect onto the resulting surface pattern. For example, the pattern will reflect the intrinsic isotropic characteristics of the material as in the case reported for Pt(111) by Michely et al. [115]. In the presence of diffusion anisotropy, as shown in Fig. 20 for Ag(110) [146], ripples may be observed even under normal ion incidence, or the ripple orientation may vary with temperature. In this context, temperature governs thermal activation of diffusion pathways, leading under certain conditions to isotropic or anisotropic configurations. Additional relevant conclusions extracted from Fig. 20 are that the pattern wavelength increases with temperature as a result of enhanced surface diffusion and that the pattern coarsens with sputtering time. It has also been observed that the pattern wavelength under this diffusive regime depends only slightly on the ion energy.

The *erosive regime* in metals [177] is attained only for near glancing incidence angles and at low temperatures in order to inhibit thermal surface diffusion. In this regime, the pattern formation cannot be correlated with any symmetry of the surface, such as the orientation of the crystal. This fact is shown for Cu(110) surfaces in Fig. 21 where the ion beam projection is modified with respect to the surface orientation without relevant change on the pattern characteristics [147]. This implies that IBS patterns can be aligned in directions that are not thermodynamically favorable, this fact being one of the major potentials of surface nanostructuring by IBS in contrast with other techniques such as MBE. In this regime, the ripples are parallel

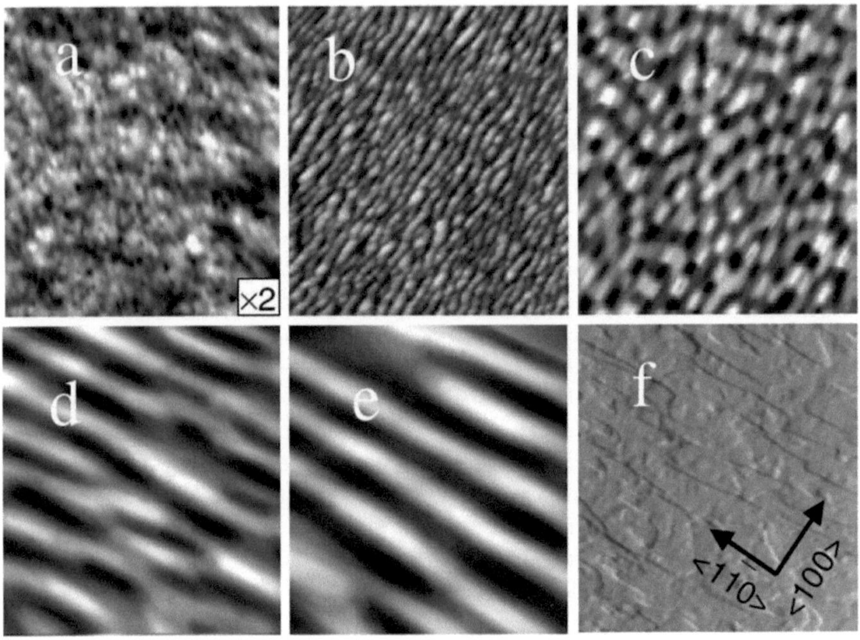

Fig. 20 STM images (350 × 350 nm^2) of Ag(110) surfaces bombarded at normal incidence with 1 keV Ar^+ ions for 15 min at (**a**) 160 K, (**b**) 230 K, (**c**) 270 K, (**d**) 290 K, (**e**) 320 K and (**f**) 350 K (taken from [177])

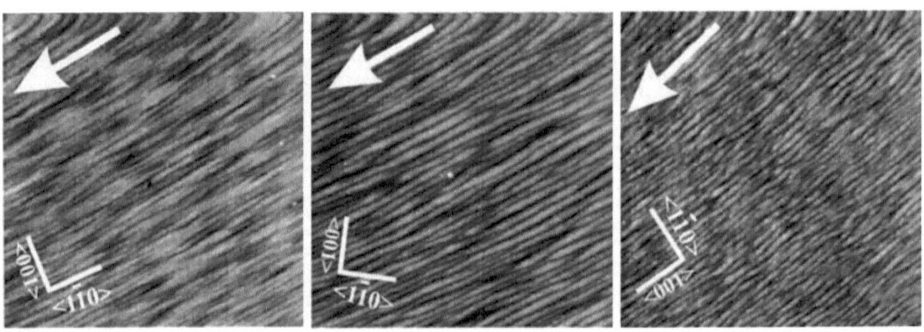

Fig. 21 STM images (400 × 400 nm^2) of Cu(110) surfaces bombarded at 70° with respect to the surface normal and 1 keV Ar^+ ions for 15 min at different angles between the beam projection and the surface orientation (adapted from [177])

to the beam direction; their wavelength depends linearly on the ion energy and both wavelength and roughness increase with fluence [177].

Finally, ion bombardment of metals does not always results in the formation of a regular pattern on the surface. In this case, the surface may additionally display kinetic roughening or not depending on the conditions. An example of these results was obtained by Krim et al. [101] on Fe surfaces bombarded with 5 keV Ar^+ ions at $\theta = 25°$.

The different parameter windows for pattern formation, smoothening or roughening have been studied by Chason et al. [31] for Cu(001) surfaces as a function of ion dose and target temperature. In this scheme, the surface remains flat for low ion fluxes and high temperatures whereas it roughens for high fluxes and low temperatures. For intermediate temperatures (200–350 K), the diffusive regime dominates and, finally, high ion fluxes and high temperatures imply an erosive regime.

3.5 Pattern Formation in Thin Metal Films by IBS

In the previous section, we have seen that ripples can also be produced by IBS on single-crystal metals for which ES energy barriers [54,150] play such a crucial role, influencing the stability of the ripple morphology even at room temperature [177]. However, a different scenario has to be considered for IBS nanostructuring of thin metallic films. These systems are usually polycrystalline and mainly formed by grains that are randomly oriented, and with sizes that are usually much smaller than the film thickness [85,180]. For such systems, the existence of ES barriers becomes improbable due to the lack of well-defined atomic steps at the surface [85]. These facts support the effective existence of an isotropic surface diffusivity rather than an anisotropic one. In this sense, thin metal films subject to IBS nanostructuring processes would be akin to the case of amorphous or amorphizable surfaces. Two groups have reported their findings for such systems.

Karmakar and Ghose have produced ripple structures on Co, Cu, Ag, Pt and Au thin film (thickness in the 30–200 nm range) surfaces irradiated at 80° by 16.7 keV Ar^+ ions [90]. They found that the ripple wavelength as a function of the target element was qualitatively consistent with the behavior predicted by Makeev et al. [112]. Under this approach, λ depends mainly on the lateral and longitudinal straggling widths of the ion cascade, as well as on the angle of incidence. Karmakar and Ghose obtained these values from the computer code SRIM [201] and found qualitative agreement between the theoretical and the experimental behaviors. However, there are sizeable quantitative differences. This is not new, quantitative disagreement between calculated and measured λ values having been previously pointed out by different groups [93,57,73]. Another interesting finding refers to the stability of the induced ripple patterns at room temperature [90] in comparison with those produced on single crystals [177]. This fact is an indication that the thermally activated diffusion energy barriers in thin polycrystalline films are comparatively higher than in single crystal metal surfaces, which agrees with the assumption on the absence of ES barriers. In fact, measurements of activation energies for adatom surface diffusion on various polycrystalline materials are consistent with this result [142]. The studies of Stepanova and coworkers consisted in the irradiation by a 1.2 keV Ar^+ ion beam of a 50-nm-thick Cu film deposited on glass or silicon substrates [162,163]. When the irradiation angle was oblique, at $\theta = 82°$, a clear ripple pattern was induced. These experiments allowed to discard any possible influence of the underlying substrate (either glass or silicon) in terms of composition and crystalline structure on the production of ripple patterns on the top metal thin film. Another set of experiments was performed under the same conditions but at normal incidence. In this case, a self-assembled network of $\approx$20-nm-sized ring-like Cu features was observed on a SiO_2 substrate after 1.2 keV Ar^+ beam etching of the Cu–substrate interface [162].

The previous studies suggest that IBS of interfaces might be a reasonable process to fabricate metal nanopatterns on non-metallic substrates. Also, the specific properties of these thin-film polycrystalline metal systems make them suitable for study through the same approach that is employed to understand IBS nanopatterning of amorphous/amorphizable surfaces.

4 Theoretical Approaches

As seen in the previous sections, typically the IBS-induced nanopatterns fully evolve in macroscopic time and length scales (minutes and several microns, respectively). It is at these scales where, e.g., interaction among ripples can be seen to lead to order improvement with fluence or, rather, to eventual disorder in heights. Although detailed knowledge on the phenomenon of sputtering is rapidly and consistently developing through (microscopic) MD type of studies (see, e.g., [15,131] and references therein, and Section 2), the scales that are currently reachable to these methods remain in the 1 ms and 50 nm ranges. Monte Carlo (MC) and continuum methods (CM) can probe larger scales; so these are the approaches that we will consider in what follows.

4.1 Sigmund's Theory of Sputtering

As mentioned in Section 5, in a classic work Sigmund analyzed the kinetic transport theory of the sputtering process [153]. Assuming an infinite medium, he found that, in the elastic collision regime at the energies of a few kiloelectron Volts where electronic stopping is not dominating, the deposited energy can be approximated by a Gaussian distribution near its maximum. Specifically, the density of energy spread out in the bulk by an ion with kinetic energy E is given by

$$\varepsilon_s(\mathbf{r}') = E N_s \, e^{-\frac{x'^2+y'^2}{2\mu^2}} e^{-\frac{(z'+a)^2}{2\sigma^2}}, \tag{5}$$

where the origin of the $\mathbf{r}' = (x', y', z')$ coordinate system is placed at the impact point of the ion within the surface; $\hat{\mathbf{z}}'$ is aligned along the ion beam direction, and $\hat{\mathbf{x}}'$ and $\hat{\mathbf{y}}'$ belong to the perpendicular plane to $\hat{\mathbf{z}}'$, (see Fig. 22). In Eq. (5) $N_s = \left[(2\pi)^{3/2}\sigma\mu^2\right]^{-1}$ is a normalization constant and, due to the initial momentum of the ion, the maximum of energy deposition occurs at a distance a along the ion trajectory inside the bulk. The longitudinal and transversal straggling widths of the

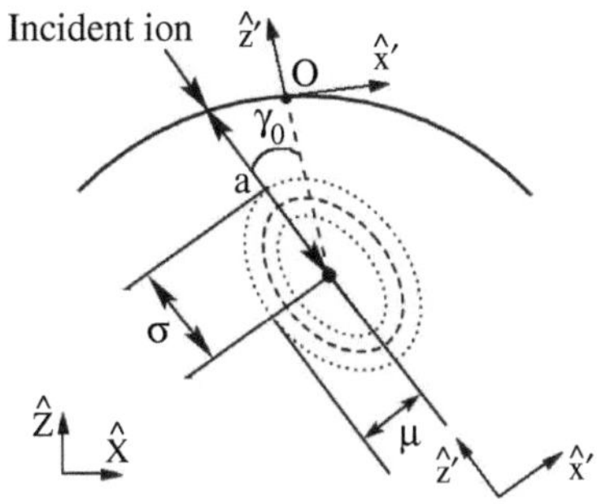

Fig. 22 Two-dimensional profile of the distribution of energy deposition, Eq. (5). σ and μ are the transversal and longitudinal widths of the Gaussian, respectively. The reference frames considered in the text are also shown. Adapted from [112]

distribution are σ and μ, respectively. If mass differences between substrate and ion are not too large, the quality of the approximation is reasonable and Eq. (5) provides a good approximation for polycrystalline and amorphous targets [153]. Despite the fact that a depends on the microscopic details of the interactions between the ion and the bulk, at intermediate energies (10 – 100 keV), it is usually considered to be proportional to E. Within this range of energies, σ, μ and a are of the order of a few nanometers. This point will be relevant when information about the evolution of the topography is obtained.

As we noted above, correlation of the sputtering yield with the surface geography is a crucial issue. In [154], Sigmund showed that the topography of the surface can indeed influence the magnitude of the rate of erosion and provided an analytical description that describes the increase of yield for geometries different from the flat morphology. It is assumed that the speed of erosion at a point O on the surface is proportional to the amount of energy deposited there by the ions, with a proportionality constant Λ that is characteristic of the substrate and depends on the atomic density of the target n_v, the atomic energy of connection in the surface U_0, and a proportional constant C_0 which is related to the square of the effective radius of the potential of effective interaction according to

$$\Lambda = \frac{3}{4\pi^2 n_v U_0 C_0}. \tag{6}$$

Under these hypotheses, it is possible to obtain the mathematical expression for the rate of volume eroded at O. This is simply given by

$$V_O = \Lambda \int_R \Phi(\boldsymbol{r}')\varepsilon_s(\boldsymbol{r}')dR. \tag{7}$$

The integral extends to the region $\mathcal{R}$ where the impact of the ions contributes to energy deposition at O. The term $\Phi(\boldsymbol{r}')$ represents the local flux. We have specified its dependence with $\boldsymbol{r}'$ to show the corrections due to the local geometry to the homogenous flux Φ_0. From (7), Sigmund derived the rate of erosion for diverse artificial geometries, such as a flat surface followed by an inclined plane or a vertex, in the case of normal incidence.

The previous description, as already observed by Sigmund [154] and briefly noted in Section 2.5 above, implies the occurrence of a morphological instability. Let us suppose that we irradiate a certain surface with an homogenous flow, as shown in Fig. 23. We can verify that distances $\overline{OA}$ and $\overline{OB}$ are smaller than $\overline{O'A'}$ and $\overline{O'B'}$ due to the geometry of the interface. This implies that, for the considered energy distribution, the penetrations of ions at A and B induce large energy deposition at O than the impacts on A' and B' at O'. As the rate of erosion is proportional to the deposited energy, erosion is faster at O than at O'. Thus, valleys are excavated more quickly than crests, amplifying initial differences in heights. Sigmund suggested that an alternative process that flattens the surface must exist and proposed atomic migration as a mechanism to correct this instability.

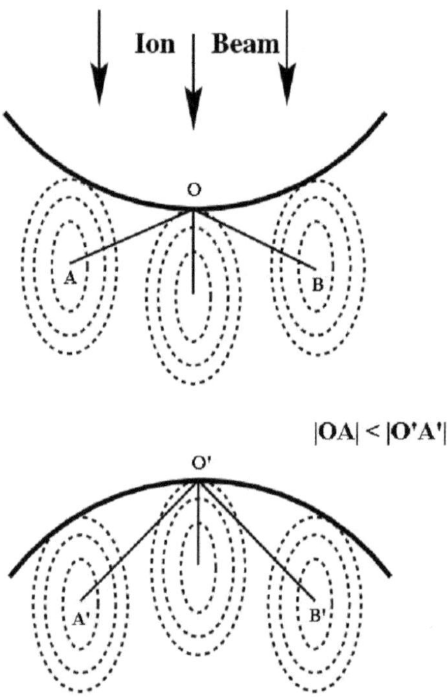

Fig. 23 Sketch of deposited energy for two different profiles. The energy deposited at O is larger than at O'. This induces more erosion in surface valleys than in crests, which produces a morphologic instability in the system. Adapted from [112]

The theory of Sigmund is the basis for most of the later continuum approaches to the dynamics of surfaces undergoing IBS. In principle, as a *theory of sputtering*, and as mentioned in Section 3.1, it is known to have limitations. Within the range of energies we are considering, two of the most conspicuous ones refer to: (i) the behavior of the sputtering yield for large incidence angles and (ii) crystalline targets. Related to the former, Sigmund's theory predicts an steady increase of the yield with incidence angle, not being able to account for the well-known decrease of S at glancing incidence due to more efficient projectile reflection and channeling effects at the surface. Nevertheless, recent continuum studies [35] have shown that improved distributions accounting for this effect (see in [35] and [129]) do not modify qualitatively the morphological predictions to be derived below from the energy distribution (5). Similarly, channeling effects limit in principle the applicability of Sigmund's theory to crystalline solids [point (ii)] by inducing sizeable departures from the energy distribution as described by (5). Thus, different distributions must be considered, as done, e.g., in [61] for the particular case of Cu targets. However, as seen in this work, again these modifications do not alter the qualitative morphological implications of the continuum theories to be described below, which justifies (at least for temperatures at which ES barriers do not dominate surface diffusion [177]) the strong similarities in nanopattern formation by IBS on metals, as compared with amorphizable targets.

As a *morphological theory*, again Sigmund's [154] has some limitations: it does not predict the alteration of the morphology during the process for scales much larger than penetration depth for ions; as surface diffusion is not considered, the wavelength of the pattern needs to be of the order of the length scales appearing in the energy distribution (5); the effects of surface shadowing or redeposition are not considered; it does not predict the time evolution of the morphology and how it affects the rate of erosion. Thus, additional physical mechanisms and a more detailed description of the surface height are needed in order to derive a morphological theory with an increased predictive power.

4.2 Monte Carlo Type Models

(Kinetic) MC methods have a long and successful tradition in the context of IBS [200,201]. For our morphological purposes, they have found increased application during the last decade. Their main advantage is that, in principle, they allow to make a more direct connection to microscopic relaxation mechanisms than, e.g., continuum approaches. In addition, they do not require analytical approximations, lending themselves rather to numerical simulations that can reach reasonably large scales (around 10 s and around 100 monolayers of erosion). However, as compared with CM, dependencies of observables with physical parameters are less direct, accessible scales are shorter and universal (i.e., substrate-/ion-independent) properties are harder to assess unambiguously. Actually, the two types of approaches—MC and CM—to some extent complement each other so that, although our focus will be mostly on CM, we consider shortly some of the main results of the most recent MC studies.

In order to access large scales as mentioned, MC methods consider physical relaxation rules in which microscopic details—such as specific interactions among ions and target atoms—are coarse-grained out and replaced by effective dynamical rules. These have to represent, on the one hand, the events leading to erosion by sputtering and, on the other hand, competing relaxation mechanisms such as surface diffusion. For the sputtering processes, starting with the study of [45] many

MC models consider an erosion rule that is either a discretization [45] or, rather, a direct implementation [81,173,17,187,188,32] of Sigmund's continuum law for energy deposition due to linear collision cascades (with corrections for large local incidence angles [161,163]). Actually, this energy deposition rule has been extended even to MC studies of different physical processes, such as, e.g., laser-induced jet chemical etching [118]. Nevertheless, other MC models do consider more microscopic (albeit still approximate) rules for erosion, such as the BCA similar to that employed in SRIM simulations [200,201], as done in the works by Koponen and collaborators [96,97,98,99,100]. Or else, still different approaches to the sputtering events consider dynamical creation of surface adatoms/advacancies [29,126,164].

As for the additional surface relaxation mechanisms, most of the works mentioned restrict their nature to surface diffusion (there being additional physical possibilities, such as redeposition and viscous flow) for adatoms and/or advacancies. The rates for diffusion vary among the different studies, being either dependent on the energies of the initial and final states—within a local equilibrium approximation— or, rather, following temperature-activated Arrhenius laws; see a recent overview in [31].

Although there are some differences in the morphological dynamics predicted by the various MC models, there is agreement in a number of basic properties, such as (i) the occurrence of a morphological instability at small/intermediate times that is well described by CM; (ii) at oblique ion incidence, the nonlinear evolution of the ripples as reflected in their non-uniform transverse motion and (iii) the saturation (stabilization) of the surface roughness at sufficiently long times. Differences among various models do occur in their predictions: for instance, in some models the dominant pattern wavelength remains fixed in the course of time [45,96,97,98,99,100,32], while in other models a consistent wavelength coarsening phenomenon occurs [81,162,187,188,31,163] with properties (such as the coarsening exponent value) that depend on physical parameters such as the substrate temperature. Then, there are properties—such as the the kinetic roughening of the interface height fluctuations at long times—about which some of these studies do not extract conclusions (if they do not focus on such large-scale issues), while other works reach different quantitative results (regarding, e.g., the values of the scaling exponents, although most predict their occurrence).

4.3 Continuum Descriptions

The wide separation in time scales between the microscopic events that take place in the target during the irradiation process and the large-scale response of its surface morphology recall a similar behavior between microscopic and collective motion in the case of, e.g., fluid dynamics. Thus, it is natural to expect [46] that some continuum description in the spirit of, say, the Navier–Stokes equations might be appropriate in the case of nanopattern formation by IBS. The advantage of continuum descriptions, if available, is that they provide compact descriptions of complex physical phenomena. Moreover, these frameworks are frequently more efficient computationally for the study of large scale properties, whose generic properties can be faithfully described by them [102]. Nevertheless, for the type of nonequilibrium phenomena which are our present focus, the relevant continuum equations are at most times nonlinear, their study requiring either analytical approximations or numerical simulations.

4.3.1 Dynamics of the Surface Height

In order to put forward a continuum description of these nonequilibrium systems, Sigmund's theory [153,154] constitutes an ideal starting point. This is so because Eq. (7) already provides an explicit expression for the most relevant kinetic quantity describing the sputtering problem, namely, the local velocity of erosion. In turn, one may add different contributions to the right-hand side of this equation if additional physical mechanisms contribute to the local variation of the target height, and this has indeed been done in order to improve the physical description. Still, for theoretical work, (7) is technically a very complex expression. The insight of BH [13] is that, under reasonable physical approximations, (7) becomes a closed time (t) evolution equation for a single physical field, the target height $h(x, y, t)$ above point (x, y) on a reference plane. It is within such (simple) single-field approach that most of the recent continuum descriptions have circumscribed themselves. However, we will see in this section that this program encounters consistency problems—that question its physical applicability—unless the dynamics of additional independent fields are taken into account. The evolution of these fields—e.g., the density of species that diffuse on the surface—also modifies the local height velocity. This does not mean that a description by a closed equation for the height field is inappropriate but, rather, that to correct such description, one has to necessarily take into account for its derivation physical mechanisms that complement Sigmund's formula. In this section we recall the main results obtained along BH's pioneering "single-field" height approach, while the next section will (mostly) consider recent developments that incorporate the dynamics of additional physical fields to the continuum description.

Bradley and Harper's Theory

BH [13] derived a partial differential equation to describe the evolution of the morphology assuming that the variation in the height of the surface is smooth when seen at a scale that is comparable with the average penetration length of the ions. By expanding the target height as a function of the surface geometry up to linear order in the surface curvatures, they obtained a closed evolution equation for the surface height. In the following, we recall the main steps of the derivation.

Define a local coordinate system with the $\hat{\mathbf{z}}$ axis oriented along the surface normal at O and the $\hat{\mathbf{x}}$ and $\hat{\mathbf{y}}$ axes lying on the perpendicular plane to $\hat{\mathbf{z}}$ (i.e., in the tangent plane through O) where $\hat{\mathbf{x}}$ is aligned along the projection of the ion beam onto this plane as shown in Fig. 22.

Assuming that the function which represents the height of the surface, $z(x, y)$, is single valued and varies smoothly, we can approximate the height in the neighborhoods of O as

$$z(x, y) = -\frac{1}{2}\left(\frac{x^2}{R_x} + \frac{y^2}{R_y}\right), \tag{8}$$

where R_x and R_y are the principal radii of curvature. The simple formula (8) is enough to our purposes for two different reasons. First, for smooth surfaces the average penetration length a is smaller than the radii of curvature and only ions penetrating the target a distance smaller than a from the point O contribute appreciably to erosion there (since energy deposition decays with distance as a Gaussian). Second, it is assumed that the surface principal directions—along which curvature is maximized or minimized—are aligned along the $\hat{\mathbf{x}}$ and $\hat{\mathbf{y}}$ axes. This holds at least for the cases in which structures are aligned either parallel or perpendicular to the projection of the incident beam (which is the usual case, as observed in previous sections).

We can obtain the local speed of erosion at O using Eq. (7). To this end, one needs to change coordinates in Eq. (5) to the (x, y) frame in Fig. 22 and take into account that the local flux of ions at any point of the surface, $\mathbf{r} = (x, y, z)$, is related to the homogenous flux, Φ_0, by $\Phi(\mathbf{r}) = \Phi_0 \cos\gamma(\mathbf{r})$, where $\gamma(\mathbf{r})$ is the angle between the beam direction and the local normal direction. By Taylor expanding to linear order in the curvature radii and using Eq. (7), BH obtained an expression for the erosion rate at O that reads

$$V_O = S_0\left[\Gamma_0(\gamma_0) + \frac{\Gamma_x(\gamma_0)}{R_x} + \frac{\Gamma_y(\gamma_0)}{R_y}\right], \tag{9}$$

where the functions Γ_0, Γ_x and Γ_y depend on Φ_0, E, Λ, the local angle of incidence γ_0 and the distances a, σ and μ of Eq. (5).

Once the dependence of the erosion rate with the local morphology has been obtained, we can obtain a continuum equation for the local surface height, $Z = h(X, Y)$, in the laboratory system of reference $(\hat{\mathbf{X}}, \hat{\mathbf{Y}}, \hat{\mathbf{Z}})$ sketched in Fig. 22. Thus, we define the component $\hat{\mathbf{Z}}$ to be perpendicular to the initial surface, $\hat{\mathbf{X}}$ as parallel to the projection of the ion beam onto the initial flat surface and $\hat{\mathbf{Y}}$ perpendicular to $\hat{\mathbf{X}}$ and $\hat{\mathbf{Z}}$. Within our linear approximation (see [112]) we can assume $\partial_t h \simeq -V_O$ and $\theta = \gamma_0 + \partial_X h$, where θ is the angle between the ion beam and the normal to the uneroded surface, and using (9) one obtains [13]

$$\frac{\partial h}{\partial t} = -v_0 + \gamma_x \frac{\partial h}{\partial X} + \nu_x \frac{\partial^2 h}{\partial X^2} + \nu_y \frac{\partial^2 h}{\partial Y^2}, \tag{10}$$

with

$$v_0 = S_0\Gamma_0(\theta), \quad \gamma_x = S_0\partial_\theta\Gamma_0(\theta), \quad \nu_x = S_0\Gamma_x(\theta), \quad \nu_y = S_0\Gamma_y(\theta), \tag{11}$$

where any nonlinear terms have been neglected and we have used $R_x^{-1} = -\partial_X^2 h$ and $R_y^{-1} = -\partial_Y^2 h$. Here, S_0 and Γ_0 are always positive and, therefore, the surface height decreases at constant speed $S_0\Gamma_0$.

In order to get information about the evolution of the surface, we will assume the following perturbation with spatial frequencies k_X and k_Y, and amplitude A at time t:

$$h(X, Y, t) = A\ \mathrm{e}^{\mathrm{i}(k_X X + k_Y Y) + \omega t}. \tag{12}$$

If we substitute this expression into (10), we obtain the real part of the dispersion relation, ω. This reads

$$\mathcal{R}e\ \omega(\mathbf{k}) = -\nu_x k_X^2 - \nu_y k_Y^2, \tag{13}$$

and the imaginary part is

$$\mathcal{I}m\ \omega(\mathbf{k}) = \gamma_x k_X. \tag{14}$$

The imaginary part of $\omega(\mathbf{k})$ indicates the magnitude and direction of the speed of transverse in-plane motion of the disturbances. Since Γ_0 is an increasing function of θ [13], the perturbation moves in the $-\hat{\mathbf{X}}$ direction (that is, upstream with respect to the projection of the beam). On the other hand, $\mathcal{R}e\ \omega(\mathbf{k})$ describes the rate at which the amplitude of a perturbation with wave-vector $\mathbf{k}$ grows or decays with time. Given that at small angles of incidence [13] $\nu_x(\theta) < 0$ and $\nu_y(\theta) < 0$, any spatial perturbations on the initial surface grow exponentially in time. Since $\nabla^2 h$ is positive at the bottom of the valleys, the rate of erosion is larger at these points than at the top of the crests where $\nabla^2 h < 0$. Therefore, the negative signs of ν_x and ν_y are the mathematical expression of Sigmund's morphological instability. At these small incidence angles, moreover, $\nu_x(\theta) < \nu_y(\theta)$ and perturbations grow faster along the $\hat{\mathbf{X}}$ axis than along the $\hat{\mathbf{Y}}$. Hence, at these angles, the ripple crests are perpendicular to the projection of the ion beam onto the surface as observed experimentally. For the critical angle θ_c, one precisely has $\nu_x(\theta_c) = \nu_y(\theta_c)$ while for larger angles $\nu_y(\theta) < \nu_x(\theta)$ and the ripple crests align with the ion beam projection.

These results predict some of most basic experimental features of IBS ripples but, without an additional mechanism which stabilizes the system, disturbances of arbitrarily small length scales would increase exponentially without bound. On the other hand, from Eq. (10) one would expect the ripple wavelengths to be of the order of the distances involved in the description, that is to say, of the order of the penetration length a. This does not happen; rather, the experimental ripple wavelengths are frequently almost two orders of magnitude larger than a. In order to solve these problems, BH incorporate the effects of the surface diffusion of thermal origin by introducing an analogous term to that derived by Mullins [121] for isotropic surfaces. With this aim, a term $-B\nabla^2\nabla^2 h$ is included into (10) to obtain the following linear equation which describes the evolution of h

$$\frac{\partial h}{\partial t} = -v_0 + \gamma_x \frac{\partial h}{\partial X} + \nu_x \frac{\partial^2 h}{\partial X^2} + \nu_y \frac{\partial^2 h}{\partial Y^2} - B\nabla^2\nabla^2 h. \tag{15}$$

Proceeding as in Eqs. (12)–(14), the real part of the dispersion relation modifies into

$$\mathcal{R}e\ \omega(\mathbf{k}) = -\nu_x k_X^2 - \nu_y k_Y^2 - B\mathbf{k}^4, \tag{16}$$

that is negative for wave vectors of sufficiently large k (i.e., short length scales). However, there is a band of unstable long-wavelength modes for which (16) is positive. The orientation and wavelength of the observed ripple structure corresponds to that mode which maximizes (16); this is $2\pi(2B/\nu_x)^{1/2}$ if the ripples are oriented along the $\hat{\mathbf{X}}$ axis as a result of $\nu_x < \nu_y$ or $2\pi(2B/\nu_y)^{1/2}$ if the ripples are oriented along the $\hat{\mathbf{Y}}$ axis as a consequence of $\nu_y < \nu_x$. With the aim of obtaining an order of magnitude estimation of this quantity, BH deduced that for normal incidence, a sputtering yield of two atoms per ion, $a \simeq \sigma \simeq \mu \simeq 10$ nm, and a typical value $B \simeq 2 \times 10^{-22}$ cm^4 s^{-1}, the value of the linear wavelength is $\lambda_l \simeq 5$ μm, in reasonable agreement with the typical distance between ripples observed experimentally.

If surface diffusion is thermally activated, as effectively occurs for high temperatures and small flows, it is easy to obtain how the wavelength of the pattern varies with the temperature, T. Since $\nu_{x,y}$ do not depend on T and assuming that B verifies $B \sim (1/T)exp(-\Delta E/k_B T)$, where ΔE is the energetic barrier to activate surface diffusion and k_B is the Boltzman's constant, the linear wavelength λ_l must verify $\lambda_l \sim (1/T^{1/2})exp(-\Delta E/2k_B T)$. On the other hand, it is also possible to obtain a relationship between the ripple wavelength and the kinetic energy and flux of ions, E and Φ_0, respectively. Assuming that a, σ and ν are proportional to E and independent of Φ_0, one gets $\lambda_l \sim E^{-1/2}$ and $\lambda_l \sim \Phi_0^{-1/2}$.

Working still within a linear approximation to the surface height in a similar spirit to the introduction of surface diffusion in (15), some alternative physical mechanisms have been proposed, specially in order to account for the lack of pattern (ripple) formation at either low temperatures [28] or small angles of incidence [19]. Thus, a term of the form $-F|k|h(k,t)$ added to the (Fourier transformed) right-hand side of Eq. (15) can describe the effect of viscous flow in the bulk onto the surface dynamics, with F being a coefficient that depends on the surface-free energy and bulk viscosity [121,122]. For low enough temperatures at which thermal surface diffusion is hampered, bulk viscous flow could dominate the surface dynamics even to the extent of preventing ripple formation for oblique angle incidence [28] or dot formation for rotating substrates [40]. Alternatively, the fact that a fraction of the sputtered atoms that move close to the surface are recaptured has led to arguing [19] for the effective description of such a process by inclusion of a term with the form $+\nu_{recc}\partial_x^2 h$, where ν_{recc} would be a positive, angle-dependent coefficient. We will see later that direct knock-on sputtering events also would have effectively an analogous continuum description, thus corresponding to stabilizing mechanisms that can counterbalance Sigmund's instability.

Higher-Order Corrections

As we have seen, BH's linear theory already predicts many important features of the nanopatterning process, but leaves out a number of additional properties. For instance, it can not predict saturation of the ripple amplitude or their nonuniform lateral motion. In order to account for these features, one needs to generalize the BH theory, the most natural procedure being the extension of their perturbative expansion in surface derivatives in (5–7) up to higher order. In this process, one needs to approximate the surface height locally through a Taylor expansion, as BH did, only that the order of approximation is higher. Along the way, not only do nonlinear contributions arise, but also corrections appear to previous linear terms. These thus provide contributions of erosive origin to linear relaxation mechanisms such as transverse pattern motion and surface diffusion. This program has been carried out in [44,111,112]. The result is the following equation (the coefficients λ with various indices appearing in this and the following height equations must not be confused with the pattern wavelength):

$$\begin{aligned}\frac{\partial h}{\partial t} = &-v_0 + \gamma_x \frac{\partial h}{\partial x} + \Omega_1 \frac{\partial^3 h}{\partial x^3} + \Omega_2 \frac{\partial^3 h}{\partial x y^2} + \xi_x \frac{\partial h}{\partial x}\frac{\partial^2 h}{\partial x^2} + \xi_y \frac{\partial h}{\partial x}\frac{\partial^2 h}{\partial y^2} \\ &+ \nu_x \frac{\partial^2 h}{\partial x^2} + \nu_y \frac{\partial^2 h}{\partial y^2} + \frac{\lambda_x}{2}\left(\frac{\partial h}{\partial x}\right)^2 + \frac{\lambda_y}{2}\left(\frac{\partial h}{\partial y}\right)^2 \\ &- D_{xy}\frac{\partial^4 h}{\partial x^2 y^2} - D_{xx}\frac{\partial^4 h}{\partial x^4} - D_{yy}\frac{\partial^4 h}{\partial y^4},\end{aligned} \qquad (17)$$

where all the coefficients depend on the experimental parameters from Sigmund's theory [112], as in BH's equation (10), which now becomes a linear low-order approximation of Eq. (17). In the first line of (17), and except for the constant average velocity v_0, all terms provide linear and nonlinear contributions to the transverse ripple motion. The second line reflects the dependence of the sputtering yield with the local curvatures (as in BH's equation) and slopes, while the third line contains (linear) surface diffusion terms that include both thermal (as in BH) and erosive contributions. This type of contributions breaks in general the $x \to -x$ symmetry while it does not alter the symmetry under $y \to -y$. The dependence of all parameters on the incidence angle is such that rotational in-plane symmetry is restored for normal incidence and Eq. (17) becomes the celebrated Kuramoto-Sivashinsky (KS) equation:

$$\frac{\partial h}{\partial t} = -v_0 + \nu \nabla^2 h - D \nabla^4 h + \frac{\lambda_0}{2}(\nabla h)^2, \qquad (18)$$

where we have used that, for $\theta = 0$, $D_{xx} = D_{yy} = D_{xy}/2$, $\nu_x = \nu_y \equiv \nu$, $\lambda_x = \lambda_y \equiv \lambda_0$, $\gamma_x = \xi_x = \xi_y = \Omega_1 = \Omega_2 = 0$ [112].

The main morphological implications of Eqs. (17) and (18) have been discussed elsewhere [112]. Here, we briefly mention the main conclusions. First of all, from the point of view of these equations, the behavior predicted by BH appears as a short-time transient, and all BH predictions—on the dependence of, e.g., the ripple structure with incidence angle—still carry over here. Moreover, for temperatures at which surface diffusion is activated, the dependences of the ripple wavelength with energy, temperature and flux are as in BH, at least for short to intermediate times. If T is low enough that the only contribution to surface diffusion is from erosive origin, these dependencies are modified and compare better with experiments at low temperatures (see [112] and Section 3 above). For intermediate to long times, the nonlinear terms in Eqs. (17) and (18) are such that the exponential growth of the ripple amplitude is stabilized, yielding to a much slower power-law increase of, e.g., the surface roughness with time. At normal incidence the surface displays kinetic roughening for long time and length scales, and this is also the case at oblique incidence as long as the nonlinearities $\lambda_{x,y}(\theta)$ have the same signs. Otherwise, Eq. (17) features *cancellation modes* as first identified in the anisotropic KS equation, that is the following particular case of Eq. (17) in which the propagative terms with coefficients $\xi_{x,y}$ and $\Omega_{1,2}$ (being arguably irrelevant to the asymptotic limit) are simply neglected [44,144]:

$$\frac{\partial h}{\partial t} = -v_0 + \gamma_x \frac{\partial h}{\partial x} + \nu_x \frac{\partial^2 h}{\partial x^2} + \nu_y \frac{\partial^2 h}{\partial y^2} + \frac{\lambda_x}{2}\left(\frac{\partial h}{\partial x}\right)^2 + \frac{\lambda_y}{2}\left(\frac{\partial h}{\partial y}\right)^2 - D_{xy}\frac{\partial^4 h}{\partial x^2 y^2} - D_{xx}\frac{\partial^4 h}{\partial x^4} - D_{yy}\frac{\partial^4 h}{\partial y^4}. \qquad (19)$$

Cancellation modes are height Fourier modes with wave vector in the unstable band, for which the nonlinear terms cancel each other, leaving the system nonlinearly unstable and inducing ripples which are oriented in an oblique direction that is parallel neither to the x nor to the y axis [144].

Equation (18) has, on the other hand, the capability (beyond linear equations such as that of BH) of predicting both "dot" and "hole" production, depending on the sign of the nonlinear parameter λ, which in turn depends on the characteristics of Sigmund's distribution [87]. However, conspicuous nonlinear features still remain beyond description by the KS equation (18) and its generalizations (17) and (19). First, the "dot" or ripple structures described by these nonlinear equations are characterized by a dominant wavelength that remains fixed in time and does not coarsen for any parameter values, so that experiments in which coarsening occurs cannot be accounted for. A stronger limitation is that the patterns described by (17), (18) and (19) disorder in heights to the extent that the size of ordered domains of ripples or dots essentially restricts to a single structure. There is thus no lateral ordering of the dots or ripples. Actually, the KS equation is known as a *paradigm of spatio-temporal chaos* in the wider field of Non-Linear Science [119].

Following the program sketched at the beginning of this paragraph, a natural step is to carry on still further the perturbative study of Sigmund's local velocity of erosion pioneered by BH. At the present stage, one could close Eq. (18) (we now consider normal incidence for simplicity) by including nonlinear terms that are quadratic in the height field and fourth order in space derivatives, reaching an equation of the form (in reference frame comoving with the eroded surface)

$$\frac{\partial h}{\partial t} = \nu \nabla^2 h - D\nabla^4 h + \lambda_1 (\nabla h)^2 + \lambda_2 \nabla^2 (\nabla h)^2. \tag{20}$$

Indeed, such a higher-order generalization has been performed [94]. However, the expressions of λ_1 and λ_2 in terms of Sigmund's parameters are such that the coefficients of the two nonlinearities in Eq. (20) have the same signs for all physical parameter values. Unfortunately, as shown shortly after [26,95], this introduces cancellation modes that seriously question the mathematical validity of (20) for our physical system. Writing the equation in Fourier space, we get

$$\frac{\partial h(\mathbf{k})}{\partial t} = (-\nu k^2 - Dk^4) h(\mathbf{k}) + (\lambda_1 - \lambda_2 k^2) FT\left[(\nabla h)^2\right], \tag{21}$$

where *FT* denotes Fourier Transform. For the unstable mode $\mathbf{k}_0$ (that indeed occurs physically, see the experiments of IBS of Pd(001) in [94,26]) such that $k_0^2 = \lambda_1/\lambda_2$, the nonlinear terms may cancel each other, so that the amplitude of this mode explodes exponentially. We are seemingly left with a matter-of-principle limitation, namely, the simpler theoretical approach that is limited to studying Sigmund's local velocity of erosion (at sufficiently high-linear and nonlinear orders) meets mathematical limitations before being able to cover for the various experimental features of the patterns we wish to study.

In the following paragraph, we will take a wider viewpoint in which the dynamics is more complete, in the sense that the surface height will be coupled to an additional physical field describing the flux of adsorbed material that diffuses on the near-surface layer. This procedure will be seen to provide an improved description solving some of the above physical and mathematical shortcomings. Before doing so, we note that additional studies exist that focus on the evolution of the height field only. Thus, the standard result in the field of Pattern Formation (see references in [136]) that a linear damping in the KS equation induces the appearance of ordered patterns has directly led to the proposition in [59] of (a modified) damped KS equation to describe the experiments described in Section 3. This modified equation actually becomes the standard damped KS equation after an appropriate nonlocal time transformation [178]. Although the natural anisotropic generalization has been duly proposed [179], these equations unfortunately do not allow much improvement in the continuum description of nanopatterning by IBS for several reasons: (i) there is no connection to phenomenological parameters (the equations are not *derived* from any model but are, rather, argued for on a phenomenological basis); (ii) damped generalizations of the KS equation seem (see [27] and references therein) *not* to allow for wavelength coarsening for any parameter values, which leaves out many of the experimentally observed patterns of these descriptions and (iii) similarly, at long times the fluctuations of the PSD function predicted by any of the damped generalizations of the KS equation are cut-off and do not show the type of power-law-like behavior that is seen, even in experiments in which there is no coarsening [195,199]. In contrast, a more successful extension of the BH-type approach has been its generalization in [35] to surfaces with steep slopes. Recall that starting with BH's, all continuum approaches mentioned (in contrast, e.g., with MC models) work within a small-slope approximation that allows neglection of higher-order terms in equations such as (17–20). Remarkably, a suitable generalization for arbitrary slope values has been seen in [35] to lead to a nonlinear equation whose traveling wave (shock) solutions compare well with experiments on the motion of the walls of pits excavated by a FIB. These results possibly provide important clues for a more complete model of nanopatterning by IBS.

4.3.2 Coupling to Diffusive Surface Species

Due to physical and mathematical considerations, the previous section leaves us with the need to enlarge the continuum description of nanopatterning by IBS. The expectation is that, by incorporating the dynamics of additional physically relevant fields, the effective height equation to be eventually derived improves its formal properties and its predictive power. Perhaps, we could compare the situation with related fields such as, e.g., the growth of thin films by physical or chemical vapor deposition (CVD) techniques. In CVD, for instance (see references in [46]), the standard continuum description arises precisely from the coupling between the local (growth) velocity and the dynamics of the concentration field of the diffusing species that eventually will stick to the growing aggregate.

There have been various attempts in the IBS context to combine the surface dynamics as predicted by a BH-type equation, with the evolution of relevant surface species, such as adatoms, addimers and surface traps (see [53,30] and references therein). However, in this approach no explicit feedback mechanism is provided from the dynamics of such species onto the local variation of the surface height so that the dynamics cannot be described by a closed system of equations. On the other extreme, there is a recent proposal in which a full Navier–Stokes (thus, highly coupled) formulation is proposed to describe ripple transverse motion onto a glass surface [2]. Trying to reach a balance between complexity and completeness in the physical description, one can seek for a formulation that, while simpler than a full hydrodynamic model, still provides an explicit coupling between the surface topography and the evolution of the relevant diffusive fields. For the case of ripple dynamics in the different (macroscopic!) context of aeolian sand dunes [42], such is the spirit of the so-called "hydrodynamic" approach, in which one sets up a system of coupled equations that describe the height of the eroded substrate profile, h, and the thickness of a mobile surface layer, R. Although there are relevant differences between both physical systems—the size of the structures is roughly seven orders of magnitude larger in aeolian ripple formation than in IBS systems—the similarity between the mechanisms of diffusion and erosion and the analogous form and behavior of the patterns in both systems seem to suggest that these processes could be modeled by similar formalisms.

This program has been followed in [6,7,25,123,125,126], where the subsequent equations are proposed to describe the evolution of the fields h and R:

$$\partial_t R = (1-\phi)\Gamma^{ex} - \Gamma^{ad} - \nabla \cdot \mathbf{J}, \tag{22}$$

$$\partial_t h = -\Gamma^{ex} + \Gamma^{ad}. \tag{23}$$

In (22) and (23), $\Gamma^{ex}(R,h)$ and $\Gamma^{ad}(R,h)$ are, respectively, rates of atom excavation from and addition to the immobile bulk, $(1-\phi) = \bar{\phi}$ measures the fraction of eroded atoms that become mobile, and the third term in Eq. (22) describes motion of mobile material onto the surface. In this way, local redeposition is allowed if $\bar{\phi} \neq 0$ [104], while the viscous near-surface layer R is provided with a dynamics of its own [175].

In [6,7], a linear dependence of Γ^{ex} and Γ^{ad} with the local geometry of the surface is considered to study the linear stability of this system. These studies reveal that depending on the values of the parameters, the ripple orientation could be aligned in any target direction, one of the limitations of this model being that the model is not well defined in the absence of redeposition $\bar{\phi} = 0$.

More detailed mechanisms of erosion and addition are explicited in [25,123,125], where nonlinear terms for the rate of excavation are introduced. These additional mechanisms are seen below to induce richer pattern dynamics than previous models for the surface height only. Unlike Aste and Valbusa's and aeolian sand ripples models, in [123,125] the momentum in the direction of the projection of the beam directly transmitted from ions to superficial atoms is neglected. Rather, a diffusive term for mass transport onto the surface is considered that, in the case of isotropic amorphous materials, is given by $\mathbf{J} = -D\nabla R$, where D is a thermally activated constant. Another feature of [25,123,125] is to consider a nonzero amount of mobile material, R_{eq}, even in the absence of excavation ($\Gamma^{ex} = 0$) or redeposition ($\bar{\phi} = 0$) which could be thermally induced. This term allows us to write the rate of addition in similar form to the Gibbs–Thompson expression for surface relaxation via evaporation–condensation, namely,

$$\Gamma^{ad} = \gamma_0 \left[R - R_{eq}(1 - \gamma_{2x}\partial_x^2 h - \gamma_{2y}\partial_y^2 h) \right], \tag{24}$$

where γ_0 is the mean nucleation rate for a flat surface (on the xy plane) and γ_{2x}, $\gamma_{2y} \geq 0$ describe variations of the nucleation rate with surface curvatures (and are positive if we assume that nucleation events are more likely in surface valleys than in protrusions). Note that, in (24), the full thickness of the mobile layer is affected by the shape of the surface.

In [123,125], an expression similar to the nonlinear generalization (17) of BH's local erosion velocity is considered for the rate at which material is sputtered from the bulk. If the beam direction is in the xz plane, we have,

$$\Gamma_{ex} = \alpha_0 \left[1 + \alpha_{1x}\partial_x h + \alpha_{2x}\partial_x^2 h + \alpha_{2y}\partial_y^2 h + \alpha_{3x}(\partial_x h)^2 + \alpha_{3y}(\partial_y h)^2 \right.$$
$$\left. -(\partial_x h)(\alpha_{4x}\partial_x^2 h + \alpha_{4y}\partial_y^2 h)\right], \tag{25}$$

where parameters reflect the dependence of Γ_{ex} on the *local* shape of the surface and are functions of the angle of incidence, ion and substrate species, ion flux, energy, and other experimental parameters just as the coefficients of Eq. (17).

4.3.3 Oblique Incidence

In a weakly nonlinear analysis of this system, the length-scale separation we have already mentioned between the timescales associated with erosive and diffusive events is seen to play a crucial role. Thus, in particular the diffusive field R relaxes much faster than the target height h, making it possible within a multiple-scale formulation to solve perturbatively for the dynamics of the former and derive a closed effective equation for the evolution of the latter which, in the case of oblique incidence [123,125], reads

$$\partial_t h = \gamma_x \partial_x h + \sum_{i=x,y} \left[-\nu_i\, \partial_i^2 h + \lambda_i^{(1)}\, (\partial_i h)^2 + \Omega_i \partial_i^2 \partial_x h + \xi_i\, (\partial_x h)(\partial_i^2 h)\right]$$
$$+ \sum_{i,j=x,y} \left[-K_{ij}\partial_i^2\partial_j^2 h + \lambda_{ij}^{(2)}\, \partial_i^2(\partial_j h)^2\right], \tag{26}$$

where these new coefficients are related to those of Eqs. (22–25) [123,125].

As in the nonlinear continuous theories shown in the previous section, the reflection symmetry in $\hat{\mathbf{x}}$ is broken in Eq. (26), but it is preserved in the $\hat{\mathbf{y}}$ axis. Equation (26) generalizes the height equations in [13,112,137] within BH approach to IBS; the main difference between (17) and (26) is that additional terms $\lambda_{ij}^{(2)}$ appear in the latter. These terms are important to correctly describe the evolution of the pattern as we will see below. To the best of our knowledge, Eq. (26) is new and indeed has a rich parameter space. Numerical integration within linear regime retrieves all features of the ripple structure as predicted by the BH theory, i.e., dependence of the ripple wavelength with linear terms and ripple orientation as a function of θ.

Top views of the morphologies described by Eq. (26) at different times are shown in Fig. 24 for different values of the parameters. In all these examples, the ripples are increasing their size in the course of time while disordering in heights for long distances, whereas the form of the ripples can vary appreciably depending on parameter values. In some cases, as in the first and third rows of Fig. 24, the ripples are disordered longitudinally, whereas in the second row, the ripples are longer and more straight. In the third row, the ripples are also very disordered in height and tend to group themselves in domains which contain, approximately, three ripples each. The ripple coarsening seen in Fig. 24 actually requires the presence of $\lambda_{i,j}^{(2)}$, whose magnitude and mathematically correct sign are due to describing redeposition by means of the additional field R. When the values of these coefficients increase relative to

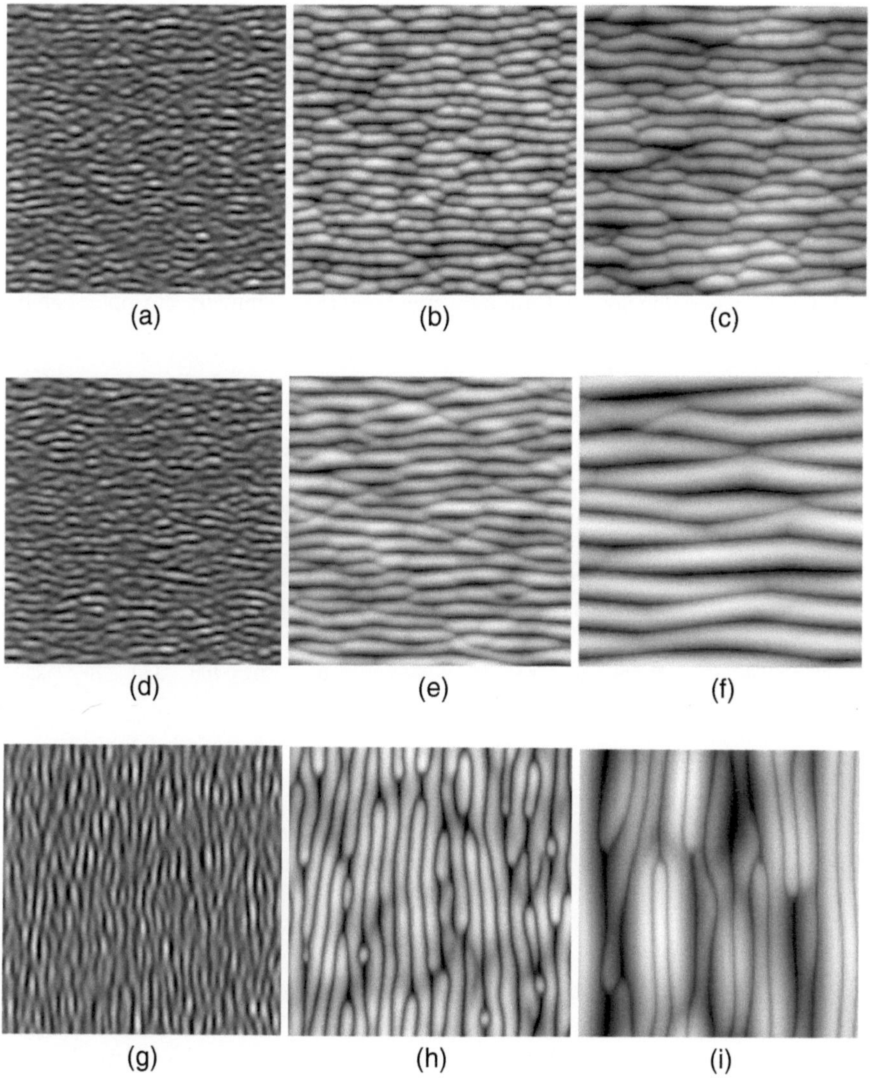

Fig. 24 (**a**), (**b**) and (**c**) Top views of morphologies described by (26) at $t = 10$ (*first column*), $t = 106$ (*second column*) and $t = 953$ (*third column*) with $\gamma_x = -0.1$, $\nu_x = 1$, $\nu_y = 0.1$, $\Omega_x = 1$, $\Omega_y = 0.5$, $\xi_i = 0.1$, $\lambda_x^{(1)} = 1$, $\lambda_y^{(1)} = 5$, $\lambda_{i,j}^{(2)} = -5$, $K_{i,j} = 1$. (**d**), (**e**) and (**f**) with the same parameters as in (**a**), (**b**) and (**c**) except $\lambda_x^{(1)} = 0.1$. (**g**), (**h**) and (**i**) with $\gamma_x = 0.1$, $\nu_x = 1$, $\nu_y = -0.95$, $\Omega_i = -0.5$, $\xi_i = 0.1$, $\lambda_x^{(1)} = 0.1$, $\lambda_y^{(1)} = 1.0$, $\lambda_{i,x}^{(2)} = -0.5$, $\lambda_{i,y}^{(2)} = -5.0$ y $K_{i,j} = 1$. The x axis is oriented vertically and the y axis horizontally

$\lambda_i^{(1)}$, coarsening stops later, and the amplitude and wavelength of the pattern also increase, analogous to the result for one-dimensional interfaces studied in [124]. The coarsening exponent n will take an *effective* value that will be larger the later coarsening stops, and may depend on (simulation) parameter values [123].

Asymmetry and transverse in-plane motion of the pattern depends on terms with an odd number of derivatives in Eq. (26). Whereas the linear terms tend to make the ripples move with a constant velocity, the nonlinear term ξ_i accounts for non-uniform movement of the ripples as observed in some experiments [79,2]. In Fig. 25, we show the evolution of a transversal section of the surface as described by Eq. (26), where the non-uniform movement and the asymmetry of ripples can be appreciated

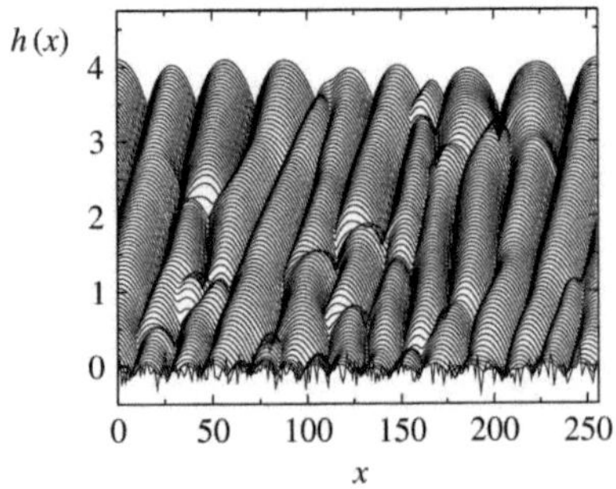

Fig. 25 Cross sections at $y = L/2$ of the two-dimensional surface given by Eq. (26) for times between $t = 0$ and $t = 1500$ separated by a time interval $\Delta t = 20$ with $\gamma_x = -2$, $\Omega_i = 0$, $\nu_x = 1$, $\nu_y = 0.1$, $\xi_i = 3.5$, $\lambda_x^{(1)} = 1$, $\lambda_y^{(1)} = 5$, $\lambda_{i,x}^{(2)} = -50$, $\lambda_{i,y}^{(2)} = -5.0$ and $\mathcal{K}_{i,j} = 1$

An example of a comparison of Eq. (26) with experiments is shown in Fig. 26 where coarsening of the ripples is clearly appreciated. As we see, Eq. (26) indeed captures essential properties of the evolution of the experimental topography.

4.3.4 Normal Incidence

In the case that the surface is bombarded perpendicularly, the in-plane asymmetry introduced by the oblique beam disappears and, for materials that do not show crystallographically privileged directions, the height equation which describes the evolution of h reads [25,125,126]

$$\partial_t h = -\nu\nabla^2 h - \mathcal{K}\nabla^4 h + \lambda^{(1)}(\nabla h)^2 + \lambda^{(2)}\nabla^2(\nabla h)^2, \tag{27}$$

where, again, the coefficients are related to the Sigmund's parameters. A stochastic generalization of this equation has also been proposed in the context of amorphous thin-film growth [140,141].

An important property of (27) is that, in the absence of redeposited material ($\bar{\phi} = 0$), $\lambda^{(1)}$ and $\lambda^{(2)}$ have the same signs, precisely as occurred in the case of the purely erosive approach [94] discussed in the previous section. This may make Eq. (27) nonlinearly unstable [141,26]. Hence, the description with a single-height

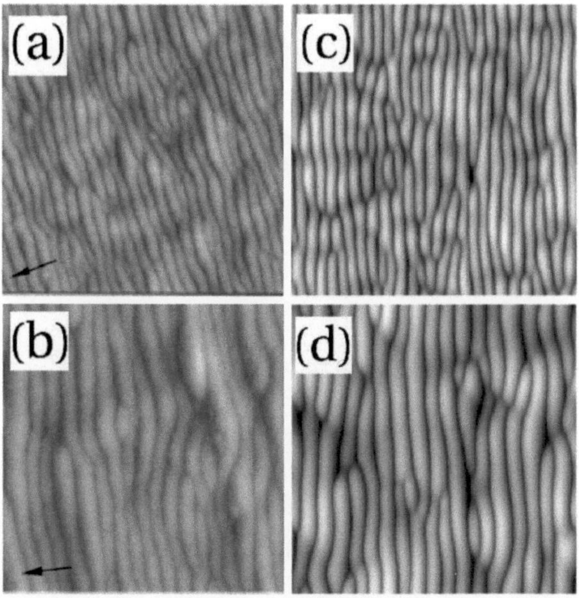

Fig. 26 $1 \times 1\ \mu m^2$ AFM views of fused SiO_2 targets irradiated under oblique incidence after 10 min (**a**) and 60 min (**b**) (taken with permission from [63]). Similar morphologies obtained by numerical integration of Eq. (26) at times t_1 (**c**) and $6t_1$ (**d**)

field seems to feature an intrinsic problem which is solved, if a fraction of redeposited material is large enough, through the two-field "hydrodynamic" approach.

In order to reduce the numberof parameters and simplify the analysis of (27), we rescale x and y, t and h by $(K/\nu)^{1/2}$, K/ν^2 and ν/λ_1, respectively, resulting in a single-parameter equation as done in the one-dimensional counterpart of (27) studied in [124]. This reads

$$\partial_t h = -\nabla^2 h - \nabla^4 h + (\nabla h)^2 - r\nabla^2(\nabla h)^2, \tag{28}$$

where $r = -(\nu\lambda^{(2)})/(K\lambda^{(1)})$ is a positive parameter which allows us to perform a numerical analysis of the complete parameter space of Eq. (27) through an appropriate rescaling.

In Figs. 27 and 28, the evolution of the height profile and the corresponding autocorrelation functions are shown for $r = 5$. Starting from an initial random distribution, a periodic surface structure with a wavelength of about the maximum of the linear dispersion relation arises and the amplitude of h increases. Then, a coarsening process occurs and dots grow in width and height, the total number of them decreasing. The apparent coarsening is quantified in the plot of $\lambda(t)$ shown in Fig. 29, where the saturation of pattern wavelength at long times can be observed. Simultaneously, with dot coarsening, the pattern increases its in-plane order, eventually leading to an hexagonal dot array as shown in Fig. 28.

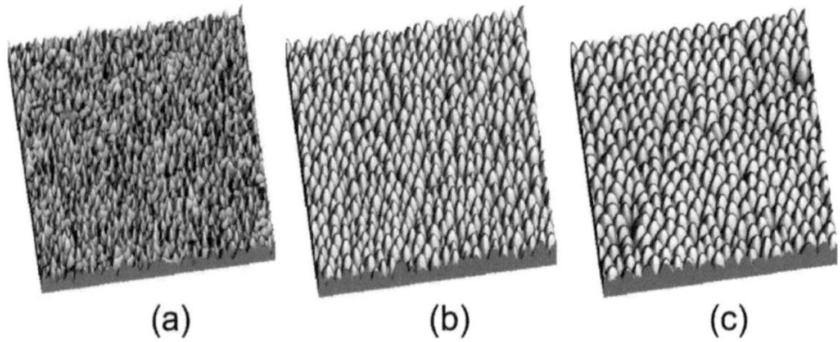

Fig. 27 Three-dimensional views of the surface obtained by numerical integration of Eq. (28) for $r = 5$ at different times: (**a**) $t = 10$; (**b**) $t = 50$ and (**c**) $t = 205$

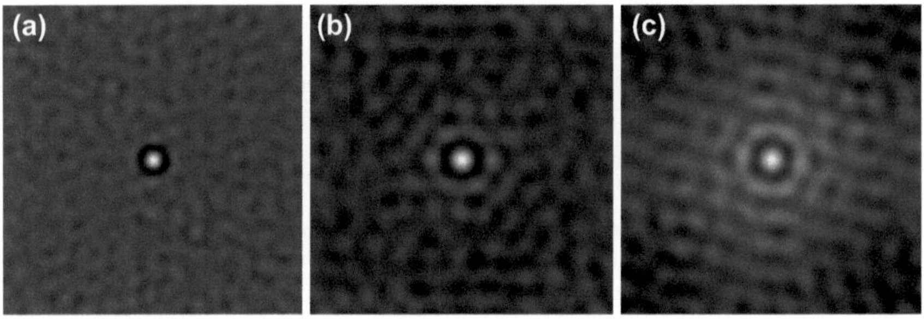

Fig. 28 Height auto-correlation corresponding to the panels of Fig. 27

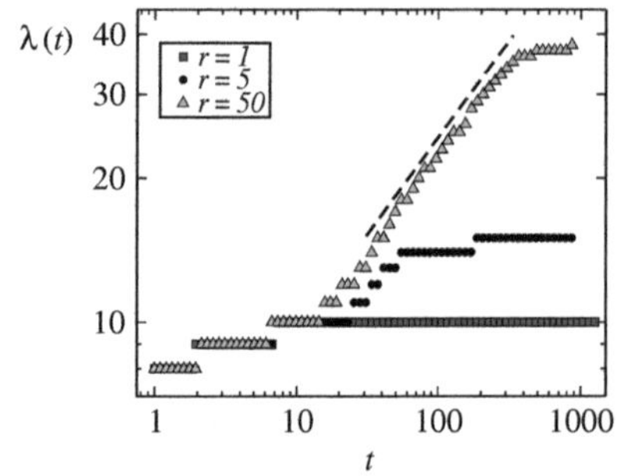

Fig. 29 Temporal evolution of the pattern wavelength, λ, given by Eq. (28) for different values of r. For $r = 50$, the *dashed line* represents the fit to a power law $\lambda \sim t^{0.40}$

As an example of comparison of this equation with specific experiments, in this paragraph we show the results recently obtained in [72] where an array of nanodots is obtained over a silicon substrate. In this work, the time evolution of nanodots is considered on Si(001) and Si(111) targets irradiated under normal incidence with Ar^+ ions of 1.2 keV. Fig. 30 shows characteristic AFM images of Si(001) and Si(111) surfaces eroded at different times. The pattern is similar in both experiments: dots of 5–7 nm height and 40–60 nm width are formed and group into short-range hexagonal order. For times larger than 20 min, the surface disorders in heights for larger distances (of the order of 6 nm in height and about 500 nm of lateral distance). This long-range disorder (kinetic roughening) increases in the course of the time. In this experiment, the rate of erosion was determined experimentally in all the processed samples by partially masking them during the sputtering process and measuring the resulting step edge height with a profilometer. It was observed that the sputtering rate (SR) was 10% higher for Si(111) than for Si(001). In order to understand the differences found experimentally between the two surface orientations, we have integrated Eq. (27) numerically considering this fact. Thus, as the model neglects target crystallinity, we have used this relative difference in SR to simulate the IBS pattern evolution in two model systems. Insets of Fig. 30a and b display the two-dimensional top-view images from the simulations for surfaces with a given SR at two different simulation times (time and length units are arbitrary). Similarly, insets of Fig. 30c and d correspond to the same simulations for a surface with a 10%

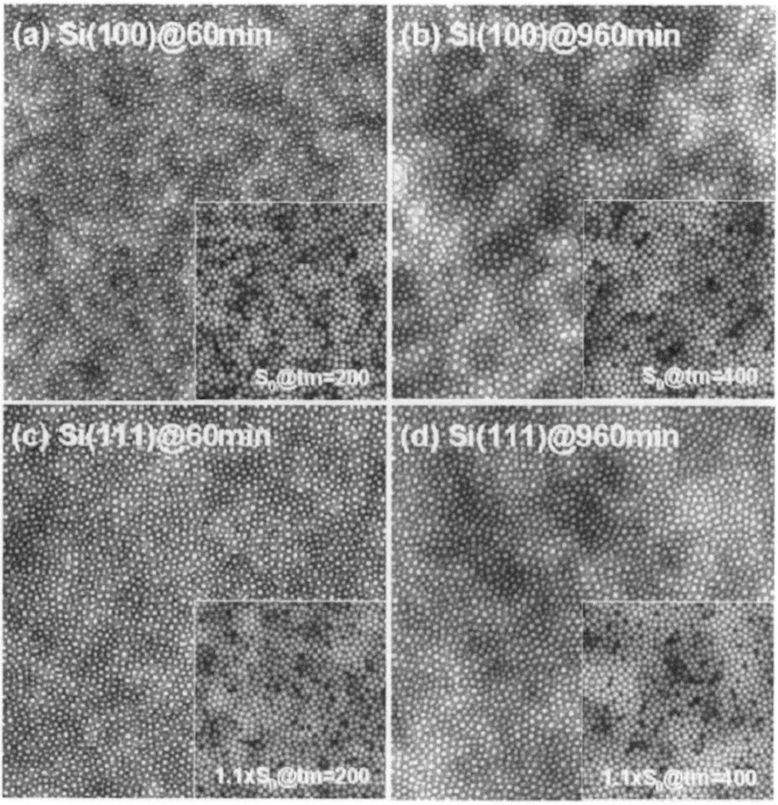

Fig. 30 AFM 3 × 3 μm^2 images of Si(001) and Si(111) surfaces sputtered with 1.2 ke V Ar^+ under normal incidence for different times (see labels). Insets: Two-dimensional views (length scales in arbitrary units) from numerical integration of Eq. (27) for different simulation times and surfaces with low (ν = 0.68, $\mathcal{K}$ = 1, $\lambda^{(1)}$ = 0.028, and $\lambda^{(2)}$ = 0.136) and high (ν = 0.75, $\mathcal{K}$ = 1, $\lambda^{(1)}$ = 0.031, and $\lambda^{(2)}$ = 0.15) erosion rates. Reprinted from [72] with permission

higher SR, thus representing the Si(111) orientation. The simulations reproduce in both cases the experimental coarsening and occurrence of a long-wavelength corrugation. In order to quantify the pattern features, the PSD functions corresponding to simulations were systematically analyzed. In Fig. 9, it is shown that the pattern wavelength saturates earlier for the surface with higher SR, but the lower SR case attains a larger final dot size. In addition, it is also shown that the correlation length also saturates earlier for a higher SR surface. Remarkably, both AFM and simulation results agree in estimating ordered domains to contain roughly three nanodots. This agreement between simulations and experimental observations allows us to conclude, e.g., that the experimental differences observed between the pattern evolution on Si(111) and Si(001) surfaces are due to their different SRs.

4.3.5 Rotating Substrate

Following the ideas of [14], in [126] we obtained the evolution equation for the surface in case the substrate is rotating simultaneously while bombarding. This reads

$$\partial_t h = -\nu_{\mathrm{r}} \nabla^2 h - K_{\mathrm{r}} \nabla^4 h + \lambda_{\mathrm{r}}^{(1)} (\nabla h)^2 + \lambda_{\mathrm{r}}^{(2)} \nabla^2 (\nabla h)^2 + \lambda_{\mathrm{r}}^{(3)} \, \nabla \cdot \left[\left(\nabla^2 h \right) \nabla h \right], \quad (29)$$

where a relationship between these parameters and those in Eq. (26) exists.

Preliminary simulations of Eq. (29) [126] show that this equation also presents interrupted coarsening and an ordered array of dots, in which the effect in the morphology evolution of the new term $\lambda_r^{(3)}$ is similar to that of $\lambda_r^{(2)}$.

4.3.6 Comparison Between Continuum Models

The continuum description of nanopattern formation by IBS is currently still open and making progress. As seen in this section, to date there are several alternative descriptions available that share some of their predictions, while differing in several other aspects. It is also true that theoretical work is not even complete yet—in terms of analyzing systematically the dependences of morphological properties with physical parameters such as, e.g., temperature or ion energy—due to two main reasons: (i) the main interfacial equations are nonlinear, thus not easily amenable to analytical solutions, while some of the physically interesting features such as the stationary pattern wavelength depend crucially on nonlinear effects and (ii) the parameter spaces of these models are frequently large, particularly in the oblique-incidence case. Actually, some of the interface equations that we have been considering seem to be new in the wider contexts of Non-Equilibrium Systems and Non-Linear Dynamics [124,27,46] even to the extent of providing examples of thus far unknown behaviors for problems of high-current interest, such as coarsening phenomena [139].

As a partial summary, and for the sake of comparison, we have collected in Table 3 the main available morphological predictions of the models that we have considered in a larger detail in the text. After comparison with the analogous

Table 3 Summary of main morphological predictions for some of the continuum models discussed in the text.

	BH [13, 61]	Anisotropic KS [44, 95, 111, 112]	Damped KS [59, 178, 179]	Nonlinear two – field model [25, 123]
λ	$\lambda(T) \sim T^{-1/2} e^{-\Delta T/2K_B T}$ $\lambda(\Phi) \sim \Phi^{-1/2}$ $\lambda(t) \sim$ const. $\lambda(E) \sim E^{-1/2}$	$\lambda(T) \sim \begin{cases} \text{const. (low } T) \\ T^{-1/2} e^{-\Delta T/2K_B T} \\ \text{(high } T) \end{cases}$ $\lambda(\Phi) \sim \begin{cases} \text{const. (low } T) \\ \Phi^{-1/2} \text{(high } T) \end{cases}$ $\lambda(t) \sim$ const. $\lambda(E) \sim \begin{cases} E \text{(low } T) \\ E^{-1/2} \text{(high } T) \end{cases}$	$\lambda(t) \sim$ const.	$\lambda(T) \sim \begin{cases} \text{const. (low } T) \\ T^{-1/2} e^{-\Delta T/2K_B T} \\ \text{(high } T) \end{cases}$ $\lambda(\Phi) \sim \begin{cases} \text{const. (low } T) \\ \Phi^{-1/2} \text{(high } T) \end{cases}$ $\lambda(t) \sim \begin{cases} \text{const. (short } t) \\ t^n \text{(intermediate } t) \\ \text{const. (large } t) \end{cases}$ $\lambda(E) \sim \begin{cases} E \text{(low } T) \\ E^{-1/2} \text{(high } T) \end{cases}$
W	NM	$W(t) \sim \begin{cases} e^{\omega t} \\ \text{(short } t) \\ t^{\beta} \\ \text{(intermediate } t) \\ \text{const.} \\ \text{(large } t) \end{cases}$	$W(t) \sim \begin{cases} e^{\omega t} \\ \text{(short } t) \\ t^{\beta} \\ \text{(intermediate } t) \\ \text{const.} \\ \text{(large } t) \end{cases}$	$W(t) \sim \begin{cases} e^{\omega t} \\ \text{(short } t) \\ t^{\beta} \\ \text{(intermediate } t) \\ \text{const.} \\ \text{(large } t) \end{cases}$
In-plane order	Disordered	Disordered	Hexagonal	Hexagonal
v	Uniform	Nonuniform	Uniform	Nonuniform
Kinetic roughening	NM	Yes	No	Yes

NM stands for *Not measurable* because there is not a well-defined saturation value of the roughness

experimental Tables 1 and 2, the two-field model leading to Eqs. (26), (27) and (29) seems to date the theoretical description that can account for a larger range of experimental behaviors from within a single framework. Naturally, more work is still needed in order to reach a still more complete description of these physical systems with a larger quantitative predictive power.

5 Applications of IBS-Patterned Surfaces

As mentioned in the Introduction, surface nanopatterning is the subject of intense research driven by the current road to miniaturization and the broad range of eventual applications. Therefore, the formation of self-organized patterns on surfaces by IBS is relevant not only from a fundamental point of view but also from a practical point of view. Whereas the understanding of the patterning process has been studied deeply from the theoretical and experimental points of view, the applications of these patterns have been scarcely addressed and a further technological effort is required for their implementation in real devices. Here, we summarize some of the applications launched in the literature.

One of the strengths of the IBS nanopatterning process comes from its universality since, as shown in Section 3, it is applicable to metallic, semiconducting or insulating surfaces. However, in comparison with other techniques, the major advantages of this technique rely on its simplicity, high output (fast) and scalable patterned area. Among the pattern characteristics, we can highlight the tunable wavelength and presence of ordering.

The first applications of IBS patterns were based on the production of microscale ripple patterns. In this case, the usage of these patterns for optical interference gratings was addressed [86]. Recently, optical activation of the surface due to the pattern formation has also been reported [128] which, apart from the surface morphology, may also be linked directly to the ion-induced defects.

Perhaps the most important and direct application of IBS patterning relies on the method based on the deposition of a given material layer on top of the target material intended to be patterned. This layer is then ion sputtered, this process inducing a nanopattern on top of it that, by continuing the sputtering process, is eventually transferred onto the target substrate. This strategy was previously used to generate a Ni dot pattern with a typical wavelength of 250–300 nm and dot height of 13–15 nm [151]. In this case, the top layer was a copolymer with self-organized polymer domains. These domains did show different SRs when ion sputtered. In this case, these Ni nanopatterns were used to separate DNA molecules.

The interest on IBS patterns was greatly increased by the report of nanodot pattern formation by Facsko et al. [55] as the distances involved were clearly in the nanometer range. Therefore, the potential application for quantum dot (QD) arrays fabrication [56] triggered further research. This route for QD fabrication is shown schematically in Fig. 31 and was experimentally proved by the successful photoemission of confined nanostructures [55]. Despite this result, there are issues to be

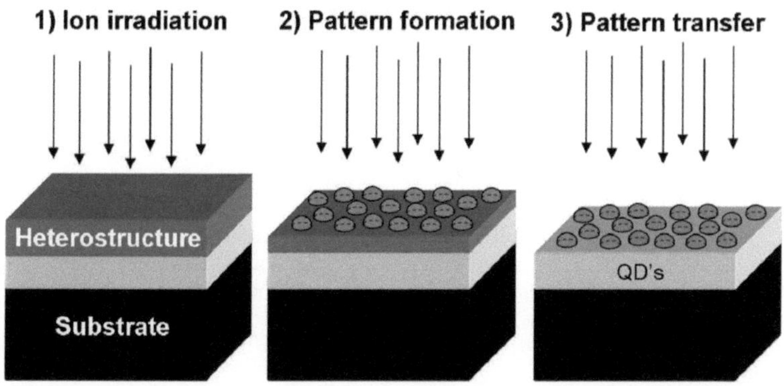

Fig. 31 Sketch of the fabrication of QD arrays by transferring the self-organized nanodot pattern induced by IBS to the interface of a semiconductor heterostructure. Adapted from [56]

improved for the final applicability, such as the crystallinity of the nanostructures, since ion bombardment induces considerable amorphization on the surface. This constraint may not play a big role if the nanodot volume is mostly crystalline. In any case, recrystallization or damage reduction may be attained by post-annealing treatments.

The IBS patterns can also be used as molds or masters to transfer or replicate the pattern to other surfaces with high functionality. This several-step process may be required when the surface cannot be processed directly, such as in the case of polymer materials, or if the desired pattern cannot be obtained in a simple way (e.g., when the diffusive regime is operating in metals). Also, this imprinting path may be interesting for reducing production costs by faster processing or avoiding the use of vacuum environments. The applicability of IBS-patterned surfaces for molding and replication has been shown by Azzaroni et al. [8] in polymer materials but the method has been successfully extended to metals [10] and even hard ceramic materials [5]. The replication method is shown schematically in Fig. 32a and consists of several sequential steps: (1) production of the IBS dot pattern on a silicon surface; (2) formation of a self-assembled monolayer of octadecyltrichlorosilane (OTS) on top of the nanostructured surface; (3) subsequent growth of the material to be patterned. This step has to be performed at mild conditions (i.e., low temperatures or under non-high energetic sputtering conditions) in order to preserve the OTS layer and (4) thanks to the antisticking properties of the OTS layer, the film deposited on top of it can be easily mechanically detached. Also, due to the extremely low thickness of the OTS layer the film surface in contact with it follows the surface morphology of the underlying nanostructured surface. Thus, the surface morphology of the detached film is the *negative* of the IBS nanostructured silicon surface. In Fig. 32b, we show the molded surface of a diamond-like carbon (DLC) film grown by radio frequency sputtering following this method. The *negative* hole structure of the IBS dot surface is observed while the two-dimensional auto-correlation function (inset) shows that the short-range hexagonal order is still preserved.

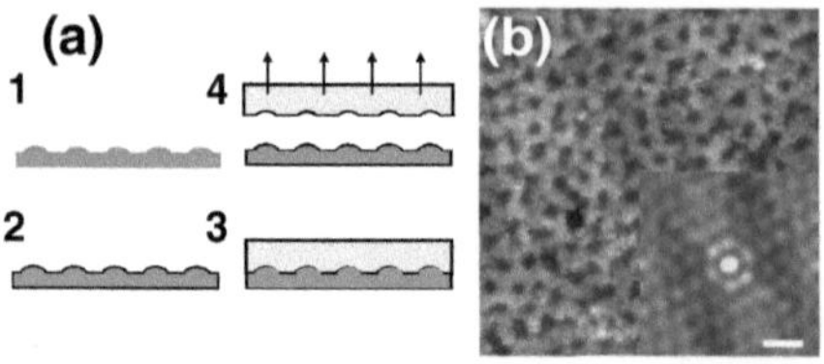

Fig. 32 (**a**) Replication method for IBS nanodot pattern transfer consisting in four steps: (1) production of the IBS pattern on Si surfaces; (2) functionalization of the nanostructured Si surface by a OTS monolayer; (3) film deposition; (4) mechanical film detachment. (**b**) 750 × 750 nm^2 AFM image of a DLC film surface molded following the method described in (**a**). Inset: Two-dimensional auto-correlation corresponding to an area of 545 × 545 nm^2. The *horizontal bar* corresponds to 63 nm

Apart from the direct exploitation of the properties induced in nanopatterned surfaces, much effort has also been devoted to the use of these surfaces as *templates* for further processing. For example, ripple patterns have been tested as substrates for alignment and manipulation of carbon nanotubes [74]. In this sense, one possible application would be to extend FIB patterning as a template technique for patterned growth of carbon nanotubes [36] to IBS nanopatterning. Also, the large range of surface roughness and characteristic wavelengths that are attainable by IBS patterns open a broad field of bioapplications. For example, surface adhesion of biological entities (proteins, DNA, etc.) may be controlled for sensing, diagnostic or biocompatiability enhancement. Moreover, the nanopit patterns could find applications as supports for catalysis [148]. Another potential application of IBS patterns is for the production of magnetic nanostructures [172]. Moroni and coworkers proved that the formation of nanoscale ripples by IBS on Co films deeply affect their magnetic properties [120]. Also, IBS patterns have been used as *growth templates* for the production of magnetic nanostructures. For example, it is well known that the magnetic properties of a superlattice formed by magnetic materials depend on the final thickness of the stack. Therefore, irregularities in the thickness due to nonconformal growth, i.e., preferential growth on the cavities (ripples or holes) or on the top of the hills, onto a patterned surface can be used to define magnetic domains [34]. Another proposal is the use of the surface topography to induce shadowed deposition under oblique precursors incidence [171]. These routes are displayed schematically in Fig. 33.

Obviously, the use of patterned surfaces for many applications is not restricted to IBS patterns. However, the presence of the, mostly undesirable, amorphous layer is a peculiarity of IBS patterns that can be used for further applications. For example, the modulation of the amorphous layer in medium-energy-induced ripples was very successfully exploited by Smirnov et al. [158] as selective doping implantation barrier. In this case, the production of ripples on silicon by nitrogen erosion leads to the formation of amorphous silicon nitride (SiN_x) regions, as shown in Fig. 34. The different nitride layer thicknesses can be used to obtain silicon regions with different degree of doping through, for instance, reactive ion-etching and As^+ implantation processes.

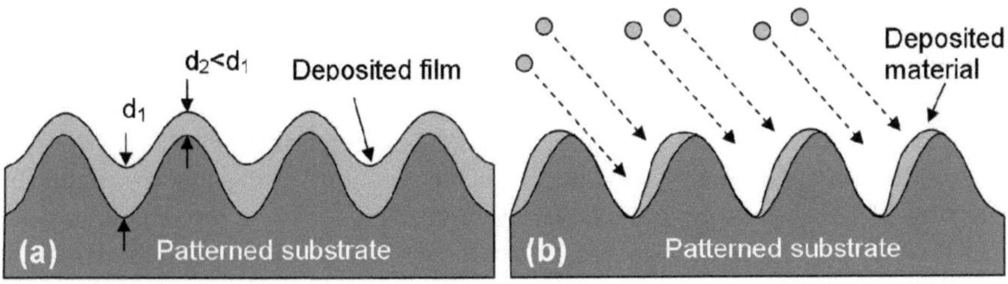

Fig. 33 Proposals for the formation of magnetic nanostructures using IBS patterned surfaces as growth templates: (**a**) nonconformal growth (the different thicknesses result in patterned magnetic domains) and (**b**) shadowed deposition

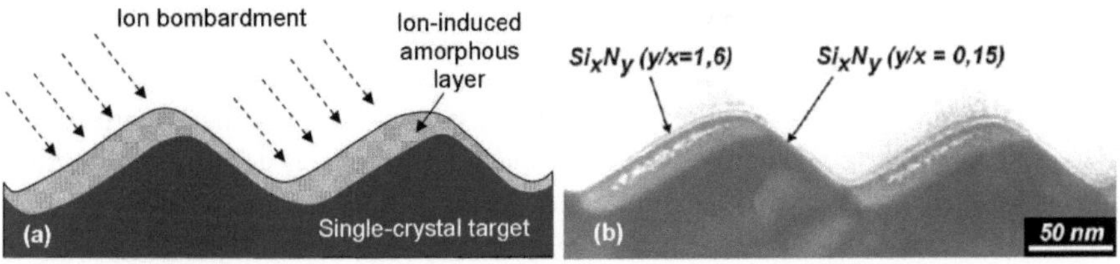

Fig. 34 (**a**) Formation of a ripple pattern and inherent modulation in the amorphized layer as a result of ion bombardment at off-normal incidence. (**b**) Cross-sectional TEM image of a ripple pattern produced by N^+ ion bombardment of silicon and the formation of different SiN_x regions that are used as implantation barriers. Figure reprinted with permission from [158]

6 Open Issues

Despite the large amount of theoretical and experimental work described above, there are still issues that remain unclear and require further research efforts. From the experimental point of view, this is especially evident for the case of nanodot patterns due to their relative novelty. For the case of nanoripples, in spite of the abundant avaliable data, there are also gaps to be filled. Perhaps, the least systematically investigated aspects concern transverse ripple propagation and the relevance of shadowing effects on the morphological evolution. The former is difficult to address due to its evident experimental complexity since in-situ, real-time monitoring of the ripple morphology has to be carried out. The latter, although quite specific, is interesting because it can provide important data for theory refinement, which requires exploring systematically the continuum models beyond the customary small-slope approximation. With regard to nanodot patterns, there is still a certain lack of systematic studies of the various targets that have proved amenable to such patterns when subject to IBS. For instance, the role of preferential sputtering on nanostructuring of hetero-semiconductors has not been unambiguously assessed yet. In addition, the effect of temperature on the nanodot morphology or the symmetry of their arrangement is not clear as different, and in some cases opposite, behaviors have been found. Another issue that will be interesting to address is the systematic comparison of the nanodot pattern properties for the same target material bombarded under normal incidence and under oblique incidence with simultaneous target rotation. In this sense, comparison is still incomplete between the IBS patterns produced with the same equipment on different materials. A related open issue is the influence of the characteristics of the ion gun optics on the pattern properties.

Another line of research with important implications in the theoretical understanding of IBS patterning is that of pattern coarsening. From the data compiled, it is clear that many systems present coarsening, but there are also others that do not. Thus, an important experimental investigation would be the study of the physical processes behind pattern coarsening. Once more, the results of these investigations would contribute to improving the theoretical models and hence deepen our understanding of IBS patterning formation. In this connection, an ambitious goal is to clarify the relationship between the more relevant parameters appearing in the continuum models with the specific properties of the target materials and the ion species. This research would indeed help to understand quantitatively the properties of different patterns found experimentally, such as, for instance, the different size and shape of dots found in Si, GaSb and InP. Moreover, it would be interesting to design specific experiments in order to verify alternative predictions from the various models. An example of this can be the experimental verification of recent proposals to produce novel nanopatterns under specific multibeam irradiation conditions [22,23,24].

Finally, from a more applied point of view, the use of IBS nanostructuring processes for technological applications remains as an important challenge since, although several specific applications have been already developed, they have not yet proved clear advantages with respect to previously existing alternative routes.

Acknowledgments We are pleased to acknowledge collaborations and exchange with a number of colleagues, in particular JM Albella, MC Ballesteros, A-L Barabási, M Camero, T Chini, M Feix, AK Hartmann, R Kree, M Makeev, TH Metzger, O Plantevin, M Varela and EO Yewande. We would like to thank specially F Alonso for his help in the ripple experiments on Si at 40 keV shown in Section 3.

Our work has been partially supported by Spanish grants Nos. FIS2006-12253-C06 (-01, -02, -03, -06) from Ministerio de Educación y Ciencia (MEC), CCG06-UAM/MAT-0040 from Comunidad Autónoma de Madrid (CAM) and Universidad Autónoma de Madrid, CCG08-CSIC/MAT-3457 from CAM and CSIC, UC3M-FI-05-007 and CCG06-UC3M/ESP-0668 from CAM and Universidad Carlos III de Madrid, and finally S-0505/ESP-0158 from CAM. RG also acknowledges financial support from the "Ramón y Cajal" program (MEC).

References

1. Alkemade PFA, Jiang ZX (2001) Complex roughening of Si under oblique bombardment by low-energy oxygen ions. J Vac Sci Technol 19: 1699–1705
2. Alkemade PFA (2006) Propulsion of ripples on glass by ion bombardment. Phys Rev Lett 96: 107602
3. Amirtharaj PM, Little J, Wood GL, Wickenden A, Smith DD (2002) The army pushes the boundary of sensor performance through nanotechnology. In Grethlein CE (ed) The AMPTIAC Newsletter, Spring 2002, Special Issue: Nanotechnology 6: 49–56
4. Andersen HH, Bay HL (1981) Sputtering by particle bombardment vol I Chapter 4: R. Behrisch (Ed.). Springer, Berlin, pp 145–218

5. Auger MA, Schilardi PL, Caretti I, Sánchez O, Benítez G, Albella JM, Gago R, Fonticelli M, Vázquez L, Salvarezza RC, Azzaroni O (2005) Molding and Replication of Ceramic Surfaces with Nanoscale Resolution. Small 1: 300–309
6. Aste T, Valbusa U (2004) Surface instabilities in granular matter and ionsputtered surfaces. Physica A 332: 548–558
7. Aste T, Valbusa U (2005) Ripples and ripples: from sandy deserts to ionsputtered surfaces. New J Phys 7: 122
8. Azzaroni O, Schilardi PL, Salvarezza RC, Gago R, Vázquez L (2003) Direct molding of nanopatterned polymeric films: Resolution and errors. Appl Phys Lett 82: 457–459
9. Azzaroni O, Fonticelli MH, Benítez G, Schilardi PL, Gago R, Caretti I, Váquez L, Salvarezza RC (2004) Direct nanopatterning of metal surfaces using self-assembled molecular films. Adv Mater 16: 405–409
10. Azzaroni O, Fonticelli M, Schilardi PL, Benítez G, Caretti I, Albella JM, Gago R, Vázquez L, Salvarezza RC (2004) Surface nanopatterning of metal thin films by physical vapour deposition onto surface-modified silicon nanodots. Nanotechnol 15: S197-S200
11. Binnig G, Quate C, Gerber Ch (1986) Atomic force microscope. Phys Rev Lett 56: 930–933
12. Bobek T, Facsko S, Kurz H, Xu M, Teichert C (2003) Temporal evolution of dot patterns during ion sputtering. Phys Rev B 68: 085324
13. Bradley RM, Harper JME (1988) Theory of ripple topography induced by ion-bombardment. J Vac Sci Technol A 6: 2390–2395
14. Bradley RM (1996) Dynamic scaling of ion-sputtered rotating surfaces. Phys Rev E 54: 6149–6152
15. Bringa EM, Johnson RE, Papaléo RM (2002) Crater formation by single ions in the electronic stopping regime: Comparison of molecular dynamics simulations with experiments on organic films. Phys Rev. B 65: 094113
16. Brown AD, Erlebacher J (2005) Temperature and fluence effects on the evolution of regular surface morphologies on ion-sputtered Si(111). Phys Rev B 72: 075350
17. Brown AD, Erlebacher J (2005) Transient topographies of ion patterned Si(111). Phys Rev Lett 95: 056101
18. Calleja M, Anguita J, García R, Birkelund K, Pérez-Murano F, Dagata JA (1999) Nanometer-scale oxidation of silicon surfaces by dynamic force microscopy: Reproducibility, kinetics and nanofabrication. Nanotechnol 10: 10044–10050
19. Carter G, Vishnyakov V (1996) Roughening and ripple instabilities on ion-bombarded Si. Phys Rev B 54: 17647–17653
20. Carter G (1999) The effects of surface ripples on sputtering erosion rates and secondary ion emission yields. J Appl Phys 85: 455–459
21. Carter G (2001) The physics and applications of ion beam erosion. J Phys D Appl Phys 34: R1-R22
22. Carter G (2004) Proposals for producing novel periodic structures by ion bombardment sputtering. Vacuum 77: 97–100
23. Carter G (2005) Surface ripple amplification and attenuation by sputtering with diametrically opposed ion fluxes. Vacuum 79: 106–109
24. Carter G (2006) Proposals for producing novel periodic structures on silicon by ion bombardment sputtering. Vacuum 81: 138–140
25. Castro M, Cuerno R, Vázquez L, Gago R (2005) Self-organized ordering of nanostructures produced by ion-beam sputtering. Phys Rev Lett 94: 016102
26. Castro M, Cuerno R (2005) Comment on "Kinetic roughening of ion-sputtered Pd(001) surface: beyond the Kuramoto-Sivashinsky model". Phys Rev Lett 94: 139601
27. Castro M, Muñoz-García J, Cuerno R, García-Hernández M, Vázquez L (2007) Generic equations for pattern formation in evolving interfaces. New J Phys 9: 102
28. Chason E, Mayer TM, Kellerman BK, McIlroy DT, Howard AJ (1994) Roughening instability and evolution of the Ge(001) surface during ion sputtering. Phys Rev Lett 72: 3040–3043

29. Chason E, Kellerman BK (1997) Monte Carlo simulations of ion-enhanced island coarsening. Nucl Instrum Methods Phys Res B 127: 225–229
30. Chason E, Erlebacher J, Aziz MJ, Floro JA, Sinclair MB (2001) Dynamics of pattern formation during low-energy ion bombardment of Si(001). Nucl Instr and Meth B 178: 55–61
31. Chason E, Chan WL (2006) Kinetic mechanisms in ion-induced ripple formation on Cu(001) surfaces. Nucl. Instrum. Methods Phys Res B 242: 232–236
32. Chason E, Chan WL, Bharathi MS (2006) Kinetic Monte Carlo simulations of ion-induced ripple formation: Dependence on flux, temperature, and defect concentration in the linear regime. Phys Rev B 74: 224103
33. Chen YJ, Wilson IH, Cheung WY, Xu JB, Wong SP (1997) Ion implanted nanostructures on Ge(111) surfaces observed by atomic force microscopy. J Vac Sci Technol B 15: 809–813
34. Chen YJ, Wang JP, Soo EW, Wu L, Chong TC (2002) Periodic magnetic nanostructures on self-assembled surfaces by ion beam bombardment. J Appl Phys 91: 7323–7184
35. Chen HH, Orquidez OA, Ichim S, Rodriguez LH, Brenner MP, Aziz MJ (2005) Shocks in ion sputtering sharpen steep surface features. Science 310: 294–297
36. Chen Y, Chen H, Yu J, Williams JS, Craig V (2007) Focused ion beam milling as a universal template technique for patterned growth of carbon nanotubes. Appl Phys Lett 90: 093126
37. Chey SJ, Van Nostrand JE, Cahill DG (1995) Surface morphology of Ge(001) during etching by low-energy ions. Phys Rev B 52: 16696–166701
38. Chini TK, Sanyal MK, Bhattacharyya SR (2002) Energy-dependent wavelength of the ion-induced nanoscale ripple. Phys Rev B 66: 153404
39. Chini TK, Okuyama F, Tanemura M, Nordlund K (2003) Structural investigation of keV Ar-ion-induced surface ripples in Si by cross-sectional transmission electron microscopy. Phys Rev B 67: 205403
40. Cirlin E-H, Vajo JJ, Doty RE, Hasenberg TC (1991) Ion-induced topography, depth resolution, and ion yield during secondary ion mass spectrometry depth profiling of GaAs/AlGaAs superlattice: effects of sample rotation. J Vac Sci Technol A 9: 1395
41. Costantini G, de Mongeot FB, Boragno C, Valbusa U (2001) Is ion sputtering always a “negative homoepitaxial deposition”?. Phys Rev Lett 86: 838–841
42. Csahók Z, Misbah C, Rioual F, Valance A (2000) Dynamics of aeolian sand ripples. Eur Phys J E 3: 71–86
43. Cuenat A, Aziz MJ (2002) Spontaneous pattern formation from focused and unfocused ion beam irradiation. Mat Res Soc Symp Proc 696: N2.8.1–N.2.8.6
44. Cuerno R, Barabási AL (1995) Dynamic scaling of ion-sputtered surfaces. Phys Rev Lett 74: 4746–4749
45. Cuerno R, Makse HA, Tomassone S, Harrington ST, Stanley HE (1995) Stochastic model for surface erosion via ion sputtering: dynamical evolution from ripple morphology to rough morphology. Phys Rev Lett 75: 4464–4467
46. Cuerno R, Castro M, Muñoz-García J, Gago R, Vázquez L (2007) Universal non-equilibrium phenomena at submicrometric surfaces and interfaces. Eur Phys J Special Topics 146: 427
47. Datta A, Wu YR, Wang YL (2001) Real-time observation of ripple structure formation on a diamond surface under focused ion-beam bombardment. Phys Rev B 63: 125407
48. Datta D, Bhattacharyya SR, Chini TK, Sanyal MK (2002) Evolution of surface morphology of ion sputtered GaAs(100). Nucl Instr and Meth B 193: 596–602
49. Datta DP, Chini TK (2004) Atomic force microscopy study of 60-keV Ar-ion-induced ripple patterns on Si(100). Phys Rev B 69: 235313
50. Datta DP, Chini TK (2005) Spatial distribution of Ar on the Ar-ion-induced rippled surface of Si. Phys Rev B 71: 235308
51. Demanet CV, Malherbe JB, van der Berg NG, Sankar V (1995) Atomic force microscopy investigation of argon-bombarded InP: Effect of ion dose density. Surf Interface Anal 23: 433–439
52. Diaz de la Rubia T, Averback RS, Benedeck R, King WE (1987) Role of thermal spikes in energetic displacement cascades. Phys Rev Lett 59: 1930–1933

53. Erlebacher J, Aziz MJ, Chason E, Sinclair MB, Floro JA (1999) Spontaneous pattern formation on ion bombarded Si(001). Phys Rev Lett 82: 2330–2333
54. Ehrlich G, Hudda FG (1966) Atomic view of surface self-diffusion: Tungsten on tungsten. J Chem Phys 44: 1039–1049
55. Facsko S, Dekorsy T, Koerdt C, Trappe C, Kurz H, Vogt A, Hartnagel HL (1999) Formation of ordered nanoscale semiconductor dots by ion sputtering. Science 285: 1551–1553
56. Facsko S, Dekorsy T, Trappe C, Kurz H (2000) Self-organized quantum dot formation by ion sputtering. Microelectron Eng 53: 245–248
57. Facsko S, Kurz H, Dekorsy T (2001) Energy dependence of quantum dot formation by ion sputtering. Phys Rev B 63: 165329
58. Facsko S, Bobek T, Kurz H, Dekorsy T, Kyrsta S, Cremer R (2001) Ion-induced formation of regular nanostructures on amorphous GaSb surfaces. Appl Phys Lett 80: 130–132
59. Facsko S, Bobek T, Stahl A, Kurz H, Dekorsy T (2004) Dissipative continuum model for self-organized pattern formation during ion-beam erosion. Phys Rev B 69: 153412
60. Fan WB, Li WQ, Qi LJ, Sun HT, Luo J, Zhao YY, Lu M (2005) On the role of ion flux in nanostructuring by ion sputter erosion. Nanotechnol 16: 1526–1529
61. Feix M, Hartmann AK, Kree R, Muñoz-García J, Cuerno R (2005) Influence of collision cascade statistics on pattern formation of ion-sputtered surfaces. Phys Rev B 71: 125407
62. Finnine I (1995) Some reflections on the past and future of erosion. Wear 186–187: 1–10
63. Flamm D, Frost F, Hirsch D (2001) Evolution of surface topography of fused silica by ion beam sputtering. Appl Surf Sci 179: 95–101
64. Frost F, Schindler A, Bigl F (2000) Roughness evolution of ion sputtered rotating InP surfaces: Pattern formation and scaling laws. Phys Rev Lett 85: 4116–4119
65. Frost F, Hirsch D, Schindler A (2001) Evaluation of AFM tips using nanometer-sized structures induced by ion sputtering. Appl Surf Sci 179: 8–12
66. Frost F, Rauschenbach B (2003) Nanostructuring of solid surfaces by ion-beam erosion. Appl Phys A 77: 1–9
67. Frost F, Fechner R, Ziberi B, Flamm D, Schindler A (2004) Large area smoothing of optical surfaces by low-energy ion beams. Thin Sol Films 459: 100–1005.
68. Frost F, Ziberi B, Höche T, Rauschenbach B (2004) The shape and ordering of self-organized nanostructures by ion sputtering. Nucl Instr and Meth B 216: 9–19
69. Frost F, Fechner R, Flamm D, Ziberi B, Frank W, Schindler A (2004) Ion beam assisted smoothing of optical surfaces. Appl Phys A 78: 651–654
70. Gago R, Vázquez L, Cuerno R, Varela M, Ballesteros C, Albella JM (2001) Production of ordered silicon nanocrystals by low-energy ion sputtering. Appl Phys Lett 78: 3316–3318
71. Gago R, Vázquez L, Cuerno R, Varela M, Ballesteros C, Albella JM (2002) Nanopatterning of silicon surfaces by low-energy ion-beam sputtering: Dependence on the angle of ion incidence. Nanotechnol 13: 304–308
72. Gago R, Vázquez L, Plantevin O, Metzger TH, Muñoz-García J, Cuerno R, Castro M (2006) Order enhancement and coarsening of self-organized silicon nanodot patterns induced by ion-beam sputtering. Appl Phys Lett 89: 233101
73. Gago R, Vázquez L, Sánchez-García JA, Varela M Ballesteros MC, Plantevin O, Albella JM, Metzger TH (2006) Temperature influence on the production of nanodot patterns by ion beam sputtering of Si(001). Phys Rev B 73: 155414
74. Granone F, Mussi V, Toma A, Orlanducci S, Terranova ML, Boragno C, Buatier de Mongeot F, Valbusa U (2005) Ion sputtered surfaces as templates for carbon nanotubes alignment and deformation. Nucl Inst. and Meth B 230: 545–550
75. Grove WR (1852) On the electrochemical polarity of gases. Phil Trans R Soc (London) B 142: 87
76. Guo LJ (2007) Nanoimprint lithography: Methods and material requirements. Adv Mater 19: 495–513

77. Habenicht S, Bolse W, Lieb KP, Reimann K, Geyer U (1999) Nanometer ripple formation and self-affine roughening of ion-beam-eroded graphite surfaces. Phys Rev B 60: R2200–R2203
78. Habenicht S (2001) Morphology of graphite surfaces after ion-beam erosion. Phys Rev B 63: 125419
79. Habenicht S, Lieb KP, Koch J, Wieck AD (2002) Ripple propagation and velocity dispersion on ion-beam-eroded silicon surfaces. Phys Rev B 65: 115327
80. Harrison DE (1988) Application of molecular-dynamics simulations to the study of ion-bombarded metal-surfaces. Crit Rev Solid State Mater Sci 14: S1–S78.
81. Hartmann AK, Kree R, Geyer U, Kölbel M (2002) Long-time effects in a simulation model of sputter erosion. Phys Rev B 65: 193403
82. Hazra S, Chini TK, Sanyal MK, Grenzer J, Pietsch U (2004) Ripple structure of crystalline layers in ion-beam-induced Si wafers. Phys Rev B 70: 121307(R)
83. Hodes G (2007) When small is different: Some recent advances in concepts and applications of nanoscale phenomena. Adv Mater 19: 639–655
84. Hofer C, Abermann S, Teichert C, Bobek T, Kurz H, Lyutovich K, Kasper E (2004) Ion bombardment induced morphology modifications on self-organized semiconductor surfaces. Nucl Inst and Meth B 216: 178–184
85. Jeffries JH, Zuo JK, Craig MM (1996) Instability of kinetic roughening in sputter-deposition growth of Pt on glass. Phys Rev Lett 76: 4931–4934
86. Johnson LF, Ingersoll KA (1979) Interference gratings blazed by ion-beam erosion. Appl Phys Lett 35: 500–503
87. Kahng B, Jeong H, Barábasi AL (2001) Quantum dot and hole formation in sputter erosion. Appl Phys Lett 78: 805–807
88. Kardar M, Parisi G, Zhang YC (1986) Dynamic Scaling of growing interfaces. Phys Rev Lett 56: 889–892
89. Karen A, Okuno A, Soeda F, Ishitani I (1991) A study of the secondary-ion yield change on the GaAs surface caused by the ion-beam-induced rippling. J Vac Sci Technol A 9: 2247–2252
90. Karmakar P, Ghose D (2004) Ion beam sputtering induced ripple formation in thin metal films. Surf Sci 554: L101–L106
91. Karmakar P, Goshe D (2005) Nanoscale periodic and faceted structures formation on Si(100) by oblique angle oxygen ion sputtering. Nucl Instr and Meth B 230: 539–544
92. Kasemo B (2002) Biological surface science. Surf Sci 500: 656–677
93. Kim J, Cahill DG, Averback RS (2003) Surface morphology of Ge(111) during etching by keV ions. Phys Rev B 67: 045404
94. Kim TC, Ghim C-M, Kim HJ, Kim DH, Noh DY, Kim ND, Chung JW, Yang JS, Chang YJ, Noh TW, Kahng B, Kim J-S (2004) Kinetic Roughening of Ion-Sputtered Pd(001) Surface: beyond the kuramoto-sivashinsky model. Phys Rev Lett 92: 246104
95. Kim TC, Ghim C-M, Kim HJ, Kim DH, Noh DY, Kim ND, Chung JW, Yang JS, Chang YJ, Noh TW, Kahng B, Kim J-S (2005) Kim et al. Reply. Phys Rev Lett 94: 139602
96. Koponen I, Hautala M, Sievänen O-P (1996) Simulations of submicrometer-scale roughening on ion-bombarded solid surfaces. Phys Rev B 54: 13502–13505
97. Koponen I, Hautala M, Sievänen O-P (1997) Simulations of roughening of amorphous carbon surfaces bombarded by low-energy Ar-ions. Nucl Instrum Methods Phys Res B 127–128: 230–234
98. Koponen I, Hautala M, Sievänen O-P (1997) Simulations of ripple formation on ion-bombarded solid surfaces. Phys Rev Lett 78: 2612–2615
99. Koponen I, Sievänen O-P, Hautala M, Hakovirta M (1997) Simulations of sputtering induced roughening and formation of surface topography in deposition of amorphous diamond films with mass separated kiloelectronvolt ion beams. J Appl Phys 82: 6047–6055
100. Koponen I, Hautala M, Sievänen O-P (1997) Simulations of self-affine roughening and ripple formation on ion bombarded amorphous carbon surfaces. Nucl Instrum Methods Phys Res B 129: 349–355

101. Krim J, Heyvaert I, Haesendonck CV, Bruynseraede Y (1993) Scanning tunnelling microscopy observation of self-affine fractal roughness in ion-bombarded film surfaces. Phys Rev Lett 70: 57–60
102. Krug J (2004) Kinetic pattern formation at solid surfaces. In: Radons G, Häussler P, Just W (eds) Collective dynamics of nonlinear and disordered systems. Springer, Berlin.
103. Kulriya PK, Tripathi A, Kabiraj D, Khan SA, Avasthi DA (2006) Study of 1.5 keV Ar atoms beam induced ripple formation on Si surface by atomic force microscopy. Nucl Instr and Meth B 244: 95–99
104. Kustner M, Eckstein W, Dose V, Roth J (1998) The influence of surface roughness on the angular dependence of the sputter yield. Nucl Instr and Meth B 145: 320–331
105. Lau GS, Tok ES, Liu R, Wee ATS, Tjiu WC, Zhang J (2003) Nanostructure formation by O_2^+ ion sputtering of Si/SiGe heterostructures. Nanotechnol 14: 1187–1191
106. Li YZ, Vázquez L, Piner R, Reifenberger R (1989) Writing nanometer-scale symbols in gold using the scanning tunneling microscope. Appl Phys Lett 54: 1424–1426
107. Lindner J, Poulopoulos P, Farle M, Baberschke K (2000) Structure and magnetism of self-organized Ni nanostructures on Cu(001). J Magn Magn Mater 218: 10–16
108. Liu ZX, Alkemade PFA (2001) Flux dependence of oxygen-beam- induced ripple growth on silicon. Appl Phys Lett 79: 4334–4336
109. Ludwig F Jr, Eddy CR Jr, Malis O, Headrick RL (2002) Si(100) surface morphology evolution during normal-incidence sputtering with 100–500 eV Ar^+ ions. Appl Phys Lett 81: 2770–2772
110. Maclaren SW, Baker JE, Finnegan WL, Loxton CM (1992) Surface roughness development during sputtering of GaAs and InP: Evidence for the role of surface diffusion in ripple formation and sputter cone development. J Vac Sci Technol A 10: 468–476
111. Makeev MA, Barabási AL (1997) Ion-induced effective surface diffusion in ion sputtering. Appl Phys Lett 71: 2800–2802
112. Makeev MA, Cuerno R, Barábasi AL (2002) Morphology of ion-sputtered surfaces. Nucl Instr Meth B 97: 185–227
113. Malherbe JB (2003) Bombardment-induced ripple topography on GaAs and InP. Nucl Instr Meth B 212: 258–263
114. Mayer TM, Chason E, Howard AJ (1994) Roughening instability and ion-induced viscous relaxation of SiO_2 surfaces. J Appl Phys 73: 1633–1643
115. Michely T, Comsa G (1991) Temperature dependence of the sputtering morphology of Pt(111). Surf Sci 256: 217–226
116. Mirkin CA (2001) Dip-pen nanolithography: automated fabrication of custom multicomponent, sub-100 nm nanometer surface architectures. MRS Bull 26: 535–538
117. Moore DF, Daniel JH, Walker JF (1997) Nano- and micro-technology applications of focused ion beam processing. Microelectron J 28: 465–473
118. Mora A, Haase M, Rabbow T, Plath PJ (2005) Discrete model for laser driven etching and microstructuring of metallic surfaces. Phys Rev B 72: 061604
119. Mori H, Kuramoto Y (1997) Dissipative structures and chaos. Springer, Berlin
120. Moroni R, Sekiba D, Buatier de Mongeot F, Gonella G, Boragno C, Mattera L, Valbusa U (2003) Uniaxial magnetic anisotropy in nanostructured Co/Cu(001): From surface ripples to nanowires. Phys Rev Lett 91: 167207
121. Mullins WW (1957) Theory of thermal grooving. J Appl Phys 28: 333
122. Mullins WW (1959) Flattening of a nearly plane solid surface due to capillarity J Appl Phys 30: 77
123. Muñoz-García J, Castro M, Cuerno R (2006) Nonlinear ripple dynamics on amorphous surfaces patterned by ion beam sputtering. Phys Rev Lett 96: 086101
124. Muñoz-García J, Cuerno R, Castro M (2006) Short-range stationary patterns and long-range disorder in an evolution equation for one-dimensional interfaces. Phys Rev E 74: 050103(R)
125. Muñoz-García J, Cuerno R, Castro M (2007) Coupling of morphology to surface transport in ion-beam irradiated surfaces: Oblique incidence. Phys Rev B 78: 205408

126. Muñoz-García J, Cuerno R, Castro M (2009) Coupling of morphology to surface transport in ion-beam irradiated surfaces: Normal incidence and rotating targets. J Phys: Condens Matter. In press
127. Murty MVR, Cowles B, Cooper BH (1998) Surface smoothing during sputtering: mobile vacancies versus adatom detachment and diffusion. Surf Sci 415: 328–335
128. Murty MVR (2002) Sputtering: The material erosion tool. Surf Sci 500: 523–544
129. Mussi V, Granonce F, Boragno C, Bautier de Mongeot, Valbusa (2006) Surface nanostructuring and optical activation of lithium fluoride crystals by ion beam irradiation. Appl Phys Lett 88: 103116
130. Nastasi MA, Mayer JW, Hirvonen JK (1996) Ion-solid interactions: Fundamentals and applications. Cambridge University Press, New York
131. Navez M, Chaperot D, Sella C (1962) Microscopie electronique – etude de l'attaque du verre par bombardement ionique. C R Acad Sci 254: 240–248
132. Nord J, Nordlund K, Keinonen J (2003) Molecular dynamics study of damage accumulation in GaN during ion beam irradiation. Phys Rev B 68: 184104
133. Oechsner H (1975) Sputtering - Review of some recent experimental and theoretical aspects. Appl Phys 8: 185–198
134. Oechsner H (ed.) (1984) Thin film and depth profile analysis. Springer, Berlin.
135. Okutani T, Shikate S, Ichimura S, Shimizu R (1980) Angular distribution of Si atoms sputtered by keV Ar^+ ions. J Appl Phys 51: 2884–2887
136. Ozaydin G, Özcan AS, Wang Y, Ludwig KF, Zhou H, Headrick RL, Siddons DP (2005) Real-time x-ray studies of Mo-seeded Si nanodot formation during ion bombardment. Appl Phys Lett 87: 163104
137. Paniconi M, Elder KR (1997) Stationary, dynamical, and chaotic states of the two-dimensional damped Kuramoto-Sivashinsky equation. Phys. Rev. E 56: 2713–2721
138. Park S, Kahng B, Jeong H, Barabási AL (1999) Dynamics of ripple formation in sputter erosion: Nonlinear phenomena. Phys Rev Lett 83: 3486–3489
139. Plücker J (1958) Observations on the electrical discharge through rarcfied gases. The London, Edimburgh and Dublin Phil Mag 16: 409
140. Politi P, Misbah C (2006) Nonlinear dynamics in one dimension: A criterion for coarsening and its temporal law. Phys Rev E 73: 036133
141. Raible M, Mayr SG, Linz SJ, Moske M, Hänggi P, Samwer K (2000) Amorphous thin-film growth: Theory compared with experiment. Europhys Lett 50: 61–67
142. Raible M, Linz SJ, Hänggi P (2001) Amorphous thin film growth: Effects of density inhomogeneities. Phys Rev E 64: 031506
143. Rossnagel SM, Robinson RS (1982) Surface diffusion activation energy determination using ion beam microtexturing. J Vac Sci Technol 20: 195–198
144. Rossnagel SM (2003) Thin film deposition with physical vapor deposition and related technologies. J Vac Sci Technol A 21: S74–S87
145. Rost M, Krug J (2005) Anisotropic Kuramoto-Sivashinsky equation for surface growth and erosion. Phys Rev Lett 75 : 3894–3897
146. Roth J (1983) Chemical sputtering. Topics Appl Phys 52: 91–146
147. Rusponi S, Boragno C, Valbusa U (1997) Ripple structure on Ag(110) surface induced by ion sputtering. Phys Rev Lett 78: 2795–2798
148. Rusponi S, Constantini G, Boragno C, Valbusa U (1998) Ripple wave vector rotation in anisotropic crystal sputtering. Phys Rev Lett 81: 2735–2738
149. Schildenberger M, Bonetti YC, Gobrecht J, Prins R (2000) Nano-pits: Supports for heterogeneous model catalysts prepared by interference lithograpy. Top Catal 13: 109–120
150. Schubert E, Razek N, Frost F, Schindler A, Rauschenbach B (2005) GaAs surface cleaning by low-energy hydrogen ion bombardment at moderate temperatures. J Appl Phys 97: 23511
151. Schwoebel RL, Shipsey EJ (1966) Step motion on crystal surfaces. J Appl Phys 37: 3682–3686

152. Seoo YS, Luo H, Samuilov VA, Rafailovic h MH, Sokolov J, Gersappe D, Chu B (2004) DNA Electrophoresis on Nanopatterned Surfaces. Nano Lett 4: 659–664
153. Shen J, Kischner J (2002) Tailoring magnetism in artificially structured materials: The new frontier. Surf Sci 500: 300–322
154. Sigmund P (1969) Theory of sputtering. I. Sputtering yield of amorphous and polycrystalline targets. Phys Rev 184: 383–416
155. Sigmund P (1973) A mechanism of surface micro-roughening by ion bombardment. J Mat Sci 8: 1545–1553
156. Soh HT, Guarini KW, Quate CF (2001) Scanning probe lithography. Kluwer, Boston.
157. Sohn LL, Willet RL (1995) Fabrication of nanostructures using atomic-force-microscope-based lithography. Appl Phys Lett 67: 1552–1554
158. Smirnov VK, Kibalov DS, Krivelevich SA, Lepshin PA, Potapov EV, Yankov RA, Skorupa W, Makarov VV, Danilin AB (1999) Wave-ordered structures formed on SOI wafers by reactive ion beams. Nucl Instr and Meth B 147: 310–315
159. Smirnov VK, Kibalov DS, Orlov OM, Graboshinikov VV (2003) Technology for nanoperiodic doping of a metal-oxide-semiconductor field-effect transistor channel using a self-forming wave-ordered structure. Nanotechnol 14: 709–715
160. Stangl J, Hol V, Bauer G (2004) Structural properties of self-organized semiconductor nanostructures. Rev Mod Phys 76: 725–783
161. Stark J (1908) Zeitsch Elecktrochem 14: 752–756
162. Stepanova M, Dew SK (2004) Surface relaxation in ion-etch nanopatterning. Appl Phys Lett 84: 1374–1376
163. Stepanova M, Dew SK, Soshnikov IP (2005) Copper nanopattern on from sputter etching a interface. Appl Phys Lett 86: 073112
164. Stepanova M, Dew SK (2006) Self-organized Cu nanowires on glass and Si substrates from suptter etching Cu/substrate interfaces. J Vac Sci Technol B 24: 592–598
165. Strobel M, Heinig K-H, Michely T (2001) Mechanisms of pit coarsening in ion erosion of fcc(111) surfaces: a kinetic 3D lattice Monte-Carlo study. Surf Sci 486: 136–156
166. Stuart RV, Wehner GK, Anderson GS (1969) Energy distribution of atoms sputtered from polycrystalline metals. J Appl Phys 40: 803–812
167. Tachi S, Okudaira S (1986) Chemical sputtering of silicon by F^+, Cl^+, and Br^+ ions - reactive spot model for reactive ion etching. J Vac Sci Technol B 4: 459–467
168. Taglauer E (1990) Surface cleaing using sputtering. Appl Phys A 51: 238–251
169. Tan SK, Liu R, Sow CH, Wee ATS (2006) Self-organized nanodot formation on InP(100) by oxygen ion sputtering. Nucl Instr and Meth B 248: 83–89
170. Tan SK, Wee ATS (2006) Self-organized nanodot formation on InP(100) by argon ion sputtering at normal incidence. J Vac Sci Technol B 24: 1444–1448
171. Teichert C (2001) Self-organization of nanostructures in semiconductor heteroepitaxy. Phys Rep 365: 335–432
172. Teichert C (2003) Self-organized semiconductor surfaces as templates for nanostructured magnetic thin films. Appl Phys A 76: 653–664
173. Terris BD, Thomson T (2005) Nanofabrication and self-assembled magnetic structures as data storage media. J Phys D: Appl Phys 38: R199-R222
174. Tok ES, Ong SW, Kang HC (2004) Dynamical scaling of sputter-roughened surfaces in 2+1 dimensions. Phys Rev E 70: 011604
175. Toma A, Buatier de Mongeot F, Buzio R, Firpo G, Bhattacharyya SR, Boragno C, Valbusa U (2005) Ion beam erosion of amorphous materials: Evolution of surface morphology. Nucl Instr and Meth B 230: 551–554
176. Umbach CC, Headrick RL, Chang K (2001) Spontaneous nanoscale corrugation of ion-eroded : The role of ion-irradiation-enhanced viscous flow. Phys Rev Lett 87: 246104
177. Vajo JJ, Doly RE, Cirlin EH (1996) Influence of O_2^+ energy, flux, and fluence on the formation and growth of sputtering-induced ripple topography on silicon. J Vac Sci Technol A 14: 2709–2720

178. Valbusa U, Boragno C, Buatier de Mongeot F (2002) Nanostructuring surfaces by ion sputtering. J Phys: Condens Matter 14: 8153–8175
179. Vogel S, Linz SJ (2005) Continuum modeling of sputter erosion under normal incidence: Interplay between nonlocality and nonlinearity. Phys Rev B 72: 035416
180. Vogel S, Linz SJ (2006) How ripples turn into dots: Modeling ion-beam erosion under oblique incidence. Europhys Lett 76: 884–890
181. Wei S, Li B, Fujimoto T, Kojima I (1998) Surface morphological modification of Pt thin films induced by growth temperature. Phys. Rev. B 58: 3605–3608
182. Witcomb MJ (1974) Prediction of apex angel of surface cones on ion-bombarded crystalline materials. J Mat Sci 9: 1227–1232
183. Wittmaack K (1984) An attempt to understand the sputtering yield enhancement due to implantation of inert-gases in amorphous solids. Nucl Instr and Meth B 230: 569–572
184. Wittmaack K (1999) Effect of surface roughening on secondary ion yields and erosion rates of silicon subject to oblique oxygen bombardment. J Vac Sci Technol A 8: 2246–2250
185. Xia YN, McClelland JJ, Gupta R, Qin D, Zhao XM, Sohn LL, Celotta RJ, Whitesides GM (1997) Replica molding using polymeric materials: A practical step toward nanomanufacturing. Adv Mater 9: 147–149
186. Xu M, Teichert C (2004) How do nanoislands induced by ion sputtering evolve during the early stage of growth? J Appl Phys 96: 2244–2248
187. Yonzon CR, Stuart DA, Zhang X, McFarland AD, Haynes CL, Van Duyne RP (2005) Towards advanced chemical and biological nanosensors? An overview. Talanta 67: 438–448
188. Yewande EO, Kree R, Hartmann AK (2005) Propagation of ripples in Monte Carlo models of sputter-induced surface morphology. Phys Rev B 71: 195405
189. Yewande EO, Hartmann AK, Kree R (2006) Morphological regions and oblique-incidence dot formation in a model of surface sputtering. Phys Rev B 73: 115434
190. Zaera F (2002) The surface chemistry of catalysis: New challenges ahead. Surf Sci 500: 947–965
191. Zalm PC (1986) Ion beam assisted etching of semiconductors. Vacuum 36: 787–797
192. Zalm PC (1994) Secondary ion mass spectrometry. Vacuum 45: 753–772
193. Zhao Y, Drotar JT, Wang GC, Lu TM (1999) Roughening in Plasma Etch Fronts of Si(100). Phys Rev Lett 82: 4882–4885
194. Zhao Y, Wang GC, Lu TM (2001) Characterization of amorphous and crystalline rough surface: Principles and applications. In Celotta R, Lucatorto T (eds.) Experimental methods in the physical sciences vol. 3. Academic Press, San Diego.
195. Ziberi B, Frost F, Tartz M, Neumann H, Rauschenbach (2004) Importance of ion beam parameters on self-organized pattern formation on semiconductor surfaces by ion beam erosion. Thin Sol Films 459: 106–110
196. Ziberi B, Frost F, Höche T, Rauschenbach B (2005) Ripple pattern formation on silicon surfaces by low-energy ion-beam erosion: Experiment and theory. Phys Rev B 72: 235310
197. Ziberi B, Frost F, Rauschenbach B, Höche T (2005) Highly ordered self-organized dot patterns on Si surfaces by low-energy ion-beam erosion. Appl Phys Lett 87: 033113
198. Ziberi B, Frost F, Rauschenbach B (2006) Pattern transitions on Ge surfaces during low-energy ion beam erosion. Appl Phys Lett 88: 173115
199. Ziberi B, Frost F, Rauschenbach B (2006) Formation of large-area nanostructures on Si and Ge surfaces during low energy ion beam erosion. J Vac Sci Technol A 24: 1344–1348
200. Ziberi B, Frost F, Rauschenbach B (2006) Self-organized dot patterns on Si surfaces during noble gas ion beam erosion. Surf Sci 600: 3757–3761
201. Ziegler JF, Biersack JP, Littmark U (1985) The stopping and range of ions in matter. Pergamon, New York.
202. Ziegler ZF (1999) IBM Research, SRIM-2006.01 (PC version), Yorktown Heights, New York. [http://www.srim.org]

Area-Selective Depositions of Self-assembled Monolayers on Patterned SiO_2/Si Surfaces

Changshun Wang and Tsuneo Urisu

Abstract Area-selective depositions of SAMs have been demonstrated on SiO_2/Si patterns, which are fabricated by synchrotron radiation (SR)-stimulated etching of SiO_2 thin films on Si(100) substrates with $SF_6 + O_2$ as reaction gases and different types of masks. The SiO_2 layer is made by thermal dry oxidation. A Co micro-pattern contact mask is fabricated by combining the photolithography and sputtering techniques, and a W submicron-pillar mask is deposited by using the focused ion beam-induced chemical vapor deposition (CVD). The SR-stimulated etching of SiO_2 thin films on Si substrates is proven to have high spatial resolution, large selectivity, anisotropic etching, and low damage. The Co and W are found to be finer materials as SR etching masks. The etched patterns exhibit three-dimensional structures and the pattern surfaces are very flat and fit for depositing well-ordered monolayers on them. A dodecene SAM is deposited on the Si surface of the patterns, and trichlorosilane-derived SAMs (octadecyltrichlorosilane, and/or octenyltrichlorosilane) are deposited on the SiO_2 of the patterns. The deposited SAMs are densely packed and well-ordered and are characterized by infrared spectroscopy, ellipsometry, and water contact angle measurements. Moreover, the surface of the octenyltrichlorosilane monolayer is changed from hydrophobic to hydrophilic by oxidizing the vinyl ($-CH=CH_2$) end groups. This patterning of SAMs on SiO_2/Si patterns will be a potential structure for fabricating novel silicon-based biosensors and in biomedical studies.

Contents

C. Wang (✉)
Department of Physics, Shanghai Jiao Tong University, 800 Dong Chuan Road, Shanghai 200240, China
e-mail: cswang@sjtu.edu.cn

Z.M. Wang (ed.), *Toward Functional Nanomaterials*, Lecture Notes in Nanoscale Science and Technology 5, DOI 10.1007/978-0-387-77717-7_11,

1 Introduction

The preparation of monolayers on solid substrates is technologically important and has been studied for many years. Apart from the monolayer prepared by the Langmuir–Blodgett method [38, 53], a great number of studies have been conducted on the self-assembled monolayer (SAM) from the aspects of structures [14, 51], formation process [4, 36], and physical properties [3, 44], The SAM is one of the promising candidates in such applications as adhesion promoters [12, 39], surface modifications [19, 24], and the fabrication of field effect transistors [31]. A pre-patterned surface of SAMs has been found to yield beneficial results for area-selective integration of biomaterials, which is technologically important to biosensor fabrications, cell studies, and tissue engineering [17, 22, 35]. The area-selective deposition and patterning of SAMs are currently being studied intensively. Patterning of SAMs has been demonstrated by using many kinds of techniques and methods, such as micro-contact printing [28], field-induced oxidation [15], photolithography [54], electrochemistry [7, 25], ultraviolet (UV) photochemistry [5, 6], and so on. The SAMs in these kinds of patterns are all within the plane of the substrate surface. However, three-dimensional (3D) structures on the substrate surface often become necessary in some practical applications of SAMs, for example, to fabricate channel structure and biomedical microdevice. Synchrotron radiation (SR)-stimulated etching has attracted considerable attention from the viewpoint of developing new surface processes, due to its unique features of high spatial resolution, perfect material selectivity, anisotropic etching, low damage, and clean etching atmosphere [37, 41, 50]. In this chapter, the area-selective depositions of SAMs are demonstrated on the patterned structures fabricated by SR-stimulated etching of SiO_2 thin films on Si substrates using several kinds of contact masks.

2 Patterning of SiO_2 Thin Films

2.1 Growth of SiO_2 Thin Films

Silicon dioxide (SiO_2) thin films are the basic components for microelectronics and other microtechnologies. It can be used in many applications as a barrier material during impurity implants or diffusion, an electrical isolator of semiconductor devices, a component in metal-oxide semiconductor (MOS) transistors, and a dielectric interlayer in multilevel metallization structures. There are several ways to form

SiO_2 thin layers, such as by thermal oxidation of silicon, deposition of a thin film by chemical vapor deposition (CVD), spin-on glass, and so on.

Thermal oxidation. Thermal oxidation of silicon is easily achieved by heating the substrate to temperatures typically in the range of 900–1200°C. The atmosphere in the furnace where oxidation takes place can either contain pure oxygen or water vapor. Both of these molecules diffuse easily through the growing SiO_2 layer at these high temperatures. Oxygen arriving at the silicon surface can then combine with silicon to form silicon dioxide. The chemical reactions that take place are either

$$Si + O_2 \rightarrow SiO_2 \tag{1}$$

for the so-called "dry oxidation" or

$$Si + 2H_2O \rightarrow SiO_2 + 2H_2 \tag{2}$$

for "wet oxidation". The dry oxidation is the standard reaction for the formation of thin oxide film. Its oxide growth is rather slow and easily controllable, while the growth kinetics for wet oxidation is about ten times faster than that of dry oxidations. The wet oxidation is the process used for the thick field oxides.

Chemical vapor deposition. Whenever the SiO_2 layers are required without oxidizing the Si, the oxide in CVD is often used to deposit SiO_2 on top of the substrates by using suitable gases and deposition conditions. There are several ways, and one is

$$SiH_2Cl_2 + 2NO_2 \rightarrow SiO_2 + 2HCl + 2N_2 \tag{3}$$

which only occurs at high temperatures (~900°C). While this reaction was used until about 1985, another reaction was offered by

$$Si(C_2H_5O)_4 \rightarrow SiO_2 + 2H_2O + C_2H_4 \tag{4}$$

This reaction can be induced at relatively low temperatures (~720°C) and produces high-quality oxides.

Spin-on glass. A kind of polymeric suspension of SiO_2 dissolved in a suitable solvent is dropped on a rapidly spinning wafer. The centrifugal forces spread a thin viscous layer of the stuff on the wafer surface, which upon heating solidifies into SiO_2.

In this section, we introduce the patterning of SiO_2 layers on Si substrates for area-selective depositions of SAMs. The amorphous SiO_2 (a-SiO_2) layers are grown on crystalline silicon (c-Si) substrates by thermal dry oxidation. On exposure to oxygen at sufficiently high temperatures, a silicon surface is oxidized producing SiO_2. The thickness of the SiO_2 layer is determined by the temperature and the oxidation time. The thermal oxidation process of silicon has generally been assumed to occur through two processes in series: the diffusion of oxidizing species through the oxide film already formed and the chemical reaction at the SiO_2/Si interface.

Initially, the growth of silicon dioxide is a surface reaction only. However, after the SiO_2 thickness begins to build up, the arriving oxygen molecules must diffuse through the growing SiO_2 layer to get to the silicon surface in order to react. As the oxidation proceeds, the oxidizing species must diffuse through a progressively thicker oxide layer, in order to arrive at the Si–SiO_2 interface and react to form more SiO_2. The oxidation rate is expected to decrease with time. A popular model for the oxide growth kinetics was proposed by Deal and Grove [8]. According to the Deal–Grove model, oxide thickness increases linearly with time for relatively small oxidation times. This is the linear regime, in which the rate-limiting process is the interfacial oxidation reaction. At longer oxidation times, the limiting process is the diffusion of oxidant, and the oxide thickness increases parabolically with time. This is the parabolic regime. This model is generally valid for temperatures between 700 and 1300°C, partial pressures between 0.2 and 1.0 atmospheres, and oxide thicknesses between 0.03 and 2 microns for both wet and dry oxidations. According to this model, about 46% of the silicon surface is consumed during the oxidation and the oxide thickness can be predicted prior to growth. For very thin oxides, the Deal–Grove model is inadequate and an exponential law is found experimentally for the dependence of the oxide thickness on time for very short times. The growth rate of oxide film deviates from the linear-parabolic kinetics. Moreover, the detailed kinetics of oxide growth is influenced by many other factors, for example, the crystallographic orientation of the Si surface, the mechanical stress in the oxide, the substrate doping, and the initial condition of the Si surface. To interpret the complicated initial regime, many modifications in the Deal–Grove model have been proposed for the thermal oxidation of silicon [18, 23, 52]. Here, we only focus on the growth of SiO_2 layers with several hundred nanometers in thickness.

A single-crystal Si(100) wafer (n-type, 80–120 ohm.cm) is cut into pieces in the size of 12 mm × 10 mm. The chips are first pre-cleaned in acetone, ethanol, and deionized water with an ultrasonic bath for 5 min, respectively, then immersed into a solution of concentrated H_2SO_4 and 30% H_2O_2 (70:30 v/v) at 110°C for 10 min to remove the contamination, and finally treated with a 2.5% hydrogen fluoride (HF) solution for 2 min to remove the native oxide. The SiO_2 thin films are formed on the freshly cleaned Si surface by annealing at 900°C for 10 h in a dry oxygen atmosphere. The thickness of the SiO_2 layer on the Si(100) surface measured with an ellipsometer is about 180 ± 10 nm. By changing the annealing temperature and the period of oxidation, SiO_2 films with different thicknesses can be obtained.

2.2 Fabrication of Mask

To make an SR-etched pattern of silicon dioxide on silicon surface, a suitable mask should be used, which should possess sufficient resistibility for SR irradiation and should be easily removable after the etching process. Here, three kinds of contact masks on SiO_2 surface are fabricated for SR-stimulated etching. One is a Co micropattern fabricated by combining the sputtering and photolithography techniques, the second is a Co milli-pattern made by directly sputtering Co thin film on SiO_2 surface through a steel stencil mask, and the third is a W submicron pillar fabricated by a focused ion beam.

Co micro-pattern mask. In order to apply the SR-etched SiO_2/Si patterns to the area-selective deposition of SAM, the etching mask should have several characteristics: (1) possess a large absorption coefficient around the photon energy of ~100 eV, where SiO_2 has a high photodecomposition rate due to the excitations of its Si 2p core electron [1], (2) show a sufficient resistance against the SR etching, (3) have a suitable process to efficiently dissolve the mask material, which does not damage the monolayers deposited before removing the mask, and (4) exhibit small grain sizes, which are necessary to the fine pattern. Considering these demands and referring to the absorption coefficient of metals [10], Co is selected as the candidate for the etching mask material. The absorption coefficient of Co is shown in Fig. 1.

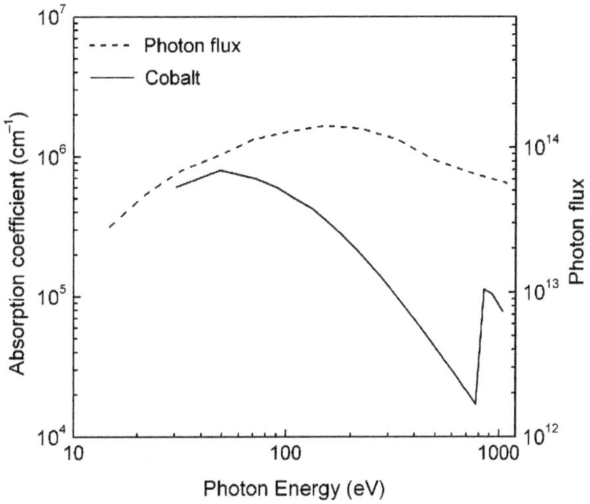

Fig. 1 The calculated absorption coefficients of the Co thin film in the vacuum UV region and the photon flux (Photons/s·mrad·0.1 λ·100 mA) of the beamline 4A2 in UVSOR

Figure 2 shows the sequence for the Co micro-pattern mask fabrication. The silicon wafer with silicon dioxide layers is first pretreated with a primer and coated with a photoresist of 1-methoxy-2-propyl acetate by a spin coater. The thickness of photoresist can be controlled by adjusting the rotating speed. The coated wafer is baked at 70°C for about 15 min to accelerate the solidification process of photoresist. A photomask that contains the wanted structure is aligned and placed above the coated wafer with a photolithography unit, which projects the pattern on the mask to the resist on the wafer. The wafer and mask are then exposed to UV radiation from an intense mercury arc lamp. This causes exposure to the photoresist in places not protected by opaque regions (chromium) of the photomask. The wafer with the exposed photoresist is put into a liquid of tetramethyl ammonium hydroxide to develop the resist pattern on the SiO_2 surface. Then, a Co thin layer is deposited on the resist pattern by a sputtering machine (SMH-2304RE). The wafer with Co layer is finally washed by acetone in an ultrasonic bath to get a Co pattern mask on

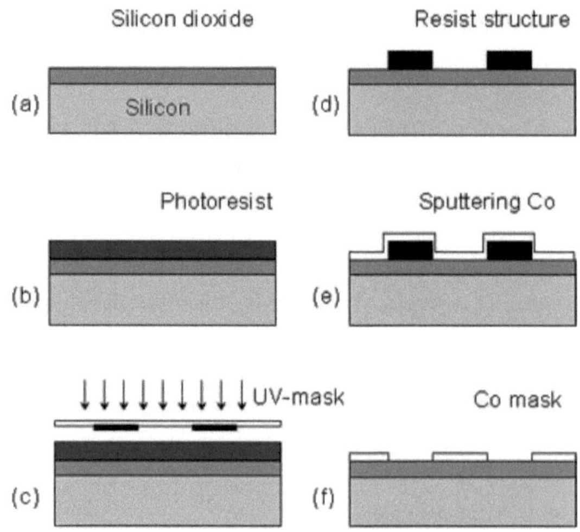

Fig. 2 The sequence for Co micro-pattern masks fabrication. (**a**) Thermal growth of SiO_2 layer on silicon substrates, (**b**) Spin coating photoresists on SiO_2 surface, (**c**) Exposure to UV radiation through a photomask, (**d**) Development, (**e**) Sputtering Co thin film, (**f**) Removing the remaining photoresist

SiO_2 surface. The size of the Co mask is about 12 mm × 12 mm. The minimum opening width of the Co mask is about 2 μm. A portion of the Co mask on SiO_2 surface, measured with a CCD camera, is shown in Fig. 3 [49]. The thickness of the Co mask measured with a step profile meter is about 165 nm.

Fig. 3 CCD image of a portion of the Co mask on SiO_2 surface

Co milli-pattern Mask. Sputtering is a common technique used to deposit thin films on the substrate surface. The basic process and setup are shown in Fig. 4. The substrate is placed in a low-pressure chamber between two electrodes. The electrodes are driven by an RF power source, which induces a plasma and ionization of the gas (e.g., argon) between the electrodes. A target made of the material to be deposited is bombarded by energetic ions which will dislodge atoms of the target,

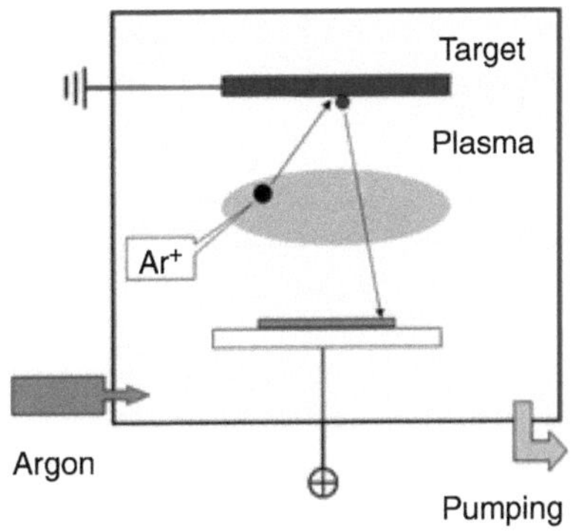

Fig. 4 The schematic figure for sputter deposition process

i.e., sputter them off. The dislodged atoms will have substantial kinetic energies, and some will fly to the substrate to be coated and stick there. The ions necessary for the bombardment of the target are extracted from argon plasma burning between the target and the substrate. This ion bombardment will liberate atoms from the target, which issue forth in all directions.

The cobalt milli-pattern mask is fabricated by directly sputtering Co thin film on the SiO_2 surface using a steel mask with an opening width of about 2 mm. This large-scale pattern is designed only for the sake of the measurement of the deposited SAMs subsequently.

W nano-pillar mask. The focused ion beam-induced CVD (FIB-CVD) is a common technique used to repair electronic devices and photomasks with very high resolution [45]. It also has the capability to grow 3D structures with a variety of materials at nanoscale resolution [13, 26, 43]. FIB systems usually use a finely focused beam of gallium ions (Ga^+) that can be operated for observing microscopic images and processing sample surfaces. Secondary electrons are generated when the primary Ga^+ beam irradiates the substrate surfaces. A local deposition can be performed by spraying compound gas on a substrate surface near the irradiation area. Secondary ions contribute to the decomposition of compound gas and split it into gas and solid parts. The gas part is evacuated in vacuum, while the solid part builds up on the substrate surface. Thus, a maskless deposition can be performed selectively on the irradiation areas. The method of using FIB-CVD to deposit tungsten 3D structures has been reported by Koops et al. [20], where $W(CO)_6$ was used as the source material to deposit tungsten-containing conductive structures. Ishida et al. have also demonstrated the fabrication of tungsten structures and improved the sidewall morphology on 3D tungsten structures with a milling process [16]. The deposited pillar is a mixture of amorphous tungsten, gallium, carbon, and oxygen, which is revealed by diffraction and element analysis. In the present case, a tungsten pillar mask for SR is directly made on the SiO_2 surface by FIB-CVD (SII Nano Technology Inc.) with gallium ions and $W(CO)_6$ source gas. The pillar mask is deposited when the gas pressure, acceleration voltage, and ion current are 1×10^{-3} Pa, 30 kV, and 0.5 pA 0.5 pA respectively. A portion of the tungsten mask on SiO_2

surface is shown in Fig. 5 [50]. The transverse width and longitudinal height of the tungsten pillar are about 80 nm and 160 nm, respectively.

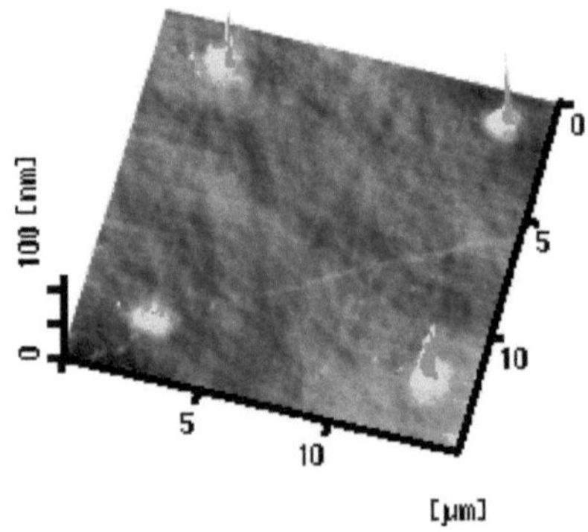

Fig. 5 AFM quasi-3D image of the tungsten nano-pillar

2.3 SR-Stimulated Etching

Microfabrication technology, a powerful tool in the manufacturing of various types of silicon-based biomedical microdevices, has made a considerable impact on the recent biotechnological researches [21, 33, 34]. In order to form a functional microstructure on a substrate, it is usually necessary to etch the thin films previously deposited and/or the substrate itself. There are some obvious requirements for most etching processes.

(1). Anisotropic etching: the etching process proceeds straight down, not in the lateral direction.
(2). Large selectivity: the selectivity is the ratio of the etching rate of the target material being etched to the etching rate of other materials.
(3). High spatial resolution to get small feature size.
(4). Reasonable etching rate.
(5). Low damage to the layer where it stops.
(6). Perfect homogeneity and reproducibility, and so on.

Actually, up to now there is no single technique that meets all the requirements for all situations. Etching process thus is almost always a search for the best compromise, and new etching techniques are introduced all the time. In general, there are two classes of etching processes: wet or chemical etching and dry etching. In wet etching, where the material is dissolved when immersed in a chemical solution, etching reactants come from a liquid source. In dry etching, where the material is sputtered or dissolved using reactive ions or a vapor phase etching agent, etching reactants come from a gas or vapor phase source and are typically ionized. Atoms or ions from the gas are the reactive species that etch the exposed film. Dry etching usually has higher degree of anisotropy, less selectivity, and lower etching rate than wet etching. The dry etching technologies can be roughly classified into three types: reactive ion etching (RIE), sputter etching, and vapor phase etching.

In RIE, the substrate is placed inside a reactor in which several gases are introduced. Plasma is struck in the gas mixture using an RF power source, breaking the gas molecules into ions. The ions are accelerated toward, and react at the surface of the material being etched, forming another gaseous material. This is known as the chemical part of RIE. There is also a physical part which is similar in nature to the sputtering deposition process. If the ions have high enough energy, they can knock atoms out of the material to be etched without a chemical reaction. Sputter etching is essentially RIE without reactive ions. The systems used are very similar in principle to sputtering deposition systems. The big difference is that the substrate is now subjected to the ion bombardment instead of the material target used in sputter deposition. Vapor phase etching is another dry etching method. In this process the wafer to be etched is placed inside a chamber, in which one or more gases are introduced. The material to be etched is dissolved at the surface in a chemical reaction with the gas molecules. The two most common vapor phase etching technologies are silicon dioxide etching using HF and silicon etching using xenon difluoride (XeF_2), both of which are isotropic in nature.

Fine patterning techniques in etching have been of great demand for the microfabrication. For example, ultra large-scale integrated circuits have been developed extensively in the past 10 years, and the size of one element has been greatly reduced. However, it is difficult to further reduce its size using the conventional method. Therefore, a new process has been desired in the manufacturing of microelectronic devices. Strong soft X-ray from an undulator in a storage ring is expected to solve this requirement [30]. SR is an ideal light source for the study of photochemical reactions because of various characteristics in the soft x-ray region such as high intensity, continuity of wavelength, and small divergence. Most gas molecules and solids used in semiconductor processes have a large cross section in the soft x-ray region and can be electronically excited at the core or valence levels by SR. In this section, the SR-stimulated etching of SiO_2/Si substrates is examined with SF_6 + O_2 as the reaction gas, and pattern structures of SiO_2 thin films on Si surface are fabricated by SR-stimulated etching with contact masks.

The SR etching of SiO_2/Si(100) substrate was conducted at the beam line 4A2 of the SR facility (UVSOR) in the Institute for Molecular Science of Japan. The electron beam energy is about 0.75 MeV. The emitted beam with horizontal acceptance angle of 11 mrad is bent by 6° and focused to the sample surface by a bent-elliptical mirror of 490-mm length. The mirror is set at 2.345 m downstream from the light source. The SR beam used for the etching had a broad energy range of 10–500 eV, which covered the core electron binding energy of Si [1]. The schematic processes for SR-stimulated etching of SiO_2/Si substrates are shown in Fig. 6. The detailed structure of the apparatus used for SR etching can be found in an earlier paper [11]. A large pressure difference between the etching chamber and the beam line is sustained by using a differential vacuum pumping system. The etching experiment is performed with a mixture of SF_6 and O_2 as the reaction gas at room temperature. The sample is set normal to the incident SR beam. Since no window is used, the

Fig. 6 The schematic processes of SR-stimulated etching with contact masks

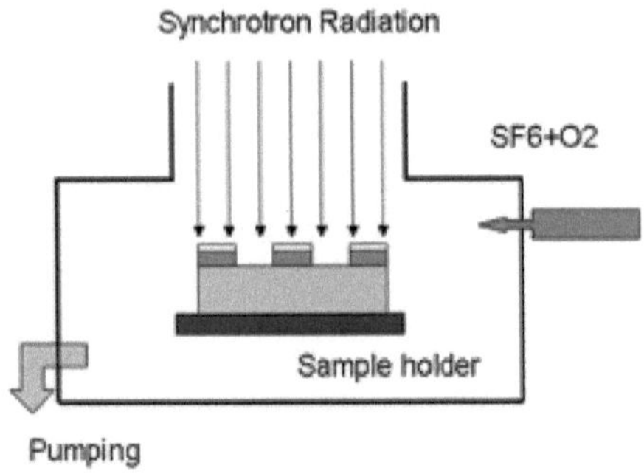

SR light irradiates the sample surface directly. The beam diameter on the sample surface is about 9 mm and the beam current of the storage ring is about 200 mA. The calculated spectrum distribution of the beam is also shown in Fig. 1. The total photon flux is about 4.2×10^{16} photons/s for 100 mA ring current. In the etching experiment, the irradiation dose of the SR beam is about 10000 mA min and the gas pressure of SF_6 and O_2 is about 0.05 torr and 0.002 torr respectively.

In the experiment, the SiO_2 is found to be effectively etched by SR radiation with a mixture of $SF_6 + O_2$ as the reaction gas at room temperature. The etching takes place only in the area irradiated with SR and proceeds in the direction of the incident beam, as shown in Fig. 7. In this figure, the SiO_2 thin film on the silicon substrate is directly irradiated with SR in the atmosphere of $SF_6 + O_2$ mixture gas without using any mask. No noticeable etching of SiO_2 is found when the sample surface is exposed only to SR at room temperature, without the introduction of SF_6 gas into the etching chamber.

Fig. 7 CCD image of a portion of the SR-etched structure without using any mask

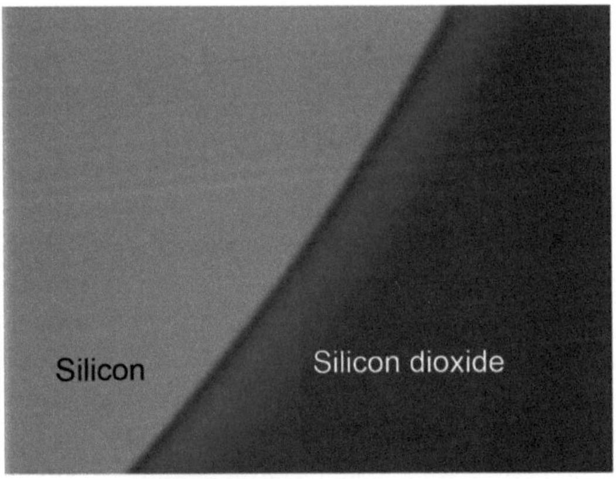

Figure 8 shows an SR-etched pattern with cobalt mask observed by a scanning electron microscopy (SEM) (JEOL, JSM-5510). The effect of interference of the SR beams results in the lower intensity of SR in some region, and a thin SiO_2 layer existed on this region due to insufficient etching. This problem can be solved by extending the period of irradiation or increasing the etching rate. Figure 9 shows

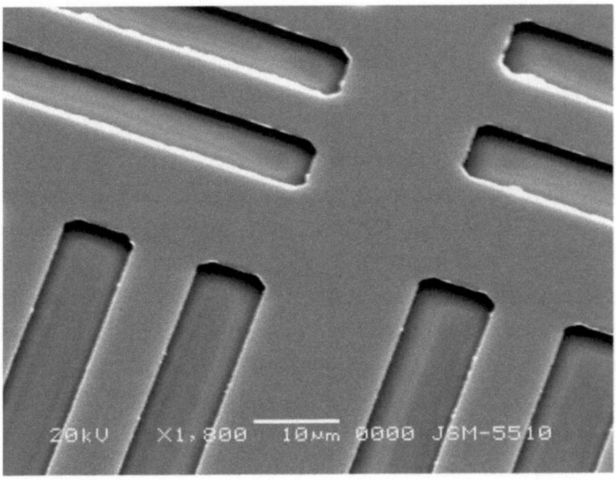

Fig. 8 SEM image of the SR-etched micro-pattern with a cobalt mask

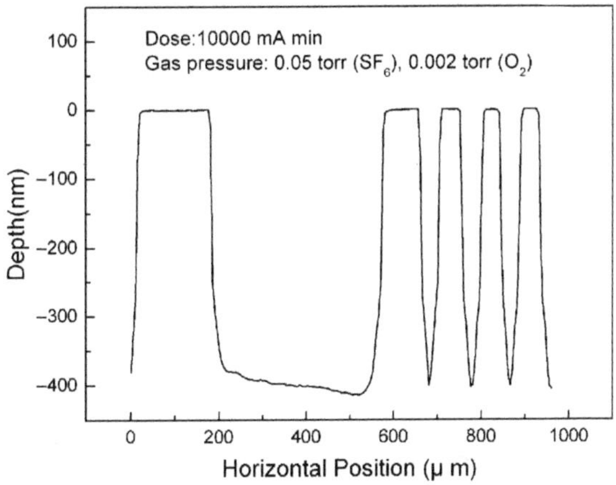

Fig. 9 The cross-section profile of a pattern after the SR-stimulated etching

a cross-section profile measured by a step profile meter. From this figure, an etching rate of 2.6 nm per 100 mA min was observed at 0.05 torr of SF_6 gas pressure in the etching chamber. After the SR etching, the Co surface looks flat and uniform, indicating that the material of Co possessed large resistibility against the SR etching.

The etching rate depends on the surface intensity of the SR beam, the surface density of the reaction molecules, and the temperature of the sample substrate. It increases with increasing gas pressure of SF_6 in the range where the attenuation of beam intensity due to the gas absorption is negligible. In the region of higher gas pressure, the beam intensity is attenuated by the absorption of the reaction gas and the etching rate saturates or decreases [11, 46]. In the present case, a thin steel tube is inserted along the beam line in the etching chamber to decrease the gas absorption, and the obtained etching rate is larger than their reported value, 0.47 nm $(100\ \text{mA min})^{-1}$ at 0.05 torr of SF_6 gas pressure in the etching chamber. The etching rate decreases when the gas pressure of SF_6 is over 0.15 torr in this etching system.

In addition, the etching rate is found to increase with decreasing substrate temperature [42, 48], which is quite different from that in plasma etching [9]. In the plasma etching, the etching rate decreases with increasing temperature since the rate-determining process is the surface thermal reaction between F atoms and the surface. In SR-stimulated etching, surface electronic excitation and photochemical reactions play important roles, which will be discussed later. One possible interpretation for the temperature dependence is that the lifetime or generation rate of the reaction center depends on the temperature, whether or not the etching rate can be increased by optimizing the reaction gas pressure and decreasing the substrate temperature. Figure 10 shows the observed temperature dependence of the etching rate. An etching rate of 3.6 nm per 100 mA min is achieved at –30C. In this experiment, dry ice is used to reduce the sample temperature, and the sample temperature is controlled via the current of a heater attached to the rear of the sample.

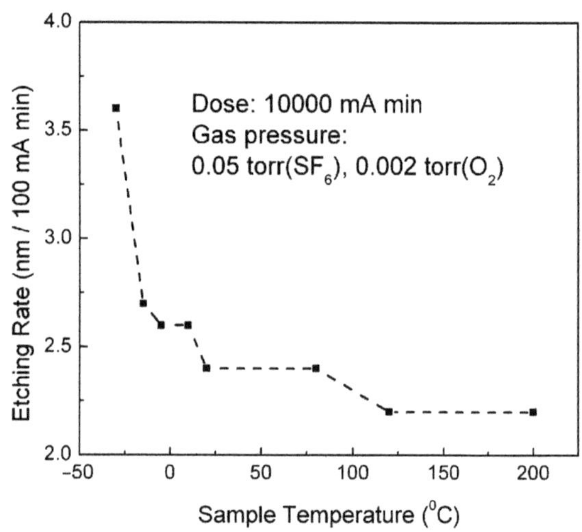

Fig. 10 Temperature dependence of the SR etching rate

Another important characteristic is the etching selectivity between Si and SiO_2. The etching process stops completely at the interface of SiO_2/Si if a little of O_2 is added to the reaction gas. The SR irradiation with flowing $SF_6 + O_2$ does not etch the Si crystal. The reaction mechanisms are extremely complex for etching process since many kinds of fragment radicals, atoms, and ions are produced by excitation, and there are a large number of possible chemical reactions. This material selectivity between Si crystal and SiO_2 can be explained by the reaction center quenching mechanism [1]. It has been reported that an amorphous SiO_2 (α-SiO_2) film continuously decomposes and evaporates when it is irradiated with SR at elevated substrate temperatures [2]. There are two main types of reaction mechanisms in photo-excited etching at room temperature. One is a gas-phase photochemical reaction in which photoexcitation of the surface does not take part in the reaction and the etching reaction occurs spontaneously. The other is a surface photochemical reaction in which photo-irradiation of the surface is necessary to the etching reaction. The former etching reaction proceeds in either the irradiated or the non-irradiated region, and the latter progresses anisotropically only in the irradiation region. It has been reported that the etching of SiO_2 is due to the surface excitation reaction that is the interaction between the excited states of the Si–O bonding and F ion or radical, while the etching of Si is induced by the gas-phase reaction. The etching of Si is constrained by adding O_2 to the SF_6 gas. The O radicals adsorbed on the surface quench the radicals in reaction gases and stop the gas-phase reaction. As a result, high etching selectivity can be obtained between Si and SiO_2 layers. These results show a different aspect from the results of plasma etching of Si/$Si0_2$ by SF_6 and O_2 gas mixture [29], in which the etching rate of Si is ten times higher than that of SiO_2.

The experimental results suggest that photo-excitation of material surfaces and the following chemical reactions play important roles in the SR-stimulated etching. In the irradiation of SR, the SiO_2 molecule is excited by high-energy photons to a reactive or excited state, and SF_6 molecules produce F ions. The excited SiO_2 reacts with the F ions, and then the resulting compounds decompose and evaporate. Therefore, the surface excitation is dominant in the above SR-excited etching SiO_2/Si system. Anisotropy and selective etching are achieved due to surface photochemical reaction. This property is practical in fabricating extremely fine patterns with a high aspect ratio.

The morphologic character of the etched surface is an important parameter for applying this pattern template [27]. Figure 11 shows a top-view image and a line profile of the SR-etched surface measured by an atomic force microscope (AFM) (SII, SPI3800N/SPA400) in dynamic force mode with a 20 μm scanner and a SI-DF40 (spring constant = 42 N/m) cantilever. The roughness (Ra) with the line profile is 0.05 nm. The maximum difference of height in the observed surface is less than 0.2 nm and the sloping angle between the two positions in the observed surface is less than 1.2°. These data show that the SR-etched surface is a very flat surface and is fit for many kinds of applications.

After the SR etching, the Co mask can be easily removed by a dilute (~ 0.01 N) HNO_3 solution. The roughness of SiO_2 surface beneath the Co mask after the SR etching is also an important parameter while considering a possible deposition of monolayers on the surface. Figure 12a and b shows the AFM quasi-3D images of the SiO_2 surface formed by thermal oxidation and the SiO_2 surface beneath the Co mask after the SR etching, respectively. From these images, it can be seen that the SiO_2 surface beneath the Co mask after the SR etching is slightly rougher than the SiO_2 surface made by thermal oxidation, which is mainly due to the irradiation of the permeated SR beam through the Co mask. It is found that the damage can be effectively lessened by increasing the thickness of the Co mask or reducing the period of SR irradiation (by increasing the etching rate). Figure 13 shows a SEM image of the SR-etched micro-pattern after removing the Co mask. Patterning SiO_2 thin film on the Si surface in micro-scale is achieved by SR-stimulated etching.

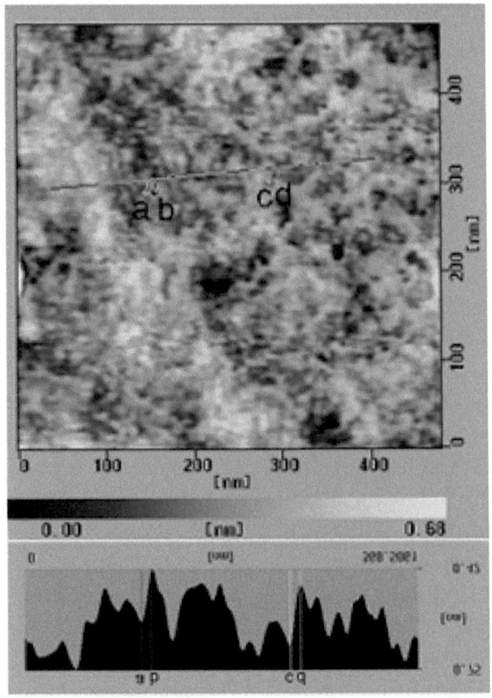

	ΔZ [nm]	D [nm]	φ [°]
ab	0.1665	9.3749	1.0176
cd	0.1895	9.3749	1.1582

Fig. 11 AFM top view image of 0.5 μm × 0.5 μm area and a line profile of the SR-etched surface. The roughness (Ra) with the line profile is 0.05 nm. D, ΔZ, and φ are the transverse distance, the longitudinal depth, and the sloping angle between position a and b, or c and d, respectively

Patterning SiO_2 thin film on the silicon surface in nano-scale is also demonstrated by SR-stimulated etching with a tungsten nano-pillar mask. The etching condition is the same as in the above experiments. Figures 14 and 15 show the SEM and AFM images of the SR-etched pattern with the tungsten pillar mask, respectively. The etched surface looks very flat, and no undercutting occurs in the etching process. The hollow appearance around the tungsten pillar is maybe due to the damage of silicon crystal during the fabrication of the tungsten mask. The damaged silicon is likely to be etched by the SR irradiation with flowing SF_6 + O_2 mixture gas. From these images, it can be concluded that the unique features of SR-stimulated etching such as high spatial resolution, high material selectivity, and anisotropy etching are proven. These patterns made by SR-stimulated etching of SiO_2 thin films on Si substrates exhibit 3D structures, which will be useful in the fabrication of many kinds of silicon-based microdevices.

In summary, SR-stimulated etching of SiO_2 layers on Si(100) surface has been investigated with SF_6 + O_2 as the reaction gas and several kinds of contact masks. The SR irradiation with flowing SF_6 and O_2 can effectively etch the SiO_2 thin films. The etching process stops completely at the interface of SiO_2/Si(100), which follows the surface photochemical reaction and has the merits of high spatial resolution,

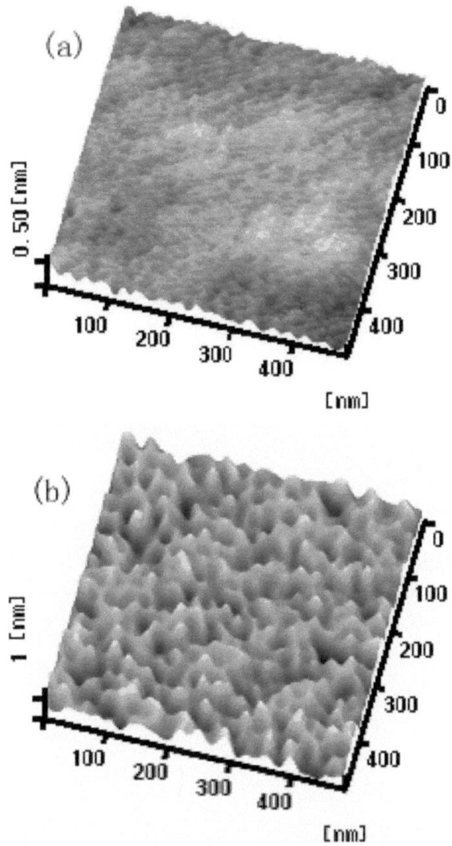

Fig. 12 (**a**) AFM quasi-3D images of 0.5 μm × 0.5 μm area of the SiO_2 surface formed by the thermal oxidation of the Si(100) surface and (**b**) the SiO_2 surface beneath the Co contact mask after the SR etching

Fig. 13 SEM image of the SR-etched micro-pattern of SiO_2 thin film on Si substrate after the removal of Co mask

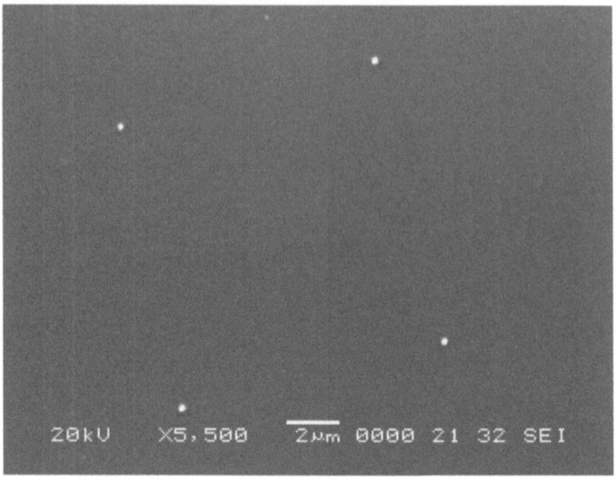

Fig. 14 SEM image of the SR-etched pattern with a tungsten pillar mask

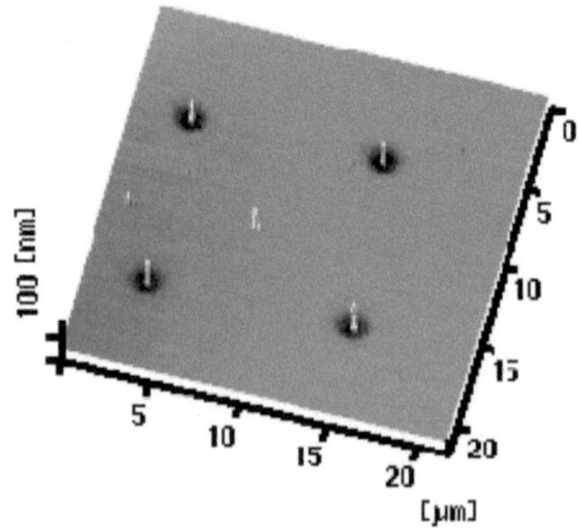

Fig. 15 AFM quasi-3D image of the SR-etched pattern with a tungsten pillar mask

large selectivity, anisotropic etching, low damage, etc. An etching rate of 3.6 nm per 100 mA min is achieved by optimizing the reaction gas pressure and decreasing the etching temperature. Co and W are proven to be finer materials as SR etching masks. Several 3D patterns of SiO_2/Si, from milliscale to nanoscale, have been fabricated with different types of masks.

3 Area-Selective Deposition of SAMs on SiO_2/Si(100) Patterns

In this section, the area-selective depositions of SAMs are demonstrated on patterns of SiO_2/Si(100) fabricated by SR-stimulated etching of SiO_2 thin films on Si substrates. The intention of this attempt is to construct a channel structure modified with SAMs for the fabrication of novel biosensors, as shown in Fig. 16. This modified pattern is provided with steady and selected hydrophilic and/or hydrophobic surfaces, which can be used for the area-selective integration of biomaterials and for building the living environment of biomolecules [40, 55]. According to the description in the last section, several patterns of SiO_2/Si have been obtained with different

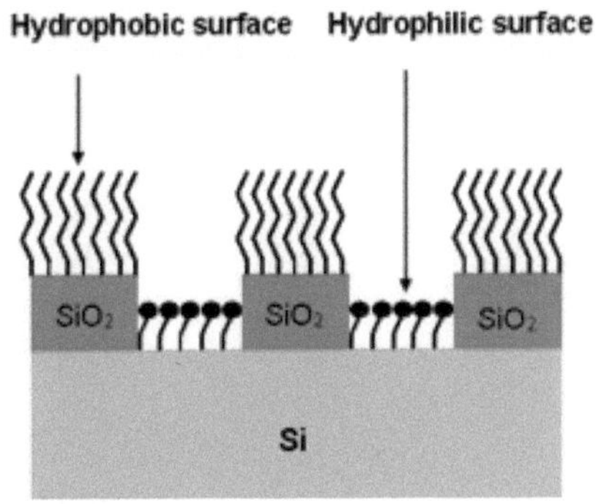

Fig. 16 Patterning of SAMs on an SR-etched pattern

types of masks. These patterns exhibit a 3D structure. As an example of the area-selective deposition of SAMs on patterned structures, let us consider the deposition of SAMs on the SiO_2/Si milli-pattern made by SR-stimulated etching with a Co contact mask [47].

We first examine the deposition of dodecene SAM on the SR-etched region. Approximately 3 mL of neat dodecene (Sigma Chemical Co.) is put in a small glass flask and is deoxygenated with dry nitrogen for 30 min. The SR-etched surface is washed with organic and aqueous chemicals and treated with a dilute HF solution to form an H-terminated Si surface, and then the etched wafer is immediately blown dry with nitrogen and put into the deoxygenated dodecene for 2 h at 200°C, while slowly bubbling pure nitrogen through the dodecene liquid. Subsequently, the wafer is taken out from the flask, rinsed three times with petroleum ether, washed in dichloromethane for 5 min, and dried in a stream of nitrogen. Then, octadecyltrichlorosilane (OTS) is deposited on the SiO_2 surface of SiO_2/Si pattern. Before deposition, the Co mask on the SiO_2 surface is removed by dipping the sample into a dilute (~0.01 N) HNO_3 solution for 10 min. This removing process does not destroy the dodecene SAM, for it is known to be stable when in contact with aqueous acid at room temperature [51]. The deposition is performed by immersing the wafer coated partially with dodecene monolayer in a 2 mM solution of OTS (Tokyo Kasei Kdgyo Co.) in toluene for 1 h below 25°C, followed by washing thoroughly with toluene, ethanol, and deionized water. For the convenience of measurement, the SAMs deposited on the milli-pattern are characterized by a Fourier transform infrared spectrometer (JASCO, FT/IR-670) in the transmission mode using a Tri-Glycine-Sulfate (TGS) detector at 4 cm^{-1} resolution, an ellipsometer, and water contact angle (WCA) measurements.

The dodecene SAM is deposited on Si surface by the hydrosilylation reaction of dodecene molecules with the H-terminated silicon surface [36]. The SAM qualities are dependent on the temperatures of depositing solution. In the temperature range of 25–200°C, it is found that the CH_2-peak intensity increases with increasing temperatures and the peak position is almost constant, independent of peak intensity.

Figure 17 shows the infrared transmission spectra of the dodecene SAM on Si surface prepared at 200°C. The observed peaks of d^+ (2850 ± 1 cm^{-1}) and d^- (2918 ± 1 cm^{-1}) are assigned to the symmetric and asymmetric stretching vibrations of methylene (-CH_2) groups, respectively; the peak of r^- (2959 ± 1 cm^{-1}) is assigned to the asymmetric stretching vibrations of methyl (–CH_3) groups. The thickness of the deposited layer measured by the ellipsometer is 1.4 (±0.1) nm. In the measurement of the SAM thickness with an ellipsometer, a value of 1.46 is used for the SAM refractive index at 633 nm. The WCA of the monolayer surface is found to be 110 (±2)°.

Fig. 17 Infrared spectra of the deposited dodecene SAM on Si surface

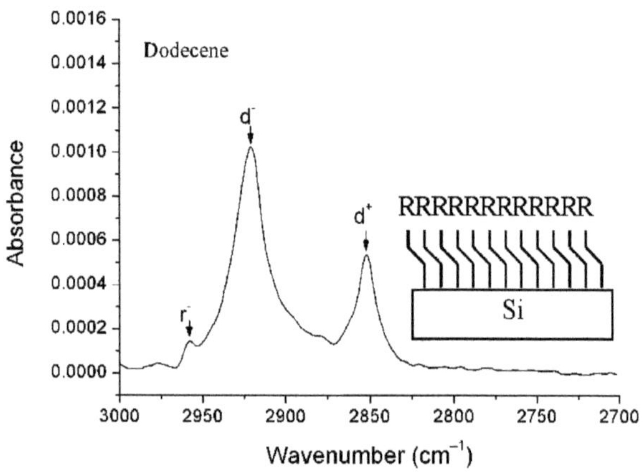

The OTS monolayer is deposited on SiO_2 by the silanization reaction of alkyltrichlorosilanes with the OH-terminated SiO_2 surface. Figure 18 shows the infrared spectra of OTS SAM on SiO_2 surface. The peaks of d^+, d^-, and r^- are 2851 cm^{-1}, 2919 cm^{-1}, and 2958 cm^{-1}, respectively. The thickness and the WCA of the deposited OTS monolayer are 2.5 (±0.1) nm and 110 (±2)°, respectively. For the deposition of OTS SAM on OH-terminated SiO_2 surface, there are two important controlling parameters to insure good deposition [4, 44]. First, the temperature of the reactive bath has to be maintained below a critical temperature T_c. The critical temperature for depositing OTS is found to be about 29°C. The hydrocarbon chains are ordered and closely packed when the monolayers are prepared below T_c, whilst disordered monolayers with low molecular density are formed above T_c. Second, the substrate surface has to be pre-hydrated with a thin layer of water. The role of this surface water is twofold: (1) it allows the transformation of chlorosilane head groups (-$SiCl_3$) into trisilanol groups [-$Si(OH)_3$] by hydrolysis, and (2) it provides a fluid substrate for the molecules, which gradually adsorbs on the surface, to rearrange laterally in the plane of monolayers. A cleaned SiO_2 surface is naturally hydroxyl-rich and hydrophilic. Moreover, the deposition of OTS SAM does not affect the precursor of dodecene SAM, since no changes in its WCA are found after the deposition of OTS monolayer.

Fig. 18 Infrared spectra of the deposited OTS SAM on SiO_2 surface

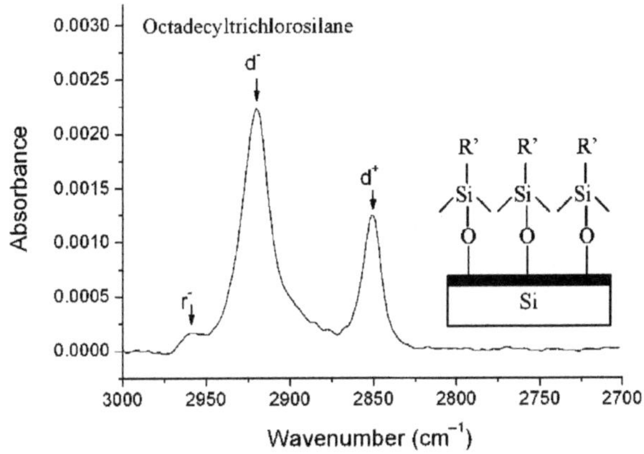

The observed peak positions for OTS and dodecene SAMs are close to the values measured on alkane crystals but far below the positions for liquid alkanes, which are at 2856 and 2928 cm^{-1}, respectively [32]. The SAM should have a thickness (d) corresponding to the extended chain length of the molecule, depending on the number of carbon atoms in the chain. Considering the small incline of SAM on the substrate surface, the measurement values of thickness are in good agreement with the formula $d\ (C_n) = 0.126\ (n - 1) + 0.478$ nm [51], where n is the number of carbon atoms in the alkyl chain, valid for hydrocarbon chains oriented perpendicular to the solid substrate and extended in their all-trans conformation. The values of WCA are consistent with a top layer of methyl ($-CH_3$) groups, as it should be if the alkyl chains are densely packed. From these characteristics, we conclude that the deposited SAMs are made of alkyl chains in their all-trans conformation, are nearly perpendicular to the substrate, and have a densely packed molecular architecture.

The morphologic character of the SiO_2/Si pattern surface has been examined in the section of SR-stimulated etching, which is an important parameter for depositing SAMs on this pattern. The relationship between the surface roughness and the peak position of the CH_2 vibrations are investigated as a function of deposition temperature for the case of dodecene SAM deposition on Si (100). With increasing substrate roughness, the -CH_2 vibration peaks shift to higher wave numbers. The magnitude of the shift for the deposition on a rough substrate is up to 6 cm^{-1}, which corresponds to a change from a condensed, almost all-trans conformational phase to a liquid-like, gauche-disordered phase [27]. The peak shifts of the CH_2 vibrations are an indicator of the amount of gauche-conformational disorder present in aliphatic SAMs. On polished substrates, although increased temperatures lead to a slightly more ordered SAM, the layers are in almost an all-trans conformational phase independent of the coverage. The experimental results also show that the surfaces of SR-stimulated etched pattern are very flat and fit for depositing well-ordered monolayers on them.

To acquire the hydrophilic surface as shown in Fig. 16, the OTS deposition is changed to OTTS (octenyltrichlorosilane, Aldrich Chemical Co.) deposition. The deposition of OTTS SAM on the SiO_2 surface of SiO_2/Si pattern is very similar in principle to the OTS deposition. The only difference is that the critical temperature for OTTS deposition is approximately – 6°C whereas it is approximately 29°C for OTS deposition. The thickness and the WCA of the deposited OTTS monolayer are measured to be 1.1 (±0.1) nm and 108 (±2)°, respectively. The infrared peaks for the symmetric and asymmetric stretching vibrations of its methylene groups are 2849 cm^{-1} and 2918 cm^{-1}, respectively. The surface of the OTTS monolayer can be oxidized by the $KMnO_4$/crown ether complex (dicyclohexano-18-crown-6) [25]. This surface oxidation is performed by immersing the wafer coated partially with OTTS monolayer into a 5 mM solution of the complex in analytical-grade benzene for about 48 h in a closed vessel, followed by several rinses with the pure benzene solvent, 3 min of sonication in pure benzene, 30 min of immersion in 5% aqueous HCl, and a final thorough rinse with purified flowing water. The vinyl (–$CH{=}CH_2$) end groups in OTTS monolayer are changed into carboxylic (COOH) end groups by the surface oxidation. After the oxidation, the WCA of the monolayer surface is about 30^0, this means the surface characteristic of the OTTS monolayer is transformed from hydrophobic to hydrophilic, which is an important issue when considering the further applications of the area-selective deposition of SAMs.

In summary, the area-selective deposition and patterning of SAMs are demonstrated on 3D pattern structures. The SR-stimulated etching of SiO_2 thin films on Si substrates is employed to make the template for area-selective deposition, for it has unique features such as high spatial resolution, high material selectivity between Si and SiO_2, anisotropy etching, clean etching atmosphere, and a very flat etched surface. SAMs are prepared on the Si surface of the SR-etched pattern by the hydrosilylation reaction of dodecene with the H-terminated surface at 200°C, and also on the SiO_2 surface of the SR-etched pattern by the silanization reaction of OTS or OTTS with the OH-terminated surface below their critical temperatures. These deposited SAMs are well ordered and densely packed, as evidenced from infrared spectroscopy, ellipsometer and WCA measurements. Moreover, the surface of the OTTS monolayer is changed from hydrophobic to hydrophilic by oxidizing the vinyl (–$CH{=}CH_2$) end groups. This work provides a method for area-selective deposition of SAMs and offers a technique for patterning of SAMs on SR-etched patterns. The patterning of the SAMs on the SR-etched pattern is a useful technique to make a microdevice for many kinds of practical applications, such as biosensor fabrications, cell studies, and tissue engineering applications.

Acknowledgments This work was supported by the National Natural Science Foundation of China (10675083), the Grant-in-Aid for Scientific Research from the Ministry of Education, Culture, Sports, Science and Technology of Japan, and the Collaboration program of the Graduate University for Advanced Studies.

References

1. Akazawa H, Takahashi J, Utsumi Y, Kawashima I, Urisu T (1991) Photostimulated evaporation of SiO_2 and Si_3N_4 films by synchrotron radiation and its application for low-temperature cleaning of Si surface. J. Vac. Sci. Technol. A 9: 2653–2661

2. Akazawa H (1995) Synchrotron-radiation-stimulated evaporation and defect formation in a-SiO_2. Phys. Rev B 52: 12386–12394
3. Balgar T, Bautista R, Hartmann N, Hasselbrink E (2003) An AFM study of the growth kinetics of the self-assembled octadecylsiloxane monolayer on oxidized silicon. Surf. Sci. 532: 963–969
4. Brzoska JB, Shahidzadeh N, Rondelez F (1992) Evidence of a transition temperature for the optimum deposition of grafted monolayer coatings. Nature 360: 719–721
5. Carroll GT, Wang DN, Turro NJ, Koberstein JT (2006) Photochemical micropattering of carbohydrates on a surface. Langmuir 22: 2899–2905
6. Chen MS, Dulcey CS, Chrisey LA, Dressick WJ (2006) Deep-UV photochemistry and patterning of (aminoethylaminomethyl) phenethylsiloxane self-assembled monolayers. Adv. Funct. Mater. 16: 774–783
7. Chi QJ, Zhang JD, Ulstrup J (2006) Surface microscopic structure and electrochemical rectification of a branched alkanethiol self-assembled monolayer. J Phys. Chem. B 110: 1102–1106
8. Deal BE, Grove AS (1965) General relationship for the thermal oxidation of silicon. J. Appl. Phys. 36: 3770–3778
9. Flamm DL, Donnelly VM, Mucha JA (1981) The reaction of fluorine atoms with silicon. J. Appl. Phys. 52: 3633–3639
10. Henke BL, Lee P, Tanaka TJ, Shimabukuro RL, Fujikawa BK (1982) Low-energy x-ray interaction coefficients: Photoabsorption, scattering, and reflection. Atom. Data Nucl. Data. Tables 27: 1–144
11. Hirano S, Yoshigoe A, Nagasono M, Mase K, Ohara J, Nonogaki Y, Takeda Y, Urisu T (1998) Ultrahigh-vacuum reaction apparatus to study synchrotron-radiation-stimulated processes. J. Synchrotron Rad. 5: 1363–1368
12. Hiremath R, Varney SW, Swift JA (2004) Selective growth of a less stable polymorph of 2-iodo-4-nitroaniline on a self-assembled monolayer template. Chem. Commun. 23: 2676–2677
13. Igaki JY, Kanda K, Haruyama Y, Ishida M, Ochiai Y, Fujita JI, Kaito T, Matsui S (2006) Comparison of FIB-CVD and EB-EVD growth characteristics. 83: 1225–1228
14. Iimura K, Nakajima Y, Kato T (2000) A study on structures and formation mechanisms of self-assembled monolayers of *n*-alkyltrichlorosilanes using infrared spectroscopy and atomic force microscopy. Thin Solid Films 379: 230–239
15. Inoue A, Ishida T, Choi N, Mizutani W, Tokumoto H (1998) Naometer-scale patterning of self-assembled monolayers films on native silicon oxide. Appl. Phys. Lett. 73: 1976–1978
16. Ishida M, Fujita J, Ichihashi T, Ochiai Y, Kaito T, Matsui S (2003) Focused ion beam-induced fabrication of tungsten structures. J. Vac. Sci. Technol. B 21: 2728–2731
17. Ito E, Nakamura F, Kanai K, Ouchi Y, Seki K, Hara M (2006) Characterization of COOH-terminated self-assembled monolayers and adsorption efficiency of DNA molecules studied by x-ray photoelectron spectroscopy and near-edge x-ray absorption fine structure spectroscopy. Jpn. J. Appl. Phys. 45: 409–412
18. Kageshima H, Shiraishi K, Uematsu M (1999) Universal theory of Si oxidation rate and importance of interfacial Si emission. Jpn. J. Appl. Phys. 38: L971–L974
19. Komeda T, Namba K, Nishioka Y (1997) Self-assembled-monolayer film islands as a self-patterned-mask for SiO_2 thickness measurement with atomic force microscopy. Appl. Phys. Lett. 70: 3398–3400
20. Koops HWP, Weiel R, Kern DP, Baum TH (1988) High-resolution electro-beam induced deposition. J. Vac. Sci. Technol. B 6: 477–481

21. Leoni L, Desai TA (2004) Micromachined biocapsules for cell-based sensing and delivery. Adv. Drug Deliv. Rev. 56: 211–229
22. Loaiza OA, Campuzano S, Lopez-Berlanga M, Pedretro MA, Pingarron JM (2005) Development of a DNA sensor based on alkanethiol self-assembled monolayer-modified electrodes. Sensors 5: 344–363
23. Lora-Tamayo A, Dominguez E, Lora-Tamayo E, Llabres J (1978) A New model of the thermal growth of a silicon dioxide layer. Appl. Phys. 17: 79–84
24. Maboudian R, Ashurst WR, Carraro C (2000) Self-assembled monolayers as anti-stiction coatings for MEMS: characteristics and recent developments. Sen. Actuators 82: 219–223
25. Maoz R, Cohen SR, Sagiv J (1999) Nanoelectrochemical patterning of monolayer surfaces: toward spatially defined self-assembly of nanostructures. Adv. Mater. 11: 55–61
26. Matsui S, Kometani R (2007) Three-dimensional nanostructure fabrication by focused-ion-beam chemical vapor deposition and its applications. IEICE Trans. Electron. E90C: 25–35
27. More SD, Graaf H, Baune M, Wang CS, Urisu T (2002) Influence of substrate roughness on the formation of aliphatic self-assembled monolayers (SAMs) on silicon (100). Jpn. J. Appl. Phys. 41: 4390–4394
28. Mrksich M, Whitesides GM (1995) Patterning self-assembled monolayers using microcontact printing: A new technology for biosensors? Trends Biotechnol. 13: 228–235
29. Ninomiya K, Suzuki K, Nishimatsu S, Okada O (1987) Role of sulfur atoms in microwave plasma etching of silicon. J. Appl. Phys. 62: 1459–1468
30. Ogawa T, Mochiji K, Ochiai I, Yamamoto S (1994) Low-temperature synchrotron-radiation-excited etching of silicon dioxide with sulfur hexafluoride adsorption. J. Appl. Phys. 75: 4680–4685
31. Park YD, Kim DH, Jang Y, Hwang M, Lim JA, Cho K (2005) Low-voltage polymer thin-film transistors with a self-assembled monolayer as the gate dielectric. Appl. Phys. Lett. 87: 243509
32. Parikh AN, Allara DL, Azouz IB, Rondelez F (1994) An Intrinsic relationship between molecular structure in self-assembled n-alkylsiloxane monolayers and deposition temperature. J. Phys. Chem. 98: 7577–7590
33. Popat KC, Robert RW, Desai TA (2002) Characterization of vapor deposited thin silane films on silicon substrates for biomedical microdevices. Surf. Coat. Technol. 154: 253–261
34. Raiteri R, Grattarola M, Butt HJ, Skladal P (2001) Micromechanical cantilever-based biosensors. Sens. Actuator B 79: 115–126
35. Sheller NB, Petrash S, Foster MD, Tsukruk VV (1998) Atomic force microscopy and X-ray reflectivity studies of albumin adsorbed onto self-assembled monolayers of hexadecyltrichlorosilane. Langmuir 14: 4535–4544
36. Sieval AB, Demirel AL, Nissink JWM, Linford MR, van der Maas JH, de Jeu WH, Zuilhof H, Sudholter EJR (1998) Highly stable Si-C linked functionalized monolayers on the silicon (100) surface. Langmuir 14: 1759–1736
37. Takahashi JI, Utsumi Y, Urisu T (1991) Material selectivity in synchrotron radiation-stimulated etching of SiO_2 and Si. J. Appl. Phys. 70: 2958–2962
38. Takizawa M, Kim YH, Urisu T (2004) Deposition of DPPC monolayers by the Langmuir-Blodgett method on SiO_2 surfaces covered by octadecyltrichlorosilane self-assembled monolayer islands. Chem. Phys. Lett. 385: 220–224
39. Tero R, Takizawa M, Li YJ, Yamazaki M, Urisu T (2004) Deposition of phospholipid layers on SiO_2 surface modified alkyl-SAM islands. Appl. Surf. Sci. 238: 218–222
40. Tirrell M, Kokkoli E, Biesalski M (2002) The role of surface science in bioengineered material. Surf. Sci. 500: 61–68
41. Urisu T, Kyuragi H (1987) Synchrotron radiation-excited chemical-vapor deposition and etching. J. Vac. Sci. Technol. B 5: 1436–1440
42. Urisu T, Kyuragi H, Utsumi Y, Takahashi J, Kitamura M (1989) Synchrotron radiation stimulated semiconductor processes: Chemical vapor deposition and etching. Rev. Sci. Instum. 60: 2157–2159

43. Utke I, Luisier A, Hoffmann P, Laub D, Buffat PA (2002) Focused-electron-beam-induced deposition of freestanding three-dimensional nanostructures of pure coalesced copper crystals. Appl. Phys. Lett. 81: 3245–3247
44. Vuillaume D, Boulas C, Collet J, Allan G, Delerue C (1998) Electronic structure of a heterostructure of an alkylsiloxane self-assembled monolayer on silicon. Phys. Rev. B 58: 16491–16498
45. Wagner A, Levin JP, Mauer JL, Blauner PG, Kirch SJ, Longo P (1990) X-ray mask repair with focused ion beams. J. Vac. Sci. Technol. B 8: 1557–1564
46. Wang CS, More SD, Wang ZH, Yamamura S, Nonagaki Y, Urisu T (2003) Patterning SiO2 thin films using synchrotron radiation stimulated etching with a Co contact mask. J. Vac. Sci. Technol. B 21: 818–822
47. Wang CS, Pan X, Sun CY, Urisu T (2006) Area-selective deposition of self-assembled monolayers on SiO_2/Si(100) patterns. Appl. Phys. Lett. 89: 233105
48. Wang CS, Urisu T (2003) Synchrotron radiation stimulated etching of SiO2 thin films with a Co contact mask for the area-selective deposition of self-assembled monolayer. Jpn. J. Appl. Phys. 42: 4016–4019
49. Wang CS, Urisu T (2005) Synchrotron radiation stimulated etching SiO_2 thin films with a contact cobalt mask. Appl. Surf. Sci. 242: 276–280
50. Wang CS, Zhang XQ, Urisu T (2006) Synchrotron-radiation-stimulated etching of SiO2 thin films with a tungsten nano-pillar mask. J. Synchrotron Rad. 13: 432–434
51. Wasserman SR, Tao YT, Whitesides GM (1989) Structure and reactivity of alkylsiloxane monolayers formed by reaction of alkyltrichlorosilanes on silicon substrates. Langmuir 5: 1074–1087
52. Watanabe T, Tatsumura K, Ohdomari I (2006) New linear-parabolic rate equation for thermal oxidation of silicon. Phys. Rev. Lett. 96: 196102
53. Wu AP, Talham DR (2000) Photoisomerization of azobenzene chromophores in organic/inorganic zirconium phosphonate thin films prepared using a combined Langmuir-Blodgett and self-assembled monolayer deposition. Langmuir 16: 7449–7456
54. Yamazaki T, Imae T, Sugimura H, Saito N, Hayashi K, Takai K (2005) Photolithographic patterning of dendrimer monolayers and pattern-selective adsorption of adsorption of linear macromolecules. J. Nanosci. Nanotechno 5: 1792–1800
55. Zhang SG (2003) Fabrication of novel biomaterials through molecular self-assembly. Nat. Biotechnol. 21: 1171–1178

Virtual Synthesis of Electronic Nanomaterials: Fundamentals and Prospects

Liudmila A. Pozhar and William C. Mitchel

Abstract Increasingly, integrated logic-storage and quantum information-processing paradigms are being viewed as dominant approaches to facilitate further advances in electronics, communication, information processing, and storage technologies. Realization of these concepts is based upon understanding the formation of coherent, polarized, and especially entangled (CPE) electron spin states, ability to control their dynamics and their contributions to quantum spin–charge transport properties of small, few-atomic systems at realistic conditions at interfaces and in quantum confinement.

This chapter is focused on novel, first-principle, synergetic theoretical and computational methods designed to predict the electron spin–charge transport properties of small atomic clusters (quantum dots, or QDs) and molecules that may be used as sources of CPE electron spin states. Theoretical methods are derived from a many-body quantum theory formalism – a projection operator method by Zubarev and Tserkovnikov (ZT) – based on equilibrium, commutatorial two-time temperature Green functions (or TTGFs). The ZT approach has been widely used to predict thermodynamic and charge transport properties of bulk systems, including metals, semi- and superconductors, etc. In this chapter, the linearized version of the ZT method is generalized to include strongly spatially inhomogeneous systems, such as a single molecule or QD.

There are several significant advantages of this approach, as compared to the traditional nonequilibrium two-time thermodynamic and field-theoretical Green function (NGF) methods that are used to study electron transport at nanoscale. In particular, the TTGF method does not require introduction of the distribution functions to link the GFs to the transport coefficients (in contrast to the existing NGF-based methods), thus avoiding effects of this major uncontrolled approximation, while also significantly reducing the use of controlled approximations and calculation efforts related to derivation of a kinetic theory. Further on, the TTGFs are susceptibilities, and thus are directly related to experimentally

L.A. Pozhar (✉)
Department of Physics, University of Idaho, P.O. Box 440903, Moscow, ID, USA 83844-0903
e-mail: LPozhar@uidaho.edu

Z.M. Wang (ed.), *Toward Functional Nanomaterials*, Lecture Notes in Nanoscale Science and Technology 5, DOI 10.1007/978-0-387-77717-7_12,

assessable microscopic charge, spin and microcurrent densities. In their turn, the latter quantities are directly related to the spin states of contributing electrons. Thus, measurements of these quantities provide direct experimental information on the contributing electron spin states. It is also important that the equilibrium TTGFs have to be calculated only once for a considered system, while NGFs must be calculated for the same system every time when process conditions change. The theoretical TTGF-based formulae for the electron transport coefficients directly related to experimental data are crucial to identify propagating (or *dynamic*) CPE electron spin states.

Practical significance of the fundamental insights into electron spin–charge transport provided by analytical theoretical means is limited without accurate data on the equilibrium electronic structure of small systems, because specific predictions are very sensitive to details of the electron spin–charge density distributions of the small systems and their quantum confinement. These data can be obtained using virtual (i.e., many-body quantum theory-based computational) synthesis and evaluation of the corresponding model molecules and QDs. Such computational results used in theoretical formulae provide reliable predictions, and thus a valuable guideline for experimental synthesis of small systems with predesigned electron CPE spin states and spin–charge transport properties. In this work, GAMESS software package has been used to synthesize computationally several artificial molecules composed of semiconductor compound atoms. The data so obtained contain detailed information on the molecular structure, composition, chemistry, electron energies, spin and charge distributions, and electron wave functions (molecular orbits) necessary to identify electron spin states of interest (such as entangled spin states where electrons share the spatial portion of their wave function while localized on or propagating to different QDs: spin-based qubits), and can be further used in the ZT-based theoretical formulae to calculate the TTGFs, and thus electron spin–charge transport properties of the studied systems.

Contents

1 Introduction

Nanotechnology-enabled, integrated logic storage and quantum information-processing paradigms are being viewed as dominant approaches to facilitate further advances in electronics, communication, and information processing and storage technologies, thus providing for the continuing technological growth of the modern society [1]. Realization of these concepts is based upon detailed understanding of synthesis principles, the structure, quantum magneto-electronic and spin–charge transport properties of nanostructure units in quantum confinements that have to be used to enable quantum logic gates [2–6] and integrated logic storage elements. Nanoheterostructures (NHSs) composed of such units have long been recognized as the most realistic candidates for implementation of these novel concepts.

This chapter is focused on a description of unified theoretical and computational methods rooted in the first principles to predict the key quantum spin–charge transport and magneto-electronic properties of systems of small quantum-confined atomic clusters (quantum dots or QDs) and artificial molecules as prospective sources/carriers of correlated/polarized/coherent dynamic electronic spin states. Explicit expressions for quantum transport properties (in particular, the conductivity, magnetic and dielectric susceptibility tensors) of such QDs in ac/dc electromagnetic fields and at weak current conditions are derived from the first principles using the equilibrium two-time temperature Green function method, and rigorously related to their electronic structure, composition and chemistry, and to those of their quantum confinement. This fundamental approach permits to reveal various contributions to these coefficients, in particular those caused by reduction of symmetry in effectively periodic systems. Such explicit correlations provide a strategic insight into the dynamics of electronic spin–charge states to help select conditions and system parameters that would identify means of manipulation of the electronic spin–charge states into desirable correlations, including spin-polarized and coherent states, that may be used to realize electron spin qubits.

Practical use of the fundamental insights provided by analytical theoretical means is limited without accurate data on the electronic and magnetic properties of small QDs, because these predictions are very sensitive to details of the (quantum-confined) electron spin–charge density distributions. These details cannot be obtained from experimental studies that provide only system-smoothed characterization of small QDs composed of a few atoms. Therefore, virtual (i.e., fundamental theory-based computational) synthesis and evaluation of the corresponding model QDs and their NHSs with desirable electron spin–charge transport and magneto-electronic properties is very important. The theoretical and computational results complement each other, and together provide a valuable guideline for experimental synthesis of the small QD-based NHSs supporting correlated/polarized/coherent dynamic electronic spin–spin and/or spin–charge states. The models of the small QDs and their NHSs with predesigned spin–spin and/or spin–charge correlated states can be collected in virtual template libraries according to their structure-confinement spin–charge correlation properties that may be used before attempting particular experimental synthesis procedures to evaluate its possible outcomes.

1.1 Experimental Studies

1.1.1 Quantum Information Processing

Physical systems that may be used to realize integrated storage logic and quantum computer operations have to satisfy a number of requirements, including (i) possession of physical degrees of freedom or physical entities that would realize well-defined qubits, (ii) reliable qubit state representation, (iii) low decoherence of the entangled states, (iv) accurate quantum gate operations, and (v) availability of assertive and nondestructive quantum measurements [7]. Electron spin states have long been identified [8] as attractive systems that satisfy these conditions in principle, and in particular possess long coherence times even at room temperature [9]. The robustness of such states with regard to interactions with the environment [10–13] (that lead to decoherence of the entangled states [14–16]) satisfies the major condition of realization of quantum computing concept. Both semiconductor QD-based systems [16–18] and magnetic nanodot materials [19–22] are thoroughly investigated experimentally, theoretically, and computationally, including modeling of possible quantum computer architectures, and controlled NOT (or CNOT) gate operations [23, 24]. Spins of electrons are less subject to decoherence than their other degrees of freedom, such as the charge and momentum. A typical electron momentum relaxation time in heterostructures is of the order of picoseconds, while electron spin relaxation time [25] in such structures exceeds a nanosecond, and can reach milliseconds [26–28] for spin states of electrons localized in artificial atoms or QDs. Localization of a single electron [16, 29, 30] on a QD, preparation of its ground state [16, 31], and manipulations by single electron spins have been realized experimentally combining spin to charge conversion with real-time single electron detection techniques [28, 32–35]. Since the late 90s, the electron spin resonance (ESR) method has been increasingly used to manipulate electron spins in QD-based spin diode setups that can act as spin filters, and devices for readout or other manipulations by single electron spin states (spin memory) [8, 36, 37]. While conceptually obvious, the ESR phenomenon involving a single electron spin (that is instrumental in inducing spin flips) has been measured in a few specific cases of solids only a few years ago [38–40]. Recently, Koppens *et al.* [16] demonstrated a possibility to control a single electron spin localized in coupled semiconductor QDs using ESR method. Other approaches to manipulate rotations of electron spins include optical excitation [41] and charge manipulation methods [42–46]. However, these approaches are even more difficult to realize (at least, at present) than ESR/magnetic field-based manipulations with electron spins. The majority of the experimental realizations of electron spin-correlated/resonant/entangled states concern semiconductor QD-based systems using ESR/magnetic field-based methods. Theoretical studies described in this chapter are focused on challenging developments in analytical and computational characterization of these and other QD-based systems to provide a fundamental guidline for their subsequent experimental synthesis. In particular, the approach is focused on revealing the inner structure of the charge–spin transport processes and coefficients in QD, artificial atoms, molecules, and other strongly inhomogeneous systems that must be studied for subsequent identification of correlated, coherent, entangled or resonant spin and charge states.

1.1.2 Quantum Dots: Realization and Applications

Quantum information processing and many other quantum electronic applications mentioned below rely entirely on the technological ability to physically prepare QD-based systems with desirable electronic properties. These properties are defined by experimental techniques used for the QD preparation. The term "quantum dot" has come to denote a variety of material structures fabricated by a wide variety of techniques. The defining characteristic, and perhaps the only common feature, is that electrons are quantum mechanically restrained in all three spatial dimensions. However, the method used to fabricate the QDs can have a critical effect on the ultimate structure of the dots and on their properties. Restricting the discussion to semiconductor host materials, some of the techniques for confining electrons are briefly reviewed below.

The so-called self-assembled QD is an outgrowth of molecular beam epitaxy growth of semiconductors on mismatched substrates or buffer layers [47]. Lattice constant mismatch between the two materials results in strain in the layer being grown. Under the right conditions, growth can convert from layer-by-layer mode to island growth in the Stranski–Krastanow mode [48]. Depending on the growth conditions, the self-assembled QD dimensions can vary from 10 to 20 nm in diameter and several nanometers in height. Such QDs can be stacked on top of each other by growing alternating layers of different semiconductors with only one in the Stranski–Krastanow regime [49]. The prototypical material combination for self-assembled QDs is InAs on GaAs, but QDs in InP on InGaP and Ge on Si, among other materials, have also been demonstrated. The QDs formed in this manner are more pyramidal than spherical. Three-dimensional growth is due to relaxation of the elastic energy related to the lattice mismatch, but even then the resulting QDs are under considerable strain that strongly affects their electronic properties. In particular, piezoelectric effects are important for self-assembled III–V QDs [50]. The QDs grown by this technology rarely form into perfect lattices of QDs, but there is some short-range order that can be taken advantage of under certain circumstances. Often, as in InAs–GaAs, the QD is composed of the narrower band gap material with host being the wider band gap material. In such cases, the QDs can act as artificial atoms that are embedded in a host material. This can be either an advantage or a disadvantage, depending on the application. The host can be an active part of the ultimate device, such as in QD infrared detectors discussed below, but it can also restrict access to the dot. This technique is limited to planar structures.

An entirely different approach to realization of QDs is growth of collodial nanocrystals from solution. Precursors and nanocrystals of the desired material form are injected into a hot surfactant. Aggregation is prevented because the nanocrystals are coated with a monolayer of the surfactant [51]. QDs grown by this method can be considered free-standing since they are not constrained in any dimension, as opposed to the self-assembled QDs discussed above. The collodial technique has been used to grow nanocrystals of II–VI semiconductors such as CdSe [52] and III–V semiconductors such as GaAs and InP [53]. QDs with diameters as small as a few nanometers are possible with this technique. These materials are relatively inexpensive to grow (compared to molecular beam epitaxy), and films of the QDs can be spun onto other semiconductors or substrates for inexpensive device fabrication. The surfactant shell can be used to control the surface states of the QD and can isolate it from its environment.

Two other techniques, both based on lithography, have to be mentioned. QDs can be formed by patterning layered two-dimensional structures and then etching out pillars containing the QDs [54]. The initial two-dimensional structures can be single quantum wells, multiple quantum wells, or superlattices, which all confine electrons to two dimensions. The remaining confinement is determined by etching. The main advantage of this approach is the control over the QDs that can be obtained. The position and size of the QDs can be more precisely controlled than in any other technique. The major disadvantage is that the lateral dimensions are restricted to the limitations of deep UV or electron beam lithography. The minimum lateral dimension is probably 10–20 nm at present. The vertical dimension can now be controlled at the near atomic level by modern epitaxial growth techniques.

The last technique to be mentioned in this brief overview is electrostatically defined QDs. Here, a gate pattern is lithographically defined and deposited on a two-dimensional semiconductor structure. Voltages applied to the gates during operation then define the QD by creating regions depleted of charge carriers around the QD. The technique was originally used to define one-dimensional quantum wires [55], but has recently gained increased interest for the creation of QDs for single electron transistor and quantum computing applications [56]. As with the etched QDs, this technique provides an additional advantage: the effective dimensions of the QD can be controlled during operation by varying the gate voltage. The principal disadvantage of this technique is that the interconnects for the gates take up much more space than the QD. This restricts this technique to applications where a single QD is the active element, such as qubits for quantum computing [57] rather than applications requiring arrays of many QDs.

Due to their unique properties that can be manipulated with, QDs are being used in an ever increasing number of applications, one of the latest being quantum information processing. A full discussion of all the applications is beyond the scope of this review, but the common ones are mentioned below.

Single Electron Transistors

QDs formed by lithography or electrostatic control can be coupled to the outside by leads. Since electrons in the QD are confined, such connections are through tunnelling junctions. The conduction through such a QD is quantized and oscillates with the number of charges on the QD [58]. The quantized conduction has led to development of single electron transistors (SET) for the use in digital classical logic applications [59]. Silicone QD-based SETs have been proposed for memory applications as well [60]. Recently, due to the possibility to control and monitor the spin of individual electrons in QD SETs, these devices have been proposed for use as qubits in quantum computing application [61].

Light Emitting Diodes

Nitride-based semiconductor heterostructures are widely used for green, blue, and white light emitting diodes (LEDs). The alloy composition of the $In_{1-x}Ga_xN$ quantum well layer is adjusted to select the wavelength of emission. However, intentionally or unintentionally self-assembled InGaN QDs have been found to be responsible for emission in many cases [62]. The presence of QDs improves the emission intensity of the LEDs and is currently used in many commercial LEDs.

Thermoelectric Applications

There is increasing interest in the development of thermoelectric (TE) devices for the direct conversion of waste thermal energy into electric power. TE devices are also under development for the use in cooling of electronic and electrooptic devices. Unlike most applications, TE devices need poor thermal conductivities so that large temperature gradients can be maintained across the device. The TE figure of merit is proportional to the electrical conductivity and the Seebeck coefficient of the material, and inversely proportional to the thermal conductivity. Stacked self-assembled QD superlattices offer increased scattering of phonons from the QD interfaces with resulting improved TE performance. TE materials based on Ge/Si [63], PbTe [64], and ErAs in InGaAs [65] QD superlattices have been reported.

Numerous and increasingly important applications of various QD, artificial molecule and atom systems have generated a necessity to develop a self-consistent and unified theoretical approach to predict their electronic equilibrium and transport properties from the first principle, to avoid a multitude of rather poor models incompatible between themselves, each of which is applicable to only a particular system. In the following section, theoretical foundations of the transport theory are considered. In Section 2 of this chapter, the latest first-principle approach specifically designed to describe charge transport processes in spatially strongly nonuniform (or inhomogeneous) systems is laid out and linked to experimentally measurable or computational quantities. In Section 3, this link developed further to demonstrate virtual syntheses of several Ga–As–P artificial molecules with predesigned electronic properties.

1.2 Theoretical Foundation

Fundamental solid state electronic structure theory of the equilibrium state [66–70] and its *ab initio* computational methods, such as the density functional theory (DFT)-based techniques, full-electron Hartree–Fock-based (HF) methods, self-consistent field (SCF) approximations, configuration interaction (CI) methods, complete active space SCF (CASSCF), multiconfiguration SCF approximation (MCSCF), coupled-clusters approximation, and various half-heuristic methods [71] established for bulk solid lattices (such as effective mass theory, **k•p** method, tight binding and envelope function models, deformation potential theory, method of invariants, degenerate bands and many-band models, and their numerous modifications), have been successfully used to identify structures that exhibit new electronic, opto- and magneto-electronic properties, including quantum wells and wires [72] (QWWs) and large magnetic QDs [73]. For smaller quantum objects, these models and approximations are progressively less accurate and fail entirely for artificial atoms and molecules composed of a few to a hundred atoms or so. Thus, for such systems the Schrödinger equation is solved numerically using the already-existing, fundamental theory-based computational methods and tools (such as GAUSSIAN [74], GAMESS [75] or NWChem [76] software packages) to compute electronic energy level structure (ELS) at equilibrium and/or simulate the spin–charge transport properties (usually quasi-classically or using simplifying conditions, such as Born–Oppenheimer approximation). Relations between the ELS and the transport properties are introduced heuristically (using the theoretical approaches mentioned above), and therefore, these computational methods on their own do not allow a first-principle evaluation of spin–charge dynamics and quantum transport properties of small QDs, molecules, and atoms. These properties, however, are essential to control the development and evolution of correlated, coherent or resonant electronic spin states in quantum structures.

Existing theoretical approaches to dynamics of electronic states in artificial atoms and molecules rely on the use of *ad hoc* modifications [15, 16, 21, 36, 77–82] of various methods developed originally for much larger systems at low-temperature conditions, and tend to lead to physically incorrect parameters even for mesoscopic tunneling junctions [83]. The best of these models provide only roughly for the quantum confinement, thus producing rather inadequate consideration of the quantum confinement effects and QD–QD and QD–environment coupling. However, the latter is the major source of both decoherence and coherence [46, 84] of the electron states (see also [16 and 85]), and that of many other unique properties of strongly inhomogeneous objects and systems. Moreover, these models do not allow first-principle, self-consistent calculations of electronic transport coefficients, such as the quantum conductivity and dielectric and magnetic susceptibility tensors. Such calculations, of course, imply the use of quantum statistical mechanical methods to derive self-consistently the corresponding formulae from the quantum Liouville equation. This formidable task, however, has been accomplished self-consistently in the simplest approximation only for mesoscale systems [82]. Thus, the coupled-equation chain for the system quantum correlators was reduced to only the first (Kadanoff–Dyson) equation to derive the corresponding quantum kinetic equations and their quasi-classical versions. Similar, but half-heuristic, methods based on a quasi-classical, Boltzmann-like kinetic approach [86, 87] were applied to large QDs and nanotubes composed of thousands of atoms. However, even for mesosystems at low temperatures, the suggested local description of the quantum conductance breaks down [88, 89]. In particular, while a sophisticated half-heuristic theory of conductance developed by Landauer, Imry, Buttiker, and others provides a valuable insight into the nature of charge transport in large QDs [87, 88, 90–93], it still breaks down for smaller QDs and molecules. Attempts [94] to generalize this approach to embrace the interacting electrons' case using Keldysh's generalization of the Green's function diagrammatic technique was successful only in the zero-temperature limit, where electron–electron interactions can be viewed as elastic. In this and other half-heuristic approximations, Coulomb blockade effects have been studied by the Green function–based diagrammatic techniques [95, 96] since the 1970s. The obtained results are only qualitative and rely on various case-specific assumptions used to describe coherence between the electrons in a mesoscopic QD and those in the leads. At higher temperatures, such coherence is ignored entirely (obviously, this assumption fails for small QDs and molecules at all temperatures). These qualitative results have provided a basis for yet another type of approaches, stemming from various phenomenological Langevin (or master) equations for the nonequilibrium distributions of electrons in metallic dots [97, 98]. Even for mesoscale QDs and with the use of a number of case-specific parameters and assumptions, the Landauer-based and similar formulae for the linear conductance describe only qualitatively the major features of the quantum charge transport, such as dependence of the conductance on a particular linear combination of incoming modes (called eigenchannels) that remain invariant upon reflection on the sample. For large QDs weakly coupled to the leads via tunnel junctions, a large number of excited states contribute to the off-resonance conductance [99] and lead to the Coulomb staircase behavior of the conductance as a function of the source-drain voltage. In the quantum regime (very low temperature), however, the current depends on the small number of the available excited levels in the dot through which only a fixed number of electrons can tunnel.

Despite of being inadequate for small QDs and molecules, the explicit half-heuristic expressions for the conductivity tensor components obtained in terms of (nonequilibrium) Keldysh Green functions in this approach and closely related expressions [36, 100] are the major sources of the quantum conductivity formulae that are widely applied to small QDs and molecules. The reliability of such formulae for each considered small QD case at low temperatures is ensured by various adjustable parameters and experimental details incorporated in the formulae and specific to the considered case.

The R-matrix theory (RMT) [101] (originally due to Wigner and Eisenbud) and its recent version [102, 103, 104] (reformulated in terms of an effective, non-Hermitian Hamiltonian approach) were applied to ballistic QDs (where the electron transport is defined by "chaotic" scattering from the nonstructured dot boundaries) weakly coupled to their environment. Many assumptions and fittings to known (in particular, Landauer–Büttiker) results integrated this approach with other existing methods. However, the RMT approach is not applicable to small QDs with highly correlated electron motions and atomically nonuniform boundary conditions. The RMT, diagrammatic approach [94] and supersymmetry-based techniques [105, 106] were used to build a theory of nonlinear conductance of low-temperature mesosystems, and to explain the universality of the conductance fluctuations and its loss in the case of large ballistic QDs [107, 108].

Over the years, self-consistency [109], inelastic scattering [110], and electron correlation effects (see, for example, [88, 94 and 111]) in mesosystems were addressed to alleviate the shortcomings of the above theoretical methods. Modified methods that use yet other types of heuristic master [112], Liouville, and/or kinetic [113] equations have also been developed for this purpose. None of the suggested improvements delivers a first-principle, self-consistent quantum statistical mechanical description of quantum transport in small QDs/molecules at finite temperatures. Even for mesosystems, general quantum statistical mechanical methods [114] used by Mahan [82] remain the only rigorous approaches to the quantum spin–charge transport. Recently, a modified linear-response method of quantum statistical mechanics was introduced by Lang et al. [115] to develop a sophisticated approach for calculations of the linear contribution to the quantum conductivity tensor of an inhomogeneous system in a weak electromagnetic field. While this approach provides an insight into the problem of quantum charge transport in inhomogeneous systems, it introduces unphysical and nonuniquely defined operators describing particle charge–field interactions, and also does not lead to explicit analytical expressions for the quantum conductivity and susceptibility tensors. In contrast, rigorous, more "traditional" linear-response theories of the quantum transport by Baranger and Stone [116], and recently, by Pozhar [117], do provide first-principle, self-consistent methods applicable to strongly inhomogeneous systems, such as small QDs and molecules. Moreover, the latter approach that uses the equilibrium two-time temperature Green functions offers additional advantages of utilization (if necessary) of computational and experimental data, as will be discussed in the following sections. While a complete, first-principle, self-consistent quantum statistical mechanical approach to quantum transport in small QDs and molecules at room temperatures is necessary to predict details of essentially quantum phenomena, such as quantization of the conductivity, coherent charge–spin transport, etc., the major features of such phenomena can be self-consistently described already within the linear approximation with respect to the external fields.

2 Linear Response Theory of Charge Transport in Small Systems in External Electro-Magnetic Fields

Consider a system of N particles each of which possesses the mass m and charge e in a weak, time-dependent electromagnetic field of intensities $\mathbf{E}(\mathbf{r}, t)$ and $\mathbf{H}(\mathbf{r}, t)$:

$$\begin{aligned} \mathbf{E}(\mathbf{r}, t) &= -\frac{1}{c}\frac{\partial \mathbf{A}(\mathbf{r}, t)}{\partial t} - \nabla_{\mathbf{r}}\varphi(\mathbf{r}, t), \\ \mathbf{H}(\mathbf{r}, t) &= \mathbf{curl}\,\mathbf{A}(\mathbf{r}, t), \end{aligned} \tag{1}$$

where bold symbols denote physical vectors; $\mathbf{r}$ and t are a position vector and time, respectively; $\nabla_{\mathbf{r}}$ denotes the gradient operator with respect to $\mathbf{r}$; c is the speed of light; and $\mathbf{A}(\mathbf{r}, t)$ and $\varphi(\mathbf{r}, t)$ are the vector and scalar potentials of the (weak) electromagnetic field in an arbitrary gauge, respectively. The total Hamiltonian $\mathcal{H}$ of the system has the form:

$$\mathcal{H} = \frac{1}{2m}\sum_{i=1}^{N}\left(\mathbf{p}_i - \frac{e}{c}\mathbf{A}(\mathbf{r}_i, t)\right)^2 + H_{\text{int}} + e\sum_{i=1}^{N}\varphi(\mathbf{r}_i, t), \tag{2}$$

where $\mathbf{p}_i = \frac{\hbar}{i}\nabla_{\mathbf{r}}$ is the momentum operator of the particle i, and H_{int} is the operator that includes interparticle interactions and interactions with the particles of the environment. The magnetization of the system is assumed to be relatively small, so that the contribution to the Hamiltonian (2) due to interactions of particle spins with the magnetic field is neglected [118]. The second-quantized representation of the Hamiltonian (2) is

$$\begin{aligned} \mathcal{H} =& \frac{1}{2m}\int d\mathbf{r}\,\psi^{+}(\mathbf{r})\left(\frac{\hbar}{i}\nabla_{\mathbf{r}} - \frac{e}{c}\mathbf{A}(\mathbf{r}, t)\right)^2\psi(\mathbf{r}) \\ &+ H_{\text{int}} + e\int d\mathbf{r}\,\psi^{+}(\mathbf{r})\psi(\mathbf{r})\varphi(\mathbf{r}, t), \end{aligned} \tag{3}$$

where $\psi^{+}(\mathbf{r})$ and $\psi(\mathbf{r})$ are the quantum field operators related to the creation and annihilation operators of the charge carriers $a^{+}_{\mathbf{k}\sigma}$ and $a_{\mathbf{k}\sigma}$, respectively, by the standard definitions:

$$\begin{aligned} \psi^{+}(\mathbf{r}) &= \frac{1}{\sqrt{V}}\sum_{\mathbf{k},\sigma} e^{-i\,\mathbf{kr}}\delta_{\sigma\,\sigma_z}a^{+}_{\mathbf{k}\sigma}, \\ \psi(\mathbf{r}) &= \frac{1}{\sqrt{V}}\sum_{\mathbf{k},\sigma} e^{i\,\mathbf{kr}}\delta_{\sigma\,\sigma_z}a_{\mathbf{k}\sigma}, \end{aligned} \tag{4}$$

with $\mathbf{k}$ being the wave vector, σ_z denoting the value of the z-component of the spin, and $\delta_{\sigma\,\sigma_z}$ being the Kronecker delta.

The current density operator at a position $\mathbf{r}$ in the system at a time t is defined as the Fréchet derivative of the Hamiltonian with respect to $\mathbf{A}(\mathbf{r}, t)$, $\mathbf{j}(\mathbf{r}, t) = -e\frac{\delta\mathcal{H}}{\delta\mathbf{A}(\mathbf{r},t)}$, and can be obtained from Eq. (3):

$$\mathbf{j}(\mathbf{r}, t) = \frac{e\hbar}{2mi}\left\{\psi^+(\mathbf{r})\nabla_\mathbf{r}\psi(\mathbf{r}) - [\nabla_\mathbf{r}\psi^+(\mathbf{r})]\,\psi(\mathbf{r})\right\} - \frac{e^2}{mc}\mathbf{A}(\mathbf{r}, t)\psi^+(\mathbf{r})\psi(\mathbf{r}). \quad (5)$$

In the absence of the fields, the current density operator is defined by

$$\mathbf{j}_0(\mathbf{r}) = \frac{e\hbar}{2mi}\left\{\psi^+(\mathbf{r})\nabla_\mathbf{r}\psi(\mathbf{r}) - [\nabla_\mathbf{r}\psi^+(\mathbf{r})]\,\psi(\mathbf{r})\right\}, \quad (6)$$

and the charge density operator is

$$\rho(\mathbf{r}) = e\psi^+(\mathbf{r})\psi(\mathbf{r}) = en(\mathbf{r}), \quad (7)$$

where $n(\mathbf{r})$ denotes the number density of the charge carriers. Using the definitions (6) and (7), one can rewrite the Hamiltonian (3) in the form:

$$\mathcal{H} = H_0 + H_1 + H_2, \quad (8)$$

where

$$H_0 = \frac{1}{2m}\int d\mathbf{r}\,\psi^+(\mathbf{r})\left[\frac{\hbar}{i}\nabla_\mathbf{r}\right]^2\psi(\mathbf{r}) + H_{int} \quad (9)$$

is the unperturbed Hamiltonian in the absence of the electromagnetic field,

$$H_1 = -\frac{1}{c}\int d\mathbf{r}\,\mathbf{j}_0(\mathbf{r})\cdot\mathbf{A}(\mathbf{r}, t) + \int d\mathbf{r}\,\rho(\mathbf{r})\varphi(\mathbf{r}, t) \quad (10)$$

is the perturbation Hamiltonian linear in the electromagnetic field potentials, and

$$H_2 = \frac{e}{2mc^2}\int d\mathbf{r}\rho(\mathbf{r})\mathbf{A}^2(\mathbf{r}, t) \quad (11)$$

is the second-order perturbation Hamiltonian [the dots "·" in Eqs. (10) and (11) denote the inner product]. Due to the assumed weakness of the magnetic field, the contribution (11) to the Hamiltonian (8) can be neglected, so that the resulting linearized Hamiltonian takes the form:

$$\mathcal{H} = H_0 + H_1 \quad (12)$$

For a system described by the Hamiltonian (12), one can obtain the expectation value $< \mathbf{O}(\mathbf{r}) >_0$ of an observable $\mathbf{O}$ using a trivial extention of the linear response theory to include spatially inhomogeneous systems [this is possible due to a quasi-local nature [119] of the field operators (4)]:

$$< \mathbf{O}(\mathbf{r},t) >=< \mathbf{O}(\mathbf{r}) >_0 + \int_{-\infty}^{\infty} dt' << \mathbf{O}(\mathbf{r},t)|H_1(t') >>_0. \tag{13}$$

where $<< \mathbf{O}(\mathbf{r},t)H_1(t') >>_0$ is the equilibrium retarded, two-time temperature Bogoliubov–Tyablikov Green function (TTGF) and $<\mathbf{O}(\mathbf{r},t)>= Tr(\wp\mathbf{O})$. Here *Tr* denotes the trace of an operator, and the statistical operator $\wp$ is the solution of the quantum Liouville equation with the Hamiltonian (12) and the equilibrium initial condition $\wp(t=-\infty) \equiv \wp_0 = e^{-\beta H_0}/Tr\, e^{-\beta H_0}$, $\wp_0$ is the equilibrium or steady state statistical operator, $\beta=1/k_B T$ is the reciprocal temperature and k_B is the Boltzmann constant. The TTGF of any two operators **O** and **K** is defined by the expression

$$<<\mathbf{O}(\mathbf{r},t)|\mathbf{K}(\mathbf{r}',t')>>_0= \vartheta(t-t')\frac{1}{i\hbar} < [\mathbf{O}(\mathbf{r},t), \mathbf{K}(\mathbf{r}',t')]>_0, \tag{14}$$

where $\vartheta(t-t')$ is the step function and the brackets [. . .,. . .] denote the commutator.

Applying Eq. (13) to the nonequilibrium charge density operator $\rho(\mathbf{r},t) = e^{iH_0t/\hbar}\rho(\mathbf{r})e^{-iH_0t/\hbar}$ and using Eq. (10), one can derive the charge conservation equation for a system in a state described by the Hamiltonian (12):

$$\begin{aligned}<\rho(\mathbf{r},t) >=&\rho_0(\mathbf{r}) - \frac{1}{c}\int d\mathbf{r}' \int_{-\infty}^{\infty} dt' << \rho(\mathbf{r},t)|\mathbf{j}_0(\mathbf{r}')>>_0 \cdot\mathbf{A}(\mathbf{r}',t') \\ &+ \int d\mathbf{r}' \int_{-\infty}^{\infty} dt' <<\rho(\mathbf{r},t)|\rho(\mathbf{r}')>>_0 \varphi(\mathbf{r}',t'),\end{aligned} \tag{14a}$$

where $\rho_0(\mathbf{r})$ is the charge density at the equilibrium or steady state. Similarly, applying Eq. (13) to the current density operator $\mathbf{j}(\mathbf{r},t)$ of Eq. (5), using Eq. (10) and noticing that in the absence of the field $< \mathbf{j}_0(\mathbf{r}) >_0= 0$, one can derive the current density conservation equation,

$$\begin{aligned}<\mathbf{j}(\mathbf{r},t)>=& - \frac{1}{c}\int d\mathbf{r}' \int_{-\infty}^{\infty} dt' << \mathbf{j}(\mathbf{r},t)|\mathbf{j}_0(\mathbf{r}')>>_0 \cdot\mathbf{A}(\mathbf{r}',t') \\ &+ \int d\mathbf{r}' \int_{-\infty}^{\infty} dt' <<\mathbf{j}(\mathbf{r},t)|\rho(\mathbf{r}')>>_0 \varphi(\mathbf{r}',t') - \frac{e}{mc}\rho_0(\mathbf{r})\mathbf{A}(\mathbf{r},t).\end{aligned} \tag{15}$$

Note here, that the current density operator of Eq. (5) satisfies the microscopic charge conservation (or continuity) equation:

$$\frac{\partial \rho(\mathbf{r})}{\partial t} = -div\,\mathbf{j}(\mathbf{r}, t).$$

2.1 Conservation Equations for the Space–Time Fourier Transforms of the Charge and Current Densities

From Eqs. (14a) and (15), the corresponding conservation equations for the space–time Fourier transforms of the expectation values of the charge and current densities $\rho(\mathbf{k}, \omega)$ and $\mathbf{j}(\mathbf{k}, \omega)$, respectively, are:

$$\begin{aligned}\rho(\mathbf{k}, \omega) =& en_0(\mathbf{k})\delta(\omega) - \frac{1}{c}\sum_{\mathbf{l}} << \rho_{\mathbf{k}} \mid \mathbf{j}_{-\mathbf{l}} >>_{0,\omega} \cdot \mathbf{A}(\mathbf{l}, \omega) \\ &+ \sum_{\mathbf{l}} << \rho_{\mathbf{k}} \mid \rho_{-\mathbf{l}} >>_{0,\omega} \varphi(\mathbf{l}, \omega),\end{aligned} \tag{16}$$

$$\begin{aligned}\mathbf{j}(\mathbf{k}, \omega) =& -\frac{e^2}{mc}\sum_{\mathbf{l}} n_0(\mathbf{k}-\mathbf{l})\mathbf{A}(\mathbf{l}, \omega) - \frac{1}{c}\sum_{\mathbf{l}} << \mathbf{j}_{\mathbf{k}} \mid \mathbf{j}_{-\mathbf{l}} >>_{0,\omega} \cdot \mathbf{A}(\mathbf{l}, \omega) \\ &+ \sum_{\mathbf{l}} << \mathbf{j}_{\mathbf{k}} \mid \rho_{-\mathbf{l}} >>_{0,\omega} \varphi(\mathbf{l}, \omega),\end{aligned} \tag{17}$$

where V is the system volume, ω is the frequency, and $\mathbf{k}$ is the wave vector, $\delta(\omega)$ denotes the Dirac's delta function, the double brackets $<< \ldots | \ldots >>_{0,\omega}$ denote the space–time Fourier-transforms of the corresponding TTGFs, $n_0(\mathbf{k}) = \frac{1}{V} < \sum_{\mathbf{q},\sigma} a^{+}_{(\mathbf{q}-\mathbf{k}),\sigma} a_{\mathbf{q},\sigma} >_0$ is the space Fourier transform of the equilibrium charge number density, and the summations run over the wave vectors $\mathbf{l}$ ($-\infty < \mathbf{l} < \infty$). Therefore, the dependence of $\rho(\mathbf{k}, \omega)$ and $\mathbf{j}(\mathbf{k}, \omega)$ on the wave vectors is defined by the corresponding space–time Fourier transforms of the microscopic charge and current density operators,

$$\rho_{\mathbf{l}} = \frac{e}{\sqrt{V}} \sum_{\mathbf{q},\sigma} a^{+}_{\mathbf{q}-\mathbf{l},\sigma} a_{\mathbf{q},\sigma}, \tag{18}$$

$$\mathbf{j}_{\mathbf{l}} = \frac{e\hbar}{m\sqrt{V}} \sum_{\mathbf{q},\sigma} (\mathbf{q} - \frac{\mathbf{l}}{2}) a^{+}_{\mathbf{q}-\mathbf{l},\sigma} a_{\mathbf{q},\sigma}, \tag{19}$$

where σ denotes the z-components of the spin.

The homogeneous (or "bulk") system case is recovered from the conservation equations (16) and (17) upon a consideration that in this case each of the sums containing the Fourier transforms of the TTGFs reduces to only one nonzero term with

$\mathbf{l}=\mathbf{k}$, and that the charge number density is constant. The equations (16) and (17) generalize the conservation equations of Refs. [120] to include the inhomogeneous system case.

2.1.1 The Generalized Susceptibility and Microcurrent–Microcurrent TTGFs

The space–time Fourier transform $\chi_{\alpha\beta}(\mathbf{k}, \mathbf{l}, \omega)$ of the generalized susceptibility tensor χ is defined by the following expression:

$$\chi_{\alpha\beta}(\mathbf{k}, \mathbf{l}, \omega) = \frac{e^2}{mc} n(\mathbf{k} - \mathbf{l})\delta_{\alpha\beta} + << \mathbf{j}_{\mathbf{k}}^{\alpha} \mid \mathbf{j}_{-\mathbf{l}}^{\beta} >>_{0,\omega}, \tag{20}$$

where the indices α and β denote Cartesian components of the microcurrents, and $\delta_{\alpha\beta}$ is the Kronecker's symbol. In the linear approximation studied here, the Fourier transform of the microcurrent–microcurrent TTGF, $<< \mathbf{j}_{\mathbf{k}}^{\alpha} \mid \mathbf{j}_{-\mathbf{l}}^{\beta} >>_{0,\omega}$, is a bilinear form of the vectors $\mathbf{k}$ and $\mathbf{l}$. Therefore, it can be expressed in terms of the corresponding second-rank tensors that are proportional to the unit matrix $\mathbf{I}$ and the tensor $\mathbf{kl}$, with coefficients that are scalar functions of the absolute values $|\mathbf{k}|\equiv k$, $|\mathbf{l}|\equiv l$, and the inner product $\mathbf{k}\cdot\mathbf{l}$,

$$\begin{aligned} << \mathbf{j}_{\mathbf{k}}^{\alpha} \mid \mathbf{j}_{-\mathbf{l}}^{\beta} >>_{0,\omega} &= \chi^{lon}(\mathbf{k}, \mathbf{l}, \omega)\frac{\mathbf{k}_{\alpha}\mathbf{l}_{\beta}}{kl} \\ &+ \left[C(\mathbf{k}, \mathbf{l}, \omega)\delta_{\alpha\beta} - \frac{\mathbf{k}_{\alpha}\mathbf{l}_{\beta}}{kl} \right] \chi^{tr}(\mathbf{k}, \mathbf{l}, \omega), \end{aligned} \tag{21}$$

where the space–time Fourier transform of the generalized scalar longitudal susceptibility can be immediately identified (the term "longitudal" is retained for the fact that this quantity reduces to the scalar longitudal susceptibility in the case of a homogeneous system),

$$\chi^{lon}(\mathbf{k}, \mathbf{l}, \omega) = \frac{<< \mathbf{k}\cdot\mathbf{j}_{\mathbf{k}} | \mathbf{l}\cdot\mathbf{j}_{-\mathbf{l}} >>_{0,\omega}}{kl}, \tag{22}$$

and where the space–time Fourier transform of the generalized scalar transversal susceptibility $\chi^{tr}(\mathbf{k}, \mathbf{l}, \omega)$ and the scalar function $C(\mathbf{k}, \mathbf{l}, \omega)$ can be determined from the identity

$$\begin{aligned} &\sum_{\alpha,\beta} << \mathbf{j}_{\mathbf{k}}^{\alpha} \mid \mathbf{j}_{-\mathbf{l}}^{\beta} >>_{0,\omega} \left[C(\mathbf{k}, \mathbf{l}, \omega)\delta_{\alpha\beta} - \frac{\mathbf{k}_{\beta}\mathbf{l}_{\alpha}}{kl} \right] \\ &= \sum_{\alpha,\beta} \chi^{lon}(\mathbf{k}, \mathbf{l}, \omega)\frac{\mathbf{k}_{\alpha}\mathbf{l}_{\beta}}{kl} \left[C(\mathbf{k}, \mathbf{l}, \omega)\delta_{\alpha\beta} - \frac{\mathbf{k}_{\beta}\mathbf{l}_{\alpha}}{kl} \right] \\ &+ \left[3C^2(\mathbf{k}, \mathbf{l}, \omega) - 2\,\frac{\mathbf{k}\cdot\mathbf{l}}{kl} C(\mathbf{k}, \mathbf{l}, \omega) + \frac{(\mathbf{k}\cdot\mathbf{l})^2}{k^2 l^2} \right] \chi^{tr}(\mathbf{k}, \mathbf{l}, \omega), \end{aligned} \tag{23}$$

that is obtained using Eq. (21). Once again, the term "transversal" is retained because the above susceptibility reduces to the scalar transversal susceptibility in the case of a homogeneous system.

Substituting $\chi^{lon}(\mathbf{k}, \mathbf{l}, \omega)$ from Eq. (22) in Eq. (23), one can prove that the first term on the right-hand side (r.h.s.) of Eq. (23) is equal to

$$C(\mathbf{k}, \mathbf{l}, \omega)\frac{\mathbf{k}\cdot\mathbf{l}}{k^2 l^2} << \mathbf{k}\cdot\mathbf{j}_\mathbf{k}|\mathbf{l}\cdot\mathbf{j}_{-\mathbf{l}} >>_{0,\omega} - \frac{(\mathbf{k}\cdot\mathbf{l})^2}{k^3 l^3} << \mathbf{k}\cdot\mathbf{j}_\mathbf{k}|\mathbf{l}\cdot\mathbf{j}_{-\mathbf{l}} >>_{0,\omega} .$$

By a convenient choice of the scalar function $C(\mathbf{k}, \mathbf{l}, \omega)$ (that does not effect a generality of this consideration),

$$C(\mathbf{k}, \mathbf{l}, \omega) = \frac{\mathbf{k}\cdot\mathbf{l}}{kl}, \tag{24}$$

this term can be set to zero. With this choice of $C(\mathbf{k}, \mathbf{l}, \omega)$, the r.h.s. of Eq. (23) becomes equal to $2\frac{(\mathbf{k}\cdot\mathbf{l})^2}{k^2 l^2}\chi^{tr}(\mathbf{k}, \mathbf{l}, \omega)$. The left-hand side (l.h.s.) of Eq. (23) can be transformed further by the use of the identity $[\mathbf{k} \times \mathbf{j}_\mathbf{k}] \cdot [\mathbf{l} \times \mathbf{j}_{-\mathbf{l}}] = (\mathbf{k}\cdot\mathbf{l})(\mathbf{j}_\mathbf{k}\cdot\mathbf{j}_{-\mathbf{l}}) - (\mathbf{k}\cdot\mathbf{j}_{-\mathbf{l}})(\mathbf{l}\cdot\mathbf{j}_\mathbf{k})$, where the square brackets $[\ldots\times\ldots]$ denote the vector cross product. In particular, dividing this identity by kl and considering the TTGF $<< [\mathbf{k} \times \mathbf{j}_\mathbf{k}]| \cdot [\mathbf{l} \times \mathbf{j}_{-\mathbf{l}}] >>_{0,\omega}$ (or formally, applying the operation $<< \ldots >>_{0,\omega}$ to the above identity), one can prove that the l.h.s. of Eq. (23) is equal to $\frac{1}{kl} << [\mathbf{k} \times \mathbf{j}_\mathbf{k}]| \cdot [\mathbf{l} \times \mathbf{j}_{-\mathbf{l}}] >>_{0,\omega}$. Thus, Eq. (23) reduces to the following form:

$$\frac{1}{kl} << [\mathbf{k} \times \mathbf{j}_\mathbf{k}]| \cdot [\mathbf{l} \times \mathbf{j}_{-\mathbf{l}}] >>_{0,\omega} = 2\frac{(\mathbf{k}\cdot\mathbf{l})^2}{k^2 l^2}\chi^{tr}(\mathbf{k}, \mathbf{l}, \omega). \tag{25}$$

From Eq. (25) the scalar transversal susceptibility $\chi^{tr}(\mathbf{k}, \mathbf{l}, \omega)$ can be easily obtained:

$$\chi^{tr}(\mathbf{k}, \mathbf{l}, \omega) = \frac{kl}{2(\mathbf{k}\cdot\mathbf{l})^2} << [\mathbf{k} \times \mathbf{j}_\mathbf{k}] \cdot [\mathbf{l} \times \mathbf{j}_{-\mathbf{l}}] >>_{0,\omega} . \tag{26}$$

Substituting $C(\mathbf{k}, \mathbf{l}, \omega)$ from Eq. (24) in Eq. (21), one can determine the TTGF $<< \mathbf{j}^\alpha_\mathbf{k}|\mathbf{j}^\beta_{-\mathbf{l}} >>_{0,\omega}$:

$$<< \mathbf{j}^\alpha_\mathbf{k} \mid \mathbf{j}^\beta_{-\mathbf{l}} >>_{0,\omega} = \chi^{lon}(\mathbf{k}, \mathbf{l}, \omega)\frac{\mathbf{k}_\alpha\mathbf{l}_\beta}{kl} + \chi^{tr}(\mathbf{k}, \mathbf{l}, \omega)\left[\frac{(\mathbf{k}\cdot\mathbf{l})}{kl}\delta_{\alpha\beta} - \frac{\mathbf{k}_\alpha\mathbf{l}_\beta}{kl}\right], \tag{27}$$

where the scalar susceptibilities are defined by Eqs. (22) and (26). The expressions (22), (23), and (27) generalize the corresponding Zubarev's formulae of Refs. [120] to the case of spatially inhomogeneous systems. They reduce to the corresponding Zubarev's expressions in the homogeneous system case.

2.1.2 The Longitudal Sum Rule

Considerations similar to those discussed above in Section 2.1.1 lead to explicit analytical expressions for the Fourier transforms of the TTGFs featuring in Eqs. (16) and (17). In particular, the linearity of the considered approximation suggests that

$$<< \mathbf{j}_{\mathbf{k}}^{\alpha} \mid \rho_{-\mathbf{l}} >>_{0,\omega} = << (\mathbf{k}\cdot\mathbf{j}_{\mathbf{k}})|\rho_{-\mathbf{l}} >>_{0,\omega} \frac{\mathbf{k}_{\alpha}}{k^2}, \tag{28}$$

$$<< \rho_{\mathbf{k}} \mid \mathbf{j}_{-\mathbf{l}}^{\beta} >>_{0,\omega} = << \rho_{-\mathbf{k}}|(\mathbf{l}\cdot\mathbf{j}_{-\mathbf{l}}) >>_{0,\omega} \frac{\mathbf{l}_{\beta}}{l^2}, \tag{29}$$

where the Fourier transforms $<< (\mathbf{k}\cdot\mathbf{j}_{\mathbf{k}})|\rho_{-\mathbf{l}} >>_{0,\omega}$ and $<< \rho_{-\mathbf{k}}|(\mathbf{l}\cdot\mathbf{j}_{-\mathbf{l}}) >>_{0,\omega}$ are scalar functions of k, l, and $\mathbf{k}{\cdot}\mathbf{l}$. Using Eqs. (28) and (29), one can transform the conservation equations (16) and (17) to the form:

$$\begin{aligned}\rho(\mathbf{k},\omega) = {} & e\sum_{\mathbf{l}} n_0(\mathbf{k})\delta(\omega)\delta_{\mathbf{kl}} - \frac{1}{c}\sum_{\mathbf{l}} << \rho_{\mathbf{k}} \mid (\mathbf{l}\cdot\mathbf{j}_{-\mathbf{l}}) >>_{0,\omega} \frac{1}{l^2}\mathbf{l}\cdot\mathbf{A}(\mathbf{l},\omega)\\ & + \sum_{\mathbf{l}} << \rho_{\mathbf{k}} \mid \rho_{-\mathbf{l}} >>_{0,\omega} \varphi(\mathbf{l},\omega),\end{aligned} \tag{30}$$

$$\begin{aligned}\mathbf{j}(\mathbf{k},\omega) = {} & -\frac{e^2}{mc}\sum_{\mathbf{l}} n_0(\mathbf{k}-\mathbf{l})\mathbf{A}(\mathbf{l},\omega)\\ & - \frac{1}{c}\sum_{\mathbf{l}}\left\{\chi^{lon}(\mathbf{k},\mathbf{l},\omega)\frac{\mathbf{kl}}{kl} + \chi^{tr}(\mathbf{k},\mathbf{l},\omega)\left[\frac{(\mathbf{k}\cdot\mathbf{l})}{kl}\mathbf{I} - \frac{\mathbf{kl}}{kl}\right]\right\}\cdot\mathbf{A}(\mathbf{l},\omega)\\ & + \sum_{\mathbf{l}} << (\mathbf{k}\cdot\mathbf{j}_{\mathbf{k}})|\rho_{-\mathbf{l}} >>_{0,\omega} \frac{\mathbf{k}}{k^2}\varphi(\mathbf{l},\omega),\end{aligned} \tag{31}$$

where the generalized scalar susceptibilities are defined by Eqs. (22) and (26), and $n_0(\mathbf{k}-\mathbf{l})$ is the $\mathbf{k}$–$\mathbf{l}$ mode of the equilibrium charge number density. Further calculations of the Fourier transforms of the TTGFs in the r.h.s. of Eqs. (30) and (31) can be achieved by the route similar to that suggested in Refs. [120], namely, assuming pairwise additivity of interparticle interactions, one can prove that $\rho_{\mathbf{k}}$ commutes with the interaction Hamiltonian H_{int} of Eq. (2), and that the commutator of $\rho_{\mathbf{k}}$ and the kinetic part of the Hamiltonian (2) are equal to $\hbar(\mathbf{k}\cdot\mathbf{j})$. Therefore, the equation of motion for the operator $\rho_{\mathbf{k}}$ takes the form:

$$\dot{\rho}_{\mathbf{k}} = \frac{1}{i\hbar}[\rho_{\mathbf{k}}, H_0] = -i\,(\mathbf{k}\cdot\mathbf{j}_{\mathbf{k}}), \tag{32}$$

where H_0 is the Hamiltonian (9), and $\dot{\rho}_{\mathbf{k}}$ denotes the Fourier transform of the time derivative of the charge density operator. Using Eq. (32), one can calculate the commutator $\frac{1}{i\hbar}[\rho_{\mathbf{k}}, \dot{\rho}_{-\mathbf{l}}]$ to obtain a generalization of *the longitudal sum rule* applicable to the case of spatially inhomogeneous systems:

$$\frac{1}{i\hbar}[\rho_{\mathbf{k}}, \dot{\rho}_{-\mathbf{l}}] = \frac{e}{m}(\mathbf{l} \cdot \mathbf{k})\rho_{\mathbf{k}-\mathbf{l}}, \tag{33}$$

where $\rho_{\mathbf{k}-\mathbf{l}} = \frac{e}{V}\sum_{\mathbf{q},\sigma} a^{+}_{\mathbf{q}-(\mathbf{k}-\mathbf{l}),\sigma} a_{\mathbf{q},\sigma}$. This sum rule reduces to the longitudal sum rule of Refs. [120] for spatially homogeneous systems when **l**=**k**,

$$\frac{1}{i\hbar}[\rho_{\mathbf{k}}, \dot{\rho}_{-\mathbf{k}}] = \frac{Ne}{mV}k^2, \tag{34}$$

with N being the charge carrier number operator, $N = \sum_{\mathbf{q},\sigma} a^{+}_{\mathbf{q},\sigma} a_{\mathbf{q},\sigma}$, and the summations running over the wave vectors **q** and z-components of the spin, σ.

2.2 The Charge Conservation Equation in Terms of the Electric Field Intensity

Using Eqs. (32), (28), and (29), one can transform Eq. (30) to the form:

$$\begin{aligned}\rho(\mathbf{k}, \omega) = e\sum_{\mathbf{l}} n_0(\mathbf{k})\delta(\omega)\delta_{kl} + \frac{i}{c}\sum_{\mathbf{l}}\frac{1}{l^2} << \rho_{\mathbf{k}} \mid \dot{\rho}_{-\mathbf{l}} >>_{0,\omega} (\mathbf{l}\cdot\mathbf{A}(\mathbf{l}, \omega))\\ + \sum_{\mathbf{l}} << \rho_{\mathbf{k}} \mid \rho_{-\mathbf{l}} >>_{0,\omega} \varphi(\mathbf{l}, \omega).\end{aligned} \tag{35}$$

Recovering the Fourier image of $<< \rho_{\mathbf{k}} \mid \dot{\rho}_{-\mathbf{l}} >>_{0,\omega}$, $<< \rho_{\mathbf{k}} \mid \dot{\rho}_{-\mathbf{l}} >>_{0,\omega} = \int_{-\infty}^{\infty} dt\, e^{i\omega t} << \rho_{\mathbf{k}} \mid \dot{\rho}_{-\mathbf{l}}(t) >>_0$, integrating by parts in the r.h.s. of this equation, and noticing that $\frac{1}{i\hbar} < [\rho_{\mathbf{k}}, \rho_{-\mathbf{l}}] >_0 = 0$, one can prove that

$$<< \rho_{\mathbf{k}} \mid \dot{\rho}_{-\mathbf{l}} >>_{0,\omega} = i\omega << \rho_{\mathbf{k}} \mid \rho_{-\mathbf{l}} >>_{0,\omega}\,. \tag{36}$$

Substituting this result in Eq. (35), one recovers the charge conservation equation in the form:

$$\begin{aligned}\rho(\mathbf{k}, \omega) &= e\sum_{\mathbf{l}} n_0(\mathbf{k})\delta(\omega)\delta_{\mathbf{kl}}\\ &+ \sum_{\mathbf{l}}\frac{1}{l^2} << \rho_{\mathbf{k}} \mid \rho_{-\mathbf{l}} >>_{0,\omega} \left\{-\frac{\omega}{c}(\mathbf{l}\cdot\mathbf{A}(\mathbf{l}, \omega)) + l^2\varphi(\mathbf{l}, \omega)\right\}.\end{aligned} \tag{37}$$

The term in the curly brackets in the r.h.s. of Eq. (37) can be easily found using the fact [121] that induced charges screen only the longitudal component of the electrical field. In terms of the space–time Fourier transforms, this can be written as $\mathbf{l} \cdot \mathbf{D}(\mathbf{l}, \omega) = \mathbf{l} \cdot \mathbf{E}(\mathbf{l}, \omega)$, where $\mathbf{E}(\mathbf{l}, \omega)$ is the space–time Fourier transform of the electric field $\mathbf{E}(\mathbf{r},t)$ from Eq. (1) and $\mathbf{D}(\mathbf{l}, \omega)$ is the induced electric field. Thus, applying the space–time Fourier transformations to Eq. (1) and finding the inner product $\mathbf{l} \cdot \mathbf{E}(\mathbf{l}, \omega)$, one derives:

$$i\mathbf{l} \cdot \mathbf{D}(\mathbf{l}, \omega) = -\frac{\omega}{c}(\mathbf{l} \cdot \mathbf{A}(\mathbf{l}, \omega)) + l^2 \varphi(\mathbf{l}, \omega). \tag{38}$$

This result allows rewriting Eq. (37) in the form that does not contain explicitly the field potentials:

$$\begin{aligned} \rho(\mathbf{k}, \omega) &= e \sum_{\mathbf{l}} n_0(\mathbf{k}) \delta(\omega) \delta_{\mathbf{kl}} \\ &+ \sum_{\mathbf{l}} \frac{i}{l^2} << \rho_{\mathbf{k}} \mid \rho_{-\mathbf{l}} >>_{0,\omega} \mathbf{l} \cdot \mathbf{D}(\mathbf{l}, \omega). \end{aligned} \tag{39}$$

This equation generalizes the corresponding Zubarev's result to the inhomogeneous system case, and reduces to Zubarev's charge conservation equation upon consideration that for a spatially homogeneous system $<< \rho_{\mathbf{k}} \mid \rho_{-\mathbf{l}} >>_{0,\omega} = << \rho_{\mathbf{k}} \mid \rho_{-\mathbf{k}} >>_{0,\omega} \delta_{kl}$.

2.2.1 The Polarization Vector and the Tensor of the Dielectric Susceptibility

Following Nozieres and Pines, the induced field intensity $\mathbf{D}(\mathbf{r},t)$ linear in the external field $\mathbf{E}(\mathbf{k},\omega)$ can be written in a general form that satisfies the causality condition:

$$\mathbf{D}(\mathbf{r}, t) = \int_0^{\infty} dt' \int d\mathbf{r}' [\delta(\mathbf{r} - \mathbf{r}')\delta(t - t')\mathbf{I} + \mathbf{F}(\mathbf{r}', t')] \cdot \mathbf{E}(\mathbf{r} - \mathbf{r}', t - t'), \tag{40}$$

where $\mathbf{F}(\mathbf{r}', t')$ is a (unknown) tensor of the second order, and the dummy variables of integration $\mathbf{r}'$ and t' run over the space and time domains, respectively. Introducing formally the (unknown) second order Cartesian tensor of the dielectric susceptibility $\boldsymbol{\varepsilon}(\mathbf{r}', t') = [\delta(\mathbf{r} - \mathbf{r}')\delta(t - t')\mathbf{I} + \mathbf{F}(\mathbf{r}', t')]$ and applying the space–time Fourier transformations to Eq. (40) one can derive a simple correlation between the Fourier transforms of the induced and applied electrical fields,

$$\mathbf{D}(\mathbf{l},\omega) = \boldsymbol{\varepsilon}(\mathbf{l},\omega) \cdot \mathbf{E}(\mathbf{l},\omega), \tag{41}$$

that holds for any physical system in a weak electromagnetic field. [In this equation $\boldsymbol{\varepsilon}(\mathbf{l}, \omega)$ is the Fourier transform of the dielectric susceptibility tensor.] The difference between the fields $\mathbf{D}(\mathbf{r},t)$ and $\mathbf{E}(\mathbf{r},t)$ at a position $\mathbf{r}$ at a time t is characterized by the polarization vector $\mathbf{P}(\mathbf{r}, t)$,

$$\mathbf{P}(\mathbf{r}, t) = \frac{1}{4\pi} [\ \mathbf{D}(\mathbf{r}, t) - \mathbf{E}(\mathbf{r}, t)\ ]. \tag{42}$$

The induced charge density $\rho_{ind}(\mathbf{r}, t) = -\nabla_{\mathbf{r}}.\mathbf{P}(\mathbf{r}, t)$ is defined by this polarization vector, so that the Fourier transform $\rho_{ind}(\mathbf{k}, \omega)$ is

$$\rho_{ind}(\mathbf{k}, \omega) = -\frac{i}{4\pi}\mathbf{k} \cdot [\ \mathbf{D}(\mathbf{k}, \omega) - \mathbf{E}(\mathbf{k}, \omega)\ ]\,. \tag{43}$$

This induced charge density can be immediately found from Eq. (39):

$$\begin{aligned}\rho_{ind}(\mathbf{k}, \omega) &= \rho(\mathbf{k}, \omega) - e\sum_{\mathbf{l}} n_0(\mathbf{k})\delta(\omega)\delta_{\mathbf{kl}} \\ &= \sum_{\mathbf{l}} \frac{i}{l^2} << \rho_{\mathbf{k}} \mid \rho_{-\mathbf{l}} >>_{0,\omega}\ \mathbf{D}(\mathbf{l}, \omega) \cdot \mathbf{l}.\end{aligned} \tag{44}$$

Combining Eqs. (41), (43) and (44), and using considerations similar to those that have lead to explicit expressions for the generalized scalar susceptibilities (22) and (26), one can derive the dielectric susceptibility tensor in terms of the TTGFs $<< \rho_{\mathbf{k}} \mid \rho_{-\mathbf{l}} >>_{0,\omega}$. Thus, the linear (in vectors **k** and **l**) contribution to the tensor $\boldsymbol{\varepsilon}(\mathbf{l}, \omega)$ that is of the major interest here can be written as follows:

$$\boldsymbol{\varepsilon}(\mathbf{k}, \omega) = \sum_{\mathbf{l}} \left\{ \varepsilon^{lon}(\mathbf{k}, \mathbf{l}, \omega)\frac{\mathbf{kl}}{kl} + \varepsilon^{tr}(\mathbf{k}, \mathbf{l}, \omega)\left[\frac{(\mathbf{k}\cdot\mathbf{l})}{kl}\mathbf{I} - \frac{\mathbf{kl}}{kl}\right]\right\},$$

where $\varepsilon^{lon}(\mathbf{k}, \mathbf{l}, \omega)$ and $\varepsilon^{tr}(\mathbf{k}, \mathbf{l}, \omega)$ are scalars that may depend only on the absolute values k, l and the inner product $(\mathbf{k}\cdot\mathbf{l})$. In its turn, the field $\mathbf{E}(\mathbf{k},\omega)$ can be decomposed in two contributions that are parallel and orthogonal to the wave vector **k**:

$$\mathbf{E}(\mathbf{k}, \omega) = \frac{\mathbf{k}}{k^2}(\mathbf{k}\cdot\mathbf{E}(\mathbf{k}, \omega)) + \frac{1}{k^2}[\mathbf{k}\times\mathbf{E}(\mathbf{k}, \omega)\times\mathbf{k}]\,.$$

Substituting the above two expressions into Eq. (41) (where **l** is changed to **k**) one can derive the following expression:

$$\begin{aligned}\mathbf{k}\cdot\mathbf{D}(\mathbf{k}, \omega) =& \sum_{\mathbf{l}} \left[\varepsilon^{lon}(\mathbf{k}, \mathbf{l}, \omega) - \varepsilon^{tr}(\mathbf{k}, \mathbf{l}, \omega)\right]\frac{k}{l} \\ &\times \left\{\frac{(\mathbf{l}\cdot\mathbf{k})}{k^2}(\mathbf{k}\cdot\mathbf{E}(\mathbf{k}, \omega)) + \frac{\mathbf{l}\cdot[\mathbf{k}\times\mathbf{E}(\mathbf{k}, \omega)\times\mathbf{k}]}{k^2}\right\} \\ &+ \sum_{\mathbf{l}} \varepsilon^{tr}(\mathbf{k}, \mathbf{l}, \omega)\frac{(\mathbf{l}\cdot\mathbf{k})}{kl}(\mathbf{k}\cdot\mathbf{E}(\mathbf{k}, \omega)).\end{aligned} \tag{45}$$

The induced magnetic field $\mathbf{B}(\mathbf{r}, t)$ is related to the curl of the electric field, $\nabla_{\mathbf{r}} \times \mathbf{E}(\mathbf{r}, t)$, by the Maxwell's equation $\nabla_{\mathbf{r}} \times \mathbf{E}(\mathbf{r}, t) = -\frac{1}{c}\frac{\partial \mathbf{B}(\mathbf{r},t)}{\partial t}$. The corresponding equation for the respective space–time Fourier transforms reads:

$$[i\, \mathbf{k} \times \mathbf{E}(\mathbf{k}, \omega)] = \frac{i\omega}{c}\mathbf{B}(\mathbf{k}, \omega). \tag{46}$$

Therefore, $\mathbf{l} \cdot [\mathbf{k} \times \mathbf{E}(\mathbf{k}, \omega) \times \mathbf{k}] = \frac{\omega}{c}\mathbf{l} \cdot [\mathbf{B}(\mathbf{k}, \omega) \times \mathbf{k}]$. Taking into account an already mentioned fact [121] that only the longitudal component of the electrical field defines the induced charge, one comes to a conclusion that the term proportional to $\frac{\omega}{c}\mathbf{l} \cdot [\mathbf{B}(\mathbf{k}, \omega) \times \mathbf{k}]$ in the r.h.s. of Eq. (45) should be small and can be neglected in the linear approximation, thus leading to the reduced form of Eq. (45):

$$\mathbf{k} \cdot \mathbf{D}(\mathbf{k},\omega) = \varepsilon(\mathbf{k},\omega)(\mathbf{k} \cdot \mathbf{E}(\mathbf{k},\omega))\,, \tag{47}$$

where the scalar dielectric susceptibility $\varepsilon(\mathbf{k},\omega)$ is defined by the "longitudal" component of the tensorial dielectric susceptibility (that remains unknown), $\varepsilon(\mathbf{k}, \omega) = \sum_{\mathbf{l}} \varepsilon^{lon}(\mathbf{k}, \mathbf{l}, \omega)\frac{(\mathbf{l}\cdot\mathbf{k})}{k\,l}$. Therefore, in the linear approximation with regard to the (weak) electromagnetic fields, the problem of determnation of the dielectric susceptibility reduces to determination of a scalar quantity $\varepsilon(\mathbf{k}, \omega)$ (albeit a function of all of the wave vectors).

Substituting the induced intensity of Eq. (47) into Eq. (43) one derives the following expression for the induced charge:

$$\rho_{ind}(\mathbf{k}, \omega) = \frac{1}{4\pi}\left[\frac{1}{\varepsilon(\mathbf{k}, \omega)} - 1\right] i\, \mathbf{k} \cdot \mathbf{D}(\mathbf{k}, \omega). \tag{48}$$

Using this result in Eq. (44) one obtains the desirable explicit expression for the space–time Fourier transform of the scalar dielectric susceptibility in terms of the space–time Fourier transform of the equilibrium/steady state charge density – charge density TTGFs:

$$\varepsilon^{-1}(\mathbf{k}, \omega) = 1 + 4\pi \sum_{\mathbf{l}} \frac{1}{l^2} << \rho_{\mathbf{k}} | \rho_{-\mathbf{l}} >>_{0,\omega}\,. \tag{49}$$

In the case of spatially homogeneous systems where $<< \rho_{\mathbf{k}} \mid \rho_{-\mathbf{l}} >>_{0,\omega} = << \rho_{\mathbf{k}} \mid \rho_{-\mathbf{k}} >>_{0,\omega} \delta_{\mathbf{kl}}$, this expression reduces to its homogeneous system counterpart [120],

$$\varepsilon^{-1}(\mathbf{k}, \omega) = 1 + \frac{4\pi}{l^2} << \rho_{\mathbf{k}} | \rho_{-\mathbf{k}} >>_{0,\omega}\,. \tag{50}$$

Note here, that Eq. (49) is obtained by neglecting the term proportional to $\frac{\omega}{c}\mathbf{l} \cdot [\mathbf{B}(\mathbf{k}, \omega) \times \mathbf{k}]$ in Eq. (45), and thus is an approximate equation even in the framework of the linear response theory developed here, while its homogeneous system counterpart, Eq. (50), is an exact equation of the linear response theory. More general approximations leading to a derivation of explicit expressions for the tensorial dielectric susceptibility of Eq. (41) can be developed by solving the system of Eqs. (41), (43) and (44) without the neglect in Eq. (45) of the term proportional to $\mathbf{l} \cdot [\mathbf{B}(\mathbf{k}, \omega) \times \mathbf{k}]$.

2.2.2 The Charge Density Conservation Equation in Terms of the Field

Using Eq. (49) one can transform Eq. (39) to the form that contains explicitly the electric field intensity $\mathbf{E}(\mathbf{k},\omega)$:

$$\rho(\mathbf{k}, \omega) = e \sum_{\mathbf{l}} n_0(\mathbf{k})\delta(\omega)\delta_{kl} + \sum_{\mathbf{l}} \frac{i}{l^2} \frac{<< \rho_{\mathbf{k}} \mid \rho_{-\mathbf{l}} >>_{0,\omega}}{\left\{1 + 4\pi \sum_{\mathbf{s}} \frac{1}{s^2} << \rho_{\mathbf{l}} \mid \rho_{-\mathbf{s}} >>_{0,\omega}\right\}} \mathbf{l} \cdot \mathbf{E}(\mathbf{l}, \omega). \tag{51}$$

Thus, the Fourier transform of the expectation value of the charge density is expressed in terms of the microscopic charge density–charge density TTGFs of an inhomogeneous system. Due to the use of the approximate Eq. (49) in derivation of Eq. (51), the latter equation is an approximation that holds only in weak fields. In the case of homogeneous systems when $<< \rho_{\mathbf{k}} \mid \rho_{-\mathbf{l}} >>_{0,\omega} = << \rho_{\mathbf{k}} \mid \rho_{-\mathbf{k}} >>_{0,\omega} \delta_{kl}$ Eq. (51) reduces to the charge conservation equation of Refs. [120].

2.3 *The Current Density Conservation Equation*

The TTGFs of microcurrents appearing in Eqs. (22) and (31) can be obtained using Eq. (32):

$$<< \mathbf{k} \cdot \mathbf{j}_{\mathbf{k}} | \mathbf{l} \cdot \mathbf{j}_{-\mathbf{l}} >>_{0,\omega} = << \dot{\rho}_{\mathbf{k}} | \dot{\rho}_{-\mathbf{l}}) >>_{0,\omega}, \tag{52}$$

$$<< \mathbf{k} \cdot \mathbf{j}_{\mathbf{k}} | \rho_{-\mathbf{l}} >>_{0,\omega} = i << \dot{\rho}_{\mathbf{k}} | \rho_{-\mathbf{l}}) >>_{0,\omega} . \tag{53}$$

Substituting Eq. (52) into the r.h.s. of Eq. (22), one can derive the longitudal susceptibility in the form:

$$\chi^{lon}(\mathbf{k}, \mathbf{l}, \omega) = \frac{<< \dot{\rho}_{\mathbf{k}} | \dot{\rho}_{-\mathbf{l}} >>_{0,\omega}}{kl}. \tag{54}$$

Further transformations of the TTGFs in Eqs. (52) and (54) can be achieved by the use of the partial integration procedure similar to that that leads to Eq. (36). Thus,

$$<< \dot{\rho}_{\mathbf{k}} | \dot{\rho}_{-\mathbf{l}} >>_{0,\omega} = \omega^2 << \rho_{\mathbf{k}} | \rho_{-\mathbf{l}} >>_{0,\omega} - \frac{1}{i\hbar} < [\rho_{\mathbf{k}}, \dot{\rho}_{-\mathbf{l}}] >_0, \tag{55}$$

and using the longitudal sum rule (33), one can rewrite Eq. (55) in the form:

$$<< \dot{\rho}_{\mathbf{k}}|\dot{\rho}_{-\mathbf{l}} >>_{0,\omega} = \omega^2 << \rho_{\mathbf{k}}|\rho_{-\mathbf{l}} >>_{0,\omega} - \frac{e^2}{m}(\mathbf{k}\cdot\mathbf{l})\, n_0(\mathbf{k}-\mathbf{l}). \tag{56}$$

Substituting the TTGF of Eq. (56) into Eq. (54) one can obtain the following expression for the space–time Fourier transform of the longitudal susceptibility:

$$\chi^{lon}(\mathbf{k},\mathbf{l},\omega) = \frac{1}{kl}\left\{\omega^2 << \rho_{\mathbf{k}}|\rho_{-\mathbf{l}}) >>_{0,\omega} - \frac{e^2}{m} n_0(\mathbf{k}-\mathbf{l})(\mathbf{k}\cdot\mathbf{l})\right\}. \tag{57}$$

The TTGF $<< \dot{\rho}_{\mathbf{k}}|\rho_{-\mathbf{l}} >>_{0,\omega}$ of Eq. (53) can be obtained by changing the sign of the r.h.s. of Eq. (36) and replacing the wave vectors –**l** by **k** and **k** by –**l**, respectively:

$$<< \dot{\rho}_{\mathbf{k}} \mid \rho_{-\mathbf{l}} >>_{0,\omega} = -i\omega << \rho_{\mathbf{k}} \mid \rho_{-\mathbf{l}} >>_0 . \tag{58}$$

Substituting the TTGFs from Eqs. (53) and (58), and the longitudinal susceptibility of Eq. (57) into Eq. (31) one obtains the following form of the current density conservation equation:

$$\begin{aligned}
\mathbf{j}(\mathbf{k},\omega) = &-\frac{e^2}{mc}\sum_{\mathbf{l}} n_0(\mathbf{k}-\mathbf{l})\mathbf{A}(\mathbf{l},\omega) \\
&-\frac{1}{c}\sum_{\mathbf{l}}\frac{1}{k^2 l^2}\left\{\omega^2 << \rho_{\mathbf{k}}|\rho_{-\mathbf{l}}) >>_{0,\omega} - \frac{e^2}{m} n_0(\mathbf{k}-\mathbf{l})(\mathbf{k}\cdot\mathbf{l})\right\}(\mathbf{l}\cdot\mathbf{A}(\mathbf{l},\omega))\mathbf{k} \\
&+\sum_{\mathbf{l}}\frac{\omega}{k^2} << \rho_{\mathbf{k}}|\rho_{-\mathbf{l}}) >>_{0,\omega}\varphi(\mathbf{l},\omega)\mathbf{k} \\
&-\frac{1}{c}\sum_{\mathbf{l}}\chi^{tr}(\mathbf{k},\mathbf{l},\omega)\left\{\frac{(\mathbf{k}\cdot\mathbf{l})}{kl}\mathbf{A}(\mathbf{l},\omega) - \frac{1}{kl}(\mathbf{l}\cdot\mathbf{A}(\mathbf{l},\omega))\mathbf{k}\right\}.
\end{aligned} \tag{59}$$

Using the space–time Fourier transform of the continuity equation (97) derived in Appendix, and Eqs. (37) and (59), one can prove the following equality:

$$\begin{aligned}
&\frac{1}{c}\chi^{tr}(\mathbf{k},\mathbf{l},\omega)\left\{\frac{(\mathbf{k}\cdot\mathbf{l})}{kl} + \frac{e^2}{m}n_0(\mathbf{k}-\mathbf{l})\right\}(\mathbf{k}\cdot\mathbf{A}(\mathbf{l},\omega)) \\
&\quad = \frac{1}{c}\left\{\chi^{tr}(\mathbf{k},\mathbf{l},\omega) + \frac{e^2}{m}\frac{(\mathbf{k}\cdot\mathbf{l})}{kl}n_0(\mathbf{k}-\mathbf{l})\right\}\frac{k^2}{kl}(\mathbf{l}\cdot\mathbf{A}(\mathbf{l},\omega)).
\end{aligned}$$

Using this equality and expressing the inner product $(\mathbf{l}\cdot\mathbf{A}(\mathbf{l},\omega))$ in terms of $(\mathbf{l}\cdot\mathbf{D}(\mathbf{l},\omega))$ from Eq. (38), the current density conservation equation (59) becomes:

$$\mathbf{j}(\mathbf{k},\omega) = \sum_{\mathbf{l}} \frac{i\omega}{k^2 l^2} << \rho_{\mathbf{k}}|\rho_{-\mathbf{l}}) >>_{0,\omega} (\mathbf{l}\cdot\mathbf{D}(\mathbf{l},\omega))\mathbf{k}$$
$$-\frac{1}{c}\sum_{\mathbf{l}}\left\{\left[\chi^{tr}(\mathbf{k},\mathbf{l},\omega)\frac{(\mathbf{k}\cdot\mathbf{l})}{kl} + \frac{e^2}{m}n_0(\mathbf{k}-\mathbf{l})\right]\right. \tag{60}$$
$$\left.\times\left[\mathbf{A}(\mathbf{l},\omega) - \frac{1}{k^2}(\mathbf{k}\cdot\mathbf{A}(\mathbf{l},\omega))\mathbf{k}\right]\right\}.$$

The transversal component of the current density is defined by the induced magnetic field $\mathbf{B}(\mathbf{r},t)$ that in its turn is defined by the transversal component of the vector potential $\mathbf{A}(\mathbf{r},t)$ via the second equation of the system (1). The corresponding equation for the space–time respective Fourier transforms reads:

$$\mathbf{B}(\mathbf{k},\omega) = [i\mathbf{k}\times\mathbf{A}(\mathbf{k},\omega)]. \tag{61}$$

Note, that the space–time Fourier transform of the induced magnetic field $\mathbf{B}(\mathbf{r},t)$ is related to the corresponding space–time Fourier transform of the curl of the electric field by the Maxwell's equation (46). The vector

$$\mathbf{B}_{\mathbf{k}}(\mathbf{l},\omega) = \frac{1}{k^2}[i\mathbf{k}\times i\mathbf{k}\times\mathbf{A}(\mathbf{l},\omega)] = \mathbf{A}(\mathbf{l},\omega) - \frac{(\mathbf{k}\cdot\mathbf{A}(\mathbf{l},\omega))}{k^2}\mathbf{k} \tag{62}$$

appearing in the r.h.s. of Eq. (60) defines the component of the vector potential which is orthogonal to the wave vector k. This vector can be written in the form:

$$\frac{1}{k^2}[i\mathbf{l}\times\mathbf{B}(\mathbf{l},\omega)] + \frac{(\mathbf{l}\cdot\mathbf{A}(\mathbf{l},\omega))}{l^2}\mathbf{l} - \frac{(\mathbf{k}\cdot\mathbf{A}(\mathbf{l},\omega))}{k^2}\mathbf{k},$$

where Eq. (61) has been used. The inner product of this vector with the wave vector $\mathbf{k}$ is zero, from which one can derive the following relation:

$$\mathbf{k}\cdot\mathbf{A}(\mathbf{l},\omega) = \frac{1}{l^2}\mathbf{k}\cdot[i\mathbf{l}\times\mathbf{B}(\mathbf{l},\omega)] + \frac{(\mathbf{l}\cdot\mathbf{k})}{l^2}(\mathbf{l}\cdot\mathbf{A}(\mathbf{l},\omega)). \tag{63}$$

Using this equality and Maxwell's Eq. (46) in conjunction with Eq. (60), one can obtain a representation of the current conservation equation in the form:

$$\mathbf{j}(\mathbf{k},\omega) = i\omega\sum_{\mathbf{l}}\frac{<< \rho_{\mathbf{k}} \mid \rho_{-\mathbf{l}} >>_{0,\omega}}{k^2 l^2}(\mathbf{l}\cdot\mathbf{D}(\mathbf{l},\omega))\mathbf{k}$$
$$-\frac{1}{c}\sum_{\mathbf{l}}\frac{1}{l^2}\left[\chi^{tr}(\mathbf{k},\mathbf{l},\omega)\frac{(\mathbf{k}\cdot\mathbf{l})}{kl} + \frac{e^2}{m}n_0(\mathbf{k}-\mathbf{l})\right] \tag{64}$$
$$\times\left[\mathbf{I} - \frac{\mathbf{kk}}{k^2}\right]\cdot\{[i\mathbf{l}\times\mathbf{B}(\mathbf{l},\omega)] - \mathbf{l}(\mathbf{l}\cdot\mathbf{A}(\mathbf{l},\omega))\}.$$

In the case of a homogeneous system, the sums over $\mathbf{l}$ in the r.h.s. reduce to one term ($\mathbf{l}$=$\mathbf{k}$) each, so Eq. (64) reduces to that of Refs. [120].

Further progress toward expressing Eq. (64) in terms of the intensity $\mathbf{E}(\mathbf{k},\omega)$ is less straightforward. In the homogeneous system case, the counterpart of Eq. (64) can be immediately transformed to an expression that does not include the vector potential $\mathbf{A}(\mathbf{k},\omega)$ explicitly. For the inhomogeneous systems, however, the presence of the summation over $\mathbf{l}$ in the second term in the r.h.s. of Eq. (64) does not allow similar exclusion of the vector potential without an additional condition. A general form of such a condition is the Lorentz anzatz widely used in electrodynamics to ascertain that the Maxwell's equations in vacuum are gauge-invariant:

$$i\mathbf{l}\cdot\mathbf{A}(\mathbf{l},\omega)=\frac{i\omega}{c}\varphi(\mathbf{l},\omega). \tag{65}$$

Using this anzatz in Eq. (38) one can express the longitudal component of the vector potential in terms of the longitudal component of the intensity $\mathbf{D}(\mathbf{l},\omega)$:

$$i\mathbf{l}\cdot\mathbf{D}(\mathbf{l},\omega)=\frac{i\omega}{c}\left\{1-\frac{c^2l^2}{\omega^2}\right\}i\mathbf{l}\cdot\mathbf{A}(\mathbf{l},\omega).$$

Recovering $\mathbf{l}\cdot\mathbf{A}(\mathbf{l},\omega)$ from this equation and substituting it in Eq. (64), one obtains the current conservation equation in the form that does not contain explicitly the field potentials:

$$\begin{aligned}\mathbf{j}(\mathbf{k},\omega)=&\sum_{\mathbf{l}}\frac{i\omega}{k^2l^2}<<\rho_{\mathbf{k}}\mid\rho_{-\mathbf{l}}>>_{0,\omega}(\mathbf{l}\cdot\mathbf{D}(\mathbf{l},\omega))\mathbf{k}\\&-\sum_{\mathbf{l}}\frac{1}{l^2}\left[\chi^{tr}(\mathbf{k},\mathbf{l},\omega)\frac{(\mathbf{k}\cdot\mathbf{l})}{kl}+\frac{e^2}{m}n_0(\mathbf{k}-\mathbf{l})\right]\\&\times\left[\mathbf{I}-\frac{\mathbf{kk}}{k^2}\right]\cdot\left\{\frac{1}{c}[i\mathbf{l}\times\mathbf{B}(\mathbf{l},\omega)]+\frac{i\mathbf{l}}{\omega\left(1-\frac{c^2l^2}{\omega^2}\right)}(\mathbf{l}\cdot\mathbf{D}(\mathbf{l},\omega))\right\}.\end{aligned} \tag{66}$$

2.4 The Longitudal Conductivity

Using Maxwell's equation (46) and Eqs. (47) and (49) one can express the current density of Eq. (66) in the form that contains explicitly only the intensity $\mathbf{E}(\mathbf{l},\omega)$:

$$\begin{aligned}\mathbf{j}(\mathbf{k},\omega)=&\frac{i\omega}{4\pi}[1-\varepsilon(\mathbf{k},\omega)]\frac{\mathbf{kk}}{k^2}\cdot\mathbf{E}(\mathbf{l},\omega)\\&-\sum_{\mathbf{l}}\frac{i}{\omega l^2}\left[\chi^{tr}(\mathbf{k},\mathbf{l},\omega)\frac{(\mathbf{k}\cdot\mathbf{l})}{kl}+\frac{e^2}{m}n_0(\mathbf{k}-\mathbf{l})\right]\\&\times\left[\mathbf{I}-\frac{\mathbf{kk}}{k^2}\right]\cdot\left\{[\mathbf{l}\times\mathbf{l}\times\mathbf{E}(\mathbf{l},\omega)]+\frac{\varepsilon(\mathbf{l},\omega)}{\left(1-\frac{c^2l^2}{\omega^2}\right)}\mathbf{l}(\mathbf{l}\cdot\mathbf{E}(\mathbf{l},\omega))\right\}.\end{aligned} \tag{67}$$

This allows straightforward identification of the diagonal tensor of longitudal conductivity in terms of the dielectric susceptibility,

$$\sigma^{lon}(\mathbf{k},\omega)\frac{\mathbf{kk}}{k^2} = \frac{i\omega}{4\pi}\{\,1-\varepsilon(\mathbf{k},\omega)\,\}\frac{\mathbf{kk}}{k^2}, \tag{68}$$

or explicitly, in terms of the equilibrium charge–charge TTGFs using Eq. (49),

$$\sigma^{lon}(\mathbf{k},\omega) = \frac{i\omega\sum_{\mathbf{l}}\frac{1}{l^2} << \rho_{\mathbf{k}}|\rho_{-\mathbf{l}} >>_{0,\omega}}{1+4\pi\sum_{\mathbf{l}}\frac{1}{l^2} << \rho_{\mathbf{k}}|\rho_{-\mathbf{l}} >>_{0,\omega}}. \tag{69}$$

A possibility to establish closed explicit expressions [such as Eq. (69)] for the tensor of longitudal conductivity depends entirely upon the availability of a closed explicit expression for the dielectric susceptibility tensor. As discussed in Section 2.2.1, Eq. (49), used in derivation of Eq. (69), is an approximate equation that holds only in weak electromagnetic fields. An immediate consequence of this approximation is simplicity and diagonality of the tensor of longitudal conductivity. Together with the linearization of the Hamiltonian (2) to the form (12), the above approximation of the dielectric susceptibility tensor leads to the longitudal conductivity (69) that is defined completely by the equilibrium microscopic charge density–charge density TTGFs. Other approximations for the dielectric susceptibility tensor (still linear in the external fields) can be developed using Eqs. (41), (43), and (44). Such approximations will lead to more sophisticated (and more complex) explicit expressions for the longitudal conductivity tensor that will also become nondiagonal. Development of such approximations is postponed to future publications.

Despite the simplicity of Eq. (49), the dielectric susceptibility tensor (49) and the tensor of the longitudal conductivity (69) still retain their dependence on the entire set of the microscopic charge density TTGFs. This dependence reflects spatial inhomogeneity of the system. For the homogeneous systems, only one term with **l**=**k** in each of the sums over **l** in Eqs. (49) and (69) survives, reducing these expressions to the form derived and discussed in Refs. [120].

2.5 *Transversal Conductivity*

2.5.1 The Induced Magnetic Moment and Magnetic Susceptibility

For the majority of experimentally studied cases, the generalized susceptibility $\chi^{tr}(\mathbf{k},\omega)$ of spatially homogeneous systems depends only on the wave vector **k** and is negative, so that the quantity $\{\chi^{tr}(\mathbf{k},\omega) + \frac{e^2}{m}n_0\,\}$ is small and describes only an insignificant diamagnetic effect derived originally by Landau. [This situation changes only in superconducting systems due to existence of a gap in the spectrum of the elementary excitations.] Similarly, one can expect that at standard conditions and in weak electromagnetic fields, for inhomogeneous systems with small magnetization, the quantity $\{\chi^{tr}(\mathbf{k}, \mathbf{l}, \omega)\frac{(\mathbf{k}\cdot\mathbf{l})}{kl} + \frac{e^2}{m}n_0(\mathbf{k}-\mathbf{l})\}$ also remains small, albeit nonnegligible. Thus, in the considered linear approximation with regard to the fields, the only manifestation of the spatial inhomogeneity effects is a dependence of this quantity on both **k** and **l** wave vectors.

Similar to electrical properties of a system of particles in an external field that are described by the polarization vector, magnetic properties of such a system in a magnetic field are characterized by the vector of induced magnetic moment $\mathbf{M}(\mathbf{r},t)$ whose space–time Fourier transform is:

$$\mathbf{M}(\mathbf{k}, \omega) = \frac{1}{4\pi}[\mathbf{B}(\mathbf{k}, \omega) - \mathbf{H}(\mathbf{k}, \omega)]. \tag{70}$$

The magnetic moment defines the transversal contribution to the current density described by the second term in the r.h.s. of Eq. (67),

$$\mathbf{j}^{tr}(\mathbf{k}, \omega) = c[i\mathbf{k} \times \mathbf{M}(\mathbf{k}, \omega)]. \tag{71}$$

Upon a consideration similar to that leading to Eq. (41), one can show that due to the causality condition, in the linear approximation

$$\mathbf{B}(\mathbf{k}, \omega) = \mu(\mathbf{k}, \omega) \cdot \mathbf{H}(\mathbf{k}, \omega), \tag{72}$$

where $\boldsymbol{\mu}(\mathbf{k},\omega)$ is the second-rank tensor of magnetic susceptibility. Using arguments similar to those described in Section 2.1.1, in the linear approximation, this tensor can be transformed to the form:

$$\mu(\mathbf{k}, \omega) = \sum_{\mathbf{l}} \left\{ \mu^{lon}(\mathbf{k}, \mathbf{l}, \omega)\frac{\mathbf{kl}}{kl} + \mu^{tr}(\mathbf{k}, \mathbf{l}, \omega)\left[\frac{(\mathbf{k}\cdot\mathbf{l})}{kl}\mathbf{I} - \frac{\mathbf{kl}}{kl}\right]\right\}, \tag{73}$$

where $\mu^{lon}(\mathbf{k}, \mathbf{l}, \omega)$ and $\mu^{tr}(\mathbf{k}, \mathbf{l}, \omega)$ are scalar longitudinal and transversal magnetic susceptibilities that have to be determined and that depend only on the absolute values k, l, and the inner product $(\mathbf{k}\cdot\mathbf{l})$. Substituting Eq. (73) in Eq. (72), one can rewrite Eq. (72) in the form:

$$\begin{aligned}\mathbf{B}(\mathbf{k}, \omega) = &\sum_{\mathbf{l}} \{\mu^{lon}(\mathbf{k}, \mathbf{l}, \omega) - \mu^{tr}(\mathbf{k}, \mathbf{l}, \omega)\}\frac{\mathbf{k}}{kl}(\mathbf{l}\cdot \mathbf{H}(\mathbf{k}, \omega)) \\ &+ \sum_{\mathbf{l}} \mu^{tr}(\mathbf{k}, \mathbf{l}, \omega)\frac{\mathbf{k}\cdot\mathbf{l}}{kl}\mathbf{H}(\mathbf{k}, \omega).\end{aligned} \tag{74}$$

From this equation one obtains:

$$[i\mathbf{k} \times \mathbf{H}(\mathbf{k}, \omega)] = \mu^{-1}(\mathbf{k}, \omega)\,[i\mathbf{k} \times \mathbf{B}(\mathbf{k}, \omega)], \tag{75}$$

where the scalar magnetic susceptibility $\mu(\mathbf{k}, \omega)$ is

$$\mu(\mathbf{k}, \omega) = \sum_{\mathbf{l}} \mu^{tr}(\mathbf{k}, \mathbf{l}, \omega) \frac{\mathbf{k} \cdot \mathbf{l}}{k\, l}. \tag{76}$$

For the following analysis, it is convenient to use the current density conservation equation in the form (60). The second term in the r.h.s. of Eq. (60) (that is, the space–time Fourier transform of the transversal component of the expectation value of the induced current density) can be written in terms of the vector potential $\mathbf{A}(\mathbf{k}, \omega)$, and then expressed in terms of the magnetic susceptibility $\mu(\mathbf{k}, \omega)$ as follows:

$$\begin{aligned} \mathbf{j}^{tr}(\mathbf{k}, \omega) &= \sum_{\mathbf{l}} \frac{i}{ck^2} \left[\chi^{tr}(\mathbf{k}, \mathbf{l}, \omega) \frac{(\mathbf{k} \cdot \mathbf{l})}{k l} + \frac{e^2}{m} n_0(\mathbf{k} - \mathbf{l}) \right] [\mathbf{k} \times \mathbf{k} \times \mathbf{A}(\mathbf{l}, \omega)] \\ &= \frac{c}{4\pi} \left[1 - \mu^{-1}(\mathbf{k}, \omega) \right] [i\mathbf{k} \times \mathbf{B}(\mathbf{l}, \omega)] . \end{aligned} \tag{77}$$

The first line in the r.h.s. of Eq. (77) is obtained using Eqs. (71), (75), and (76), while the second line also uses Eq. (70). Comparing these two lines and using Eq. (61) again, one can establish the following explicit expression for the magnetic susceptibility $\mu(\mathbf{k}, \omega)$:

$$\mu^{-1}(\mathbf{k}, \omega) = 1 + \frac{4\pi}{c^2 k^2} \sum_{\mathbf{l}} \left[\chi^{tr}(\mathbf{k}, \mathbf{l}, \omega) \frac{(\mathbf{k} \cdot \mathbf{l})}{k l} + \frac{e^2}{m} n_0(\mathbf{k} - \mathbf{l}) \right]. \tag{78}$$

2.5.2 Explicit Expression for the Transversal Conductivity

Using the second line in the r.h.s. of Eq. (77) to replace the second term in the r.h.s. of Eq. (67) and using Eq. (46), one can rewrite Eq. (67) in the form:

$$\begin{aligned} \mathbf{j}(\mathbf{k}, \omega) = &\frac{i\omega}{4\pi} \left[1 - \varepsilon(\mathbf{k}, \omega) \right] \frac{\mathbf{k}}{k^2} \mathbf{k} \cdot \mathbf{E}(\mathbf{k}, \omega) - \frac{c^2 k^2}{4\pi i \omega} \left[1 - \mu^{-1}(\mathbf{k}, \omega) \right] \\ &\frac{1}{k^2} \left[\mathbf{k} \times \mathbf{k} \times \mathbf{E}(\mathbf{k}, \omega) \right] . \end{aligned} \tag{79}$$

From this equation, $\sigma^{tr}(\mathbf{k}, \omega)$ can be identified as the factor with $\frac{1}{k^2} [\mathbf{k} \times \mathbf{k} \times \mathbf{E}(\mathbf{k}, \omega)]$, and thus related to the space–time Fourier transform of the scalar magnetic susceptibility,

$$\sigma^{tr}(\mathbf{k}, \omega) = \frac{c^2 k^2}{4\pi i \omega} \{ 1 - \mu^{-1}(\mathbf{k}, \omega) \} . \tag{80}$$

As follows from Eq. (80), the transversal component of the conductivity tensor is completely defined by the equilibrium/steady state charge density and the microcurrent density TTGFs appearing in the explicit expression (78) for the magnetic susceptibility $\mu(\mathbf{k}, \omega)$. With this result, one can reduce Eq. (67) to the *form* that formally coincides with that specific to the homogeneous system case

$$\begin{aligned}\mathbf{j}(\mathbf{k}, \omega) &= \sigma^{lon}(\mathbf{k}, \omega)\frac{\mathbf{kk}}{k^2} \cdot \mathbf{E}(\mathbf{k}, \omega) \\ &\quad - \sigma^{tr}(\mathbf{k}, \omega)\frac{1}{k^2}[\mathbf{k} \times \mathbf{k} \times \mathbf{E}(\mathbf{k}, \omega)].\end{aligned} \tag{81}$$

However, Eqs. (68) and (80) for the scalar longitudal and transversal conductivities, respectively, appearing in the conservation equation (81) for the linear quantum current density of inhomogeneous systems differ significantly from their homogeneous case counterparts. In particular, the space–time Fourier transforms of both quantum conductivities now depend upon the entire (infinite) set of the space–time Fourier transforms of the TTGFs of the microscopic charge and current densities via Eqs. (69) and (78). Of course, simplicity of Eq. (81) originates from the use of the linear approximation that is justified by weakness of the external electromagnetic field.

2.5.3 Quantum Conductivity of Homogeneous Systems

Noticing that in the case of homogeneous system the sums over **l** in the r.h.s. of Eq. (69) reduce to one term each (**l**=**k**), one can immediately derive an expression for the linear contribution to the space–time Fourier transform of the longitudal quantum conductivity of homogeneous systems in terms of the space–time Fourier transform of the equilibrium/steady state microscopic charge density–charge density TTGF:

$$\sigma^{lon}(\mathbf{k}, \omega) = \frac{\frac{i\omega}{k^2} << \rho_{\mathbf{k}}|\rho_{-\mathbf{k}} >>_{0,\omega}}{1 + 4\pi\frac{1}{k^2} << \rho_{\mathbf{k}}|\rho_{-\mathbf{k}} >>_{0,\omega}}. \tag{82}$$

An explicit expression for the linear contribution to the space–time Fourier transform of the transversal conductivity in terms of the space–time Fourier transform of the equilibrium/steady state microcurrent–microcurrent TTGF can be obtained in a similar manner from Eqs. (26), (76), (78), and (80):

$$\sigma^{tr}(\mathbf{k}, \omega) = \frac{1}{i\omega}\frac{\frac{1}{2k^2} << [\mathbf{k} \times \mathbf{j}_{\mathbf{k}}] \cdot [\mathbf{k} \times \mathbf{j}_{-\mathbf{k}}] >>_{0,\omega} + \frac{e^2}{m}n_0}{1 + \frac{4\pi}{c^2k^2}\left\{\frac{1}{2k^2} << [\mathbf{k} \times \mathbf{j}_{\mathbf{k}}] \cdot [\mathbf{k} \times \mathbf{j}_{-\mathbf{k}}] >>_{0,\omega} + \frac{e^2}{m}n_0\right\}}. \tag{83}$$

Expressions (82) and (83) coincide with those derived in Refs. [120].

2.6 Calculations of the Equilibrium TTGFs

To calculate the space–time Fourier transforms of the dielectric susceptibility, Eq. (49), the generalized susceptibility, Eqs. (54) and (57), the longitudal and transversal conductivities, Eqs. (69) and (80), and the magnetic susceptibility, Eq. (78), one has to calculate the equilibrium/steady state microscopic charge and current density TTGFs appearing in these expressions. While no diagrammatic technique exists for these TTGFs, one can use a method established by Zubarev and Tserkovnikov (ZT) [122]. In the framework of this method one has to choose a set of operators that would serve to form and calculate physically meaningful, or irreducible, projections of the TTGFs. In this section such calculations are demonstrated for the case of the Fourier transform of the equilibrium charge density TTGF, $\langle\langle\rho_{\mathbf{k}}|\ \rho_{-\mathbf{l}}\rangle\rangle$. In this case, a convenient and sufficient set of operators on which such projections are realized consists of the Fourier transforms of the charge density, $\rho_{\mathbf{k}}$, and its time derivative, $\dot{\rho}_{-\mathbf{l}}$. Using this set, one can obtain from Eq. (2.15) of Ref. [122] the following relations:

$$((\rho_{\mathbf{k}},\ \rho_{-\mathbf{l}}))_{0,\omega} = \frac{(\rho_{\mathbf{k}},\rho_{-\mathbf{l}})_0}{\hbar\omega - \{(i\hbar\dot{\rho}_{\mathbf{k}},\rho_{-\mathbf{l}})_0 - ((i\hbar\dot{\rho}_{\mathbf{k}},-i\hbar\dot{\rho}_{-\mathbf{l}}))_{1,\omega}\}\,(i\hbar\rho_{\mathbf{k}},\rho_{-\mathbf{l}})_0^{-1}}, \tag{84}$$

$$((i\hbar\dot{\rho}_{\mathbf{k}},\ i\hbar\dot{\rho}_{-\mathbf{l}}))_{1,\omega} = \frac{(i\hbar\dot{\rho}_{\mathbf{k}},-i\hbar\dot{\rho}_{-\mathbf{l}})_1}{\hbar\omega - \frac{((i\hbar)^2\ddot{\rho}_{\mathbf{k}},-i\hbar\dot{\rho}_{-\mathbf{l}})_1}{(i\hbar\dot{\rho}_{\mathbf{k}},-i\hbar\dot{\rho}_{-\mathbf{l}})_1} + \frac{(i\hbar\dot{\rho}_{\mathbf{k}},\rho_{-\mathbf{l}})_0}{(\rho_{\mathbf{k}},\rho_{-\mathbf{l}})_0} - \frac{(((i\hbar)^2\ddot{\rho}_{\mathbf{k}},(i\hbar)^2\ddot{\rho}_{-\mathbf{l}}))_{2,\omega}}{(i\hbar\dot{\rho}_{\mathbf{k}},-i\hbar\dot{\rho}_{-\mathbf{l}})_1}}, \tag{85}$$

that couple the space–time Fourier transforms of the Kubo charge density relaxation function $((\rho_{\mathbf{k}},\ \rho_{-\mathbf{l}}))_{0,\omega}$ to the higher order Kubo charge density relaxation functions, $((i\hbar\dot{\rho}_{\mathbf{k}},\ i\hbar\dot{\rho}_{-\mathbf{l}}))_{1,\omega}$ and $\left(((i\hbar)^2\ddot{\rho}_{\mathbf{k}},\ (i\hbar)^2\ddot{\rho}_{-\mathbf{l}})\right)_{2,\omega}$. Subscripts 1 and 2 here denote the irreducible parts of these relaxation functions, their initial values (at ω=0), and the initial values of their time derivatives in the sense of Eq. (2.12) of [122]. The inner products within parentheses in Eq. (84), such as $(\rho_{\mathbf{k}},\ \rho_{-\mathbf{l}})_0$, are the initial values of the corresponding relaxation functions, such as $((\rho_{\mathbf{k}},\ \rho_{-\mathbf{l}}))_{0,\omega}$. These relaxation functions are related to the space–time Fourier transforms of the corresponding TTGFs by Eq. (1.15) of [122],

$$((\rho_{\mathbf{k}},\ \rho_{-\mathbf{l}}))_{0,\omega} \equiv \frac{1}{\hbar\omega}\{\langle\langle\rho_{\mathbf{k}}|\ \rho_{-\mathbf{l}}\rangle\rangle_{0,\omega} + (\rho_{\mathbf{k}},\ \rho_{-\mathbf{l}})_0\}. \tag{86}$$

Noticing that $(i\hbar\dot{\rho}_{\mathbf{k}},\ \rho_{-\mathbf{l}})_0 = 0$ and $((i\hbar)^2\ddot{\rho}_{\mathbf{k}},\ -i\hbar\dot{\rho}_{-\mathbf{l}})_1 = 0$, and using Eq. (86), one can obtain from Eqs. (84) and (85) the following expressions,

$$\langle\langle\rho_{\mathbf{k}}|\ \rho_{-\mathbf{l}}\rangle\rangle_{0,\omega} = \frac{(i\hbar\dot{\rho}_{\mathbf{k}},i\hbar\dot{\rho}_{-\mathbf{l}})_{1,\omega}}{\hbar\omega - ((i\hbar\dot{\rho}_{\mathbf{k}},-i\hbar\dot{\rho}_{-\mathbf{l}}))_{1,\omega}\,(\rho_{\mathbf{k}},\rho_{-\mathbf{l}})_0^{-1}}, \tag{87}$$

where

$$((i\hbar\dot{\rho}_{\mathbf{k}},i\hbar\dot{\rho}_{-\mathbf{l}}))_{1,\omega} = \frac{(i\hbar\dot{\rho}_{\mathbf{k}},-i\hbar\dot{\rho}_{-\mathbf{l}})_1}{\hbar\omega - \frac{(((i\hbar)^2\ddot{\rho}_{\mathbf{k}},(i\hbar)^2\ddot{\rho}_{-\mathbf{l}}))_{2,\omega}}{(i\hbar\dot{\rho}_{\mathbf{k}},-i\hbar\dot{\rho}_{-\mathbf{l}})_1}}. \tag{88}$$

Using the longitudal sum rule, Eq. (33) of Section 2.1, one can prove that $(i\hbar\dot{\rho}_{\mathbf{k}}, i\hbar\dot{\rho}_{-\mathbf{l}})_1 = \frac{e\hbar^2}{m\sqrt{V}}(\mathbf{l}\cdot\mathbf{k})\langle\rho_{\mathbf{k}-\mathbf{l}}\rangle_0$, where $(\mathbf{l}\cdot\mathbf{k})$ is the inner product of the wave vectors and e and m is the charge and mass of the charge carrier, such as an electron. The relaxation function $\left(\left((i\hbar)^2\ddot{\rho}_{\mathbf{k}}, (i\hbar)^2\ddot{\rho}_{-\mathbf{l}}\right)\right)_{2,\omega}$ in Eq. (88) can be evaluated using Eq. (2.14b) of Ref. [122],

$$\lim_{\hbar\omega\to\infty} \hbar\omega\left(\left((i\hbar)^2\ddot{\rho}_{\mathbf{k}}, (i\hbar)^2\ddot{\rho}_{-\mathbf{l}}\right)\right)_{2,\omega} = \left((i\hbar)^2\ddot{\rho}_{\mathbf{k}}, (i\hbar)^2\ddot{\rho}_{-\mathbf{l}}\right)_2, \tag{89}$$

and the initial value in the r.h.side can be evaluated using an explicit expression for the Hamiltonian of a system of interest. On the other hand, using Eq. (2.12) of Ref. [122], one can relate this initial value to $(\rho_{\mathbf{k}}, \rho_{-\mathbf{l}})_0$:

$$\begin{aligned}\left((i\hbar)^2\ddot{\rho}_{\mathbf{k}}, (i\hbar)^2\ddot{\rho}_{-\mathbf{l}}\right)_{2,\omega} = {} & \left\langle(i\hbar)^2\ddot{\rho}_{\mathbf{k}}, -i\hbar\dot{\rho}_{-\mathbf{l}}\right\rangle_0 - \langle[i\hbar\dot{\rho}_{\mathbf{k}}, \rho_{-\mathbf{l}}]\rangle_0 (\rho_{\mathbf{k}}, \rho_{-\mathbf{l}})_0^{-1} \\ & \langle[\rho_{\mathbf{k}}, -i\hbar\dot{\rho}_{-\mathbf{l}}]\rangle_0 .\end{aligned} \tag{90}$$

Noticing that $\langle[\rho_{\mathbf{k}}, -i\hbar\dot{\rho}_{-\mathbf{l}}]\rangle_0 = \langle[i\hbar\dot{\rho}_{\mathbf{k}}, \rho_{-\mathbf{l}}]\rangle_0 = \frac{e\hbar^2}{m\sqrt{V}}(\mathbf{l}\cdot\mathbf{k})\langle\rho_{\mathbf{k}-\mathbf{l}}\rangle_0$ for any inhomogeneous system, one can determine $(\rho_{\mathbf{k}}, \rho_{-\mathbf{l}})_0$ from Eq. (90). Thus, while in less sophisticated projection operator methods the initial value $(\rho_{\mathbf{k}}, \rho_{-\mathbf{l}})_0$ is considered as an adjustable parameter that has to be known from experiment or other considerations, the ZT method allows increasingly accurate approximations of this value using a self-consistent procedure.

Alternatively, the initial value $(\rho_{\mathbf{k}}, \rho_{-\mathbf{l}})_0$ can be found computationally for any given system using GAMESS or similar software upgraded by addition of a simple module for calculations of such inner products from the computational data provided by the software. In a similar fashion, one also can obtain $\left\langle[(i\hbar)^2\ddot{\rho}_{\mathbf{k}}, -i\hbar\dot{\rho}_{-\mathbf{l}}\right\rangle_0$ and similar expectation values from the computational data.

Note also, that for a given Hamiltonian the quantity $\left\langle[(i\hbar)^2\ddot{\rho}_{\mathbf{k}}, -i\hbar\dot{\rho}_{-\mathbf{l}}\right\rangle_0$ can be related analytically to the inner product of the charge density fluctuations $(\delta\rho_{\mathbf{k}}, \delta\rho_{-\mathbf{l}})_0$. Thus, introducing the notations $E_{k,-l}$ and R_k, so that $E_{k,-l}^2 = \dfrac{\left\langle[(i\hbar)^2\ddot{\rho}_{\mathbf{k}}, -i\hbar\dot{\rho}_{-\mathbf{l}}\right\rangle_0}{\frac{e\hbar^2}{m\sqrt{V}}(\mathbf{l}\cdot\mathbf{k})\langle\rho_{\mathbf{k}-\mathbf{l}}\rangle_0}$ and $i\hbar\dot{\rho}_{\mathbf{k}} = \dfrac{e\hbar^2}{m\sqrt{V}}(\mathbf{l}\cdot\mathbf{k})\langle\rho_{\mathbf{k}-\mathbf{l}}\rangle_0 R_{\mathbf{k}}$, one can rewrite Eq. (90) in the form:

$$(\rho_{\mathbf{k}}, \rho_{-\mathbf{l}})_0^{-1} = \frac{1}{\frac{e\hbar^2}{m\sqrt{V}}(\mathbf{l}\cdot\mathbf{k})\langle\rho_{\mathbf{k}-\mathbf{l}}\rangle_0} E_{\mathbf{k},-\mathbf{l}}^2 - (R_{\mathbf{k}}, R_{-\mathbf{l}})_2 . \tag{91}$$

Using this result and Eq. (90) one can obtain from Eq. (87) a tractable expression for the charge density–charge density TTGF to the second order in the density fluctuations,

$$\langle\langle \rho_{\mathbf{k}} | \, \rho_{-\mathbf{l}} \rangle\rangle_{0,\omega} = \frac{\frac{e\hbar^2}{m\sqrt{V}}(\mathbf{l}\cdot\mathbf{k})\,\langle \rho_{\mathbf{k}-\mathbf{l}}\rangle_0}{(\hbar\omega)^2 - \frac{e\hbar^2}{m\sqrt{V}}(\mathbf{l}\cdot\mathbf{k})\,\langle\langle R_{\mathbf{k}},\, R_{-\mathbf{l}}\rangle\rangle_{2,\omega} - E^2_{\mathbf{k},-\mathbf{l}}}. \qquad (92)$$

From Eq. (92), it becomes clear that the major contribution to the charge density–charge density TTGF, and thus to the longitudal quantum conductivity of Eq. (54), is defined by the expectation value of the "local" charge density. At the same time, both of the initial values of the higher order charge density TTGFs and the TTGFs of the density fluctuations [included in the denominator of Eq. (92)] can contribute dramatically in the case of strongly spatially inhomogeneous systems. Due to this, computations of TTGFs of Eq. (92) are not trivial and require specific method and algorithm developments. Calculations of the microcurrent density TTGFs appearing in the explicit expressions (49), (54), (57), (69), (80), and (78) for the transport coefficients can also be completed using the ZT method in a fashion similar to that of the charge density–charge density TTGFs. Note again that once the TTGFs are expressed in terms of the initial values of the corresponding quantities (microscopic charge or current densities) and their fluctuations, the latter fluctuations can be obtained from relatively accurate computational data provided by GAMESS, NWChem, and similar quantum chemistry software upgraded by the corresponding "analytical" modules. Thus, the problem of prediction of the susceptibilities and conductivity of the spatially inhomogeneous systems of any nature (including small QDs and molecules) is reduced to a tractable computational problem using the existing analytical method (ZT) and software. In particular, $\langle\langle R_{\mathbf{k}},\, R_{-\mathbf{l}}\rangle\rangle_{2,\omega}$ and $E_{\mathbf{k},-\mathbf{l}}$in Eq. (92) can be expressed in terms of electronic energy level values and correlators of the charge density and spin density fluctuations using a relatively straightforward generalization of the ZT method [122], similar to the case of homogeneous systems considered by Zubarev and Tzercovnikov.

The explicit expressions for the quantum susceptibilities and conductivities derived above are applicable to inhomogeneous systems of any nature and degree of inhomogeneity, including atomic and molecular clusters, quantum dots and wells, artificial atoms, etc. They reduce to their counterparts specific to the homogeneous system case when inhomogeneity can be neglected. These expressions have been obtained in the framework of the linear response theory formulated in terms of the equilibrium/steady state TTGFs. The fact that derivation of such explicit expressions has become possible by the use of Zubarev–Tserkovnikov two-time Green function formalism, while it has not been achieved via many other routes, including the recent technique [115] specifically developed to solve the problem of the quantum conductivity of strongly inhomogeneous systems, confirms yet again the power and flexibility of the ZT approach.

Reduction of the conservation equation (67) for the space–time Fourier transform of the expectation value of the linear quantum current density to Eq. (81) has become possible by keeping only the linear contribution with respect to the fields (or technically, the wave vectors) to the second term in the r.h.s. of Eq. (67). This linearization may not be applicable to strongly inhomogeneous systems subject to moderate and strong electromagnetic fields (such as ferromagnetic quantum dots).

Note also that Eq. (31) for the space–time Fourier transform of the expectation value of the quantum current density does not use any assumptions or conditions concerning correlations between the scalar and vector field potentials and, therefore, is the most general equation for the linear contribution to the quantum current density in any inhomogeneous system with small nonlinear effects. This equation provides a starting point for any possible developments, including the use of particular gauges for the field potentials. All results derived above reduce to their counterparts specific to the homogeneous system case if system inhomogeneity is negligibly small.

From the obtained results, it follows that the linear contributions to the quantum charge transport in an inhomogeneous system in a weak electromagnetic field can be entirely accounted for in terms of the microscopic equilibrium/steady state charge and microcurrent density TTGFs. These and nonequilibrium TTGFs of various observables have been intensively studied in numerous publications such as Refs. [120, 122–125]. It has been shown that the TTGFs of increasing complexity (n-operator TTGFs, where n runs from 2 to ∞) satisfy a system of coupled *algebraic* equations [122] that can be decoupled to any required accuracy by the use of the generalized continued fraction formalism. A general structure of the explicit expressions for the TTGFs so obtained is given in terms of the energy spectrum of the charge carriers specific to the equilibrium state of a system and various charge–spin density correlation functions, and correlation functions of their fluctuations. This methodology has been intensively used to solve numerous general and particular problems of statistical mechanics and quantum field theory. The TTGFs appearing in the conservation equations for the charge and current densities and in the explicit expressions for the quantum susceptibilities and conductivity derived above can be obtained using this methodology. Therefore, explicit calculations of the required TTGFs are entirely feasible at present, provided the electron energy spectrum and correlators of the charge and spin densities and their fluctuations specific to the equilibrium state of the system are available. Such calculations will reveal all details of the correlated/collective motion in a system, including those leading to coherent and resonant motions of the subsystems, such as a few electrons, thus providing necessary information on expectation values of the coherent states, such as those that can be used for realization of qubits and their processing.

In the following section, accurate computations of electronic energy levels and charge–spin density distributions necessary to determine the charge–spin correlation functions and those of their fluctuations are demonstrated for several artificial molecules composed of semiconductor compound atoms.

3 Virtual Synthesis of Small Artificial Molecules with Predesigned Electronic Properties

As demonstrated in previous sections, the problem of prediction of the charge–spin transport coefficients of strongly inhomogeneous systems and that of their thermodynamic counterparts [126] can be reduced to the problem of calculation of the corresponding equilibrium/steady state TTGFs in terms of equilibrium or steady state physically measurable quantities, such as electronic energies, charge and spin density distributions and their fluctuations, and various correlations of the charge–spin density distributions and their fluctuations. These quantities can be predicted analytically, computationally, or simply measured experimentally. While such data can be obtained in the framework of analytical equilibrium quantum statistical mechanics, many details of particular systems will not be possible to include due to extreme analytical difficulties such inclusion would cause. However, such details are extremely important, in particular for understanding of mechanisms leading to the development of collective modes or coherent evolution of the subsystems. Therefore, to develop practically valuable theoretical predictions, these data should be obtained by means of the first-principle quantum computations or simulations. This means solving the Schrödinger equations for the systems of interest. Numerical solutions of the Schrödinger equation for various systems (or virtual synthesis of such systems) are also of major importance for material scientists, chemists, and engineers as such solutions allow to probe various routes of synthesis of artificial molecules with desirable equilibrium properties. In this section, such computations are demonstrated in the case of several artificial molecules composed of Ga, As, and P atoms, including those synthesized at conditions reflecting quantum confinement.

In the majority of studies to date, such as those of Ref. [127], formation and growth of artificial molecules did not involve spatial restrictions applied to the centers of mass of participating atoms. The structure of the molecules is obtained by a minimization of the total atomic cluster energy by solving the corresponding Schrödinger equation (without any spatial constraints applied to the clusters' atomic coordinates) using well-established algorithms based on first-principle theoretical methods, such as the density functional theory (DFT), Hartree–Fock (HF), multi-configuration self-consistent field (MCSCF), and similar methods. For the majority of the considered cases, numerical data on the structure, stoichiometry, and electronic, optical, and magnetic properties are specific only for artificial molecules synthesized in the absence of spatial constraints (below called "vacuum"), and differ significantly from those specific to molecules (composed of the same type and numbers of atoms) formed in quantum confinement or on surfaces. At the same time, actual formation of both artificial and "natural" molecules and other atomic/molecular structures follows primarily the "heterogeneous" nucleation route, which becomes even more important in view of the use of the quantum confinement effects to manipulate properties of artificial molecules and larger objects, such as nanowires and quantum dots (QDs) [128]. Nucleation, growth, and formation of molecules from the small atomic clusters in quantum confinement are dramatically affected by the presence of confinement atoms that not only stabilize the artificial molecules that could not be stabilized at unconfined conditions, but also affect their electronic energy spectra, geometry, charge–spin density distributions, and other equilibrium properties derived from those.

Manipulations with confinement structure and conditions offer an attractive way to obtain artificial molecules and larger objects with predesigned electronic, and by virtue of the theoretical approach discussed in Section 2, charge–spin transport properties. Unfortunately, the task of mimicking of quantum confinement details is very computationally demanding even for small atomic clusters. A sophisticated and computationally feasible alternative is offered by modeling the molecular synthesis in quantum confinement by means of building "template," or predesigned artificial molecules whose major parameters (such as the ratios of the covalent radii of atoms in the clusters, clusters' shape and stoichiometry) are chosen to resemble those of clusters that are expected to form in some "typical" quantum confinements. This approach has been proven [129] to deliver both insightful and practical results confirmed by experimental data and simulations. Below, this approach is used to synthesize virtually several artificial nonstoichiometric molecules of tetrahedral symmetry composed of Ga, As, and P atoms. Such molecules resemble smaller versions of pyramidal QDs currently fabricated experimentally (see Section 1 for details and references). Moreover, they may examplify possible outcomes of experimental synthesis in the near future when further miniaturization advances are made.

3.1 Pyramidal Artificial Molecules of Ga with As and P

The electronic energy level structure and charge–spin density distributions of several predesigned and vacuum pyramidal molecules composed of Ga, As, and P atoms has been investigated by means of progressively accurate HF-based methods, including the restricted and restricted open-shell HF (RHF and ROHF, respectively), configuration interactions (CI), complete active space self consistent field (CASSCF), and MCSCF methods as realized by the GAMESS software package [75, 130]. For the predesigned molecules, the form, stoichiometry, and atomic covalent radii have been derived from the bulk structure of GaAs in a manner similar to that described in [129], and have been manipulated with to reflect a possible influences of quantum confinement. The obtained results confirm that the electronic energy level structure (ELS), the direct optical transition energy (OTE), and charge–spin density distributions (CDD and SDD, respectively) of the virtually synthesized molecules are sensitive to manipulations with the covalent radii and stoichiometry. Thus, a very small displacement (in the range of 10^{-2} Å) of atoms (that occurs in the predesigned molecules) from their respective positions in vacuum molecules leads to a significant change in the OTE of the predesigned molecules (by tenths of electron Volt) and restructuring of their ELSs in the highest occupied–lowest unoccupied molecular orbital (HOMO–LUMO) regions. The latter signifies a tendency toward the development of the valence and conduction level bands. In agreement with previous results [129] and other data available in literature, the OTEs of these molecules are in the range of several electron Volts and can be manipulated up to 100% in the studied cases by manipulations with the composition and covalent radii of the atoms comprising the molecules defined by the quantum confinement. [In practice, the latter manipulations can be realized by the corresponding adjustment of parameters of the quantum confinement.]

Tetrahedral symmetry fragments of zincblende units of GaAs bulk crystalline lattices have provided rich material for virtual design of stable molecules without dangling bonds. Larger Ga–As clusters of this type have also been observed and synthesized in numerous experimental studies (see Ref. [129] for references and discussions). [While passivation of dangling bonds by saturating them with hydrogen atoms is aften used in experiments to produce stable atomic clusters, it significantly affects [131] the synthesized molecules' ELSs, CDDs, and SDDs, making such models inapplicable to molecular synthesis in quantum confinement that usually does not provide for hydrogen atoms.] The condition of the absence of dangling bonds reflects results of "natural" synthesis of stable molecules and largely defines both stoichiometry and shape of the synthesized molecule, and spatial constraints defined by the tetrahedral symmetry of the bulk Ga–As lattice and applied to the atomic coordinates provide for the major effects of quantum confinement. In Ref. [129] and other publications, applications of this condition and spatial constraints have lead to virtual synthesis of almost perfect predesigned pyramidal molecules. These molecules are composed of ten Ga or In atoms providing the pyramid "scaffolds" and four As atoms inside of the pyramids. The Ga covalent radius of 1. 26 Å has been chosen at 300 K, and that of As, 1.18 Å, at 298.5 K. In the study discussed below, one or two As atoms in the $Ga_{10}As_4$ clusters so predesigned were substituted by P atoms (without changing positions of the centers of mass of all atoms) to form the corresponding $Ga_{10}As_3P$ and $Ga_{10}As_2P_2$ predesigned molecules (Fig. 1).

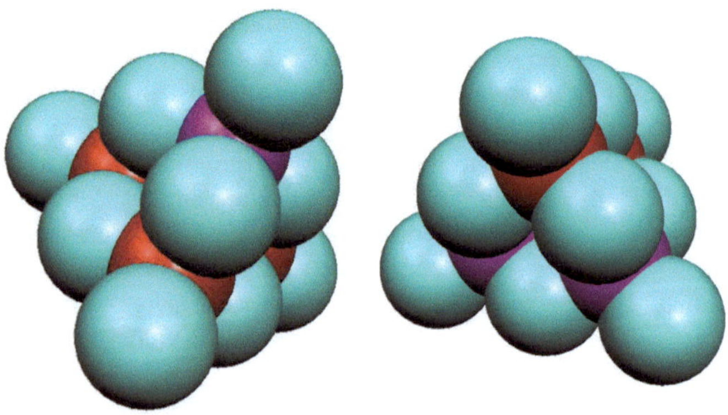

Fig. 1 Left to right: predesigned pyramidal clusters $Ga_{10}As_3P$ and $Ga_{10}As_2P_2$ of Ga (*blue*), As (*red*) and P (*purple*) atoms. All dimensions are to scale

This means that the P atoms in the predesigned molecules were assigned the covalent radii of As. The corresponding molecules formed in the absence of spatial constraints, or vacuum molecules, have been obtained from the predesigned ones on relaxation of the spatial constraints applied to the centers of mass of the atoms while using the RHF/ROHF/CI/CASSCF/MCSCF energy minimization procedure. In this minimization procedure, each of the atoms of the predesigned clusters has been randomly displaced from its initial position and the total cluster energy; forces acting on each atom and the fundamental vibration frequencies of the molecule have been calculated for each of such displacements. The process was repeated until the coordinates of the atomic centers that provided the minimum of the total energy were identified, thus indicating the formation of and identifying the geometry of the synthesized molecule.

Geometrically, the vacuum counterpart of the $Ga_{10}As_3P$ predesigned molecule practically coincides with its predesigned "parent," with the atomic centers in the vacuum molecule being displaced by a few hundredths of Angstrom from their respective positions in the predesigned molecule. This is not surprising, because the Ga and As covalent radii of the predesigned molecules coincide with the corresponding covalent radii specific to the bulk crystalline structure of Ga–As. These computational resulsts suggest that at the "vacuum" conditions the covalent radii change only slightly, in agreement with available experimental data. However, as demonstrated by the OTE, CDD, and SDD data discussed below, this very small change in the covalent radii leads to a significant change in the electronic properties.

The form of the "vacuum" cluster $Ga_{10}As_2P_2$ depicted in Fig. 2 deviates significantly from that of the original predesigned pyramid. One of the Ga atoms has moved inside of the original pyramidal structure, and both P atoms shifted to accommodate this atom and to allow two other Ga atoms move closer together. This distortion of the initial pyramid is likely to be caused by the difference in size between P atoms and As atoms they substituted. Because P atoms are smaller in size than As ones, the replacement of two As atoms by two P atoms without changing the interatomic separation distances creates a large "extra" space inside of the predesigned $Ga_{10}As_2P_2$ molecule. Due to the spatial constraint applied to the atomic coordinates within the predesigned molecule formed in quantum confinement, the electronic orbits had to change their shape to accommodate for the replacement of As atoms by P atoms. In the corresponding vacuum molecule where the spatial constraints were relaxed, the energy minimization procedure has lead to distortion of the original pyramidal shape.

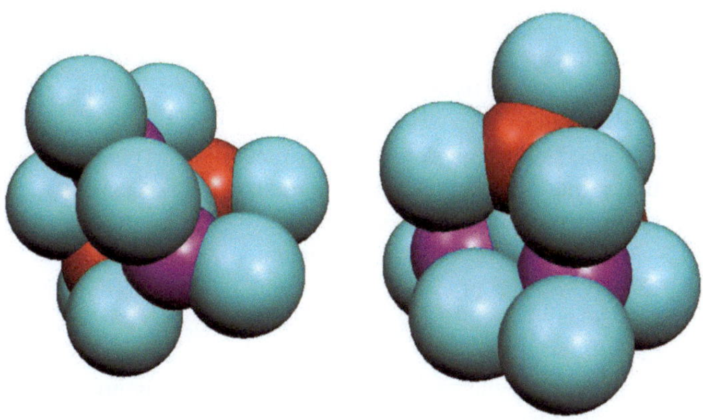

Fig. 2 Two views of the irregular vacuum cluster $Ga_{10}As_2P_2$. Ga atoms are *blue*, As *red*, and P *purple*. All dimensions are to scale

As has been mentioned already, the ELSs, CDDs, and SDDs of the predesigned and vacuum molecules discussed above have been obtained by RHF/ROHF/CI/CASSCF/MCSCF methods as realized by the GAMESS software package [75, 130]. The stable predesigned molecules have been developed to model molecular synthesis in quantum confinement that supports the pyramidal geometry of these molecules. Computationally, and in this study, this corresponds to the spatial constraints applied to the centers of mass of the atoms to support the pyramidal geometry of the predesigned molecules while minimizing their total energy. This approach has been used previously to synthesize virtually small clusters that possess desirable electronic energy spectrum properties and delivered results compatible with experimental findings [129, 132]. The information on the cluster geometry and composition can be used to identify appropriate conditions for experimental nucleation of similar molecules in confinement or on surfaces. This approach provides an insight into the details of synthesis of artificial molecules in quantum confinement while avoiding the computationally demanding procedure of minimization of the total energy of the entire "cluster plus surface/confinement atoms" system with all of the participating confinement atoms included (totaling to at least several hundred atoms in the case of the considered small molecules). Such a "complete" minimization procedure involving from several hundreds to many thousands of confinement atoms is not computationally feasible at this time due to the existing hardware limitations that significantly restrict power of the CI/CASSCF/MCSCF software.

The ground state energies and OTEs of the studied molecules are collected in Table 1. The minimum complete active spaces (CAS) that provide for convergence of the energy minimization procedures include at least 6 electrons and 8 orbits (6×8 for the molecules with two P atoms, and 6×9 and 8×8 for the predesigned and vacuum $Ga_{10}As_3P$ molecules, respectively). As demonstrated previously [129, 132] for other small molecules and for the predesigned $Ga_{10}As_3P$ molecule (see Table 1) synthesized in this work, smaller CASs, such as 4×6, do not allow to predict accurately the ground state energies of artificial molecules composed of semiconductor compound atoms.

Similar to previously studied "pure" Ga–As clusters [129], the composite predesigned and vacuum clusters are stable ROHF/CI/CASSCF/MCSCF ground state triplets, with the exception of the vacuum molecule $Ga_{10}As_2P_2$ whose ground state is the the RHF/CI singlet (MCSCF computations for this molecule are not completed yet). However, unlike the degenerate triplets of the "pure" Ga–As clusters of previous studies, these triplets are entirely nondegenerate due to the presence of the P atoms breaking the tetrahedral symmetry of the original Ga–As predesigned pyramids. The ROHF/RHF HOMO–LUMO energy difference and the OTEs, in the case of CI/CASSCF/MCSCF procedures, are in the range from about 2 eV to 4 eV for all of the molecules, save for the vacuum RHF/CI singlet $Ga_{10}As_2P_2$. This latter result has been expected though, given a total loss of symmetry of this molecule as compared to the rest of the molecules. Although there is no data in literature for the artificial molecules studied in this work, the above findings correlate very well with numerous data available for small Ga-rich Ga–P molecules [133]. As expected, the CI/CASSCF/MCSCF OTEs largely improve ROHF/RHF data on the

Table 1 Energy data for the studied clusters

Predesigned clusters	Ground state energy, (–1)×1 Hartree	HF HOMO–LUMO energy difference,eV	CI/CASSCF/MCSCF optical transition energy,eV	CAS type
$Ga_{10}As_3P$	MCSCF Triplet:			
	2596.312563811	3.4776	1.9430	6×9
	2596.289733633		1.8591	4×6
$Ga_{10}As_2P_2$	MCSCF Triplet: 2596.609992140	3.5102	1.9983	6×8
$Ga_{10}As_3P$ (VC)	MCSCF Triplet:			
	2596.325927751	3.4858	4.2702	8×8
$Ga_{10}As_2P_2$ (VC)	CI Singlet:			
	2596.697688143	6.5034	3.2450	6×8

VC denotes vacuum molecules.

HOMO–LUMO energy differences. However, the OTE of the CI singlet $Ga_{10}As_2P_2$ seem too small, so the MCSCF data for this molecule have to be obtained before drawing any conclusions concerning the OTE of this molecule.

Comparison of the OTEs of the predesigned and the corresponding vacuum molecules confirms previous findings [129, 132] that almost an infinitesimal change (3×10^{-2} Å, on average) in coordinates of the clusters' atoms can lead to a sizable change in the OTEs. A larger change (in the range of 0.4 Å for a few atomic coordinates) in the case of $Ga_{10}As_2P_2$ molecule leads to a dramatic change in the structure and, consequently, in the OTE: the ground state of this vacuum molecule changes to the RHF/CI singlet, and the RHF HOMO–LUMO energy difference of this vacuum molecule becomes almost about 1.5 times larger than that of its predesigned "parent".

Sensitivity of the RHF/ROHF/CI/CASSCF/MCSCF energy data to even the smallest changes in the cluster structure is further demonstrated by the ELSs and is typical for all studied molecules in this and previous works. The major effects of the quantum confinement on the ELS (as revealed by the studies of the predesigned molecules) result in displacement of all of the occupied orbits in the ROHF/RHF HOMO region toward HOMO, and the corresponding displacement of the CI/CASSCF/MCSCF almost fully occupied orbits toward the highest of them (in recognition of almost full occupation or unoccupation of these orbits, in the discussion below, the abbreviations HOMO and LUMO are sometimes used for both ROHF/RHF and CI/CASSCF/MCSCF cases). The energy difference between the HOMO and the nearest occupied orbit for the studied molecules decreases significantly as compared to that of the corresponding vacuum molecules. Because this and other occupied orbits in the vicinity of the HOMO group closer to each other and to the HOMO in the case of the predesigned molecules, one concludes that the quantum confinement effects lead to the formation of the electron level substructures that will transform into valence bands in the case of larger clusters or molecules. Similar tendency to close grouping is observed for unoccupied orbits in the vicinity of the LUMO for the studied predesigned molecules, as compared to that of their vacuum counterparts. This indicates a tendency to formation of the conduction band in larger clusters/molecules. In correspondence with tendencies reported earlier for small artificial molecules of semiconductor compound atoms, deeper lying occupied orbits up to about 50 (calculated from the HOMOs) in the studied cases are affected by the confinement and contribute to the valence configuration. Deeper lying occupied orbits are not much affected by the confinement in the studied cases. The data on the charge–spin density distributions in the valence-forming orbits (totaling to about 50 orbits in the studied cases) define the major features of the total charge–spin density distributions of the molecules, and the fluctuations of all the charge–spin-involving distributions, that have to be incorporated into the analytical formulae for the equilibrium TTGFs to predict the charge–spin transport coefficients of these molecules. [Obviously, it is almost impossible to develop analytical models to calculate the CDDs, SDDs, their fluctuations and correlations with such fine details incorporated. Moreover, the models would have to be changed from one molecule case to another.] In Figs. 3 to 12 below the HOMO and LUMO orbits and the CDDs and SDDs are depicted for several of the studied molecules.

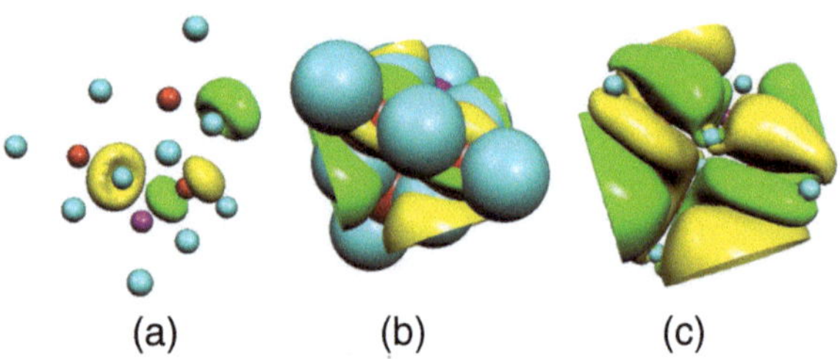

Fig. 3 The highest almost fully occupied CI/CASSCF/MCSCF orbit of the predesigned $Ga_{10}As_3P$ molecule. The surfaces of both signs (positive is *green*, negative *yellow*) correspond to portions (cuts) of the maximum (minimum) values (not shown) of 0.120292 (–0.120292) in arbitrary units: (**a**) cut 0.05; (**b**) and (**c**) cuts 0.005. Ga atoms are *blue*, As *red*, and P *violet*. In (**a**) and (**c**) atomic dimensions are reduced

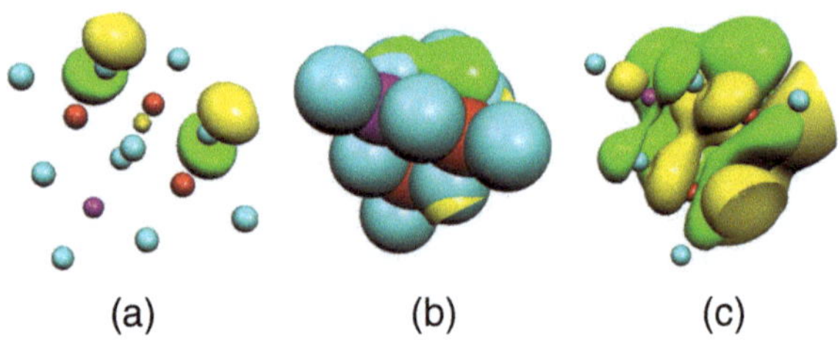

Fig. 4 The highest almost fully unoccupied CI/CASSCF/MCSCF orbit of the predesigned $Ga_{10}As_3P$ molecule. The surfaces of both signs (positive is *green*, negative *yellow*) correspond to portions (cuts) of the maximum (minimum) values (not shown) of 0.196151 (–0.131281) in arbitrary units: (**a**) cut 0.05; (**b**) and (**c**) cuts 0.005. Ga atoms are *blue*, As *red*, and P *violet*. In (**a**) and (**c**) atomic dimensions are reduced

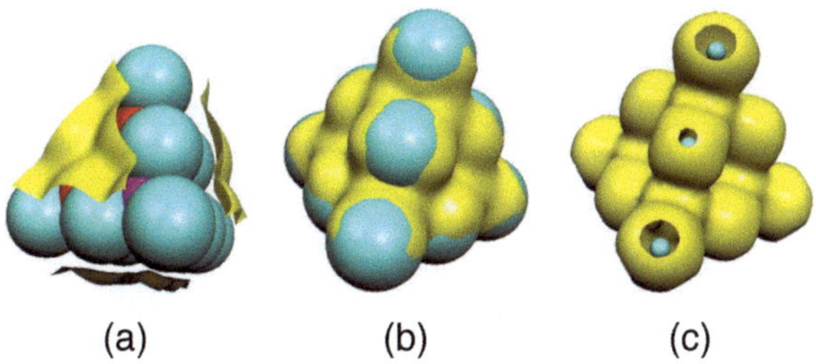

Fig. 5 The CI/CASSCF/MCSCF charge density distribution (CDD) of the predesigned $Ga_{10}As_3P$ molecule. The charge surfaces (*yellow*) correspond to portions (cuts) of the CDD maximum value (not shown) of 10.4268 in arbitrary units: (**a**) cut 0.01; (**b**) cut 0.03, and (**c**) cut 0.05. Ga atoms are *blue*, As *red*, and P *violet*. In (**c**) atomic dimensions are reduced

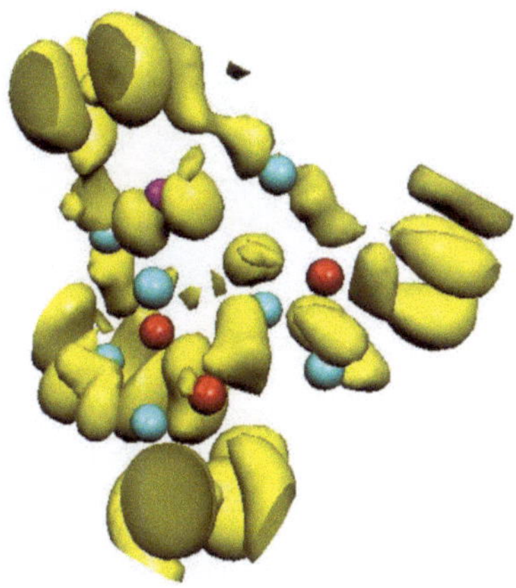

Fig. 6 The CI/CASSCF/MCSCF spin density distribution (SDD) of the predesigned $Ga_{10}As_3P$ molecule. The SDD surfaces (*yellow*) correspond to portion (cut) 0.001 of the SDD maximum value (not shown) of 0.0277249 in arbitrary units. Ga atoms are blue, As *red*, and P *violet*. Atomic dimensions are reduced

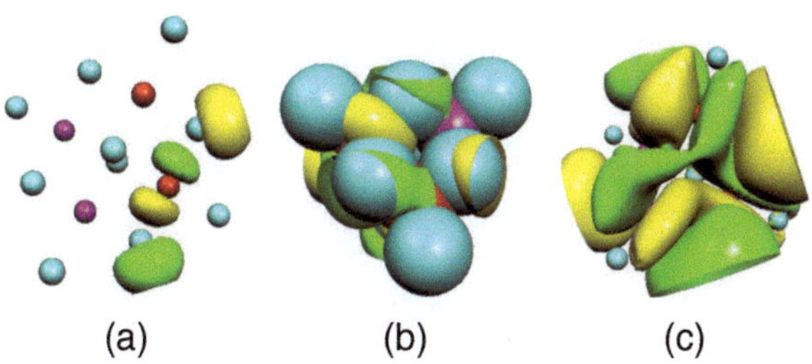

Fig. 7 The highest almost fully occupied CI/CASSCF/MCSCF orbit of the predesigned $Ga_{10}As_2P_2$ molecule. The surfaces of both signs (positive is *green*, negative *yellow*) correspond to portions (cuts) of the maximum (minimum) values (not shown) of 0.116968 (–0.123702) in arbitrary units: (**a**) cut 0.05 and (**b**) and (**c**) cuts 0.005. Ga atoms are *blue*, As *red*, and P *violet*. In (**a**) and (**c**) atomic dimensions are reduced

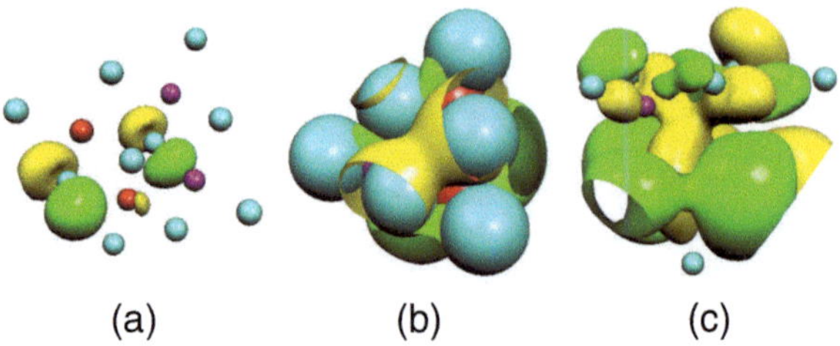

Fig. 8 The highest almost fully unoccupied CI/CASSCF/MCSCF orbit of the predesigned $Ga_{10}As_2P_2$ molecule. The surfaces of both signs (positive is *green*, negative *yellow*) correspond to portions (cuts) of the maximum (minimum) values (not shown) of 0.241172 (–0.147003) in arbitrary units: (**a**) cut 0.05 and (**b**) and (**c**) cuts 0.003. Ga atoms are *blue*, As *red*, and P *violet*. In (**a**) and (**c**) atomic dimensions are reduced

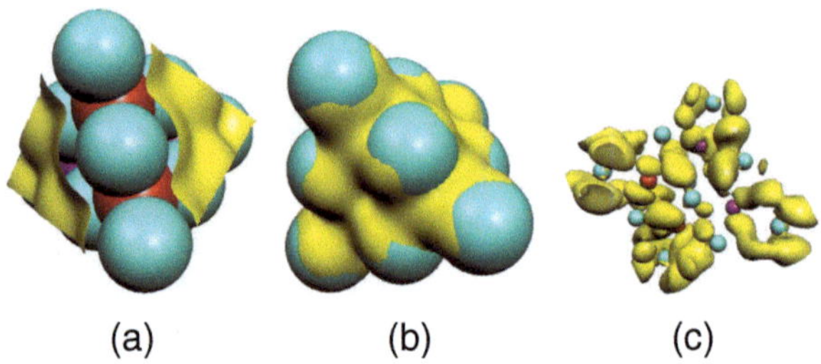

Fig. 9 The CI/CASSCF/MCSCF charge and spin density distributions (CDD and SDD, respectively) of the predesigned $Ga_{10}As_2P_2$ molecule. The CDD and SDD surfaces (*yellow*) correspond to portions (cuts) of the corresponding maximum values (not shown) of 10.4026 and 0.024302, respectively, in arbitrary units: (**a**) CDD cut 0.05; (**b**) CDD cut 0.003; and (**c**) SDD cut 0.001. Ga atoms are *blue*, As *red*, and P *violet*. In (**c**) atomic dimensions are reduced

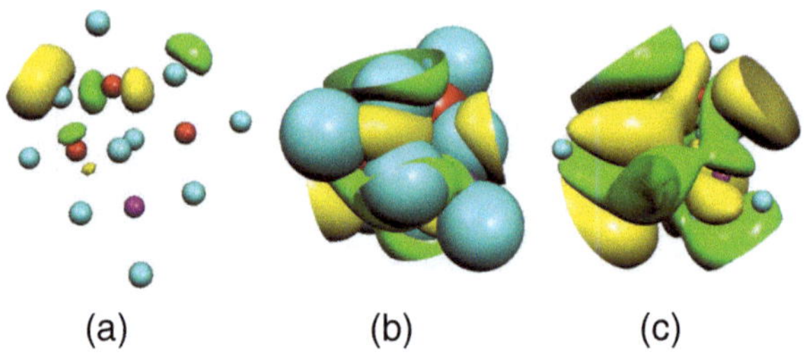

Fig. 10 The highest almost fully occupied CI/CASSCF/MCSCF orbit of the vacuum $Ga_{10}As_3P$ molecule. The surfaces of both signs (positive is *green*, negative *yellow*) correspond to portions (cuts) of the maximum (minimum) values (not shown) of 0.130274 (–0.207202) in arbitrary units: (**a**) cut 0.05 and (**b**) and (**c**) cuts 0.005. Ga atoms are *blue*, As *red*, and P *violet*. In (**a**) and (**c**) atomic dimensions are reduced

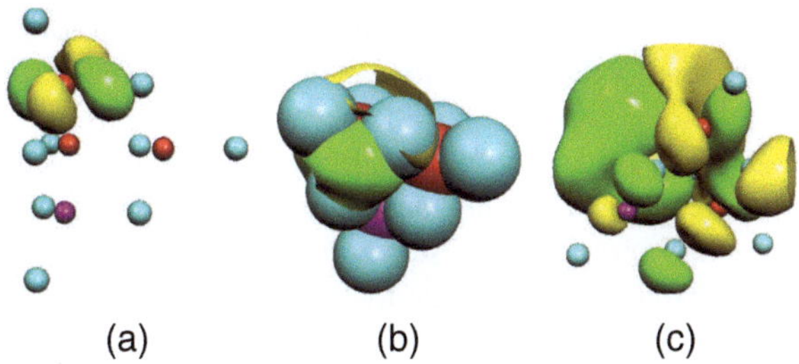

Fig. 11 The highest almost fully unoccupied CI/CASSCF/MCSCF orbit of the vacuum $Ga_{10}As_3P$ molecule. The surfaces of both signs (positive is *green*, negative *yellow*) correspond to portions (cuts) of the maximum (minimum) values (not shown) of 0.144046 (–0.187724) in arbitrary units: (**a**) cut 0.05 and (**b**) and (**c**) cuts 0.005. Ga atoms are *blue*, As *red*, and P *violet*. In (**a**) and (**c**) atomic dimensions are reduced

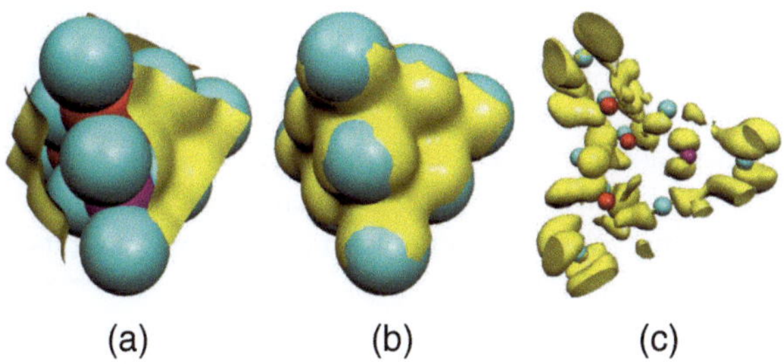

Fig. 12 The CI/CASSCF/MCSCF charge and spin density distributions (CDD and SDD, respectively) of the vacuum $Ga_{10}As_3P$ molecule. The CDD and SDD surfaces (*yellow*) correspond to portions (cuts) of the corresponding maximum values (not shown) of 10.4828 and 0.0245593, respectively, in arbitrary units: (**a**) CDD cut 0.05; (**b**) CDD cut 0.003; and (**c**) SDD cut 0.001. Ga atoms are *blue*, As *red*, and P *violet*. In (**c**), atomic dimensions are reduced

Thus, similar to other cases of small artificial molecules, the highest almost fully occupied orbit of the $Ga_{10}As_3P$ predesigned molecule (Fig. 3c) is both bonding for some groups of atoms and antibonding for other groups. The major contributions to this orbit come from a 2π-orbit of an As atom and two hybrid *sp*-orbits centered near two Ga atoms (Fig. 3a). To a large degree, the orbit is contained within the molecule (Fig. 3b). The lowest almost fully unoccupied orbit of this molecule is depicted in Fig. 4. The structure of this orbit is defined by numerous small contributions (Fig. 4c), with the major contributions coming from two 2π-orbits of two Ga atoms and one *sp*-hybrid orbit centered near an As atom (Fig. 4a). The LUMO also is almost entirely contained within the molecule.

The CDD of this molecule is pictured in Fig. 5 and reflects the tetrahedral symmetry of the molecule. From analysis of contributions of deeper (up to 50th) lying occupied orbits, one can conclude, in correspondence with analysis available in literature for such small nonstoichiometric molecules [132, 134], that the valence properties of this molecule differ dramatically from those of similar stoichiometric ones. In particular, the octet rule is no longer applicable to such nonstoichiometric molecules, and one has to compute contributions from several to several tenths of occupied orbits to decipher the type of the valence and chemical properties of such molecules, as well as their charge and spin density distributions.

As demonstrated in Fig. 5a, the CDD surfaces of this molecule reach into the space beyond that occupied by the molecule. The "surface" of this molecule, similar to those of other small molecules of semiconductor compound atoms [129, 132, 134], is negatively charged, although the total charge on the molecule is zero. The SDD of this molecule (Fig. 6) is not remarkable, with its maximum value very small and leading contributions provided by electronic spins of Ga atoms and of P atom (electrons of As atoms do not contribute nearly as much).

It is very informative to compare the electronic properties of the $Ga_{10}As_3P$ predesigned molecule detailed above (Fig. 3) to those of the predesigned molecule $Ga_{10}As_2P_2$. The highest almost fully occupied and the lowest almost fully unoccupied CI/CASSCF/MCSCF orbits of this latter molecule are represented in Figs. 7 and 8. The nature of these orbits and the major contributions to them resemble closely those of the predesigned molecule $Ga_{10}As_3P$. This indicates that P atoms do not paticipate noticably in the formation of these orbits in both molecules. The P atoms do influence, however, the details of the ELSs and the OTE values. In particular, the latter molecule with two P atoms seems more stable (its ground state is lower in energy than that of the molecule with one P atom), and its OTE is slightly larger than that of the other molecule.

This observation signifies that the electronic properties of small predesigned Ga–As–P molecules may be fine-tuned by sophisticated manipulations with concentration of P atoms. The CDD and SDD of the predesigned $Ga_{10}As_2P_2$ molecule (Fig. 9) are very similar both in form and in strength to those of the corresponding $Ga_{10}As_2P$ one. In both cases, P atoms contribute substantially to SDDs, while As atoms do not.

Electronic properties of the studied vacuum $Ga_{10}As_3P$ molecule are demonstrated in Figs. 10, 11, and 12. The loss of the tetrahedral symmetry of this molecule results in a slight decrease in its ground state energy and significant increase of its OTE as compared to that of the corresponding predesigned molecule (Table 1).

While the nature and major contributions to the highest almost fully occupied orbit of this molecule (Fig. 10) still resembles those of the predesigned molecules (with slightly increased contribution of As atoms, two of which now contribute), the nature of the lowest almost fully unoccupied orbit differs significantly from those of the predesigned molecules. In particular, the major contribution to this orbit comes from the (antibonding) 3*d*-orbit of an As atom. The ground state of the vacuum molecule is lower and the OTE larger than those of the corresponding predesigned molecule. This, of course, confirms expectations based on physical reasoning that, ideally, molecules synthesized without any spatial constraints applied to their atoms realize the absolute minimum of their total energy (ground state), while "constrained" molecules realize only the conditional minimum of their total energy. Correspondingly, the OTEs of vacuum molecules are also larger than those of their predesigned counterparts. The CDD and SDD of the vacuum $Ga_{10}As_3P$ molecule are represented in Fig. 12 and are similar to those of the predesigned molecules, thus confirming a common tendency that As and P atoms do not contribute significantly to the formation of the collective charge distributions while contributing to the formation of the collective spin distributions of the studied molecules.

From the above results it follows that addition of P atoms to Ga-rich Ga–As atomic clusters may lead to a significant change in equilibrium electronic properties of the clusters (studies on the MCSCF electronic properties of the vacuum molecule $Ga_{10}As_2P_2$ still continue). Comparison with the equilibrium electronic properties of the $Ga_{10}As_4$ molecules of [129] shows that a substitution of one As atom by P atom in the predesigned or vacuum $Ga_{10}As_4$ molecules (with the covalent radii of As assigned to the P atom) leads to a significant increase (over two times) in the OTE and decrease in the ground state energies, thus considerably stabilizing the molecule. This observation may be helpful in the development of new experimental synthesis strategies to finely tune the equilibrium electronic properties of artificial molecules composed of Ga, As, and P atoms, and thus manipulate their electronic transport properties, in synergy with results of Section 2.

Summary

First-principle quantum statistical mechanical methods provide powerful tools for predictions of the electronic transport properties of strongly spatially inhomogeneous systems, such as nanosystems, quantum dots, artificial atoms, and molecules, etc. Among such methods, Zubarev–Tzerkovnikov's approach and its modifications laid out in Section 2 provide the best first-principle method for such predictions. The strong feature of this method that makes it outstanding among other first-principle quantum statistical mechanical approaches concerns tractability and simplicity of its theoretical predictions that allow derivation of explicit expressions for the transport coefficients in terms of the equilibrium TTGFs, as opposed to a number of other methods that utilize nonequilibrium TTGFs. In addition to analytical calculations, this provides an opportunity to determine the TTGFs computationally, once the necessary software is developed and interfaced with the existing quantum software packages. The reason is that the equilibrium state of a system is unique, so such TTGFs have to be calculated/computed only once and can be used to predict all transport properties (see Section 2, Refs. [117] and [126] and references therein), while the nonequilibrium TTGFs have to be calculated or computed anew for any change in a nonequilibrium process in the same system. Moreover, experimental and computational data and simulations of small nonequilibrium systems are ambiguous and orders of magnitude less accurate than those of small equilibrium systems, while the equilibrium TTGFs can be obtained computationally or derived from experimental data with high accuracy.

In Section 2, it has been demonstrated how to express analytically the equilibrium TTGFs in terms of physically meaningful, measurable or computable quantities, such as the electronic energies, charge density distributions, charge and spin correlators of increasing order, and those of the charge and spin density fluctuations. These data can be predicted theoretically for various (general or specific) models using methods of equilibrium statistical mechanics, computed for the cases of interest by solving the Schrödinger equation(s), simulated by means of equilibrium quantum molecular dynamics (MD) or derived from experimental data. The easiest route is to use the first-principle "quantum chemistry" software (such as GAMESS, NWChem, and GAUSSIAN packages) to solve the Schrödinger equation numerically, as it was demonstrated in Section 3, and then calculate all the desirable quantities from the obtained data on electronic orbits, CDDs and SDDs. At this time, however, one can only compute in this fashion the charge density distributions and electronic energy levels, due to the "design" of the software used. To compute the charge and spin fluctuations and their correlations necessary to calculate the charge transport coefficients with the accuracy beyond the zeroth approximation of the ZT method, additional interface modules have to be developed and appended to the corresponding software packages. The work on such modules is greatly helped by the availability of analytical representations of a range of the equilibrium TTGFs of increasing order appearing in the transport coefficients of increasing accuracy in terms of the measurable (computable) quantities. Such analytical studies are in progress.

The obtained results show that theoretical predictions of the transport properties of nanosystems, artificial molecules, etc., in synergy with their computational studies, provide a feasible and sophisticated way to obtain useful and insightful data to guide experimental synthesis of such systems. In particular, virtual design of numerous small molecules composed of semiconductor compound atoms confirms once again the importance and usefulness of the effects of quantum confinement in manipulation of the equilibrium, and thus transport, electronic properties of systems of utmost importance for modern technology of electronic materials.

Acknowledgments Support of the National Science Foundation through the grants DMR No. 0340613 and No. 0647356 is appreciated.

Appendix

Using Eqs. (32) and (18) one can derive the following equation for the longitudal component of the space–time Fourier transform of the current density:

$$\mathbf{k}\cdot\mathbf{j}(\mathbf{k},\omega) = -\frac{e^2}{mc}\sum_{\mathbf{l}} n_0(\mathbf{k}-\mathbf{l})\,\mathbf{k}\cdot\mathbf{A}(\mathbf{l},\omega) - \frac{i}{c}\sum_{\mathbf{l}} << \dot{\rho}_{\mathbf{k}} \mid \mathbf{j}_{-\mathbf{l}} >>_{0,\omega} \cdot\mathbf{A}(\mathbf{l},\omega)$$
$$+\, i\sum_{\mathbf{l}} << \dot{\rho}_{\mathbf{k}} \mid \rho_{-\mathbf{l}} >>_{0,\omega}\, \varphi(\mathbf{l},\omega). \tag{93}$$

Recovering the Fourier image of $<< \dot{\rho}_{\mathbf{k}} \mid \mathbf{j}_{-\mathbf{l}} >>_{0,\omega}$, $<< \dot{\rho}_{\mathbf{k}} \mid \mathbf{j}_{-\mathbf{l}} >>_{0,\omega} = \int_{-\infty}^{\infty} dt\, e^{i\omega t} << \dot{\rho}_{\mathbf{k}}(t) \mid \mathbf{j}_{-\mathbf{l}} >>_0$, and integrating by parts in the r.h.s. of this equation, one can prove that

$$<< \dot{\rho}_{\mathbf{k}} \,|\, \mathbf{j}_{-\mathbf{l}} >>_{0,\omega} = -\frac{1}{i\hbar} < [\rho_{\mathbf{k}}, \mathbf{j}_{-\mathbf{l}}] >_0 - i\omega << \rho_{\mathbf{k}} \,|\, \mathbf{j}_{-\mathbf{l}} >>_0 . \tag{94}$$

Substituting the Fourier transform (94) into Eq. (93), using the conjugated Eqs. (36) and (16), and noticing that the system is uncharged (total charge is equal to zero), one can recover Eq. (93) in the form:

$$\mathbf{k}\cdot\mathbf{j}(\mathbf{k}, \omega) = -\frac{e^2}{mc} \sum_{\mathbf{l}} n_0(\mathbf{k}-\mathbf{l})\, \mathbf{k}\cdot\mathbf{A}(\mathbf{l}, \omega) + \frac{1}{c\hbar} \sum_{\mathbf{l}} < [\rho_{\mathbf{k}}, \mathbf{j}_{-\mathbf{l}}] >_0 \cdot\mathbf{A}(\mathbf{l}, \omega) + \omega\rho(\mathbf{k}, \omega). \tag{95}$$

Using Eqs. (18) and (19), one can express the commutator $[\rho_{\mathbf{k}}, \mathbf{j}_{-\mathbf{l}}]$ as follows:

$$\frac{1}{\hbar}[\rho_{\mathbf{k}}, \mathbf{j}_{-\mathbf{l}}] = \frac{e}{mV}\mathbf{k}\rho_{\mathbf{k}-\mathbf{l}}. \tag{96}$$

With this result, and because of $< \rho_{\mathbf{k}-\mathbf{l}} >_0 = en_0(\mathbf{k} - \mathbf{l})$, one can prove that the first two terms in Eq. (95) sum up to zero, so that the space–time Fourier transform of the continuity equation for the Fourier transform of the local charge density is recovered:

$$\mathbf{k} \cdot \mathbf{j}(\mathbf{k}, \omega) = \omega\rho(\mathbf{k}, \omega). \tag{97}$$

Therefore, in addition to the continuity equation for charge density operator, a similar continuity equation for its local expectation value (i.e., the local charge density) holds. It follows from Eq. (97) that the induced charge density is determined by the induced current density. However, the converse statement is not correct, as only the longitudinal component of the induced current density is determined by the induced charge density.

References

1. P. Zoller P, T. Beth, D. Binosi, R. Blatt, H. Briegel, D. Bruss, T. Calarco, J.I. Cirac, D. Deutsch, J. Eisert, A. Ekert, C. Fabre, N. Gisin, P. Grangiere, M. Grassi, S. Haroche, A. Imamoglu, A. Karison, J. Kempe, L. Kouwenhoven, S. Kroll, G. Leuchs, M. Lewenstein, D. Loss, N. Lutkenhouse, S. Massar, J.E. Mooij, M.B. Plenio, E. Polzik, S. Popescu, G. Rempe, A. Sergienko, D. Suter, J. Twamley, G. Wendin, R. Werner, A. Winter, J. Wrachtrup, and A. Zellinger, Euro. Phys. J. D **36**, 203 (2005).
2. G. Chen, Z.J. Diao, J.U. Kim, A. Neogi, K. Urtekin, and Z.G. Zhang, Int. J. Quant. Inform. **4**, 233 (2006).
3. S. Bandyopadhyay, Superlattices and Microstructures, **37**, 77 (2005).
4. T. Radtke and S. Fritzsche, Comp. Phys. Commun. **173**, 91 (2005).
5. R.G. Mani, W.B. Johnson, V. Narayanamutri, V. Privman, and Y.-H. Zhang, Physica E **12,** 152 (2002).
6. G. Burkard and D. Loss, Phys. Rev. B **59**, 2070 (1998); D.P. DiVincenzo, Phys. Rev. A **51**, 1015 (1994).
7. D. Loss and D.P. DiVincenzo, Phys. Rev. A V. **57**(1), 120–126 (1998).

8. G. Prinz, Phys. Today **48**, 58 (1995); G. Prinz, Science **282**, 1660 (1998).
9. T. Gaebel, M. Domhan, I. Popa, C. Wittmann, P. Neumann, F. Jelezko, J.R. Rabeau, N. Stavrias, A.D. Greentree, S. Praver, J. Meijer, J. Twamley, P.R. Hemmer, and J. Wrachtrup, Nature (Physics) **2**, 408 (2006).
10. D.D. Awschalom and J.M. Kikkawa, Phys. Today 52, 33 (1999); J.M. Kikkawa and D.D. Awschalom, Phys. Rev. Lett. **80**, 4313 (1998); J.M. Kikkawa and D.D. Awschalom, Nature (London) **397**, 139 (1999).
11. R. Fiederling, M. Kelm, G. Reuscher, W. Ossau, G. Schmidt, A. Waag, and L.M. Molenkamp, Nature (London) **402**, 787 (1999).
12. Y. Ohno, D.K. Young, B. Beschoten, F. Motsukura, H. Ohno, and D.D. Awschalom, Nature (London) **402**, 790 (1999).
13. F.G. Monson and M.L. Roukes, J. Magn. Magn. Mater. **198,** 632 (1999).
14. D.P. DiVincenzo, Science **270**, 255 (1995).
15. D.P. DiVincenzo and D. Loss, J. Magnet. Magnet. Mater. **200**, 202 (1999).
16. F.H.L. Koppens, C. Buizert, K.J. Tielrooij, I.T. Vink, K.C. Nowack, T. Meunier, L.P. Kouwenhoven, and L.M.K. Vandersypen, Nature **442**, 766 (2006); D. Kitchens, A. Richardella, J.-M. Tang, M.E. Flatte, and A. Yazdani, Nature **442**, 436 (2006).
17. L.J. Sham and J. Magnet. Magnet. Mater. **200**, 219 (1999).
18. G.M. Jones, B.H. Hu, and C.H. Young, Appl. Phys. Lett. **88**, 192102(3) (2006); G.M. Jones, B.H. Hu, and C.H. Young, Appl. Phys. Lett. **89**, 073106 (3) (2006).
19. R. Skomsky, J. Zhou, A.Y. Istomin, A.F. Starace, and D.J. Sellmyer, J. Appl. Phys. **97**, 10R511(3) (2005).
20. R. Skomsky, A.Y. Istomin, A.F. Starace, and D.J. Sellmyer, Phys. Rev. A **70**, 062307(4) (2004).
21. X. Hu and S. Das Sarma, Phys. Rev. A **61**, 062301 (19) (2000).
22. A. Mizel and D.A. Lidar, Phys. Rev. B **70**, 115310 (13) (2004).
23. V.W. Scarola and S. Das Sarma, Phys. Rev. A **71**, 032340 (12) (2005).
24. B. Zhou, R. Tao, S.-Q. Shen, and J.-Q. Liang, Phys. Rev. A **66**, 010301 (R) (2002).
25. M. Wagner, U. Merkt, and A.V. Chaplik, Phys. Rev. B **45**, 1951 (1992).
26. T. Fujizawa, D.G. Austing, Y. Tokura, Y. Hirayama, and S. Tarucha, Nature **419**, 278 (2002).
27. R. Hanson, B. Witkamp, L.M.K. Vandersypen, L.H. Willems van Beveren, J.M. Elzerman, and L.P. Kouwenhoven, Phys. Rev. Lett. **91**, 196802 (4) (2003).
28. J.M. Eizerman, R. Hanson, L.H. Willems van Beveren, B. Witkamp, L.M.K. Vandersypen, and L.P. Kouwenhoven, Nature **430**, 431 (2004).
29. J.M. Eizerman, R. Hanson, J.S. Greidanus, L.H. Willems van Beveren, S. De Franceschi, L.M.K. Vandersypen, S. Tarucha, and L.P. Kouwenhoven, Phys. Rev. B **67**, 161308 (R) (2003).
30. M. Bauer, P. Hawrylak, K. Hinzer, S. Fafard, M. Korkusinski, Z.K. Wasilewski, O. Stern, and A. Forchel, Science **291**, 451 (2001).
31. M. Atatűre, J. Dreiser, A. Badolato, A. Högele, K. Karrai, and A. Imamoglu, Science **312**, 551 (2006).
32. R. Schleser, E. Ruh, T. Ihn, K. Ensslin, D.C. Driscoll, and A.C. Gossard, Appl. Phys. Lett. **85**, 2005 (2004).
33. R. Hanson, L.H. Willems van Beveren, I.T. Vink, J.M. Elzerman, W.J.M. Naber, F.H.L. Koppens, L.P. Kouwenhoven, and L.M.K. Vandersypen, Phys. Rev. Lett. **94**, 196802 (4) (2005).
34. W. Lu, Z.Q. Li, L. Pfeiffer, K.W. West, and A.J. Rimberg, Nature **423**, 422 (2003).
35. L.M.K. Vandersypen, J.M. Elzerman, R.H. Schouten, L.H. Willems van Beveren, R. Hanson, and L.P. Kouwenhoven, Appl. Phys. Lett. **85**, 4394 (2004).
36. P. Recher, E.V. Sukhorukov, and D. Loss, Phys. Rev. Lett. **85**, 1962 (2000).
37. C.P. Poole, Electron Spin Resonance, 2nd edn., Wiley, New York (1983).
38. M. Xiao, I. Martin, E. Yablonovich, and H.W. Jiang, Nature (London) **430**, 435 (2004).

39. F. Jelezko, T. Gaebel, I. Popa, A. Gruber, and J. Wrachtrup, Phys. Rev. Lett. **92**, 076401(4) (2004).
40. D. Rugar, R. Badakian, H.J. Mamin, and B.W. Chui, Nature (Londin) **430**, 329 (2004).
41. A. Imamoglu, D. D. Awschalom, G. Burkard, D. P. DiVincenzo, D. Loss, M. Sherwin, and A. Small, Phys. Rev. Lett. **83**, 4204 (1999).
42. Y. Kato, R.C. Myers, A.C. Gossard, and D.D. Awschalom, Nature **427**, 50 (2004).
43. V.N Golovach, M. Borhani, and D. Loss, www.arXiv.org/con-mat/0601674 (2006).
44. Y. Tokura, W.G. Van der Wiel, T. Obata, and S. Tarucha, Phys. Rev. Lett. **96**, 047202 (4) (2006).
45. R. Hanson, L.H. Willems van Beveren, I.T. Vink, J.M. Elzerman, W.J.M. Naber, F.H.L. Koppens, L.P. Kouwenhoven, and L.M.K. Vandersypen, Physica E **34**, 1 (2006).
46. J. Nogues, J. Sort, V. Langlais, V. Skumryev, S. Sarinach, J.S. Munoz, M.D. Baro, Phys. Reports **422**, 65 (2005).
47. J. Stangl, V. Holý, and G. Bauer, Rev. Mod. Phys. **76**, 725 (2004).
48. I.N. Stranski and L. Krastanow, Sitzungsbericht d. Akad. d. Wissenschaften in Wien, Abt. IIb, Band **146**, 797 (1937).
49. Q. Xie, A. Madhukar, P. Chen, and N. P. Kobayashi, Phys. Rev. Lett. **75**, 2542 (1995).
50. G. Bester, A. Zunger, Z. Wu, and D. Vanderbilt, Phys. Rev. B **74**, 081305R (2006).
51. C. B. Murray, C. R. Kagan, and M. G. Bawendi, Annu. Rev. Mater. Sci. **30**, 545 (2000).
52. J. E. Bowen Katari, V. L. Colvin, and A. P. Alivisatos, J. Phys. Chem. **98**, 4109 (1994).
53. O. I. Mićić and A. J. Nozik, J. Lumin. **70**, 95 (1996).
54. M. A. Reed, R. T. Bate, K. Bradshaw, W. M. Duncan, W. R. Frensley, J. W. Lee, and H. D. Shih, J. Vac. Sci. Technol. B **4**, 358 (1986).
55. A. B. Fowler, A. Hartstein, and R. A. Webb, Phys. Rev. Lett. **48**, 196 (1982); T. J. Thornton, M. Pepper, H. Ahmed, D. Andrews, and G. J. Davies, Phys. Rev. Lett. **56**, 1198 (1986).
56. H. Ishikuro and T. Hiramoto, Appl. Phys. Lett. **71**, 3691 (1997); A. Aassime, G. Johansson, G. Wendin, R. J. Schoelkopf, and P. Delsing, Phys. Rev. Lett. **86**, 3376 (2001).
57. B. E. Kane, N. S. McAlpine, A. S. Dzurak, R. G. Clark, G. J. Milburn, H. B. Sun, and H. Wiseman, Phys. Rev. B **61**, 2961 (2000).
58. M. A. Kastner, Rev. Mod. Phys. **64**, 849 (1992).
59. K. K. Likharev, Proc. IEEE **87**, 606 (1999).
60. L. Guo, E. Leobandung, S. Y. Chou, Science **275**, 649 (1997).
61. B. E. Kane, N. S. McAlpine, A. S. Dzurak, R. G. Clark, G. J. Milburn, H. B. Sun, and H. Wiseman, Phys. Rev. B **61**, 2961 (2000).
62. S. K. Su, S. J. Chang, L. W. Ji, C. S. Chang, L. W. Wu, W. C. Lai, T. H. Fang, and K. T. Lam, Semicond. Sci. Technol. **19**, 389 (2003); D. Damilano, N. Grandjean, J. Massies, S. Dalmasso, J. L. Reverchon, M. Calligaro, J. Y. Duboz, L. Siozade, and J. Leymarie, Phys. Stat. Sol. A **180**, 363 (2000).
63. M. Shamsa, W. Liu, A. A. Balandin, and J. Liu, Appl. Phys. Lett. **87**, 202105 (2005).
64. T. C. Harmon, R. E. Reeder, M. P. Walsh, B. E. LaForge, C. D. Hoyt, and G. W. Turner, Appl. Phys. Lett. **88**, 243504 (2006).
65. J. M. Zide, D. O. Klenov, S. Stemmer, A. C. Gossard, G. Zeng, J. E. Bowers, D. Vashaee, and A. Shakouri, Appl. Phys. Lett. **87**, 1121102 (2005).
66. J.R. Chelikowsky, J. Phys. D **33**, R33 (2000).
67. Y. Alhassid, Rev. Mod. Phys. **72**, 895 (2000).
68. D.L. Smith and C. Mailhiot, Rev. Mod. Phys. **62**, 173 (1990).
69. E. Landsberg (ed). Quantum Theory of Real Materials, Kluwer, Boston, MA (1996).
70. A.J. Freeman, Ann. Rev. Mat. Sci. **25**, 1 (1995).
71. N.M. Ashkroft and N.D. Mermin, Solid State Physics, Holt, Rinehart&Winston, New York (1976).
72. E.L. Ivchenko and G.E. Pikus, Superlattices and Other Heterostructures, 2nd edn., Springer-Verlag, Berlin, New York (1997).
73. J.M. Tang and M.E. Flatte, Phys.Rev. Lett. **92**, 047201 (2004).

74. http://www.gaussian.com.
75. http://www.msg.ameslab.gov/GAMESS.
76. http://www.emsl.pnl.gov/docs/nwchem/nwchem.html.
77. F. Meier, J. Levy, and D. Loss, Phys. Rev. B **68**, 134417 (15) (2003).
78. V.W. Scrola and S. Das Sarma, Phys. Rev. A **71**, 032340 (12) (2005).
79. A. Mizel and D.A. Lidar, Phys. Rev. B **70**, 115310 (13) (2004).
80. F. Meier, V. Carletti, O. Gywat, D. Loss, and D.D. Awschalom, Phys. Rev. B **69**, 195315 (12) (2004).
81. J. Lehmann and D. Loss, Phys. Rev. B **73**, 045328 (10) (2006).
82. G.D. Mahan, Many Particles Physics, 2nd edn, Plenum, New York (1993).
83. C.W. Miller, Z.-P. Li, I.K. Schuller, R.W. Dave, J.M. Slaughter, and J. Akerman, Phys. Rev. B **74**, 212404 (4) (2006).
84. B.D. Cullity, Introduction to Magnetic Materials, Addison-Wesley, Reading, MA (1972).
85. M.I. Dykman, L.F. Santos, and M. Shapiro, J. Optics B **7**, Special Issue SI, S363 (2005).
86. J. Rammer, Rev. Mod. Phys. **63**, 781 (1991), and references therein.
87. S. Fujita, Introduction to Non-Equilibrium Quantum Statistical Mechanics, W.B. Saunders, Philadelphia (1986), etc.
88. S. Datta, Electronic Transport in Mesoscopic Systems, Cambridge University, Cambridge, England (1995).
89. D.K. Ferry and S.M. Goodnick, Transport in Nanostructures, Cambridge University, Cambridge, England (1997).
90. Y. Imry and R. Landauer, Rev. Mod. Phys. **71**, S306 (1999).
91. C.W.J. Beenakker, Phys. Rev. **B 44**, 1646 (1991).
92. Y. Imry, Directions in Condensed Matter Physics, Vol. 1, World Scientific, Singapore (1986).
93. Mesoscopic Phenomena in Solids, B.L. Altshuler, P.A. Lee and R.A. Webb (eds). North-Holland, Amsterdam (1991).
94. Y. Meir and N.S. Wingreen, Phys. Rev. Lett. **68**, 2512 (1992).
95. D.V. Averin and K.K. Likharev, Phys. Rev. **B 44**, 6199 (1991).
96. Nanotechnology, G. Timp (ed.), AIP, New York (1998).
97. C.W.J. Beenakker, Phys. Rev. **B 44**, 1646 (1991).
98. S.N. Datta, J. Phys. Chem. A **109**, 11417 (2005).
99. D.V. Averin and Yu. N. Nazarov, Phys. Rev. Lett. **65**, 2446 (1990); L.I. Glazman and K.A. Matveev, JETP Lett. **51**, 484 (1990).
100. A. Wacker, Phys. Reports **357**, 1 (2002)
101. G.B. Arnold, J. Low Temp. Phys. **68**, 1 (1987).
102. R.A. Jalabert, A.D. Stone, and Y. Alhassid, Phys. Rev. Lett. **68**, 3468 (1992); J.C. Cuevas, A. Martin-Rodero, and A.L. Yeyati, Phys. Rev. **B 54,** 7366 (1996).
103. Y. Alhassid, Rev. Mod. Phys. **72**, 895 (2000).
104. Y.V. Fyodorov and H.J. Sommers, J. Math. Phys. **38**, 1918 (1997).
105. O. Bohigas and M.-J. Giannoni, Mathematical and Computational methods in Nuclear Physics. Lecture Notes in Physics, Vol. **209**, J.S. Dehesa, J.M.G. Gomez, and A. Polls (eds). Springer, Berlin (1984).
106. B.L. Altshuler, D.E. Khmelnitskii, A.I. Larkin, and P.A. Lee, Phys. Rev. **B 22**, 5142 (1980).
107. P.A. Lee and A.D. Stone, Phys. Rev. Lett. **55**, 1622 (1985); E.A. Yuzbashyan, W. Happer, B.L. Altshuler, and S.B. Shastry, J. Phys. **A 36**, 2577 (2003).
108. K.B. Efetov, Adv. Phys. **32**, 53 (1983).
109. T.N. Todorov, J. Phys: Condens. Matter. **12**, 8995 (2000); M. Buttiker, Pramana J. Phys. **58**, 241 (2002); M. Brandbyge, J. Taylor, K. Stokbro, J.-L Moroz, and P. Ordejon, Phys. Rev. B **65**, 165401 (2002).
110. M. Buttiker, Phys. Rev. B **33**, 3020 (1986); J.L. D'Amato and H. Pastawski, Phys. Rev. B **41**, 7411 (1990).
111. H. van Hauten, C.W.J. Beenakker, and A.A.M. Starring, in: Single Electron Tunneling, H. Grabert, M.H. Devoret (Eds.), Plenum Press, New York, 167 (1991).

112. E.V. Sukhorukov and D. Loss, Phys. Rev. B **59**, 13054 (1999); B. Qiao, H.E. Ruda, and M.S. Zhan, Phys. Rev. A **65**, 042325(11) (2002); X.-Q. Li, W.-K. Zhang, P. Cui, J. Shao, Z. Ma, and Y. Yan, Phys. Rev. B **69**, 085315(9) (2004); H. Moya-Cessa, Phys. Reports **432**, 1 (2006).
113. B. Qiao, X.S. Xing, and H.E. Ruda, Physica A **355**, 319 (2005); B. Qiao and H.E. Ruda, Physica A **334**, 459 (2004); B. Qiao and H.E. Ruda, Physica A **333**, 197 (2004); B. Qiao and H.E. Ruda, Physica A **327**, 425 (2003).
114. E.N. Economou, Green's Functions in Quantum Physics, Springer, Berlin (1983), and references therein; possible simplifications of some of these methods and their use are discussed in S. Mukamel, Phys. Rev. E **68**, 021111(14) (2003).
115. I. G. Lang, L. I. Korovin, J. A. de la Cruz and S. T. Pavlov, Sov. Phys.–JETP **96**, 268 (2003).
116. H.U. Baranger and A.D. Stone, Phys. Rev. B **40**, 8169 (1989).
117. L.A. Pozhar, cond-mat/0502476, http://www.arXiv.org (2005); L.A. Pozhar, Mat. Res. Soc. Proc. **789**, 49 (2004); L.A. Pozhar, Mater. Res. Soc. Proc. **900E**, O03-02.1 (2006); L.A. Pozhar, Mat. Res. Soc. Proc. **789**, 49 (2004).
118. Note that these interactions have also been neglected in [115]. This makes the approach developed in [115] not applicable to systems in strong magnetic fields and to systems with significant magnetization, despite an inclusion of the corresponding vector potential term into the particle momentum operator in [115].
119. A. I. Akhieser and S. V. Peletminskii, Methods of Statistical Physics, Nauka, Moscow (1977).
120. D. N. Zubarev, Soviet Phys. Uspekhi **3**, 320 (1960/61); D.N. Zubarev, Nonlinear Statistical Thermodynamics, Plenum Press, New York (1974).
121. P. P. Ewald, Ann. de Phys. **54**, 519 (1917); P. P. Ewald, Ann. de Phys. **66**, 253 (1921); V. M. Agranovich and V. L. Ginzburg, Optical Properties of Crystalline Solids, Space Dispersion and the Theory of Excitons, Glavnaya Redaktsiya Fiz-Mat. Literatury, Moscow, (1965) (in Russian); Yu. L. Klimontovich, Statistical Theory of Non-Equilibrium Processes in Plasma, Glavnaya Redaktsiya Fiz-Mat. Literatury, Moscow, (1964) (in Russian).
122. D. N. Zubarev and Yu. A. Tserkovnikov, Trudy Matematicheskogo Instituta im. Steklova **175**, 139 (1986); Proceedings of the Steklov Institute of Mathematics **2**, 139 (1988), and references therein.
123. S. V. Tyablikov, Methods in the Quantum Theory of Magnetism, Plenum Press, New York (1967); N. N. Bogolyubov, Introduction to Quantum Statistical Mechanics, World Scientific, NJ, (1982); N. N. Bogolyubov, A. A. Logunov, A. I. Oksak, I. T. Todorov, General Principles of Quantum Field Theory, Nauka, Moscow (1987); N.N. Bogolubov, Selected Works, N. N. Bogolubov, Jr. and A. M. Kurbatov (Eds.), Gordon and Breach, New York (1991).
124. Yu. A. Tserkovnikov, Dokl. Acad. Nauk SSSR **143**, 832 (1962); Soviet Phys. Dokl. **7**, (1962/1963); G. F. Mazenko, Phys. Rev. A **7**, 209 (1973); **7**, 222 (1973); **9**, 360 (1974).
125. Yu. A. Tserkovnikov, Teoreticheskaya I Matematicheskaya Fizika, **7**, 250 (1971); English translation in Theor. Math. Phys. **7** (1971); **23**, 221 (1975); English translation in Theor. Math. Phys. 23 (1975); **26**, 77 (1976); English translation in Theor. Math. Phys. 26 (1976); **50,** 261 (1982); English translation in Theor. Math. Phys. **50** (1982); **52**, 147 (1982); English translation in Theor. Math. Phys. **52** (1982); N.N. Bogolyubov, V.V. Tolmachev, and D.V. Shirkov, New Methods in the Theory of Superconductivity, Chapman and Hall, London (1959); N.M. Plakida, in Statistical Physics and Quantum Field Theory, N.N. Bogolyubov (Ed.), Nauka, Moscow, p. 205 (1973), etc.
126. L.A. Pozhar, V.F. de Almeida, and M.Z.-C. Hu, Ceramics Transactions **137**, 101 (2003); L.A. Pozhar, E.V. Kontar, and M. Z.-C. Hu, J. Nanosci. Nanotech. **2**, 209 (2002); L.A. Pozhar, Phys. Rev. E **61**, 1432 (2000); L.A. Pozhar and K.E. Gubbins, Int. J. Thermophys. **20**, 805 (1999); L.A. Pozhar and K.E. Gubbins, Phys. Rev . E **56**, 5367 (1997); E. Akhmatskaya, B.D. Todd, P.J. Daivis, D.L. Evans, K.E. Gubbins, L.A. Pozhar, J. Chem. Phys. 106, 4684 (1997); J.K. Percus, L.A. Pozhar, and K.E. Gubbins, Phys. Rev . E **51**, 261

(1995); L.A. Pozhar and K.E. Gubbins, J. Chem. Phys. **99**, 8970 (1993), etc. [Wider list of relevant publications is available on http://bama.ua.edu/~lpozhar].
127. A.K. Singh et al., Phys. Rev. Lett. **91**, 146802 (2003); V. Kumar and Y. Kawazoe, Phys. Rev. Lett. **90**, 055502 (2003), etc.
128. P.S. Fodor, G.M. Tsoi, and L.E. Wenger, J. Appl. Phys. **91**, 8186 (2002).
129. These and other aspects of such virtual synthesis of small artificial molecules with predesigned electronic properties are a subject of intensive research, and have been discussed in numerous publications, including L.A. Pozhar, A.T. Yeates, F. Szmulowicz and W.C. Mitchel, Phys. Rev. B **74**, 085306 (2006); J. Virtual Nanoscale Sci. Technol. **14**, No. 8 (2006): http://www.vjnano.org; L.A. Pozhar, A.T. Yeates, F. Szmulowicz and W.C. Mitchel, EuroPhys. Lett. **71**, 380 (2005); Mat. Res. Soc. Proc. **829**, 49 (2005), etc.
130. M.W. Schmidt et al., J. Comput. Chem. **14**, 1347 (1993).
131. G.W. Bryant and W. Jaskolski, Mat. Res. Soc. Proc. **789**, N13.2 (2004).
132. L.A. Pozhar and G. Brown, Mater. Res. Soc. Proc. **957**, 0957-K10-04 (2007); L.A. Pozhar, G. Brown and W.C. Mitchel, Mater. Res. Soc. Proc. **959**E, 0959-M05-02 (2007); L.A. Pozhar, A.T. Yeates, F. Szmulowicz and W.C. Mitchel, Mater. Res. Soc. Proc . **891**, EE02-04.1 (2006); Ibid., **906**E, HH01-05.1 (2006).
133. E.F. Archibong, A. St-Amant, S.K. Goh, and D.S. Marynick, Chem.Phys. Lett. **361**, 411 (2002); S. Katircioğlu and H. Kökten, J. Mol. Struct.: Theochem **712**, 67 (2004); E.-L. Li, X.-M. Luo, W. Shi, and X.-W. Wang, J. Mol. Struct.: Theochem **723**, 79 (2005).
134. A.I. Boldyrev and L.-S. Wang, J. Phys. Chem. A **105**, 10759 (2001).

Index

MIX
Papier aus verantwortungsvollen Quellen
Paper from responsible sources
FSC® C105338

If you have any concerns about our products,
you can contact us on
ProductSafety@springernature.com

In case Publisher is established outside the EU,
the EU authorized representative is:
Springer Nature Customer Service Center GmbH
Europaplatz 3, 69115 Heidelberg, Germany

Printed by Libri Plureos GmbH
in Hamburg, Germany